III–V
COMPOUND
SEMICONDUCTORS

Integration with Silicon-Based Microelectronics

III–V

COMPOUND SEMICONDUCTORS

Integration with Silicon-Based Microelectronics

Edited by
TINGKAI LI
MICHAEL MASTRO
ARMIN DADGAR

CRC Press
Taylor & Francis Group
Boca Raton London New York

CRC Press is an imprint of the
Taylor & Francis Group, an **informa** business

CRC Press
Taylor & Francis Group
6000 Broken Sound Parkway NW, Suite 300
Boca Raton, FL 33487-2742

First issued in paperback 2019

ISBN-13: 978-1-4398-1522-9 (hbk)
ISBN-13: 978-0-367-38326-8 (pbk)

Library of Congress Cataloging-in-Publication Data

III-V compound semiconductors : integration with silicon-based microelectronics /
    editors, Tingkai Li, Michael Mastro, and Armin Dadgar.
        p. cm.
    "A CRC title."
    Includes bibliographical references and index.
    ISBN 978-1-4398-1522-9 (alk. paper)
    1. Compound semiconductors. I. Li, Tingkai. II. Mastro, Michael A., 1975- III.
Dadgar, Armin. IV. Title.

TK7871.99.C65I38 2010
621.3815'2--dc22   2010025720

Visit the Taylor & Francis Web site at
http://www.taylorandfrancis.com

and the CRC Press Web site at
http://www.crcpress.com

Contents

Part IV   Defect and Properties Evaluation and Characterization

Part V   Device Structures and Properties

Preface

After many decades, the true potential of integrating compound semiconductors with Si is now being realized. The two major revolutions occurring in the semiconductor industry—the integration of compound semiconductors into Si microelectronics and the fabrication of compound semiconductors on large-area Si substrates—will be addressed in this book.

After several decades of scaling, Si-based processors are approaching a number of fundamental limitations that can be addressed only with the use of compound semiconductors. To meet increasingly challenging and complex system requirements as well as remaining cost effective, it is not sufficient to use one single semiconductor material system. Therefore, major efforts have been made in recent years to combine the low-cost and well-established Si-based CMOS processing attributes with the superior performance attributes of compound semiconductors. Such a combination—marrying the best of both worlds—will enable performance superior to that achievable with compound semiconductors or Si-based CMOS alone, while retaining the affordability that is achieved with Si-based CMOS.

Conversely, large-area Si substrates can simply act as a building block to lower the fabrication cost of compound semiconductor devices. In many cases, a superior compound semiconductor technology has little or no commercial market simply because of the high cost of manufacturing. The lack of large-area substrates, e.g., GaN, GaAs, or InP, creates a cost and throughput choke hold on compound semiconductor foundries. A number of dedicated teams have already shown that compound semiconductor devices, such as GaN-based LEDs, high-power diodes and transistors, and III–V solar cells, can be manufactured on large-area Si substrates.

Straddling these two extremes is the die level insertion of compound semiconductor transistor technologies within or surrounding the Si CMOS circuitry. This approach circumvents many material incompatibility issues between compound semiconductors and Si, and has been revived under the DARPA COSMOS program (details will be provided in the book). All the technologies mentioned above are ongoing challenges to the semiconductor community. This book will outline the state of the art and set a roadmap for the future of this field.

The book gives a comprehensive overview of the scientific concepts and the recent developments in the field of III–V materials and device integration with Si microelectronics. It has been written by experts in the field and includes topics such as materials growth, characterization, device design, and device integration. When appropriate, the focus shifts to critical optical, optoelectronic, or electronic device technology.

The book is divided into five parts. Part I introduces the fundamental properties and issues of these systems to provide a basis for the remainder of the book. It describes the structure, phase diagram, and physical and chemical properties of III–V and Si materials, as well as the challenges to integrating III–V materials with Si. Part II delves into one particularly important structure, GaN, and its related alloys on silicon, which is an enabling technology for the commercial penetration of a new class of power diodes and transistors. Part III expands our analysis to include more traditional III–V materials and discusses their merits and drawbacks for device integration with Si microelectronics. It then goes on to discuss the fabrication and behavior of III–V devices integrated within Si microelectronics. Part IV discusses the properties of these III–V semiconductors and describes approaches to evaluate and characterize their attributes. It introduces novel technologies for the measurement and evaluation of material quality and device properties. Part V investigates state-of-the-art devices and components that are being implemented on Si. It focuses mainly on optical devices; LEDs; Si photonics; high-speed, high-power III–V materials and devices; and III–V solar cell devices—all within the integration of the Si framework.

This book is intended as a reference book for scientists and engineers working at the intersection of Si and compound semiconductor technology. The chapters are written in a comprehensive manner so as to be suitable for students as well as experts in this field.

Contributors

Frank Bertram
Institut fuer Experimentelle Physik
Fakultät für Naturwissenschaften
Otto-von-Guericke-Universität
    Magdeburg
Magdeburg, Germany

Jürgen Bläsing
Institut fuer Experimentelle Physik
Fakultät für Naturwissenschaften
Otto-von-Guericke-Universität
    Magdeburg
Magdeburg, Germany

John E. Bowers
Department of Electrical and
    Computer Engineering
University of California
Santa Barbara, California

Edward Yi Chang
Department of Materials Science
    and Engineering
National Chiao Tung University
Hsin Chu, Taiwan, R.O.C.

Rainer Clos
Institut fuer Experimentelle Physik
Fakultät für Naturwissenschaften
Otto-von-Guericke-Universität
    Magdeburg
Magdeburg, Germany

Armin Dadgar
Institut fuer Experimentelle Physik
Fakultät für Naturwissenschaften
Otto-von-Guericke-Universität
    Magdeburg
Magdeburg, Germany

Tyler J. Grassman
Department of Electrical and
    Computer Engineering
The Ohio State University
Columbus, Ohio

Katherine J. Herrick
Raytheon Company
Waltham, Massachusetts

Anthony A. Immorlica Jr.
Professional Engineer
Mount Vernon, New Hampshire

Hongxing Jiang
Department of Electrical and
    Computer Engineering
Texas Tech University
Lubbock, Texas

Thomas Kazior
Raytheon Company
Waltham, Massachusetts

Alois Krost
Institut fuer Experimentelle Physik
Fakultät für Naturwissenschaften
Otto-von-Guericke-Universität
    Magdeburg
Magdeburg, Germany

Jeffrey LaRoche
Raytheon Company
Waltham, Massachusetts

Jing Li
Department of Electrical and
    Computer Engineering
Texas Tech University
Lubbock, Texas

Tingkai Li
Micron Technology Inc.
Boise, Idaho

Di Liang
Department of Electrical and
    Computer Engineering
University of California
Santa Barbara, California

Jingyu Lin
Department of Electrical and
    Computer Engineering
Texas Tech University
Lubbock, Texas

Michael Mastro
Power Electronic Materials Section
Laboratory for Advanced Materials
    Synthesis
U.S. Naval Research Laboratory
Washington, District of Columbia

Srinivasan Raghavan
Materials Research Centre
Indian Institute of Science
Bangalore, India

Joan M. Redwing
Department of Materials Science
    and Engineering
The Pennsylvania State University
University Park, Pennsylvania

Steven A. Ringel
Department of Electrical and
    Computer Engineering
The Ohio State University
Columbus, Ohio

Nobuhiko Sawaki
Department of Electrical and
    Electronic Engineering
Aichi Institute of Technology
Toyota, Japan

1

Fundamentals and the Future of Semiconductor Device Technology

Michael Mastro

CONTENTS

1.1 Introduction

In a curious historical twist, the first semiconductor amplifier was built from germanium and for a time, germanium was thought to be the semiconductor of the future. In a short time, the ease of manufacturing silicon-based devices made the germanium device industry unviable. A major advance was by Jack Kilby, at Texas Instruments, who *integrated circuits* of resistors,

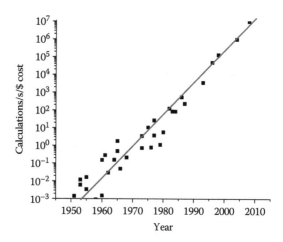

FIGURE 1.1
Exponential growth of computation power per equivalent cost of production.

capacitors, and transistors onto one die. By 1971, Hoff and Faggin designed the Intel 4004 chip with 2200 transistors. Over the past five decades, the silicon-based industry has grown into a $249 billion market (ITRS, 2007). The societal impact of the silicon-based semiconductor industry is even more important than its revenue.

Several variations of Moore's law are often cited such as the number of transistors per integrated circuit (IC) doubling approximately every 2 years. The true impact of Moore's law on society is primarily about reducing cost per function, as displayed in Figure 1.1 (Kurzweil, 2005).

There are many perspectives on the intrinsic limits to the scaling. Zhirnov et al. (2003) point out that the Shannon–von Neumann–Landauer minimum energy to switch a bit from 1 to 0 in a solid-state device is

$$E_{s|min} = \ln(2)k_BT = 0.017 \text{ eV}.$$

The Heisenberg uncertainty relationships determine the minimum distance required is

$$L_{min} = \frac{\hbar}{\sqrt{2m_eE_{s|min}}} = 1.5 \text{ nm},$$

where m_e is the electron mass, and the minimum time is

$$\tau_{min} = \frac{\hbar}{E_{s|min}} = 0.04 \text{ ps}.$$

Several limits will be approached but still not reached at the 18 nm generation: $E_s \approx 4200E_{s|min}$; $L \approx 5 \times L_{min}$; $\tau \approx 3\tau_{min}$. Perhaps a more critical limitation at

Part I

Technical Fundamentals and Challenges

the 18 nm generation is a predicted transistor density of 3.5×10^9 cm^{-2}, which would greatly exceed the thermal stability limit of silicon. Intelligent circuit design can limit the number of operating transistors at any given time, thus limiting the thermal load; however, the increase in long interconnects is creating a propagation bottleneck.

There will be a point at which the long-run exponential scaling of transistor density and inverse length will fail. Nevertheless, there are other approaches to maintain the exponential growth in computation power per equivalent cost. Novel electronic and optoelectronic devices could be integrated to provide new functionality beyond digital switches as systematic device scaling slows down.

1.1.1 Comparison of Materials and Devices

To understand the role of compound semiconductors, it is fruitful to examine why silicon is dominant. Specific advantages of silicon include the source material, silica, which is widely available, inexpensive, and easily refined into high-purity silicon. Large ingots of silicon are readily produced by the Czochralski process; 12 in. (300 mm) diameter wafers are currently in production lines and 18 in. (450 mm) diameter wafers are scheduled for introduction in the next 5–10 years. Silicon can be formed in a semi-resistive state or can be easily doped n-type with net negative charge or doped p-type with a net positive charge. Perhaps, the most important characteristic is the ability to grow a high-quality native oxide, SiO_2. This oxide forms a sharp interface with silicon and displays a low level of traps at this interface or in the oxide itself. This is critical as the upper boundary of the active channel of a CMOS transistor is defined by this silicon/SiO_2 interface. It is generally accepted that the recent introduction of a high-κ dielectric as the insulator was only accomplished by allowing a few atomic layers of SiO_2 to form at the high-κ/Si interface.

An important yet much smaller industry is the approximate $25 billion compound semiconductor industry. A sampling of applications served by compound semiconductors is given in Table 1.1 (Zhalko-Tytarenko, 2007). The advantageous cost structure of manufacturing devices with silicon necessitates that any other semiconductor material must present a specific technological advantage. Silicon with its indirect bandgap cannot be used for light-emitting devices (LEDs), which are now dominated by the GaN, GaP, and GaAs material systems. Wideband gap materials such as GaN and SiC allow a higher voltage limit in power switching devices. The high mobility of GaAs and InP allows the commercialization of transistors operating at above 100 GHz.

Direct bandgap compound semiconductors such as GaAs and GaN when fabricated into p-n junctions emit light efficiently. The near-UV, blue, and green portion of the light spectrum is produced by InGaN-based LEDs while amber, red, and IR light is generated by GaP or GaAs-based LEDs. An

TABLE 1.1

Semiconductor Materials Used for Various Commercial Device Markets

Component	Device Martial	Applications	Market ($ Billion)
Solar cell	CuGeSe, GaAs, GaInP	High-efficiency ground and satellite device	0.5
LEDs	GaN, GaP	White light illumination, flat panel display backlighting, signs, indicators	8.5
Lasers	GaAs, GaN, InP	Optical storage, data communications, telecom switches	2.5
Wireless	GaAs, InP, GaN	Mobile communication, base stations	8

appropriate combination of blue, green, and red LEDs is an efficient white light source. There is a large, rapidly growing demand for high-brightness LED white lighting for backlighting and flash photography, as well as general illumination as a replacement to the incandescent light bulb. Compound semiconductor optoelectronics are also used as LEDs in outdoor displays, warning signs, and traffic lights, and as laser diodes in CD/DVD-ROM drives.

The telecommunications market has driven major advancements in compound semiconductor laser devices. For example, fiber optic communication employs InGaP-based lasers to coincide with the absorption minima found in fiber optics between 1.3 and 2 μm. Silicon is also transparent in the IR, which encourages integration of III–V laser diodes for on-chip optical communication.

The high-power/high-voltage market is primarily served by silicon-based laterally diffused MOS (LDMOS) and high-voltage vertical FETs (HVVFET), and silicon carbide-based devices such as 600 V Schottky diodes have been in commercial production for some time. The main market segments for high-power, high-frequency transistors are defense and military applications (radar, jamming, counter-measures, guided weapons), wireless infrastructure (3G, 3G+, WiMAX/LTE base stations and backhaul), and broadcast and communication satellites (SatCom).

Most applications for semiconductor electronic devices are classified as high-speed or high-power. A rough guide to this application space is given in Figure 1.2 as a two-dimensional (2D) slice in power frequency. The power axis is actually a function of current and voltage handling capability as well as several factors that influence the selection of a certain technology for a given application including reliability (ruggedness and thermal considerations), performance (linearity, efficiency), size, cost, and legacy. Scaling the device size in general increases the speed, i.e., frequency of the device, but it is necessary to look at the particular device design as well as the underlying material parameters.

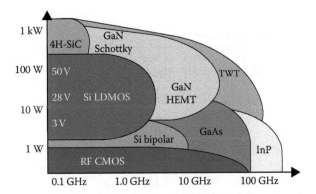

FIGURE 1.2
Power frequency diagram of the application space for several semiconductor materials. At extremely high power and/or frequency, vacuum electronics are still implemented and included for comparison.

Semiconductor electronics can be broadly classified as bipolar (or minority carrier), or unipolar (or majority carrier). Majority carrier devices are usually faster and include Schottky diode, power vertical diffused MOSFET, and JFET. Minority carrier devices usually have better on-state performance and include variants of the PIN diode, IGBT, BJT, and thyristor.

An important parameter in bipolar power devices, such as pin diodes, is the minority carrier lifetime. Indirect semiconductors such as Si and SiC intrinsically have a longer minority carrier lifetime compared to direct bandgap semiconductors such as GaN. There are several types of bipolar devices built in Si and SiC that display high blocking voltages.

Semiconductor diodes, used in switches or rectifiers, when forward biased (on-state) should have minimal voltage across the two terminals and the leakage current should be very low when reverse-biased (in the off-state). Schottky diodes have high switching speed but tend to have high leakage in the off-state. Increasing the thickness or decreasing the doping in the drift region increases the breakdown voltage but also increases the on-resistance as described by

$$R_{\text{on-sp(ideal)}} = \frac{W_D}{q\mu_n N_D} = \frac{4(V_B)^2}{\varepsilon_s E_c^3 \mu_n},$$

where
V_B is the breakdown voltage
E_c is the critical field for avalanche breakdown in a particular semiconductor

Several figures of merit (FOM) are listed in Table 1.2 (Zhang et al., 2003). In general, the large bandgap of SiC and GaN allows high-temperature

TABLE 1.2

Normalized Unipolar Power-Device Figures of Merit

Semiconductor	λ	JM $(E_c v_{sat}/\pi)^2$	MJM $\lambda^* JM$	KM $\lambda(v_{sat}/\varepsilon_r)^{1/2}$	Q_{F1} $\lambda\sigma_A$	Q_{F2} $\lambda\sigma_A E_c$	MB(Q_{F3}) $\mu\varepsilon_r E_c^3$	BHFM μE_c^2
Si	1	1	1	1	1	1	1	1
Ge	0.4	0.03	0.012	0.2	0.06	0.02	0.2	0.3
GaAs	0.33	7.1	2.4	0.45	5.2	6.9	15.6	10.8
GaP	0.53	37	20	0.7	10	40	16	5
GaN	0.87	760	655	1.6	560	6,220	650	77.8
AlN	1.7	5,120	8,700	21	52,890	2059,000	31,700	1,100
4H-SiC	3	180	540	4.61	390	2,580	130	22.9

operation without complete carrier ionization; the large critical field of SiC and GaN allows high-voltage operation (relative to maximum breakdown); the high electron mobility of an AlGaN/GaN high-electron-mobility transistor (HEMT) structure is advantageous as R_{on} is inversely proportional to mobility; the large thermal conductivity, λ, of SiC minimizes self-heating effects.

Gallium nitride–based transistors possess fundamental electronic properties that make it an ideal candidate for high-power microwave devices (Johnson et al., 2001). A number of these properties derive directly from the wide bandgap of GaN ($E_g = 3.4\,eV$), including an exceptionally high electric breakdown field (~3 MV cm^{-1}). This high breakdown field allows GaN-based devices to be biased at a high drain voltage ($V_{break-down} > 50\,V$) while maintaining low microwave added noise (NF = 0.6 dB at 10 GHz) and thus, a large dynamic range (Oxley, 2001). Furthermore, the wide bandgap of GaN allows device operation at elevated temperature (>300°C) without degradation. Additionally, GaN has a high saturation electron velocity ($v_{sat} = 2 \times 10^7$ cm s^{-1}), which is partially accountable for the high current density, I_{max} ($I_{max} \approx q n_s v_{sat}$ where $q = 1.6 \times 10^{-19}$ C, n_s = sheet charge density, v_s = electron saturation velocity), and high operating frequency as $f_t \approx v_{sat}/L_{eff}$.

Cost is a major criteria for the consumer market and there is a large demand for electronic devices that can withstand hundreds of volts. Silicon carbide is well suited for low-frequency power applications but the high device cost (due to the cost and limited size of the SiC substrate) is only appropriate for high value-added technologies. The prospect of a low-cost alternative is now entering the commercial marketplace based on GaN on Si substrate technology. The large diameter Si substrate allows the GaN devices to reach an economy of scale appropriate for high-volume commercial markets. GaN on Si technology is just now entering commercial implementation: High-power converters DC/DC and DC/AC; high-power RF amplification (L, S, C and X bands). GaN on Si Schottky diodes are now sold by Velox and International Rectifier as replacement to 600 V Si or SiC rectifiers targeted at switch mode power converters.

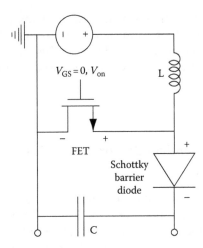

FIGURE 1.3
Boost converter where two main semiconductor components are the transistor and the rectifying diode. Current power supply technology incorporates designs (not included in the schematic) to account for the limited capability of Si-based transistors and rectifiers.

A boost converter shown in Figure 1.3 is the critical component of a switch-mode power supply. The freewheeling diode also known as a suppressor diode, catch diode, snubber diode, or flyback diode, is used to eliminate a sudden voltage spike across an inductive load when its supply voltage is abruptly removed or reduced. These diodes are almost always found in circuitry where inductive loads are opened/closed or controlled by silicon components such as a silicon laterally diffused MOS transistor. Examples include motor drivers and relay drivers. Schottky diodes are advantageous due to their low forward voltage drop and fast reverse recovery voltage. Gallium nitride on Si technology can play a role in all but the highest power variants of conversion circuits: AC/AC converters (frequency changers, cycloconverters); AC/DC converters (rectifiers, off-line converters); DC/AC converters (inverters); and DC/DC converters.

Chen et al. (2009) demonstrated monolithic integration on a Si substrate of an AlGaN/GaN lateral-field rectifier and a normally off AlGaN/GaN transistor switch. Typically, a standard AlGaN/GaN HEMT is a normally on device, which is undesirable for practical implementation. A fluorine ion plasma was used to covert the HEMT to a normally off device, and to deplete the 2DEG in the Schottky diode.

HEMTs based on the III-nitride material system have attracted interest for high-frequency electronic components operating at high-power levels. Nitride-based HEMTs can achieve power, bandwidth, and efficiency levels that exceed the performance of GaAs or SiC-based devices. The development of nitride HEMTs for compact RF and microwave electronics can provide a significant cost and performance advantage for a number of military and commercial systems including X-band radar systems, and power amplifiers for satellite and terrestrial communication (Mastro et al., 2003). Fundamentals of AlGaN/GaN HEMTs are discussed later in this chapter and implementations of various HEMT technologies are discussed throughout the book.

1.2 Fundamental Semiconductor Properties

1.2.1 Band Structure

Electrons accelerate in opposite direction to the electric field and decelerate due to collisions and lattice scattering event. This net electron motion is defined by a drift velocity. Electrons are best described by a **k**-space vector where $\mathbf{k} = 2\pi\mathbf{p}/h$, which is related to the electron momentum by $\mathbf{p} = m\mathbf{v}$ and velocity, **v** (Van Zeghbroeck, 2008). The energy of a free electron is simply the kinetic energy $m\mathbf{v}^2/2$. Rewriting the electron energy in terms of the basis vectors in **k** space gives

$$E = \frac{\hbar^2}{2m}\left(k_x^2 + k_y^2 + k_z^2\right).$$

In a semiconductor crystal, the wavefunction for the electron must satisfy the Schrodinger equation (Miller, 2000). The application of appropriate boundary conditions gives a wavefunction of the form

$$\psi(\mathbf{r}) = C \exp(i\mathbf{k}\cdot\mathbf{r}).$$

Only a limited density of states (or modes) are available for occupation for a particular wavelength (or energy) in a crystal. Especially important is the energy bandgap where no electron is predicted to exist in a pure crystal at $T = 0\,\mathrm{K}$. This forbidden band is defined by the valence band, which the last filled energy level, and the conduction band, which is first unfilled energy level, at $T = 0\,\mathrm{K}$.

The wavevector is related to the wavelength, λ, by $\mathbf{k} = 2\pi/\lambda$. It is common to approximate the wavevector at the bottom of a particular energy band (E_0) by isotropic parabolic relation as

$$k = \frac{1}{\hbar}\sqrt{2m(E - E_0)}.$$

Within an allowed energy band, electrons behave similar to free particle except their effective mass, m^*, is a fraction of the free space mass of an electron. The effective mass describes curvature in k-space by

$$m^* = \hbar k \left(\frac{\partial E}{k} \right)^{-1},$$

and several values are listed in Table 1.3 (Li, 1993).

Electrons tend to accumulate, i.e., the highest probability occupation, at the bottom of the conduction band, E_c, as described by the dispersion relationship for electrons,

$$E(k) = E_c + \frac{\hbar^2 k^2}{2m_e^*},$$

and holes,

$$E(k) = E_v - \frac{\hbar^2 k'^2}{2m_h^*}.$$

These relationships are valid for direct-gap semiconductors such as GaAs or InP where the constant energy surfaces near the conduction band minimum and the valence band maximum are spherical. In silicon and germanium, the constant energy surface near the band edge is an ellipsoidal, which requires evaluation of effective mass tensor to find

TABLE 1.3

Bandgap, Effective Mass, and Mobility of Important Semiconductors

Semiconductor	Bandgap (eV)	Effective Mass		Mobility (cm² V⁻¹ s⁻¹)	
		m_n^*/m_0	m_p^*/m_0	μ_n	μ_p
Si	1.12	$m_l = 0.98$	$m_H = 0.49$	1,500	450
		$m_t = 0.19$	$m_L = 0.16$		
Ge	0.67	$m_l = 1.64$	$m_H = 0.28$	3,900	1,900
		$m_t = 0.082$	$m_L = 0.04$		
GaN	3.36	0.27	0.8	380	20
GaP	2.25	0.82	0.6	110	75
GaAs	1.43	0.067	0.45	8,500	400
InP	1.35	0.077	0.64	4,600	150
GaSb	0.75	0.042	0.4	5,000	850
InAs	0.36	0.023	0.4	33,000	460
InSb	0.17	0.0145	0.4	80,000	1,250
ZnS	3.68	0.4	0.35	102	16
PbTe	0.31	0.17	0.2	6,000	7,000

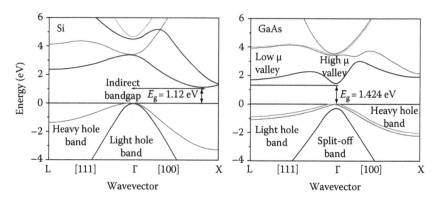

FIGURE 1.4
Band diagrams of Si and GaAs, Γ point: $k = (0,0,0)$; **X** point: $k = 2\pi a^{-1}(1,0,0)$; **L** point: $k = \pi a^{-1}(1,1,1)$.
Si is an indirect semiconductor because the maximum of the conduction band at Γ does not
align with the minimum of the conduction band near **X**. In contrast, GaAs is a direct semiconductor as the conduction band minimum and the valence band maximum align at Γ.
GaAs has a high mobility at low applied electric field due to the large curvature of conduction band at Γ; however, a low mobility valley along **L** becomes populated at high applied
field.

$$E(k) = E_c + \frac{\hbar^2}{2}\left(\frac{k_l^2}{m_l} + \frac{k_t^2}{m_t}\right).$$

Examination of the band structure of silicon, germanium, and gallium arsenide in Figure 1.4 shows a heavy-hole and light-hole valence band that are
degenerated at $k = 0$ as well as a spin–orbit split-off band located a small fraction of an eV below the top of the valence bands. The heavy hole shows a
smaller curvature and thus higher effective mass while the light hole has a
larger curvature and thus a lower effective mass.

1.2.2 Density of States

The number of available states near a certain k-space $(= 2\pi/\lambda)$ are listed in
Table 1.4 (Van Zeghbroeck, 2005). Specifically, the number of states per unit
in a 3-D bulk, a 2-D annulus with radius k and thickness $k + dk$ shell, and a
1-D line of length L are listed. The parabolic band relation is used to reex-
press this number of states in k-space as the density of states (DOS) per unit
volume per unit energy.

The Fermi–Dirac function of electrons

$$f_n(E) = \frac{1}{\left[1 + e^{(E - E_f)/k_B T}\right]},$$

and of holes

TABLE 1.4

Number of States in k-Space in $[k, k+dk]$ and Density of States in a Semiconductor Crystal in the Bulk 3D, a 2D Plane, and a 1D Line

$(E \geq E_c)$	States in $[k, k+dk]$	Density of States, $D(E)$ (or g_c)
3-D	$\dfrac{dN_{3D}}{dk} = 2\left(\dfrac{L}{2\pi}\right)^3 4\pi k^2$	$\dfrac{dN_{3D}}{dE} = \dfrac{4\pi}{h^3}(2m_n^*)^{3/2}(E-E_c)^{1/2}$
2-D	$\dfrac{dN_{2D}}{dk} = 2\left(\dfrac{L}{2\pi}\right)^2 2\pi k$	$\dfrac{dN_{2D}}{dE} = \dfrac{4\pi}{h^2}m_n^*$
1-D	$\dfrac{dN_{1D}}{dk} = 2\left(\dfrac{L}{2\pi}\right)$	$\dfrac{dN_{1D}}{dE} = \sqrt{\dfrac{2\pi m_n^*}{h^2}}(E-E_c)^{-1/2}$

$$f_p(E) = \frac{1}{\left[1+e^{(E_f-E)/k_BT}\right]},$$

describes the probability of a state being occupied by an electron. This distribution is valid for a particle that is indistinguishable, such as an electron, obeys the exclusion principle, and can only occupy a particular state (Hanson, 2007). Bringing these two concepts together, we can express the total number of filled electronic states per unit volume. The electron concentration, n, is found by integrating over energy the product of the DOS and the Fermi–Dirac function,

$$n_0 = \int_{E_c}^{\infty} f_n g_n dE,$$

which is conveniently expressed as

$$n_0 = N_c F_{1/2}(\eta)$$

The effective density of states in the conduction band is

$$N_c = 2\left(\frac{2\pi m_n^* k_B T}{h^2}\right)^{3/2}$$

and the Fermi integral of order one-half is

$$F_{1/2}(\eta) = \left(\frac{2}{\sqrt{\pi}}\right)\int \frac{\varepsilon^{1/2}d\varepsilon}{1+e^{(\varepsilon-\eta)}}$$

where

$\varepsilon = (E - E_c)/k_B T$ is the reduced energy
$\eta = -(E_c - E_f)/k_B T$ is the reduced Fermi energy
m_n^* is the scalar effective mass for electrons listed in Table 1.3

Elemental semiconductors, such as Si and Ge, have multi-valley conduction band minima. Specifically, in Si, there are six ellipsoids along the $\langle 100 \rangle$ axes, and in Ge, there are eight half-ellipsoids located at the zone boundaries of the first Brillion zone along the {111} axes (Li, 1993). For υ conduction band minima, the total density of electrons is given by

$$n_0' = \upsilon n_0 = \upsilon N_c F_{1/2}(\eta) = N_c' F_{1/2}(\eta),$$

and the effective DOS ($\upsilon_{Si} = 6$) is

$$N_c' = \upsilon N_c = 2 \left(\frac{2\pi m_{dn}^* k_B T}{h^2} \right)^{3/2},$$

where the DOS effective mass is given by

$$m_{dn}^* = \upsilon^{2/3} \left(m_l^* (m_t^*)^2 \right)^{1/3}.$$

The top of the valence bands for Si, Ge, and GaAs are nonspherical and warped. The valence bands consist of a heavy-hole band and a light-hole band described by a heavy-hole mass, m_H, and a light-hole mass, m_L, respectively (Li, 1993). In general, the split-off band can be neglected to give an expression

$$m_{dp}^* = \left(m_H^{3/2} + m_L^{3/2} \right)^{2/3}$$

for the effective density of the valence band states. A summary of the conduction and valence band parameters is given in Table 1.5 (Li, 1993). Transport is highly dependent on crystal orientation but an average conductivity effective mass for Si and Ge is defined as

$$m_{cn}^* = 3 \left(\frac{1}{m_l^*} + \frac{2}{m_t^*} \right)^{-1},$$

which is $0.26 m_0$ for Si and $0.12 m_0$ for Ge. These concepts will reappear in our discussion of injection velocity dependence on transport effective mass and ballistic conduction channels dependence on DOS effective mass.

TABLE 1.5

Density of States and Band Parameters of Si, Ge, and GaAs

	v	N'_c (cm^{-3})	Conduction Bands/m_0		DOS Eff. Mass m_{dn}/m_0
Si	6	2.75×10^{19}	$m_l = 0.98$	$m_t = 0.19$	1.084
Ge	4	1.03×10^{19}	$m_l = 1.64$	$m_t = 0.082$	0.561
GaAs	1	3.67×10^{19}	$m_n = 0.067$	—	0.067
GaN	1	2.23×10^{18}	$m_n = 0.2$	—	0.2

	N'_v (cm^{-3})	Valence Bands/m_0		DOS Eff. Mass m_{dp}/m_0
Si	1.28×10^{19}	$m_H = 0.49$	$m_L = 0.16$	0.55
Ge	5.42×10^{18}	$m_H = 0.28$	$m_L = 0.044$	0.29
GaAs	7.0×10^{18}	$m_H = 0.082$	$m_L = 0.45$	0.47
GaN	1.19×10^{18}	$m_H = 0.8$	$m_L = 0.3$	0.545

1.2.3 Mobility

Mobility describes the ease of carrier transport through the semiconductor crystal lattice. Electrons travel in a straight line until their movement is perturbed by a lattice, impurity, or some other scattering mechanism (Ilegems and Montgomery, 1972). Increasing the temperature of the lattice above 0 K induces a perturbation of the atoms from their equilibrium positions. This creates an oscillating potential field that scatters the charged carriers with the loss in energy transferred as phonons (Crouch et al., 1978). With increasing thermal energy, the frequency of this interaction increases and thus the mobility decreases with increasing temperature. In nonpolar Si, the elastic scattering dependence on *acoustic phonon* interaction can be approximated by $\mu_l \approx (m^*)^{-5/2}T^{-3/2}$, where m^* is the effective mass of the carrier (Sze, 1982).

Optical phonon scattering is significant for polar semiconductors such as GaAs. An oscillating dipole is created by the motion of negative and positive atoms in the unit cell of a polar crystal (Li, 1993). The vibration of this dipole is the polar optical-mode phonon. The interaction of conduction band electrons with optical-mode phonons manifests as a displacement of the atomic polarization and is a significant scattering mechanism for moderately doped GaAs at room temperature.

Ionized impurity scattering is an elastic process where charged carriers are deflected by the Coulomb potential of an ionized donor or acceptor. The mobility dependence on ionized impurities can be approximated by $\mu_i \approx (m^*)^{-1/2}T^{3/2}N_I^{-1}$, where N_I is the density of ionized impurities.

Scattering by *dislocations* can be viewed as a line charge with an effect similar to a charged center or as a strain field with an associated deformation potential. The mismatched growth of GaN and its related alloys on Si substrates generates density of dislocations of 10^8–10^{10} threading dislocations cm⁻³. At this level, the dislocation dominates most of the empirical properties of GaN with the mobility found to follow a $\mu_{dis} \approx (n)^{1/2}/N_{dis}$ relationship, where n is the net carrier concentration (Choi et al., 2001).

A general equation to calculate the electron mobility in a semiconductor is $\mu_n = q\tau_c/m^*$, where τ_c is the mean (reciprocal average) free time between collisions.

The effective mass, m^*, the hole or electron is related to the curvature of the conduction or valence band, respectively. Thus, a material such as InSb with a small effective mass has a large mobility, which is advantageous for high-frequency operation.

Drift is charge carrier motion in response to an electric field. Drift current, $I = qnv_d A$ is the amount of charge, n, crossing a unit plane, A. The momentum gained by an electron in an electric field is equal to the momentum lost upon collision, $m^* v_d = -qE\tau_c$, where the drift velocity is $v_d = -qE\tau_c/m^*$ and the mobility, $\mu_e = -q\tau_c/m^* = v_d/E$, is a measure of the velocity per unit of applied electric field. The current density is $J = qnv_d = q\mu_n nE$.

At low electric field, the velocity is linearly related to the electric field, $v_d = \mu_e E$ and the constant mobility factor, μ_e, is independent of the electric field. Under high electric fields, the mobility decreases and the velocity will approach a saturation velocity. At high electric field, electrons gain an effective electron temperature that is higher than the (ambient) lattice temperature. The increase in energy increases optical phonon scattering and the velocity approaches a saturation velocity, v_{sat}, as described by $v(E) = \mu E/(1 + \mu E/v_{sat})$.

An examination of the drift velocity in the III–V semiconductors in Figure 1.5 shows that the velocity peaks and then decreases rather than approaching a saturation velocity. The band diagram of GaAs in Figure 1.4 shows high-mobility valley ($\mu \approx 8000 \, cm^2 \, V^{-1} \, s^{-1}$) at the Brillouin zone center where the effective mass of the electron is $0.068m_0$. At a slightly higher energy ($\Delta E = 0.31 \, eV$), a low-mobility ($\mu \approx 100 \, cm^2 \, V^{-1} \, s^{-1}$) satellite valley, with an effective mass of the electron is $1.2m_0$, is found along the $\langle 111 \rangle$ axes. At high electric fields, this low-mobility valley is readily populated as its density of states is approximately 70 times that of the high-mobility valley (Sze, 1982).

The description of carrier transport begins to break down as the length, L, of the conduction channel approaches the average distance between scattering events. If the distance traveled is less than a mean free path between scattering events, then the electron (in theory) will accelerate similar to transport in a vacuum tube (Anantram et al., 2008). In general, the scattering mechanisms described above can be classified as either elastic or inelastic collisions. The interaction between electrons and fixed impurities are elastic as there is minimal energy transfer. Conversely, inelastic scattering arises from the collision between electrons and electrons, or electrons and quantized lattice vibrations (phonons) (Hanson, 2007). Electrons experiencing an elastic collision will maintain its phase memory while an inelastic collision will cause an electron to lose its phase coherence. Thus, it is beneficial to define

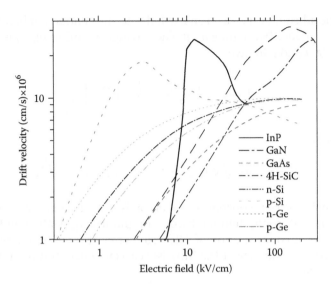

FIGURE 1.5
Carrier drift velocity dependence on applied electric field.

an average mean free length between elastic collisions, L_m, and a mean free length between inelastic collisions, L_φ. Assuming a time-independent potential, the phase of the wavefunction, $\psi(r,t) = \psi(r)e^{-Et/\hbar}$, will evolve smoothly between inelastic dephasing events.

1.2.4 Carrier Concentration

An important inherent advantage to silicon is that undoped crystal is semi-insulating, and the crystal can be easily doped n-type or p-type by doping with donor or acceptor atoms, respectively. A column-V donor atom, with sufficient thermal energy, is ionized and introduces an extra electron into the conduction band. This raises the position of the Fermi level and the electron carrier concentration in the conduction band is described by

$$n = N_c e^{(-E_c - E_f)/kT}.$$

A column-III acceptor dopant is deficient an electron compared to the host silicon atoms. The dopant will covalently bond to the silicon matrix by accepting an extra electron, which will leave a conducting hole in the valence band. This lowers the Fermi level and the hole concentration in the valence band is described by

$$p = N_v e^{(-E_f - E_v)/kT}.$$

In intrinsic silicon, every electron excited into the conduction band leaves a hole in the valence band, thus $n = p$. The intrinsic Fermi level, E_i, is obtained by equating n and p to give

$$E_f = E_i = \frac{E_c + E_v}{2} - \frac{kT}{2} \ln\left(\frac{N_c}{N_v}\right),$$

and, similarly,

$$n_i = n = p = \sqrt{N_c N_v} e^{-(E_c - E_v)/2kT} = \sqrt{N_c N_v} e^{-E_g/2kT}.$$

The intrinsic carrier concentration of Si is $1.4 \times 10^{10}\,\text{cm}^{-3}$ and of Ge is $2.0 \times 10^{13}\,\text{cm}^{-3}$ at room temperature. Comparatively, GaN yields a low intrinsic carrier concentration of $2.25 \times 10^{-10}\,\text{cm}^{-3}$ at room temperature. The temperature dependence of the intrinsic carrier concentration in GaN can be modeled as

$$n_i(T) = 1.98 \times 10^{16} \cdot T^{2/3} \cdot \exp\left(-\frac{20488}{T}\right).$$

The low intrinsic carrier concentration at room and elevated temperature makes GaN ideal for high-power and -temperature applications. The direct bandgap limits the minority carrier lifetime in GaN, thus this material is favored for high-power unipolar devices. Another important wide bandgap semiconductor material, SiC, has an indirect bandgap, which exhibits a longer minority carrier lifetime and, therefore, is favored for high-power bipolar devices.

1.2.5 Fermi Stabilization Theory

The mathematical descriptions of semiconductors do not provide an intuitive understanding as to why a particular intentionally undoped semiconductor tends be n-type, p-type, or semi-insulating. Similarly, it has been found to be difficult in changing the native conductivity of a particularly semiconductor, e.g., ZnO is n-type and can easily be doped to a highly degenerate n^{++} carrier concentration; however, p-type doping in ZnO has proven to be extremely challenging.

A relatively new understanding of defect formation and dopant activation in semiconductors has been developed based on the Fermi stabilization level concept (Khanal et al., 2007). Figure 1.6 shows the conduction and valence band edge, which defines the bandgap, relative to the vacuum level for several semiconductors (Wu et al., 2003). Numerous seemingly unrelated studies found that the Fermi level in a semiconductor always trended to a point approximately 4.9 eV from the vacuum level—the *Fermi stabilization level*.

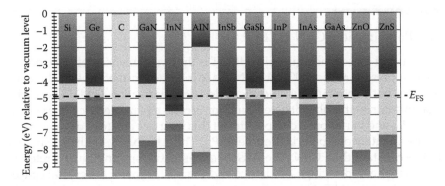

FIGURE 1.6
Position of semiconductor bandgap relative to the vacuum level. The location of the Fermi stabilization level relative to the conduction and valence band edge can be viewed as the tendency of the semiconductor to form n-type, semi-insulating, or p-type. It is energetically unfavorable to move the Fermi level approximately 800 meV above or below the Fermi stabilization level at approximately 4.9 eV from the vacuum level.

For example, any surface of a semiconductor inherently possesses broken bonds and it was found that the Schottky barrier height is often pinned at the Fermi stabilization level. Similarly, proton or neutron irradiation produces damage on the surface and within the bulk of the semiconductor, and here again the Fermi level is found to trend with increasing damage toward the Fermi stabilization level (Wu et al., 2004).

In essence, any perturbation to the semiconductor crystal, including irradiation, severed bonds on a surface, dislocations, high doping levels, and implant damage, drives an equilibrium process to generate defects (Wu et al., 2002). This native defect system has a wave function independent of the semiconductor host (Yu et al., 2005). This analysis is quite insightful to understanding why some semiconductors show superior properties for a particular carrier, e.g., Ge is an excellent p-type channel material as it is easily doped p-type with high-hole mobility.

Another viewpoint of this theory is the amphoteric defect model, which states an equilibrium process to form defects is increasingly energetically favorable as the Fermi level is driven above or below the Fermi stabilization level. Thus, an intrinsically n-type material such as InN can be heavily doped with a Mg acceptor atom but the crystal will respond by generating an equivalent level of vacancies that as a complex will passivate the acceptor atoms. The surface of any semiconductor intrinsically possesses severed bonds that reconstruct to minimize the free energy of the local system. A nanowire is particularly interesting since its properties are dominated by its surface due to its extremely large surface-to-volume ratio. For example, Simpkins et al. (2008) recently found that the surface states of a GaN nanowire create a depletion region that encompasses the conductive interior of the wire with an insulating shell.

1.3 Polarization/GaN HEMT

The basis of a III-nitride thin-film HEMT is the two-dimensional electron gas (2DEG) present at the (0001) AlGaN/GaN interface formed due to the existence of both spontaneous and piezoelectric polarization fields. The variation in strain at this abrupt heterojunction creates a piezoelectric polarization field, P_{pz}, and variation in composition at this interface creates a discontinuity in the spontaneous polarization field, P_{sp}. The AlGaN crystal has a smaller lattice constant than GaN. The AlGaN a-axis lattice spacing must stretch to match to the underlying GaN lattice and this distortion of the crystal causes the c-axis of the AlGaN layer to contract. The composition-dependent strain for a thin $Al_xGa_{1-x}N$ layer on a relaxed GaN layer is depicted in Figure 1.7a.

The III-nitride crystal consists of an alternating sequence of metal (Al, Ga, In) atoms and N atoms in the c-axis direction (Mastro et al., 2005). The strain-induced distortion of the III-nitride atoms away from their equilibrium positions creates a piezoelectric polarization as depicted in Figure 1.7b.

The spontaneous polarization exists at a discontinuity in the composition of crystal regardless of the strain state of the structure. In contrast, the piezoelectric polarization field is only generated by strain within the crystal. A common example is a 25 nm $Al_{0.3}Ga_{0.7}N$/thick-GaN where the positive fixed spontaneous and piezoelectric polarization at the interface produces a charge, which induces the accumulation of negatively charged electrons as a 2DEG. Therefore, it is vital to understand the limit or *critical thickness* that a layer can be grown and generate the full piezoelectric polarization.

The critical thickness of AlGaN on a thick GaN layer is displayed in Figure 1.8. Examination of Figure 1.8 shows that the standard HEMT 25 nm $Al_{0.3}Ga_{0.7}N$ layer is slightly below the critical thickness. In this calculation,

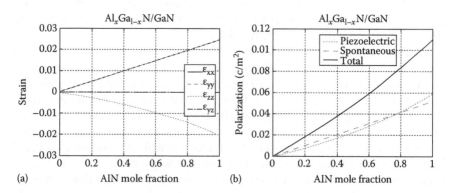

FIGURE 1.7

(a) Elastic strains in a pseudomorphic $Al_xGa_{1-x}N$ film as a function of mole fraction, x, on a relaxed GaN layer, and resultant (b) piezoelectric, spontaneous, and total polarization fields calculated using the polarization coefficients given in Vurgaftman and Meyer (2003).

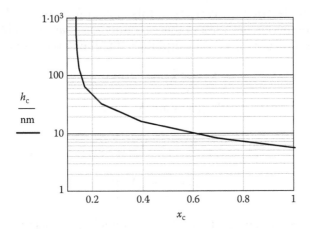

FIGURE 1.8
Critical thickness, h_c, as a function of AlN mole fraction, x_c, for an $Al_xGa_{1-x}N$ layer on relaxed GaN layer. The calculation employed Fisher model (10^{10} dislocation cm^{-3}) with Peierls barrier. Note that a 25 nm $Al_{0.3}Ga_{0.7}N$ layer or a 5 nm AlN layer on GaN layer is slightly below the critical thickness.

the dislocation density was assumed to be equal to 10^{10} dislocations cm^{-2}. The Peierls barrier, which accounts for the energy required for dislocation movement along different lattice planes, was incorporated into the Fischer theory, which accounts for the interaction between existing defects with the misfit strain field. It is quite common for all major misfit strain theories such as Matthews and Blakeslee (Kirby et al., 1979), Dodson and Tsao (Jain et al., 1995), and Fisher (Luri and Suhir, 1986) to under-predict the critical thickness of the thin film. In actuality, it is difficult to experimentally define the exact critical thickness as there is likely a small transition between the onset of relaxation and full relaxation (Vescan et al., 1993).

The high current density of the AlGaN/GaN heterostructure is chiefly a result of the large polarization-induced field and large conduction band offset in the AlGaN/GaN system. This polarization field has both a piezoelectric, strain-induced, component as well as a built-in spontaneous polarization component. The total polarization field induces a 2DEG that is approximately linearly proportional to the AlN-mole fraction $x(0 \leq x \leq 1)$ as $n_s = x(5 \times 10^{13}$ cm$^{-2})$ (Mastro et al., 2003). For the 30% Al (the composition most widely studied), a channel sheet electron density of approximately 1.5×10^{13} cm^{-2} can be realized, which is a factor of 5–10 times higher than typical GaAs or InP pHEMTs Moreover, this formation of the 2DEG at the AlGaN/GaN interface with a subsequent high electron mobility ($\mu > 1500$ cm^2 V^{-1} s^{-1}), contributes to a low on resistance (low knee voltage) as channel resistance is related to $1/(qn_s\mu E)$ at low electric field. All these factors have allowed the development of AlGaN/GaN HEMTs with high breakdown voltage at high current density that can sustain high channel operating temperature (Xu, 2000).

To properly describe the carriers in this system requires a self-consistent Schrodinger–Poisson calculation that accounts for quantum subbands. The total polarization $(P = P_{sp} + P_{pz})$ incorporates into the Poisson equation as

$$\nabla \cdot D = \nabla \cdot (\varepsilon E + P) = \rho,$$

$$\nabla \cdot [(\varepsilon(-\nabla\phi)] + \nabla \cdot P = \rho,$$

assuming $\nabla\varepsilon \to 0$, then

$$\nabla^2\phi = -\frac{\rho}{\varepsilon} + \frac{1}{\varepsilon}[\nabla \cdot P],$$

where the polarization occurring at hetero-interface results in a fixed polarization charge, ρ^{Pol}, to give

$$\nabla^2\phi = -\frac{\rho}{\varepsilon} - \frac{\rho^{Pol}}{\varepsilon}.$$

The charge, ρ, in the semiconductor is primarily composed of hole carriers, p, electron carriers, n, ionized donors, N_D^+, and ionized acceptors N_A^-; however, other effects are often incorporated, including the donor and acceptor traps (Freeman, 2003). Substituting these factors gives a functional form of the Poisson equation as

$$\nabla^2\phi = -\frac{q}{\varepsilon}\left[N_D^+ - N_A^- + p - n\right] - \frac{\rho^{Pol}}{\varepsilon},$$

which describes the formation of charged carriers in semiconductor in response to the potential field. As can be seen in Figure 1.9, the polarization field creates an enormous density of electron carriers on the GaN side of the $Al_{0.3}Ga_{0.7}N$/GaN interface even without intentional doping. This is in contrast to AlGaAs/GaAs-based HEMTs, in which carriers from the larger bandgap modulation-doped AlGaAs layer accumulate in a triangular potential on the GaAs-side of the interface to create the 2DEG.

An alternative structure is to form a HEMT based on an AlN/GaN structure. This binary layer accesses a higher theoretical polarization field (Figure 1.7) but the critical thickness is approximately 5 nm (Figure 1.8). Other complications arise such as the difficulty in forming an Ohmic source and drain contacts to the AlN. This can be eliminated by forming an (Al) GaN contact layer. Also, the thin AlN layer allows a significant interaction between the Schottky gate and the 2DEG channel. This interaction can be diminished by inserting a thin (Al)GaN cap or a high-κ dielectric.

A rather recent innovation is to flip the entire structure to form the crystal along the (000–1) N-polar direction. The equations that describe the

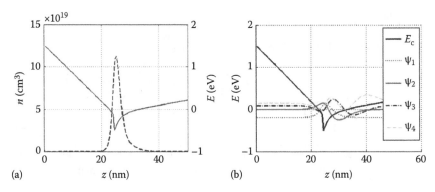

FIGURE 1.9
Standard HEMT (a) carrier concentration (dashed line) and conduction band (solid line), and (b) conduction band (solid line) and the first four subbands for a 25 nm $Al_{0.3}Ga_{0.7}N/GaN$ structure. Incorporating quantum mechanics correctly shows that the peak sheet carrier is actually slightly displaced from the $Al_{0.3}Ga_{0.7}N/GaN$ interface. A similar displacement is seen at the oxide/Si interface in inversion mode MOSFETs.

spontaneous and piezoelectric polarization are still valid except the sign is inverted. In an N-polar GaN/AlGaN/GaN structure, the polarization field at the top GaN/AlN interface creates a 2DEG in the upper GaN layer. Thus, the AlGaN layer effectively does not participate in the contacting or current transport to the Ohmic contacts. The challenge of this design is from the unstable energetics of the N-face GaN surface leads to a rougher morphology of the as-grown surface. The difference in bonding and work function of the N-polar surface requires new process recipes to be developed for the Ohmic metallization.

The default substrate for high-performance AlGaN/GaN HEMTs is silicon carbide, which has a relatively small lattice and thermal expansion mismatch with GaN, and has a high thermal conductivity that minimizes self-heating in the active device (Binari et al., 2001). The thermal expansion mismatch is especially detrimental as a simple GaN layer grown over 1 μm in thickness on Si will form a high density of cracks during cool-down from growth temperature. This tensile stress from the Si can be mitigated by step-grading AlGaN to GaN or introducing low-temperature Al(Ga)N layers within the buffer layer (Dadgar et al., 2002). These techniques are designed to introduce compressive stress during growth to mainly counteract the large tensile stress generated during cool-down from growth temperature (Figure 1.10).

The advantage of SiC over Si in thermal conductivity, as can be seen in Figure 1.11, is important for devices self-generating heat by operating at high-frequency and high-power levels. It was once assumed that the thermal resistance of the silicon wafer would be an unavoidable deterrent to GaN on Si HEMT technology; however, the impact of the Si thermal resistance on self-heating effects has been addressed by thermal design and advanced packaging (Johnson et al., 2008).

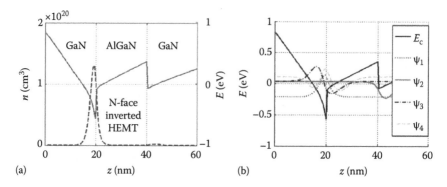

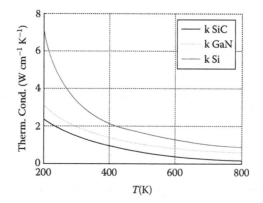

FIGURE 1.10
Inverted HEMT (a) carrier concentration (dashed line) and conduction band (solid line), and (b) conduction band (solid line) and the first four subbands for an N-face 20 nm GaN/20 nm $Al_{0.3}Ga_{0.7}N/GaN$.

FIGURE 1.11
Bulk thermal conductivity of SiC, GaN, and Si (Glassbrenner and Slack, 1964). The difference in thermal conductivity between the materials decreases with increasing temperature. Additionally, the thermal conductivity of SiC is often much lower than depicted due to the defective nature of available substrates (Nilsson et al., 1997). Note a typical GaN HEMT operating temperature is 450 K.

One important step was to thin the Si substrate to minimize the thermal resistance, R_{th}, of the substrate. Commercial products are now available where the Si substrate is thinned down to 4 nm to reduce the thermal impedance normalized to the substrate width to approximately 35°C mm W^{-1}. At this point, the thermal impedance of the package is comparable to the R_{th} from the substrate. The use of an advanced copper-flanged air cavity package has allowed commercialization of a GaN on Si HEMT product that can operate at a power density of 650 W cm^{-3} (Johnson et al., 2008).

In general, the maximum RF output power density attainable in GaN HEMTs is less than predicted due to current dispersion via deep levels and surface states (Mittereder et al., 2003). Surface states lead to a depletion of the 2DEG; however, these surface states can be minimized through the deposition of passivation films on the AlGaN surface. Deep-level traps have proved more difficult to control. It has been shown experimentally that carriers are trapped by dislocations (possibly grain boundaries) and carbon impurities (Klein et al., 2001), deteriorating the high-frequency response. This effect can be minimized but it is inherent to the nitride system as the material is defective due to its growth on mismatched substrates. This variability found in post-processing testing is limiting the introduction of this device into commercial systems. This reliability issue will be revisited later in the book.

1.3.1 Polarization in InGaN

Similar strain and polarization fields are generated for (0001) InGaN/GaN interfaces as can be seen in Figure 1.12. The active material in all blue and green optoelectronics is InGaN. The actual device structure is complex but the device essentially operates by injecting holes and electrons into the InGaN quantum well from an upper p-GaN layer and a lower n-GaN layer, respectively. There is a large positive polarization field at the upper GaN/InGaN interface and large negative polarization field at the lower InGaN/GaN interface. This creates a gradient in the conduction and valence bands, which is well known to decrease the overlap of the electron and hole wavefunctions, and consequently decrease the recombination efficiency. High InN content green light emitters are particularly plagued by the quantum-confined stark effect, which causes a blueshift in the emission peak with increasing drive current (Feezell et al., 2009).

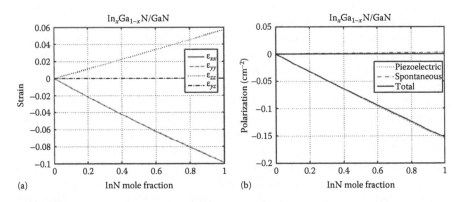

FIGURE 1.12
(a) Elastic strains and (b) polarization in a pseudomorphic In$_x$Ga$_{1-x}$N film as a function of mole fraction, x, on a relaxed GaN layer. The contribution of spontaneous polarization is minimal across the entire InGaN compositional range.

1.4 CMOS Inverter

A CMOS VLSI chip has three primary parameters: integration density, switching speed, and power dissipation. The various circuit elements are built from the two fundamental building blocks: nMOSFETs and pMOSFETs. An early enabler of LSI logic was to design the basic circuit elements to have negligible standby power. An excellent example is the inverter. The simple circuit inverts a high-voltage (logic 1) input signal to a zero out logic signal, or inverts a zero-voltage input signal to a high-voltage output signal. The design of this circuit allows a large noise margin of operation. We will describe the basic design of an inverter as a baseline for discussion of the future of Si electronics and the potential for integration of compound semiconductors within the transistor.

The classic approach to building CMOS transistors is to connect the body p-substrate terminal of an nMOSFET to the lowest voltage ground, and to connect the body n-well terminal of the pMOSFET to the highest voltage of the power supply V_{dd}. A CMOS inverter is formed by connecting the nMOSFET source to the ground and the pMOSFET source to V_{dd} as is shown in Figure 1.13. The gates of the n- and pMOSFETs are linked as the common input node, and the two drains of the n- and pMOSFETs are linked as the common output node.

The nMOSFET and pMOSFET are not symmetric devices. The lower hole mobility of the pMOSFET will intrinsically lead to a smaller drive current. An equivalent nMOSFET of equal width, has a higher carrier (in this case electron) mobility and thus higher drive current. Assuming matched channel lengths and threshold voltages, then $I_n/I_p \approx \mu_n/\mu_p$. A CMOS inverter compensates the mobility difference by increasing the pMOSFET channel width W_p relative to nMOSFET channel width W_n by the ratio $W_p/W_n = I_n/I_p$. This widening of the p-channel width results in a shift in drive currents as is displayed in step (a) in Figure 1.14.

As mentioned above, the critical feature of an ideal CMOS inverter is that only one transistor is on at steady state, thus there is no static power dissipation. Figure 1.13 and step (c) in Figure 1.14 shows illustrate the proper voltage reference levels to achieve an inverter with no static current in the steady state. The pMOSFET has the same operating principles as the nMOSFET, except all the polarities of the currents and voltages are reversed. For example, the pMOSFET is turned on by negative voltage ($-V_{dd}$) on the gate while the nMOSFET is turned on by a positive voltage (V_{dd}) on the gate.

As shown in step (b), the orientation of the pMOSFET is flipped to change the sign of the current. The gate-source voltage of the pMOSFET is referenced to the power supply (V_{dd}) while the nMOSFET is referenced to the ground (0 V). Thus, when $V_{in} = V_{dd}$, the nMOSFET is on and the pMOSFET is off. Conversely, when $V_{in} = 0$, the nMOSFET is off and the pMOSFET is on.

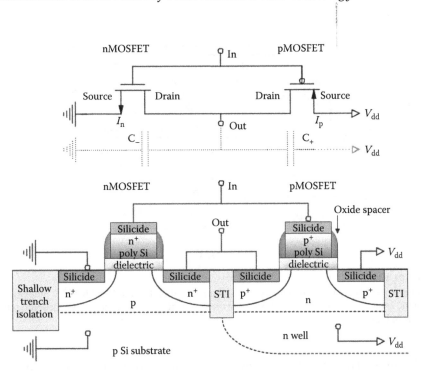

FIGURE 1.13
Circuit diagram and cross-sectional schematic of a CMOS inverter. The two gates are connected as the input node and the two drains are connected as the output node. The design of an ideal inverter implies that for any voltage on the input, one of circuits is open and thus no power will flow. The source of a pMOSFET is connected to the power supply V_{dd} in a CMOS circuit so the device conducts if the gate voltage is lower than $(V_{dd} - V_t)$. The pMOSFET n-well is contacted to V_{dd} while the nMOSFET is tied to the ground.

Figure 1.13 and step (c) in Figure 1.14 shows that the source of the nMOSFET is connected to the ground. Thus, when the nMOSFET is on, the n-channel provides a low resistance path between the ground (on the source) and the drain. Therefore, the voltage on the output node, which is connected to the drain side of the nMOSFET, is *pulled-down* to zero (ground). Conversely, when the pMOSFET channel is on, the p-channel provides a low resistance path to the supply voltage, which is connected to the source of the pMOSFET. The voltage on the output node, which is connected to the drain on the pMOSFET, is *pulled-up* to the supply voltage (V_{dd}).

A graphical analysis provides insight into the relation between shape of the n- and pMOSFET *I–V* curves and the resultant voltage transfer characteristic. Starting with a circuit with $V_{in} = V_{dd}$ and $V_{out} = 0$, when the input voltage if instantly changed to ground, $V_{in} = 0$, the nMOSFET will be off and the pMOSFET will turn on as now $V_{GSp} = -V_{dd}$. Initially $V_{SDp} = -V_{dd}$ because the source of the pMOSFET is connected to the power supply (V_{dd}) and the source voltage is still equal to the zero output voltage, $V_{out} = 0$. In the figure below, this corresponds to the top-left point on the thick dashed pMOSFET

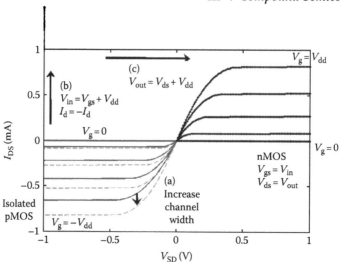

FIGURE 1.14

Drain current of an n-channel (top right) and a p-channel (bottom left) MOSFET. Steps (a,b,c) are the essence of constructing a CMOS inverter as is also shown in the cross-sectional schematic in Figure 1.13.

I–V curve. The pMOSFET flows current into output node (on the drain), which pulls up the output voltage to V_{dd}. The circuit is now steady state with $V_{out} = V_{dd}$ and $V_{in} = 0$. Similarly, the circuit can again be inverted by switching V_{in} to V_{dd} and pulling down V_{out} to zero through the nMOSFET. At the steady-state points, no current flows as $I_{DSp} = I_{DSn} = 0$.

At any input voltage between $V_{in} = 0$ and $V_{in} = V_{ds}$, the output voltage, V_{out}, is found by equating $I_n(V_{in}) = I_p(V_{in})$. Graphically (Figure 1.15), this is found as the intercept of nMOSFET and pMOSFET load curves at a common input voltage. This set of V_{out} vs. V_{in} is transfer curve of the CMOS inverter.

The shape of the nMOSFET and pMOSFET *I–V* curves and corresponding shape of the transfer curve (Figure 1.16) impacts several performance factors of the inverter circuit and, by extension, performance metrics of the entire integrated circuit. The regions of stable operation (or noise margin) are determined by the sharpness of the high-to-low transition in $V_{in}–V_{out}$ plot below. It is necessary to note that offsetting the lower mobility and thus I_{DSp} of the pMOSFET by increasing the channel width shifts the central unstable region of the transfer curve to a higher range of input voltage. This effect can be beneficially used to create a symmetric transfer curve or an asymmetric transfer curve to offset a noisy voltage source.

Substituting Ge or III–V materials as the channel layer can present quite different transistor current–voltage curves due to differences in material parameters including mobility, velocity saturation, and high-field scattering. The derivation of the CMOS inverter, which is just one of several circuits found in an IC, highlights this effect.

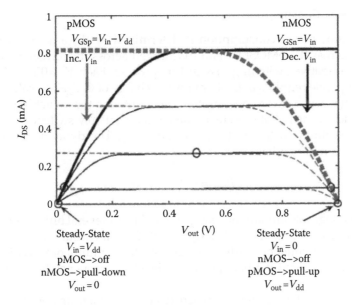

FIGURE 1.15
Load curves for nMOSFET (solid line) and pMOSFET (dashed line) of a static CMOS inverter with the power supply voltage $V_{dd}=1$ V. The circles represent steady-state operation where the intercept of $I_p(V_{in})=I_n(V_{in})$ determines the matching V_{out}.

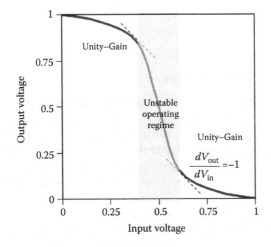

FIGURE 1.16
The unity gain points determine the unstable operating regime of the CMOS inverter. Ideally, the inverter is operating with the nMOSFET or pMOSFET in a resistive mode at low or high input voltage, respectively.

1.4.1 Power Consumption

The dynamic power consumption is determined by the amount of energy charged across the pMOSFET and discharged across the nMOSFET as well as any dissipative switching loss in this process. Each $V_{in}=0\,V \rightarrow V_{dd} \rightarrow 0\,V$ switching cycle requires a fixed amount of charge of $C_L V_{dd}^2$. A realistic measure of dynamic power consumptions accounts for the switching activity, $S_{0 \rightarrow 1}$, which measures the average rate a circuit actively switches compared to the maximum switching (clock) frequency, f. The dynamic power consumption can be written as

$$P_{dym} = C_L V_{dd}^2 S_{0 \rightarrow 1} f.$$

The unmanageable amount of heat generated from several billion transistors led to a major change in VLSI design away from a constant march toward higher clock frequency; rather designs such as multi-core architectures have increased computational power as the clock frequency has been fairly constant.

The current CMOS scaling has reached a regime where an appreciable current flows in the off-state due to the short source–drain spacing. Over the last several design generations, the off-current has increased exponentially while the on-current has remained relatively constant at approximately $1\,mA\,\mu m^{-1}$, thus the switches are no longer ideal. One effect is that the standby power is no longer negligible according to

$$P_{off} = N_G I_D (off) V_{dd},$$

where N_G is number of gates.

1.4.2 Transient Circuit Behavior

This response of this circuit is dominated by the load capacitance C_L, which consists of the input capacitance of the fan-out gates, the capacitance of the connecting wires, and the drain diffusion capacitances of the nMOSFET and pMOSFET. The propagation delay of the CMOS inverter is the time to charge through the pMOSFET ($\approx R_p C_L$) and discharge through the nMOSFET ($\approx R_n C_L$). There are a myriad of interrelated factors but the speed of the circuit can be increased by decreasing the on-resistance of the nMOSFET and pMOSFET as well as minimizing the output capacitance.

Focusing on the delay at the individual transistor, here delay is approximated by CV_{dd}/I_{Dsat}, where V_{dd} is the drain current, I_{Dsat} is the drain current at $V_{dd} = V_{DS} = V_{GS}$. For circuits dominated by intrinsic transistor delay, the total load capacitance, C, is typically assumed to equal the gate capacitance in inversion, C_{inv}. This metric focuses on the switching charge $C_{inv} V_{dd}$ and excludes the gate and other parasitic capacitances.

In a modern CMOS design, drain current does not reach I_{Dsat} (Sutherland et al., 1999). One improvement is to replace the on-current I_{Dsat} with the average (effective) switching current I_{eff} (Khakifirooz and Antoniadis, 2008). At short channel lengths, the drain induced barrier lowering (DIBL) significantly decreases I_{eff} relative to I_{Dsat}. In modern short-channel MOSFETs, the saturation drain current is essentially a linear function of gate voltage, which corresponds to a near-constant transconductance (Weste and Harris, 2005). The transistor current in saturation can be expressed as

$$I_D = WC'_{inv}(V_{GS} - V_t)v,$$

where the width W normalizes the drain current and v is approximately the effective average velocity of carriers at the energy barrier near the source (Khakifirooz and Antoniadis, 2008). The virtual source velocity v_{x0} provides a more accurate description by

$$v_{x0} = \frac{v}{1 - C'_{inv}R_SW(1+2\delta)v}.$$

This includes a correction for the source series resistance due the DIBL coefficient, $\delta = \partial V_t/\partial V_{DS}$, which is an increase in threshold voltage from the voltage drop across the source and drain series resistances.

1.4.3 Transistor Design: Linear Regime

A MOSFET above threshold and at a low source-to-drain bias has large induced channel charge per unit area at any point x given by

$$Q_i(x) = -C_{ox}(V_{GS} - V_T - V(x)),$$

where the potential, $V(x)$, varies from zero at the source to V_d at the drain. The capacitance per unit area presented by the gate dielectric is

$$C_{ox} = \frac{\varepsilon_{ox}}{t_{ox}} = \frac{\kappa\varepsilon_o}{t_{ox}},$$

where
   ε_{ox} is the oxide permittivity
   ε_o is the permittivity of free space
   κ is the relative dielectric constant
   t_{ox} is the thickness of the oxide

The gate oxide thickness is scaled down to increase the charge in the channel and current in the device. Unfortunately, the scaling of the silicon oxynitride dielectric down to 1.2 nm resulted in a significant increase in quantum-mechanical-based tunneling. Thus, the industry recently switched to a hafnium dioxide or nitrided hafnium silicate (HfSiON) process to increase the dielectric constant.

Perhaps, the greatest historical advantage of silicon for microelectronics is actually its ability to form a high-quality native oxide. The circa 2007 shift away from a (nitrided) silicon oxide dielectric was a one of the biggest challenges in the several-decade history of silicon microelectronics. Linked to this issue was the move to a metal gate electrode, which replaced the polysilicon gate electrode. It was found that Fermi level pinning from defects at the polysilicon/high-κ dielectric interface unacceptably increases the turn-on voltage. Additionally, surface phonon scattering by the high-κ dielectric limited mobility in the channel; however, metal gates are effective at screening the phonon scattering from coupling to the channel and thus improve the mobility under inversion (Chau et al., 2004). The switch away from the native oxide on Si and polysilicon gates removes two major rationalizations to why the Si channel should not be replaced with a compound semiconductor material. The higher mobility of quantum well InSb transistors results in a superior CV/I metric, which is inversely related to the switching frequency, f_t.

A low barrier height at the Schottky metal–AlInSb/InSb QW junction results in high gate leakage and resultant low I_{ON}/I_{OFF} ratio. Chau et al. (2005) showed that a high-κ dielectric is requisite to reduce drain leakage in this narrow-gap QW transistor.

The voltage difference applied between the source and the drain causes the current to flow. Current through the channel is a product of width of channel, W, mobile charge, Q, and the carrier velocity, v, which is expressed as

$$I_D = WQ_i(x)v_x(x)$$

Since current is constant across the channel, due to charge conversation, one can evaluate the current at any position along the channel. A straightforward technique is to calculate the current at $x = 0$ as

$$I_D^{@x=0} = WQ_i(0)v_x(0),$$

with the charge at $x = 0$ given as

$$Q_i(0) = -C_{ox}\left(V_{GS} - V_T - \left[V(0) = 0\right]\right).$$

This partially eliminates the drain voltage dependence; however, the drain voltage reenters through effective mobility, μ_{eff}, which is a product of the electric field along the channel,

$$E_x = \frac{(V_{DS} - 0)}{L}.$$

Substituting for the carrier drift velocity gives the above threshold current as

$$I_D = WC_{ox}(V_{GS} - V_T)\mu_{eff}E_x = \left(\frac{W}{L}\right)\mu_{eff}C_{ox}(V_{GS} - V_T)V_{DS}.$$

This equation describes the MOSFET in the linear region of the *I–V* curve, where MOSFET behaves as a voltage-controlled resistor.

1.4.4 Transistor Design: Saturation Regime

As the drain-to-source voltage is increased, a point is reached where, $V_{DS} > V_{GS} - V_t$, i.e., the voltage at the drain exceeds the voltage applied by the gate (above the threshold voltage). There is no longer an inversion layer charge near drain and the channel is *pinched off*. There is no voltage drop (for an ideal device) across the pinch-off region near the drain. The voltage drop over the induced channel (from the source to the beginning of the pinch-off region) is $V_{GS} - V_t$. Thus, the current does not increase for any reasonable increase in V_{DS} above saturation.

The electric field in the inversion layer is limited to the gate voltage as

$$E_x = \frac{\left[(V_{GS} - V_T) - 0\right]}{L}.$$

Note the contrast to the equation for the electric field in the linear regime given above. The average velocity is $v = \mu E_x$ and the since the potential field is limited and the pinch-off region is small, L can be assumed to be relatively unchanged. This creates a squared dependence on gate voltage for the drain current as

$$I_D = \left(\frac{W}{L}\right)\mu_{eff}C_{ox}(V_{GS} - V_T)(V_{GS} - V_T) = \left(\frac{W}{2L}\right)\mu_{eff}C_{ox}(V_{GS} - V_T)^2,$$

where the factor of 1/2 was introduced to account for non-linearities at large source-to-drain fields.

1.4.5 Transistor Design: Velocity Saturation

Under low electric fields, carrier velocity is proportional to the applied electric field by $v = \mu E$, where mobility, μ, is the constant of proportionality. At high fields, the carrier velocity will saturate or even, in the case of GaAs, begin to decrease. Since the electric field is the ratio of applied drain voltage over a given distance, velocity saturation is common in the short channel lengths found in modern transistors. For example, a source–drain bias of $V_{DS}=1$ V over a channel length $L=60$ nm yields an electric field of 1.67×10^5 V cm^{-1}, which is much greater than the approximate 3×10^3 critical field for electron carrier saturation in Si.

Above threshold, the charge $[Q_s \sim \exp(q\Psi_s/2kT)]$ under the gate is strongly dependent on any change in surface potential, Ψ_s. This strong inversion occurs when the surface potential exceeds

$$\Psi_s^{inv} = 2\Psi_B = 2\frac{kT}{q}\ln\left(\frac{N_a}{n_i}\right),$$

where $\Psi_B = \Psi_{fermi} - \Psi_{intrinsic}$. The incremental increase in gate field does not result in a further increase of the depletion layer width; rather the charge is contained within the inversion layer and the electrons effectively shield the underlying Si (Taur and Ning, 1998).

1.4.6 Subthreshold MOSFET

At a subthreshold voltage $(V_G < V_T)$, the charge under the gate is weakly dependent on surface potential with $Q_s \sim (\Psi_s)^{1/2}$ for $0 < \Psi_s < 2\Psi_B$. The charge density is small yet non-negligible in the weak inversion regime $(\Psi_B < \Psi_s < 2\Psi_B)$. In weak inversion, the following approximation can be made to charge by

$$Q_i = -\sqrt{\frac{\varepsilon_{si}qN_a}{2\psi_s}}\left(\frac{kT}{q}\right)\left(\frac{n_i}{N_a}\right)^2 e^{q(\Psi_s-V)/kT},$$

and gate voltage by

$$V_g = V_{fb} + \Psi_s + \frac{\sqrt{2\varepsilon_{si}qN_a\Psi_s}}{C_{ox}}.$$

The dependence on gate voltage is exponential for below threshold drain current as given by

$$I_D = \mu_{eff}C_{ox}\frac{W}{L}\left(\frac{k_BT}{q}\right)e^{q(V_{GS}-V_T)/mk_BT}\left(1-e^{qV_{DS}/k_BT}\right),$$

and inverse subthreshold slope as

$$S = \left(\frac{d(\log_{10} I_{ds})}{dV_G}\right)^{-1} = 2.3\frac{mkT}{q} = 2.3\frac{kT}{q}\left(1+\frac{C_{dm}}{C_{ox}}\right),$$

which is typically 70–100 mV per decade (Taur and Ning, 1998). The body-effect coefficient accounts for charge in the bulk by

$$m = 1 + \frac{\sqrt{\varepsilon_{Si}qN_a/4\Psi_B}}{C_{ox}} = \left(1+\frac{C_{dm}}{C_{ox}}\right) = 1 + \frac{3t_{ox}}{W_{dm}},$$

where C_{dm} is the bulk depletion capacitance.

An alternate derivation to the subthreshold (and above threshold) behavior is attained by treating the current flow through the n$^+$ source/p$^+$ channel/n$^+$ drain MOSFET as a bipolar junction transistor (BJT) (Johnson, 1973). The barrier in conduction band between the source and drain prevents carrier flow. A large drain voltage pulls the conduction band down and carriers are rapidly swept out, which is similar to an npn bipolar transistor.

A direct substitution as discussed by Lundstrom (2005) {of $V_{BE} \rightarrow \psi_S$, $W_B \rightarrow L$, $Wt_{inv} \rightarrow I_{DS}$, $T_{inv} \rightarrow (k_BT/q)/E_{sd}$, $V_{CE} \rightarrow V_{SD}$, $\psi_S = C_{ox}V_g/(C_{ox}+C_D) = V_G/m$, $E_S = qN_A/[(m-1)C_{ox}]$, $(n_i/N_A)^2 = e^{-q2\psi_B/k_BT} = e^{-q\psi_T/mk_BT}$, $D_{n-} > k_BT\mu_{eff}/q$} transforms the MOSFET drain current equation into an analogous equation for the BJT collector current density given as

$$J_c = q\frac{D_n n_i^2}{W_B N_A}\left(e^{qV_{BE}/k_BT} - e^{qV_{BC}/k_BT}\right) = q\frac{D_n n_i^2}{W_B N_A}e^{qV_{BE}/k_BT}\left(1 - e^{qV_{CE}/k_BT}\right).$$

The critical region in a BJT is the width of the base, W_b, which is similar to the channel length, L, in the MOSFET. The controlling barrier in the BJT is base emitter potential, V_{BE}, while in MOSFET, the gate voltage modifies the surface potential, which indirectly changes the band structure in the channel. This analogy between a MOSFET and a BJT is even more obvious below as we introduce additional theory to describe nanoscale FET operation by emission of carriers over a potential barrier.

1.4.7 Alternative Materials as the Active Channel

Higher mobility is a conceptually simple reason to introduce compound semiconductors as the active channel in n- and pMOSFETs. The growth of lattice mismatched semiconductors introduces defects that are deleterious to device operation. These growth-related issues are discussed elsewhere in this book. If heteroepitaxy issues can be overcome, an increase in material mobility leads to a direct increase in drain current according to the long channel (low field) current equation given as

$$I_{DS} = \frac{\mu_{eff}C_{ox}(W/L)(V_g - V_t)V_{ds}(m/2)V_{ds}^2}{1 + (\mu_{eff}V_{ds}/v_{sat}L)}.$$

In a short channel device, the drain current will saturate at a lower voltage due to velocity saturation (discussed above) according to

$$I_{dsat} = C_{ox}Wv_{sat}(V_g - V_t)\frac{\sqrt{1 + 2\mu_{eff}(V_g - V_t)/(mv_{sat}L)} - 1}{\sqrt{1 + 2\mu_{eff}(V_g - V_t)/(mv_{sat}L)} + 1},$$

and

$$V_{dsat} = \frac{2(V_g - V_t)/m}{\sqrt{1 + 2\mu_{eff}(V_g - V_t)/(mv_{sat}L)} + 1},$$

where m is the body-effect coefficient ($m \approx 1.1$–1.4) (Taur and Ning, 1998). Transport in current state-of-the-art transistors are no longer adequately described by classical drift–diffusion. Current technology is approaching ultrashort channel lengths where carrier transport is at approximately 50% of the ballistic limits. Before launching into a detailed description of one theory of ballistic transport, it is important to highlight that in ballistic transport, the current is a product of the number of conductions subbands and carrier injection from the source. The high curvature of the conduction band in a material such as GaAs translates into a higher mobility but this also leads to a decrease in DOS and resultant decrease in the number of channels for ballistic transport. A top-of-the-barrier theory is presented to aid in our understanding of alternative channel material and designs for ballistic transport.

1.5 Ballistic Transport

In ultrashort channels, the drift–diffusion model fails to account for the velocity overshoot. At high field, particularly for a steep potential gradient near the drain, a significant fraction of the carriers will have energy much higher than the thermal energy. These hot carriers are no longer in thermal equilibrium with the lattice and can exceed the bulk saturation velocity. Fischetti (1998) found that within 50 nm channels, velocity overshoot can reach high levels, which leads to a terminal saturation transconductance that exceeds the velocity saturation limited value.

At extremely short channel lengths with resultant strong electric fields, the electrons traveling in the channel do not have time to scatter and maintain equilibrium with crystal lattice and have their velocity saturate (Faux et al., 1992). In the conduction band profile of a MOSFET, electrons stream across at high energy and the average velocity exceeds the saturated velocity in the bulk, especially at drain edge. At the source edge, the electron velocity is also high but the electrons have to overcome a barrier in the conduction.

The physics at the *top of the barrier* provides a satisfactory description of the behavior of a ballistic MOSFET. This discussion follows Lundstrom (2005) and details given in Lundstrom and Guo (2005) and Datta (2005). Briefly, it is necessary to relate the density of states at the *top of the barrier* with the self-consistent potential, U_{scf}, which can be modulated by the gate. The source and drain Fermi levels are equal at equilibrium and shift under the applied bias by qV_D (Figure 1.17).

The source contact fills electrons into channel with the number of electrons in equilibrium, $N_S^0(E)$, at energy, E, given as

$$N_S^0(E) = D(E - E_C) f_s(E),$$

which is the product of the number of states, $D(E - E_C)$, and Fermi function, $f_s(E)$. The rate of the process,

$$\frac{dN(E)}{dt} = \frac{\left(N_s^{\,0}(E) - N(E)\right)}{\tau_s},$$

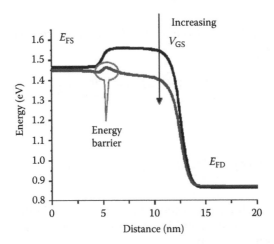

FIGURE 1.17
Energy barrier between source and drain inhibits current flow. An increase in applied gate voltage decreases the height of the barrier (Rahman et al., 2005). The source contact has a Fermi level of E_{FS} and the drain contact has a Fermi level of $E_{FD} = E_{FS} - qV_D$.

provides a description of the approach to the equilibrium value and τ_s is the characteristic time for electrons to enter or leave the device from the source contact. Similarly, the drain contact fills states into the device, according to its Fermi level, and trends to an equilibrium number given as

$$N_D^0(E) = D(E - E_C) f_D(E),$$

at a rate

$$\frac{dN(E)}{dt} = \frac{\left[N_D^0(E) - N(E)\right]}{\tau_D},$$

which is determined by the characteristic time, τ_D, and the displacement from the equilibrium value. The first derivative in time of the number of electrons in the device when both the source and the drain contacts are attached is given by

$$\frac{dN(E)}{dt} = \frac{N_S^0(E) - N(E)}{\tau_S} + \frac{N_D^0(E) - N(E)}{\tau_D} = 0,$$

which is the rate of change in electron density via the source and drain contacts. This rate is equated to zero under steady-state conditions. The total number of electrons, N, is solved by integrating over energy,

$$N = \int N(E)dE = \int [D_S(E) f_S(E) + D_D(E) f_D(E)]dE.$$

With the two Fermi levels (source and drain) and a corresponding density of states given as

$$D_S(E) = \frac{\tau_D}{\tau_S + \tau_D} D(E - E_C) = \frac{\gamma_S}{\gamma_S + \gamma_D} D(E - E_C),$$

and

$$D_D(E) = \frac{\tau_S}{\tau_S + \tau_D} D(E - E_C) = \frac{\gamma_D}{\gamma_S + \gamma_D} D(E - E_C).$$

The broadening factors, $\gamma_{S,D}$, account for broadening of energy levels due to the source or drain contact and is related to the escape time for electron to leave through contact, $\tau_{S,D}$, by

$$\gamma_S = \frac{\hbar}{\tau_S},$$

and

$$\gamma_D = \frac{\hbar}{\tau_D}.$$

Typically, the source and drain contacts are equivalent, thus

$$\gamma_S = \gamma_D = \frac{\hbar}{\tau} = \frac{\hbar}{L'/v_x},$$

where L' is length of the potential barrier.

From the electron density one can evaluate current; at steady state, the continual flow of electrons into the device from the source contact is equal to the flow of electrons out of the device through the drain. The steady-state current is

$$I(E) = \frac{q\left(N_S^0(E) - N(E)\right)}{\tau_S} = \frac{q\left(N_D^0(E) - N(E)\right)}{\tau_D}.$$

Substituting the equilibrium density of electrons entering from the source contact as well as the steady-state number of electrons at energy E gives

$$I(E) = \frac{q}{\tau_S}\left[D(E)f_S(E) - D_S(E)f_S(E) - D_D(E)f_D(E)\right],$$

$$I(E) = \frac{q}{\tau_S}\left[\frac{\tau_S + \tau_D}{\tau_S + \tau_D}D(E)f_S(E) - \frac{\tau_D}{\tau_S + \tau_D}D(E)f_S(E) - \frac{\tau_S}{\tau_S + \tau_D}D(E)f_D(E)\right],$$

$$I(E) = q\left[\frac{D(E)}{\tau_S + \tau_D}\right]\{f_S(E) - f_D(E)\},$$

$$I(E) = \frac{2q}{h}[M(E)]\{f_S(E) - f_D(E)\},$$

where the number of propagating modes that carry current is

$$M(E) = \frac{h}{2(\tau_S + \tau_D)}D(E).$$

Integrating over all energies gives the Landauer-Buttiker formula for mesoscopic devices with unity transmission, $T(E) = 1$, as

$$I_D = \frac{2q}{h}\int M(E)\left(f_1(E) - f_2(E)\right)dE,$$

where constants for this device are lumped into $M(E)$ by

$$M(E) = \pi D(E) \frac{\gamma_1 \gamma_2}{\gamma_1 + \gamma_2}.$$

The following steps will derive an analytical equation for the current–voltage relationship in a 2D planar MOSFET with one subband occupied to simplify the mathematics. The $E(k)$ dispersion is assumed to be parabolic at the top of the barrier. The 2D DOS for a planar MOSFET is

$$D(E) = WL' \frac{m^*}{\pi h^2} \Theta(E - E_c),$$

where Θ is a step function with the DOS equal zero below bottom of conduction (sub)band. The source and drain contacts are assumed to be identical and thus have the same DOS give as

$$D_S(E) = D_D(E) = \frac{D(E)}{2}.$$

Substituting the 2D DOS and changing the limits of the integral gives the number of electrons as

$$N = WL \frac{m^*}{2\pi h^2} \int_{E_C}^{\infty} \left[f_S(E) + f_D(E) \right] dE.$$

Current flows when there is a difference in Fermi levels. Integrating gives the electron density in device equal to the 2D DOS:

$$N = \frac{N_{2D}}{2} WL' \left[F_o(\eta_{F1}) - F_o(\eta_{F2}) \right],$$

where F_o is the Fermi–Dirac integral of the Fermi function with

$$\eta_{F1} = \frac{\left(E_{F1} - E_C^{FB} + q\psi_S \right)}{k_B T},$$

and

$$\eta_{F2} = \frac{\left(E_{F1} - qV_D - E_C^{FB} + q\psi_S \right)}{k_B T}.$$

describing movement of the Fermi levels. Of particular interest is the factor

$$N_{2D} = m^* \frac{k_B T}{\pi \hbar^2},$$

which is related to fundamental parameters and the effective mass material parameter, m^*. This result is startling as it states a material with a higher effective mass has a large number of states available for transport in a ballistic transistor. This is dissimilar to the drift–diffusion equations introduced earlier in the chapter, where a high effective mass is shown to result in a low mobility and thus low current in the transistor.

When only viewed from the classical regime, the low effective mass III–V materials are seen as a superior channel material to achieve higher mobilities and current in the transistor. The low effective mass also reveals a low DOS, which reduces the number of available subbands in the ballistic channel and reduces quantum capacitance in ultrashort channel MOSFETs; however, this low DOS also translates into a high electron injection velocity into a ballistic channel as will be described below (Rahman et al., 1998).

In true ballistic transport, only electrons flowing in positive direction toward the source are predicted. The electric field from the source to the drain bias and gate bias is so large that no electrons are injected from the drain. That is, no electrons are flowing with a negative velocity (or negative k-vector). To maintain equilibrium, the Fermi level at the source will be pulled down (Datta, 2005).

The current is calculated by evaluating

$$I_D = \frac{2q}{h} \int M(E) \big(f_S(E) - f_D(E) \big) dE$$

The 2D MOSFET DOS is substituted into the number of propagating modes that carry current to give

$$M(E) = \frac{h}{4\tau} \frac{m^* W L'}{\hbar}.$$

In a 2D planar MOSFET, the carriers are injected at any angle in the plane. Since the sum of over all angles, i.e., $\langle \cos \theta \rangle$ is $2/\pi$ and the energy of the ballistic carriers is $E = (1/2) m v_x^2$, then

$$\langle \tau_S \rangle = \frac{L}{\langle v_x \rangle} = \frac{L}{\sqrt{2E/m^*} \, (2/\pi)}.$$

As expected, current flows when there is a difference between two Fermi–Dirac functions according to

$$I_D = \frac{qN_{2D}}{2} W v_T \left[F_{1/2}(\eta_{F1}) - F_{1/2}(\eta_{F2}) \right],$$

with current proportional to width of MOSFET as well as the thermal injection velocity, which is calculated by

$$v_t = \sqrt{\frac{2k_B T}{\pi m^*}},$$

and is listed in Table 1.6.

Injection velocity (from source) increases with increasing source–drain bias until it saturates at the degenerate thermal injection velocity. Velocity saturates at the top of the barrier. MOSFET current is limited by the average carrier velocity at source. The carrier density

$$Q_i = C_{ox}(V_g - V_t),$$

at the source is independent of the drain voltage. The carrier velocity near the source is dependent on the energy barrier and scattering rates at the source side of the channel region as well as the thermal velocity of carrier injected from the source into the channel (Lundstrom and Guo, 2005). The thermal injection of carriers sets the upper limit current in the ballistic regime. Above threshold, the ideal MOSFET charge is

$$C_{ox}(V_{GS} - V_T) = \frac{qN}{WL}.$$

TABLE 1.6

Thermal Velocity of Prospective Material for Ballistic Transistors

Semiconductor	Bandgap (eV)	Thermal Velocity (m s^{-1} × 10^5)	
		Electrons	Holes
Si	1.12	1.23	0.77
Ge	0.67	1.90	1.02
GaAs	1.43	2.07	0.80
InP	1.35	1.94	0.67
GaSb	0.75	2.62	0.86
InAs	0.36	3.54	0.85
InSb	0.17	4.47	0.82

Rearranging the equation for the 2D DOS yields

$$N = \frac{C_{ox}WL'(V_G - V_t)}{q} = \frac{N_{2D}W}{2}L'[F_0(\eta_{FS}) - F_0(\eta_{FD})],$$

and

$$C_{ox}(V_{GS} - V_T) = \frac{N_{2D}}{2}\tilde{v}_T\left[F_0(\eta_{F1}) + F_0\left(\eta_{F1} - \frac{qV_{DS}}{k_BT}\right)\right].$$

Substituting these values into the drain current expression gives

$$I_D = WC_{ox}(V_{GS} - V_T)\tilde{v}_T\left[\frac{1 - \dfrac{F_{1/2}(\eta_{FS} - qV_{DS}/k_BT)}{F_{1/2}(\eta_{FS})}}{1 + \dfrac{F_{1/2}(\eta_{FS} - qV_{DS}/k_BT)}{F_{1/2}(\eta_{FS})}}\right].$$

The degenerate thermal injection velocity at the top of the barrier is

$$\tilde{v}_t = \sqrt{\frac{2k_BT}{\pi m^*}}\frac{F_{1/2}(\eta_{FS})}{F_0(\eta_{F1S})},$$

which equals one for non-degenerate semiconductors. The thermal injection velocity is higher for degenerate statistics.

Equivalently,

$$\langle v \rangle = \tilde{v}_t\left[\frac{1 - F_{1/2}(\eta_{FD})/F_{1/2}(\eta_{FS})}{1 + F_{1/2}(\eta_{FD})/F_{1/2}(\eta_{FS})}\right],$$

is the average velocity, which is composed of the thermal injection velocity times bias-dependent factors of Fermi levels (Lundstrom, 2005).

The behavior of the drain current is easier to visualize when non-degenerate statistics are assumed:

$$I_D = WC_{ox}(V_{GS} - V_T)v_T\left[\frac{1 - e^{-qV_{DS}/k_BT}}{1 + e^{-qV_{DS}/k_BT}}\right].$$

The previous equation indicates that at low V_{DS}, the channel is behaving as voltage-controlled resistor according to

$$I_D = \mu_{Ballistic}\frac{W}{L}C_{ox}(V_{GS} - V_T)V_{DS},$$

which is the same definition given in our previous classic drift definition of current in the linear regime; however, the classic definition of mobility is replaced by a ballistic mobility:

$$\mu_{\text{Ballistic}} = \frac{L(v_{\text{T}}/2)}{k_{\text{B}}T/q}.$$

In ballistic transport, the definition of mobility is physically not valid but it is useful for understanding the time for carrier transport across the channel (Lundstrom and Guo, 2005).

In the limit of large V_{DS}, the drain current expression equation becomes

$$I_{\text{D}} = WC_{\text{ox}}\left(V_{\text{GS}} - V_{\text{T}}\right)v_{\text{T}}.$$

Thus, in a ballistic transistor, current saturates but velocity is replaced with the thermal injection at the beginning of the channel. This replicates the drift-based saturated drain current solution, which is linear with $(V_{\text{GS}} - V_{\text{T}})$. For completeness, the drain current is proportional to $(V_{\text{G}} - V_{\text{T}})\alpha$ with $\alpha = 1$ under non-degenerate statistics and can vary up to $\alpha = 1.5$ under degenerate statistics (Lundstrom, 2005).

The detailed derivation presented above was performed to highlight the impact of a particular semiconductor's band structure (e.g., effective mass) on the DOS available for transport and the injection velocity from the source. To directly compare III–V materials to Si and Ge requires an investigation into a number of interrelated factors including the quantum capacitance. Rahman et al. (2005) found the ballistic performance of III–V materials was, in general, inferior to Si and Ge due to the low DOS. Rahman et al. (2005) used the top-of-the-barrier transport model where carriers semi-classically fill k-states at the top of the source-channel barrier. Current, I_{D}, is proportional to the charge Q and thermal injection velocity, v_{inj}. The charge, $Q = C_{\text{Gate}}(V_{\text{G}} - V_{\text{T}})$ is dependent of the gate insulator capacitance (Figure 1.18) as

$$C_{\text{G}} = \frac{C_{\text{ins}}C_{\text{Q}}}{C_{\text{ins}} + C_{\text{Q}}},$$

which is a function of the capacitance of the (dielectric) insulator, C_{ins}, and the quantum capacitance as given by

$$C_{\text{Q}} = \frac{\partial(-qn_{\text{s}})}{\partial\Psi_{\text{s}}} = q^2\left\langle D_{\text{2d}}\left(E_{\text{F}}\right)\right\rangle \sim m^*.$$

At high V_{g}, v_{inj} increases due to carrier statistics as described above. The GaAs injection velocity initially increases then decreases due to carriers

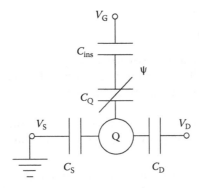

FIGURE 1.18
The gate capacitance is a series combination of insulator capacitance C_{ins}, which is determined by EOT and quantum capacitance, C_Q. Quantum capacitance depends on the 2D DOS, which is a function of the number of valleys, number of subbands occupied, and the DOS of each subband.

filling higher-energy, low-velocity L valleys. The Si v_{inj} saturates at high V_g as higher X4 valleys are populated (Rahman et al., 2005).

In III–V materials, the small conduction band DOS creates a small C_Q, which creates a small C_g and carrier density, Q at the top of the energy barrier. In contrast, Ge has a low v_{inj} but a large DOS (large C_Q) and thus a large Q, which yields an overall larger current. At low V_g, the carriers are non-degenerate and v_{inj} is flat.

In thicker semiconductor devices (100 atomic layers), the quantum effects are negligible, since the subbands are closely spaced in energy and multiple subbands can be populated at high V_g. Thus, the C_g is less dependent on the DOS. For a transistor with a thick equivalent oxide thickness (EOT), $C_{ins}(\sim 1/EOT) \ll C_Q$ and thus C_Q does not control C_g but rather the injection velocity controls I_{on}. Conversely, for a thin EOT and a thin body, $C_{ins} \gg C_Q$, therefore any degradation of C_G, i.e., lack of carriers at the top of the channel, limits I_{on}.

Thus, the advantage of low effective mass III–V channel material is only important when the body and EOT are thick. It is likely that EOT will not scale with scaling of other geometric sizes as the gate leakage will exceed a controllable amount at very thin EOT. Interestingly, for deeply scaled ballistic devices, Ge seems to offer the best injection velocity-DOS product (Rahman et al., 2005).

Nevertheless, this analysis does not include the ability to bandgap engineer III–V multilayer structures, particularly to form a 2DEG. This is one of the advantages of moving away from a Si or SiGe as the channel material. Recent experimental findings on high-speed III–V transistors will be revisited throughout the book.

1.6 Themes of Book

Within this book, there are two overreaching themes: *why* and *how* to replace silicon components with compound semiconductor components. The *why* to replace silicon can be split into three subthemes:

1. Introducing compound semiconductors to improve the functionality of silicon microelectronics, e.g., selectively depositing InSb as the active channel.

2. Parallel usage of silicon microelectronics and compound semiconductor devices to link their functionality for distinct operation, e.g., embedding high-speed InP transistors next to silicon digital circuitry.

3. Fabricating compound semiconductor on silicon-based substrates that serve merely as a large area host, e.g., GaN LEDs on 6-in or larger diameter Si substrates.

There are some technologies that can fall into one or both technologies. For example, within a CMOS processor, the integration of InGaSb lasers would be a key component toward optical communication, which would improve the functionality (1) by solving the interconnect bottleneck. Once merged optical/electronic technology matures, a number of applications will inevitably develop. Integrated optical sensors that communicate with Si digital circuits are better classified as (2) parallel usage.

The *how* to replace silicon theme is similarly split into three sub-themes:

1. All semiconductor applications are preferably made in Si (or silicon germanium). Thus, we will often introduce the established or competing Si-based technology and explain how challenging it is to use a compound semiconductor process to displace the typically mature Si manufacturing process.

2. Monolithic integration of compound semiconductor on Si is conceptually simpler as it may merely involve introducing a new tool in the process sequence. In practice, a process to deposit compound semiconductors onto silicon usually introduces defect densities that prevent acceptable operation of a device built in the compound semiconductor. Second, the compound semiconductor integration step needs to be viewed from a larger process flow viewpoint as the instantaneous and accumulative environment (thermal, pressure, stress, etc.) of the deposition process may harm other parts of the die.

3. Hybrid integration of compound semiconductors on Si circumvents many of the disadvantages of monolithic integration but raises its own issues, particularly with cost and yield.

The current interconnect bottleneck is an example of a problem that can be completely or partially solved with compound semiconductor technology. The current interconnect design consists of several ($\approx$10) layers of metals separated by low-κ insulator. According to Rent's rule: $\tau_{global} \approx R_{int}C_{int} \approx l^2 \approx 100$ ps at the 90 nm technology node. Although most wires are short, the occasional long wire must cross the chip and signals will lag the transistor frequency.

It is desirable to replace the electronic interconnects with optical interconnects. It is not clear that it is necessary to replace the short transistor-to-transistor interconnects as the delays and cross talk are minimal. Additionally, the difference in size dimensions of electronics and optical technologies are not compatible at present. Plasmonics offers an intriguing way to circumvent the diffraction limit of light and is thus projected as a short-run link in sections of the die that are highly dense. At present, there are limitations to plasmonic propagation from loss by absorption in the metal and the requirement to couple light into and out of the plasmonic components.

Large gains in RC delay and cross talk would be made by replacing the long run interconnects that are now present in multi-core die. Waveguide technology is fairly well understood and a large foot print is available or could be designed between each core on the die to accommodate each waveguide highway.

References

Anantram, M.P., Lundstrom, M.S., Nikonov, D.E. 2008. Modeling of nanoscale devices. *Proc. IEEE.* 96: 1511.

Binari, S.C., Ikossi, K., Roussos, J.A., Kruppa, W., Park, D., Dietrich, H.B., Koleske, D.D., Wickenden, A.E., Henry, R.L. 2001. *IEEE Trans. Electron Dev.* 48: 465.

Chau, R., Datta, S., Doczy, M., Doyle, B., Kavalieros, J., Metz, M. 2004. *IEEE Electron. Dev. Lett.* 25: 408.

Chau, R., Brask, J., Datta, S., Dewey, G., Doczy, M., Doyle, B., Kavalieros, J., Jin, B., Metz, M., Majumdar, A., Radosavljevic, M. 2005. *Microelectron. Eng.* 80: 1–4.

Chen, W.C., Wong, K.-Y., Chen, K.J. 2009. *IEEE Electron. Dev. Lett.* 30: 430.

Choi, W., Zhang, J., Chua, S.J. 2001. Dislocation scattering in n-GaN. *Mater. Sci. Semiconductor Process.* 4:567–570.

Crouch, R.K., Debnam, W.J., Fripp, A.L. 1978. *J. Mater. Sci.* 13: 2358.

Dadgar, A., Poschenrieder, M., Bläsing, J., Fehse, K., Diez, A., Krost, A. 2002. *Appl. Phys. Lett.* 80: 3670.

Datta, S. 2005. *Quantum Transport: Atom to Transistor*, Cambridge University Press, Cambridge, U.K.

Faux, D.A., Laux, S.E., Fischetti, M.V. 1992. *IEDM Tech Dig.* P553.

Feezell, D.F., Schmidt, M.C., DenBaars, S.P., Nakamura, S. 2009. *MRS Bull.* 34: 318–323

Fischetti, M.V. 1998. *Phys. Rev. B* 59: 4901.

Freeman, J.C. 2003. NASA technical report: Basic equations for the modeling of gallium nitride high electron mobility transistors, NASA/TM 211983.

Glassbrenner, C.J., Slack, G.A. 1964. *Phys. Rev.* 134: A1058–A1069.

Hanson, G.W. 2007. *Fundamentals of Nanoelectronics*, Prentice Hall, Upper Saddle River, NJ.

Ilegems, M., Montgomery, H.C. 1972. *J. Phys. Chem. Solids* 34: 885.

ITRS 2007. International Technology Roadmap for Semiconductors. www.itrs.net

Jain, S., Harker, A., Atkinson, A., Pinardi, K. 1995. *J. Appl. Phys.* 78: 1630.

Johnson, E.O. 1973. *RCA Rev.* 34: 80.

Johnson, J.W., LaRoche, J.R., Ren, F., Gila, B.P., Overberg, M.E., Abernathy, C.R., Chyi, J.-I., Chou, C.C., Nee, T.E., Lee, C.M., Lee, K.P., Park, S.S., Park, Y.J., Pearton, S.J. 2001. *Solid-State Electron.* 45: 205.

Johnson, J.W., Singhal, S., Hanson, A., Therrien, R., Chaudhari, A., Nagy, W., Rajagopal, P., Martin, Q., Nichols, T., Edwards, A., Roberts, J., Piner, E., Kizilyalli, I., Linthicum, K. 2008. GaN-on-Si HEMTs: From device technology to product insertion. *Mat. Res. Soc. Proc.* 1068: 3–12.

Khakifirooz, A., Antoniadis, D.A. 2008. *IEEE Trans. Electron. Dev.* 55: 1391.

Kirby, P.A., Selway, P.R., Westbrook, L.D. 1979. *J. Appl. Phys.* 50: 7.

Klein, P.B., Binari, S.C., Ikossi, K., Wickenden, A.E., Koleske, D.D., Henry, R.L. 2001. *Electron. Lett.* 37: 1550.

Khanal, D.R., Yim, J., Walukiewicz, W., Wu, J. 2007. *Nano Lett.* 7:1186–1190.

Kurzweil, R. 2005. *The Singularity Is Near: When Humans Transcend Biology*, Viking Press, New York.

Li, S. 1993. *Semiconductor Physical Electronics*, Springer, New York.

Lundstrom, M. 2005. Ballistic Nanotransistors. http://nanohub.org/resources/612

Lundstrom, M., Guo, J. 2005. *Nanoscale Transport: Device Physics, Modeling, and Simulation*, Springer, New York.

Luri, S., Suhir, E. 1986. *Appl. Phys. Lett.* 49: 140.

Mastro, M.A., Tsvetkov, D., Soukhoveev, V., Uskiov, A., Dmitriev, V., Luo, B., Ren, F., Basi, K.H., Pearton, S.J. 2003. *Solid-State Electron.* 47: 1075.

Mastro, M.A., LaRoche, J.R., Bassim, N.D., Eddy, Jr., C.R. 2005. *Microelectron. J.* 36: 705–711.

Mittereder, J.A., Binari, S.C., Klein, P.B., Roussos, J.A., Katzer, D.S., Storm, D.F., Koleske, D.D., Wickenden, A.E., Henry, R.L. 2003. *Appl. Phys. Lett.* 83: 1650.

Miller, D. 2000. *Semiconductor Optoelectronics Devices*, Stanford University Press, Stanford, CA.

Nilsson, O., Mehling, H., Horn, R., Fricke, J., Hofmann, R., Muller, S.G., Eckstein, R., Hofmann, D. 1997. *High Temperatures-High Press.* 29: 73–79.

Oxley, C.H. 2001. *Solid-State Electron.* 45: 677.

Rahman, A., Klimeck, G., Lundstrom, M. 1998. *IEEE International Electron Devices Meeting*, San Francisco, CA, pp. 615–618

Rahman, A., Klimeck, G., Lundstrom, M. 2005. *IEDM Technical Digest*, Washington, DC, p. 604.

Simpkins, B.S., Mastro, M.A., Eddy, Jr., C.R., Pehrsson, P.E. 2008. *J. Appl. Phys.* 103: 104313–104319.

Sutherland, I.E., Sproull, R.F., Harris, D.F. 1999. *Logical Effort: Designing Fast CMOS Circuits*, Morgan Kaufmann, San Francisco, CA.

Sze, S.M. 1982. *Physics of Semiconductor Devices*, John Wiley & Sons, New York.

Taur, Y., Ning, T. 1998. *Fundamentals of Modern VLSI Devices*, Cambridge University Press, Cambridge, U.K.

Van Zeghbroeck, B. 2008. *Principles of Semiconductor Devices*, Prentice Hall, Englewood Cliffs, NJ.

Vescan, L., Stoica, T., Dieker, C., Luth, H. 1993. *Mater. Res. Soc. Symp. Proc.* 298: 45.

Vurgaftman, I., Meyer, J., 2003. *J. Appl. Phys.* 94: 3675–3696.

Weste, N., Harris, D. 2005. *CMOS VLSI Design: A Circuits and Systems Perspective*, Pearson/Addison-Wesley, Boston, MA.

Wu, J., Walukiewicz, W., Yu, K.M., Ager III, J.W., Haller, E.E., Lu, H., Schaff, W.J., Saito, Y., Nanishi, Y. 2002. *Appl. Phys. Lett.* 80: 3967.

Wu, J., Walukiewicz, W., Yu, K.M., Shan, W., Ager III, J.W., Haller, E.E., Lu, H., Schaff, W.J., Metzger, W.K., Kurtz, S. 2003. *J. Appl. Phys.* 94: 6477.

Wu, J., Walukiewicz, W., Li, S.X., Armitage, R., Ho, J.C., Weber, E.R., Haller, E.E., Lu, H., Schaff, W.J., Barcz, A., Jakiela, R. 2004. *Appl. Phys. Lett.* 84: 2805.

Xu, J. 2000. PhD dissertation, University of California at Santa Barbara, Santa Barbara, CA.

Yu, K.M., Liliental-Weber, Z., Walukiewicza, W., Li, S.X., Jones, R.E., Shan, W., Ager III, J.W., Haller, E.E., Lu, H., Schaff, W.J. 2005. *Appl. Phys. Lett.* 86: 071910.

Zhang, A.P., Ren, F., Han, J., Pearton, S.J., Park, S.S., Park, Y.J., Chyi, J.-I. 2003. GaN and AlGaN high voltage power rectifiers, in *Wide Energy Bandgap Electronic Devices*, ed. Ren, F., Zolper, J.C., World Scientific, Singapore.

Zhalko-Tytarenko, A. 2007. Compound semiconductor materials: Technology, developments and markets, BCC Research Report.

Zhirnov, V.V., Cavin, R.K., Hutchby, J.A., Bourianoff, G.I. 2003. Limits to binary logic switch scaling—A gedanken model. *Proc. IEEE* 91(11): 1934–1939.

2

Challenge of III–V Materials Integration with Si Microelectronics

Tingkai Li

CONTENTS

2.1 Introduction

After several decades of scaling, Si-based processors are approaching a number of fundamental limitations that can only be addressed by the use of compound semiconductors (CS). To meet increasingly challenging and complex system requirements as well as staying cost effective, it is not enough to use one single semiconductor material system. Therefore, major efforts have been expended in recent years to combine the low cost and well-established Si-based CMOS processing attributes with the superior performance attributes of CS. Such a combination—marrying the best of both worlds—will enable performance superior to that achievable with CS and Si-based CMOS alone with CMOS affordability. The following advantages of III–V materials on Si have been realized:

1. Si substrate has high thermal and electrical conductivity, and mechanical hardness, which is better for very large-scale device integration.
2. Si substrate would significantly reduce manufacturing costs compared to sapphire, SiC, InP, or GaAs substrates.
3. Si substrate has potential wafer-size expansion, which would further reduce the manufacturing process costs.
4. III–V materials have excellent low-field and high-field electron transport properties.
5. III–V materials have ultrahigh-speed switching at very low supplied voltages.
6. Realized the ultimate vision of high switching activity factor low-voltage and high-speed III–V-based logic circuit blocks coupled with the functional density advantages provided by the Si CMOS platform.

The lack of large area of the substrates and the high price, e.g., sapphire, SiC, GaN, GaAs, or InP, create a high cost and throughput chokehold on CS foundries. In many cases, a superior CS technology has little or no commercial market simply based on the high cost of manufacturing. Based on the above advantages, Si substrate offers a low price as compared to sapphire, SiC, InP or GaAs substrates, high crystalline perfection, availability of large-size substrates, all types of conductivity, and high thermal conductivity ($1.5\,W\ cm^{-1}$). Large-area Si substrates can simply act as a building block to further lower the fabrication process cost of CS devices. The true promise of integrating CS into traditional Si microelectronics is now coming to fruition after many decades of promise. With an approach that directly integrates the CS into the CMOS wafer, only one wafer is processed to achieve a finished chip. Therefore, great efforts have been made to achieve direct integration of III–V materials systems such as GaN, GaAs, SiC and related alloys with Si substrates.

Integration of III–V CS devices with Si microelectronics can be used in four important applications in our society today. First is the high-speed devices. CS are needed to solve the "interconnect bottleneck" in which the parasitic capacitance and propagation delay of electrical signals along the copper lines actually controls the maximum frequency of next-generation processors. For example, there is a strong interest in replacing the channel in Si MOSFETs with III–V CS. A promising candidate is an InSb quantum-well structure, which has the lowest energy-delay product, an important metric for logic microprocessors. Therefore, III–V high electron mobility transistor (HEMT) and Si photonics are urgently needed for data transfer in combining systems with cell phone, TV, and Internet together. Second is the white LEDs for solid-state lighting, which promises to replace conventional light sources, with impressive economic and environmental savings. In the United States, expenditures for lighting may be reduced by $100 billion over the period 2000–2020. By the year 2020, electricity used for lighting may be cut by 50%, sparing the atmosphere 28 million metric tons of carbon emission annually. Third is the high-speed and high-power devices, which will be used in power system and hybrid cars to save energy. The III-nitride materials on Si have been demonstrated for high voltage, high power density, and high-speed device applications. Fourth is III–V solar cells, which have the highest photoelectric conversion efficiency and is future green energy source. In the last decade, heteroepitaxy of III-nitride such as GaN and III–V materials such as GaAs, InSb thin films on Si substrate have been broadly studied using a variety of tools, including molecular beam epitaxy (MBE) [Guha 1997], [Oye 2007], [Rao 2000], [Contreras 2005], chemical beam epitaxy (CBE) [Huang 2002], [Sung 2004], metalorganic vapor phase deposition (MOCVD) [Alam 2001], [Ting 2000], [Tokunaga 2000], laser ablation [Ohta 2001], and vacuum reactive deposition [Zhang 2000]. A number of inspired teams have already shown that CS devices such as high-speed devices, GaN-based LEDs, high-power transistors, and III–V solar cells, etc., can be manufactured on large-area Si substrates. However, there are still many issues that deal with integration of III–V CS devices with Si microelectronics: (1) lattice mismatch causes a high-density dislocation in III–V materials layer, which may reduce significantly usable area and decrease material quality; (2) thermal mismatch is a major problem, which causes strain, defects, and even cracks during device preparation and operation. The thick epilayers of III–V materials on Si substrates for device fabrication is difficult to be achieved without cracks; (3) lack of semi-insulating Si substrate, which may lead to parasitic capacitance effects during high-frequency operation; (4) Si diffusion into CS. During the III–V materials grown, under H_2 flow and high temperature, Si substrate may react with H_2 out–gases of Si, or react and diffuse into III–V materials, and this Si will lead to a high n-type background doping, which makes it difficult to achieve p-type doping and poor device performance; and (5) Si optical absorption reduced the permeance of LED devices on Si substrates.

Many different approaches have been developed to overcome major lattice and thermal expansion mismatches, etc., between III–V materials and Si substrate with great success as illustrated by the multitude of excellent work presented in the book and the fact that excellent device performance has been achieved, and are starting to become commercially available. In order to further fabricate high-quality electronic devices, the challenges of III–V materials integration with Si microelectronics should be addressed.

2.2 Lattice Mismatch

Based on the lattice mismatch between coated film and substrates, and film growth methods, there are three kinds of thin films grown on substrates, which is epitaxial, amorphous, and polycrystalline films, as shown in Figure 2.1. Two types of epitaxy can be defined and each deals with the advancements of sciences and technologies [Ohring 1992]. Homoepitaxy refers the case where the film and substrate are the same materials. When the epitaxial film and substrate crystal are identical, the lattice parameters are perfectly matched and there is no interfacial bond straining, which is homoepitaxy, as shown in Figure 2.1a. Such as epitaxial Si deposited on Si substrate, epitaxial GaAs deposited on GaAs substrate are the most typical examples of the homoepitaxy. Based on the theories, the homoepitaxial films are defect and stress free, and have the pure compositions of the substrates. Technically speaking, the homoepitaxial film quality also depends on the substrate surface conditions, process methods, and process parameters. The second type of epitaxy is heteroepitaxy where the film and substrate are different materials. Such as GaN deposited on Si, AlAs deposited on GaAs substrates. In the heteroepitaxy, the lattice parameters between film and substrate are unmatched. Epitaxial growth of thin films and the control of defects in thin-film heterostructures are key considerations for the next generation of microelectronic, optical, and magnetic devices. Depending on the extent of the mismatch, there are three distinct epitaxial regimes. If the lattice mismatch is very small, the heterojunction interfacial structure is essentially like homoepitaxy, in which there is a small strain in the epitaxial film. There is almost

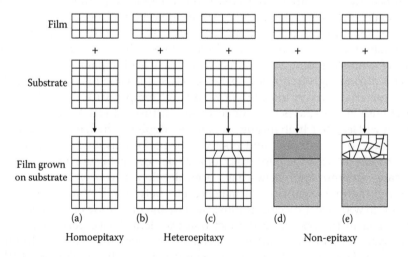

FIGURE 2.1
Three kinds of thin films deposited on substrates: epitaxial, amorphous, and polycrystalline films. (a) Matched, (b) strained, (c) relaxed, (d) amorphous, and (e) polycrystalline.

no dislocation or dislocations are generated at the film surface and glide to the interface; therefore, the Burgers vectors and planes of the dislocations are dictated by the slip vectors and glide planes of the crystal structure of the film. However, differences in film and substrate chemistry and coefficient of thermal expansion can strongly influence the electronic properties and perfection of the interface. In the well-established lattice-matching epitaxy, where the lattice misfit is much small ($\ll 7\%$), films grow pseudomorphically up to a "critical thickness" where it becomes energetically favorable for the film to contain dislocations. If the film and substrate have a larger lattice mismatch, either edge dislocation defects form at the interface, or the two lattice strain to accommodate their crystallographic difference, as shown in Figure 2.1b and c. The former situation (relaxed epitaxy with defects) generally prevails during later film formation stages irrespective of crystal structure or lattice mismatch. The latter case (strained film) is based on strained-layer heteroepitaxy. This phenomenon occurs between film-structure pairs composed of different materials having the same crystal structure. $Si_{1-x}Ge_x$ films grown on Si is a typical example of strained-layer epitaxy [Maa 2007a]. In most cases, the heteroepitaxy is a partial strained-layer epitaxy, in which there are relaxed epitaxy with defects and even cracks and strained layer. GaN films deposited on Si with AlN buffer layers is the important example [Lin-a 2007]. The dislocations generated at the edge of islands during three-dimensional (3D) growth, geometrical constraints determine the Burgers vectors of the dislocations at the film–substrate interface. For example, during 3D growth of germanium on silicon, it has been found that $90°$ dislocations with $a/2$ $\langle 110 \rangle$. Burgers vectors are created at the edge of germanium islands and lie in the (001) film–substrate interface [LeGoues 1994]. Conventional technologies maintain that lattice matching epitaxy (LME) during thin film growth is possible as long as the lattice misfit between the film and the substrate is less than 7%–8%. Smaller lattice misfit leads to smaller interfacial energy and coherent epitaxy is formed. Above this misfit (>7%–8%), growing epitaxially in the form of single crystals is by domain matching epitaxy (DME), where integral multiples of lattice planes match across the film–substrate interface, and the size of the domain equals integral multiples of planar spacing. If the film and the substrate have similar crystal structures, then the matching of planes becomes equivalent to the matching of lattice constants. In this case, it was surmised that the film will grow textured or largely polycrystalline. Accordingly, for small misfits ($\ll 7\%$–8%), the generalized DME is equivalent to the conventional LME [Narayan 2003].

If the films having a totally different lattice structure with substrates, or at certain process methods and conditions, the amorphous or polycrystalline films were deposited, as shown in Figure 2.1d and e. For example, amorphous Si can be deposited on glass at low temperature and high growth rate, and polycrystalline Si can be deposited on glass with very high flow ratio of H_2 to saline using plasma-enhanced chemical vapor deposition (PECVD) method [Joshi 2005]. Therefore, the deposited film quality depends extensively on the lattice mismatch between growth films and substrates, as well as the film growth process conditions.

The III–V materials growth on Si substrate is the typical heteroepitaxy where the film and substrate are different materials. Based on main materials and device applications, the III–V materials can be divided into two groups: III-nitrides and conventional III–V materials. Generally speaking, III–nitrides have a higher deposition temperature around 1200°C, and conventional III–V materials have a lower deposition temperature around 850°C, which related with thermal expansion mismatch that will be discussed in Section 2.3 thermal expansion mismatch. III-nitrides and its alloys have a wide band gap range (0.7 eV for InN, 3.4 eV for GaN–6.2 eV for AlN), which can used to make optoelectronics such as LEDs with various colors from UV to IR wavelength ranges of the electromagnetic spectrum, and III-nitride solar cell. The indium gallium nitride series of alloys is photo-electronically active over virtually the entire range of the solar spectrum.

III-nitrides also have high carrier mobility at high temperature, very high breakdown voltage, and excellent chemical robustness, which are important materials for high-power, high-frequency electronic devices. III-nitride technology emerged to meet not only the huge lighting and display market, but also the high-power electronics market, which are enabled by the hetero-epitaxial growth of III-nitride thin films on lattice-mismatched substrates such as silicon, sapphire, and silicon carbide, and the fabrication of high-quality III-nitride-based devices. A consequence of this mismatch is that III-nitride epitaxial layers contain a high density of threading dislocations that have been linked to the failure of lasers and the breakdown of *pn* junctions. Furthermore, it is anticipated that threading dislocations will aggravate compositional stability problems associated within segregation from III-nitride alloys such as InGaN alloys. Therefore, reducing the defect density specially for threading dislocations are critical issues to obtain high-quality III-nitride films grown on Si substrates. The band gap and lattice constants of Si, SiC, sapphire, and III-nitride materials are shown in Figure 2.2 [Weast 1990], [Strite 1994], [Mohammad 1995], [Kingery 1976], [Zamir 2001], [Kasper 1995]. From these figures, III-nitride materials have a smaller lattice mismatch with SiC(4H or 6H), but larger lattice mismatch with Si and sapphire.

Table 2.1 shows the defect densities of GaN grown on sapphire, SiC, and Si substrates with various film thickness and deposition processes [Metzger 1998], [Heinke 2000], [Chakraborty 2004], [Faleev 2005], [Li 2007], [Kim 2008]. From the table, GaN deposited on SiC substrates has the smaller defect density. Due to the better thermal mismatch of sapphire than Si, the GaN grown on Si has the more defect densities than sapphire, which will be discussed on the chapter of thermal mismatch. Among various deposition processes, GaN made by MBE has the smallest defect density. Typically, the substrate of choice is either SiC or sapphire due to the smaller lattice mismatch between III-nitrides and SiC, and better thermal mismatch between III-nitrides and sapphire, and much progress has been made in understanding the nucleation

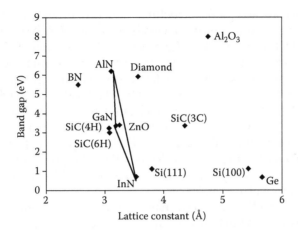

FIGURE 2.2
The band gap and lattice constants of III-nitride materials.

and growth processes on these two substrate materials such that most useful devices are currently fabricated using sapphire [Amano 1986], [Wu 1998] and silicon carbide [Ponce 1995], [Weeks 1996], [Tanaka 1996], [Edwards 1998], [Lahreche 2000].

In order to improve the working functions of Si-based electronics, reduce manufacturing costs of III–V based devices, a heterogeneous integration of III–V compound materials with Si platform is necessary. However, for III-nitride materials growth on Si substrates, the large difference in the lattice parameters of GaN ($a_{GaN} = 0.31892\,nm$) and Si ($a_{Si(111)} = 0.38403\,nm$), yields a lattice parameter mismatch 16.9% for GaN on Si(111), resulting in a high dislocation density of ~10^{10} cm^{-2} for GaN on Si(111), which is comparable to GaN on sapphire. Several technologies have been used to solve these problems.

TABLE 2.1

Defect Densities of Screw (N_s) and Edge Dislocations in Random (N_{er}) and in Grain Boundary (N_{eg})

Substrate	Growth Method	GaN (μm)	N_s (10^9 cm^{-2})	N_{er} (10^9 cm^{-2})	N_{eg} (10^9 cm^{-2})	Reference
Sapphire	MOCVD	1.0	0.44	120	23	[Metzger 1998]
Sapphire	MBE	0.7–1.5	0.01–3	5–100	—	[Heinke 2000]
Sapphire	HVPE-LEO	10.0	0.005	—	—	[Chakraborty 2004]
SiC(6H)/AlN	MOCVD-LEO	10	0.3	—	—	[Kim 2008]
SiC(6H)	HVPE	0.29	0.02	—	—	[Faleev 2005]
SiC(6H)	HVPE	31	0.72	—	—	[Faleev 2005]
Si(111)/AlN	MOCVD	2	0.47	4.8	—	[Li 2007]
Si(111)/AlN	MOCVD	1	0.49	1.9	0.26	[Li 2007]

GaN films with acceptable structures and properties have been grown on Si(111) using a various buffer layers [Follstaedt 1999], [Strittmatter 1999], [Schremer 2000], [Kaiser 2000], [Lahreche 2000], [Molina 1999], [Nikishin 1999], [Ishikawa 1999], [Liaw 2000]. Improvements in microstructure and properties were demonstrated using lateral epitaxial overgrowth [Ujiie 1989], [Kato 1991], and selective area epitaxy [Tanaka 2000], [Yang 1999]. Devices based on GaN/Si(111) layers have also been demonstrated, including LEDs [Guha 1998a], [Guha 1998b], [Tran 1999], [Yang 2000], [Dalmasso 2000], photo-detectors [Osinsky 1998], [Pau 2000], field-effect transistors [Chumbes 1999], [Egawa 2000] and the high-power devices, the total power of 368 W at 60 V with 70% drain efficiency under pulsed operation of GaN-on-Si HFET has also been reported [Therrien 2005].

The conventional III–V materials have very wide applications of photoelec-tronic devices such as LEDs, lasers, and high-efficiency solar cells, optical cir-cuit, microwave applications, and very large-scale integration circuit devices, etc. The band gap and lattice constants of Si, Ge, and conventional III–V mate-rials are shown in Figure 2.3 [Weast 1990], [Strite 1994], [Mohammad 1995], [Kasper 1995], [NCSR-2007], [Vurgaftman 2001], [Adachi 1985]. Some III–V materials such as GaAs, AlAs, GaInNAs, ZnSe, $Al_xGa_{1-x}As$, and $In_xG_{1-x}P$ alloys with certain x values, etc., have almost the same lattice constant with various band gaps. Therefore, the lattice matched compounds with various band gaps can be epitaxially grown to form a multijunction solar cell, the

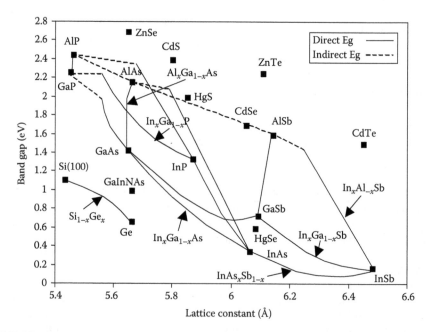

FIGURE 2.3
The band gap and lattice constants of conventional III–V materials.

top junction captures high-energy photons, while others pass through to the lower band gap junctions below, which is photo-electronically active over virtually the entire range of the solar spectrum. In this way, the high-efficiency solar cell can be made. The III–V materials of GaAs, AlAs, GaInNAs, ZnSe, $Al_xGa_{1-x}As$, and $In_xG_{1-x}P$ alloys with certain x values, etc., have almost the same lattice constant with Ge, but larger lattice mismatch with Si(100), which results in large number of threading dislocations, strain, and poor quality of the III–V materials. For GaAs growth on Si(100), the large difference in the lattice parameters of GaAs ($a_{GaAs} = 0.5653$ nm) and Si ($a_{Si(100)} = 0.5431$ nm) yields a lattice parameter mismatch 4% for GaAs on Si(100), and the growth of polar GaAs materials on nonpolar Si substrates results in larger number of defects and formation of anti-phase boundary. The realization of low defect density GaAs on Si heteroepitaxy would enable monolithic integration of III–V materials and devices with conventional Si integrated circuit technology. Unfortunately, the 4.1% lattice mismatch between GaAs and Si typically results in the nucleation of unacceptably high threading dislocation densities on the order of 10^8 cm^{-2}. GaAs and Ge, however, feature a more amenable 0.07% lattice mismatch, which allows the fabrication of high-efficiency III–V solar cells and other III–V devices on Ge substrates.

In order to further develop high-quality III–V devices integration with Si microelectronics, two steps SiGe buffer layer [Chang 2005], [Lin-b 2007], metamorphic GaAs on Si(100) substrate [Lueck 2006], selective area growth combined with thermal cycle annealing [Aitken 2008], SrTiO₃(STO) buffer [Eisenbeiser 2002], compositionally graded SiGe buffer [Lin-b 2007], [Yang 2003], and ZnSe buffer layers [Sarinanto 1998] have been used to III–V materials such as GaAs on Si(100) substrates. GaAs layers grown on Si with Ge/SiGe graded buffers have an unprecedented threading dislocation (TDD) of 1×10^6 cm^{-2} and a minority carrier lifetime of 10.5 ns. These substrates can be used as a platform for fabrication of monolithically integrated optical circuits and AlGaAs/GaAs lasers that operate under CW/RT [Currie 1998], [Carlin 2000].

2.3 Thermal Mismatch

For III–V materials growth on Si substrates, the large difference in the lattice parameters yields a lattice parameter mismatch of 16.9% for GaN on Si(111), and 4% for GaAs on Si(100), resulting in a high dislocation density of ~10^{10} cm^{-2} for GaN on Si(111) and ~10^8 cm^{-2} for GaAs on Si(100). However, the most serious problem is the large thermal mismatch between GaN or GaAs and Si. When the film and the substrate experience different rates of thermal expansion and contraction, the thin film can experience considerable tensile or compressive stress during a temperature cycle, leading to crack formation or delamination. Such crack formation can be observed on thick GaN films

TABLE 2.2

Thermal Expansion Coefficient of Si, SiC, Sapphire, and Various III–V
Materials

Materials	TEC ($\times 10^{-6}$)	Reference	Materials	TEC ($\times 10^{-6}$)	Reference
Al_2O_3	7.50 (a)	19	InN	4.00	[Strite 1994]
	8.50 (c)				
AlN	4.20 (a)	19	InP	4.60	[Weast 1990]
	5.30 (c)				
AlAs	5.20	62	Si	3.59	[Zamir 2001]
GaAs	5.40	18	$Si_{0.65}Ge_{0.35}$	3.64–5.46	[Kasper 1995]
GaN	5.59 (a)	19	SiC(3C)	2.90	[Strite 1994]
	3.17 (c)				
GaP	5.30	18	SiC(6H)	4.20 (a)	[Strite 1994]
				4.68 (c)	
Ge	5.78	23, 72	ZnO	2.90 (a)	[Strite 1994]
				4.75 (c)	

Note: (a): a axis; (c): c axis.

grown on Si with AlN buffer or GaAs films grown on Si with SiGe virtual
substrates after a 1000°C–700°C temperature drop from the growth temperature to room temperature. The GaN and GaAs films have the thermal expan-
sion coefficient (TEC) of 5.59×10^{-6} K^{-1}, and 5.40×10^{-6} K^{-1} respectively, which
are much higher than the TEC of Si virtual substrate about 3.59×10^{-6} K^{-1}
as shown in Table 2.2. The thermal mismatch leads to a large tensile stress
during cooling from the growth temperature to room temperature, often
resulting in cracked layers preventing device applications. The thermal mismatch may cause strain, defects, and even cracks during device preparation
and operation. With lattice mismatch, extra dislocations generated from the
interface with temperature changes.

The TEC of Si, SiC, sapphire, and various III–V materials are shown in
Table 2.2. From the table, III–V materials such as GaN and GaAs have smaller
thermal mismatch with SiC(6H), and large thermal mismatch with Si and
sapphire. Because the III–V materials deposited on sapphire results in com-
pressed stress other than tensile stress of III–V materials grown on Si due to
the thermal mismatch, which is an advantage of sapphire used as substrate.
On the other hand, there are AlN, InN, ZnO having the TEC between GaN
and Si, and InP, $Si_{1-x}Ge_x$ having the TEC between GaAs and Si. These materials have been used as intermeddle layer or buffer layers for GaN or GaAs
growth on Si substrates, respectively [Lin-a 2007], [Lin-b 2007], [Lueck 2006].
However, for GaN or GaAs grown on Si substrate, the TEC of the buffer layers such as AlN/AlGaN buffer layer for GaN, or SiGe for GaAs is dominated
by that of Si since the Si substrate is much thicker than the total thickness
of the buffer layer of such SiGe-graded layers plus the top Ge layer, which is
approximately 10 μm. Therefore, using intermeddle layers to solve thermal

mismatch problems is limited. The presence of cracks in thin films is not desirable for device fabrication because they can act as scattering centers for light propagation, can resist in-plane electrical current flow, and can introduce electrical shorting paths in vertical currents.

2.3.1 Stress Calculations and Nanoheteroepitaxy Theory

Temperature changes cause materials to expand and contract. During the case where temperature deformation is not permitted, an internal stress created. The internal stress created is termed as thermal stress, which can be calculated by

$$\sigma = -E\alpha(T_t - T_o) \tag{2.1}$$

where
  σ is the thermal stress
  E is the Young modules
  α is the coefficient of thermal expansion
  T_t is the final temperature
  T_o is the start temperature

In order to understand the thermal stress of III–V materials such as GaN and GaAs on Si, following considerations have been made based on reference papers [Zamir 2001], [Slack 1975], [Timoshenko 1925]. The tensile stress causes a concave bending of the film/substrate system. The stress values can be easily determined via the curvature of the sample that is proportional to the stress value. The curvature leads to a strong broadening of the Bragg peaks in x-ray diffraction of the GaN or GaAs thin film and the Si substrate. The radius of curvature (ρ) can be measured, e.g., by measuring the width of the Si(111) Bragg peak using different apertures. When a coated film is cooled after deposition, if its TEC, α_c, is larger than that of the substrate, α_s (as in the case of GaN and GaAs on Si), the coated film is under tensile strain. As a result, the uncracked film–substrate composite bends, having a radius of curvature, ρ, is derived by Timoshenko [Zamir 2001], [Timoshenko 1925] as

$$\frac{1}{\rho} = \frac{(\alpha_s - \alpha_c)(T_f - T_g)}{\left[h/2 + 2\left(E_c^* I_c + E_s^* I_s\right)\Big/h\left(1/E_c^* t_c + 1/E_s^* t_s\right)\right]} \tag{2.2}$$

where
  T_f is the final temperature after cooling
  T_g is the growth temperature
  t_c and t_s are the individual coated film and substrate thicknesses, respectively
  h is the total thickness ($h = t_c + t_s$)
  I is the moment of inertia, $I = t^3/12$
  E^* is the effective modulus of elasticity

For wide layers and plain strain conditions $E^* = E/(12 - v^2)$, where E is the Young's modulus of elasticity and v is the Poisson's ratio. Since the coated film is thin ($t_c < 0.1t_s$), the predicted in-plane normal stress in the uncracked film is uniform and is given by

$$\sigma_p = \frac{1}{\rho\left[2\left(E_c^* I_c + E_s^* I_s\right)/ht_c + E_c^* t_c/2\right]} \tag{2.3}$$

Under typical MOCVD growth conditions, the stress amount of GaN on Si is about 0.9 GPa/μm [Dadgar 2000].

After the first cracking occurs, the stress in the coated film is relaxed close to the crack faces. The stress distribution near a crack is described by the shear-lag model

$$\sigma_x = \sigma_p\left[\tanh(\beta\xi)\sinh(\beta x) - \cosh(\beta x) + 1\right] \tag{2.4}$$

where

$$\beta^2 = \frac{G_c\left(1/t_c Q_c + 1/t_s Q_s\right)}{t} \tag{2.5}$$

$$Q_c = \frac{E_c}{2\left(1 - v^2\right)} \quad \text{and} \quad Q_s = \frac{E_s}{2\left(1 - v^2\right)} \tag{2.6}$$

x is the position along the coated film between cracks ($x=0$ is the crack face position)

σ_p is the predicted stress of the uncracked layer, far from the crack, which is calculated by Equations 2.2 and 2.3

ξ is one half of the distance between cracks

G_c is the shear modulus of the coated film

Equations 2.2 through 2.4 predicts that the stress is maximum at $x=\xi$, and decreases toward $x=0$ and $x=2\xi$. The transition occurs over an edge region of width, u, such that for $x \le u$ and for $x \ge (2\xi - u)$, the local stress is reduced ($\sigma < \sigma_p$). Analogously, an uncracked sample, with lateral size $L=2\xi$, has to exhibit similar strain distribution, since free surfaces of the sample must be relaxed (as in the case of crack faces). The above analysis shows that the buffer layers can reduce the thermal stress by decreasing ($\alpha_s - \alpha_c$) value, decreasing patterned size of III–V materials can release the thermal stress, and the deformation and nanocracks can release the thermal stress.

On the other hand, Zubia [Zubia 1999] describes an approach to the heteroepitaxy of lattice mismatched semiconductors, which he calls nanoheteroepitaxy. The theory developed here shows that the 3D stress relief mechanisms that are active when an epi-layer is nucleated as an array of nanoscale islands on a compliant patterned substrate will significantly reduce the strain energy in the epi-layer and extend the critical thickness dramatically. The stress in the epitaxial layer was shown to be of the formula:

$$\sigma_{epi}(y,z) = \frac{\epsilon_T \, Y_{epi} x(y,z) e^{-\pi z/2l}}{(1 - v_{epi})} \tag{2.7}$$

$$\epsilon_T = \frac{2(la_{epi} - a_{sub}l)}{(a_{epi} + a_{sub})} = \epsilon_{epi} + \epsilon_{sub} \tag{2.8}$$

where
  ϵ_{epi}, ϵ_{sub}, and ϵ_T are the partitioned strains in the epitaxial layer, substrate, and the total lattice mismatch strain, respectively
  v_{epi} is Poisson's ratio for the epi-layer material
  Y_{epi} is Young's modulus for the epi-layer material
  $2l$ is the island diameter
  $x(y, z)$ characterizes the lateral stress distribution

From the formula (2.2 through 2.7), the stress of the epitaxial film reduces with decreasing the island diameter.

In the conventional heteroepitaxy, the strain at growth interface remains constant and strain energy grows linearly with epilayer thickness. As the thickness is above critical thickness, the strain relaxes, and dislocations eventually are created. On the other hand, it has been demonstrated both theoretically and experimentally that there exist stress relief in nanoheteroepitaxy. The three-dimensional strain in nanosize nucleus gives exponential stress/strain decay, with the decay length proportional to (and of similar magnitude to) island diameter and, therefore, the strain energy saturates at a maximum value (Figures 2.4a, b and 2.5) [Zubia 1999], [Wang-a 2006]. Figure 2.5

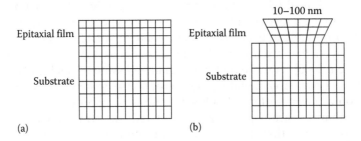

(a)　　　　　　　　　　　　(b)

FIGURE 2.4
The theory of nanoheteroepitaxy. (a) Conventional epitaxy and (b) nanoheteroepitaxy.

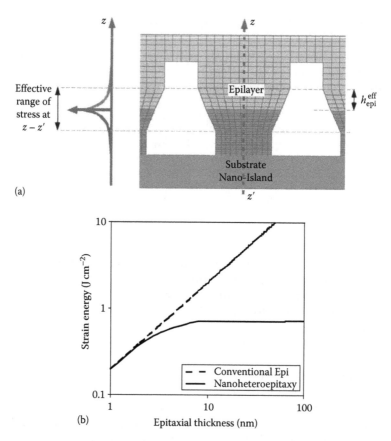

FIGURE 2.5
The 3D stress relief and compliance nanoheteroepitaxy, which combines 3D stress relief and substrate compliance, where the stress and strain decay exponentially on both sides of hetero-interface. (a) Nanoheteroepitaxy and (b) 3D stress relief by nanoheteroepitaxy.

shows the 3D stress relief and compliance by nanoheteroepitaxy. By combining the 3D stress relief and substrate compliance, the stress and strain decay exponentially on both sides of hetero-interface. Strain energy saturates when the epitaxial film thickness (h) is larger than effective film thickness (h_{eff}). According to the above nanoheteroepitaxy theory, with nano-island radii in the 10–100 nm range, it is possible to eliminate mismatch defects for hetero-epitaxial layers that are mismatched by as much as 4.2% of Ge on Si [Zubia 1999], [Zubia 2000]. In more highly mismatched systems such as GaN on Si, using the selective-area metalorganic vapor phase epitaxy (MOVPE) growth of GaN on a novel, patterned SOI substrate with nanoscale silicon islands further enhances strain partitioning and reduces strain energy. Electron microscopy reveals that the defect concentration decays rapidly away from the GaN/Si heterointerface, in agreement with NHE theory. Beyond this defected region, the GaN appears to be of high quality.

Based on above theory, the four methods can be used to solve the thermal mismatch problem of III–V materials such as GaN and GaAs on Si. (1) Buffer layers such as AlN with grading $Al_xGa_{1-x}N$ ($1 \geq x \geq 0$) for GaN growth on Si substrate, and Ge with grading $Si_{1-x}Ge_x$ ($1 \geq x \geq 0$) for GaAs growth on Si substrate. (2) Lateral nanoheteroepitaxy overgrowth of GaN or GaAs on Si. (3) Set up week points in nano-nuclear area other than GaN or GaAs film to release the thermal stress. (4) Using patterned wafer and selective growth of III–V materials on Si substrate to release the thermal stress. (5) Transfer Si on the substrates with similar thermal expansion coefficients.

2.3.2 Thermal Expansion Coefficients of III–V Materials and Substrates

Reducing the difference in TEC between III–V materials and substrate materials is a key to producing crack-free III–V films. Therefore, many scientists have been working at researches of the TEC and resultant strain for III–V materials on Si, SiC, and sapphire substrates with various buffer layers such as AlN, $Al_xGa_{1-x}N$ ($1 \geq x \geq 0$), $Si_{1-x}Ge_x$ ($1 \geq x \geq 0$), Ge, etc. Figure 2.6 plots linear TEC of III–V materials, SiGe, Ge, Si, SiC, and sapphire as a function of temperature as reported by various literature sources [Kasper 1995], [Slack 1975], [Okada 1984], [Roder 2005], [Talwar 2002], [Figge 2009], [Glazov 2000], [Dobrovinskaya 2009]. In all cases, the linear TEC increase with temperature. Note that the values from different sources do not all agree completely with each other. From Figure 2.6, the TEC of SiC has a good match with GaN and GaAs materials, especially at high temperature, the plots for a and c lattice

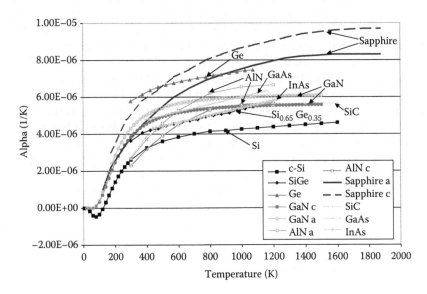

FIGURE 2.6
Linear thermal expansion coefficients of III–V materials, SiGe, Ge, Si, SiC, and sapphire as a function of temperature.

constants of AlN and GaN, it can be seen that the TEC of SiGe can be made to agree quite well with the nitride values by adjusting the Ge content of the film; ~35% Ge is fairly close. The TEC of SiGe is also between GaAs and Si in the all temperature range, which is good to be used as intermeddle layer for GaAs deposited on Si substrates.

2.3.3 Strain Calculated from Domain Mismatch Epitaxy

According to Section 2.2 LME, if the film and substrate have a larger lattice mismatch above this misfit (>7%–8%), either edge dislocation defects form at the interface, or the two lattice strain to accommodate their crystallographic difference. On the other hand, growing epitaxially in the form of single crystals can be also by DME [Narayan 2003], [Wang 2002]. In the DME, we consider the matching of lattice planes, which could be different in different directions of the film–substrate interface. In the DME framework, the film can have either a fixed or the same orientation relationship with the substrate, depending upon the nature of the misfit. The misfit is accommodated by matching of integral multiples of lattice planes, and there is one extra half plane (dislocation) corresponding to each domain. The misfit can range from being very small to very large. In the small misfit regime, the DME reduces to LME where matching of the same planes or lattice constants is considered with a misfit typically less than 7%–8%. If the misfit falls in between the perfect matching ratios of planes, then the size of the domain can vary in a systematic way to accommodate the additional misfit. In the conventional LME, the initial or unrelaxed misfit strain (ε_c) is given by $\varepsilon_c = a_f/a_s - 1$, where a_f and a_s are lattice constants of the film and the substrate, respectively. In LME, the ε_c is less than 7%–8%, which is relaxed by the introduction of dislocations beyond the critical thickness during thin film growth. In DME, the matching of lattice planes of the film d_f with those of the substrate d_s is considered with similar crystal symmetry and lattice constants. LME involves the matching of the same planes between the film and the substrate. However, in DME, the film and the substrate planes could be quite different as long as they maintain the crystal symmetry, which may involve the matching of the different planes between the film and the substrate. In DME, the initial misfit strain ($\varepsilon = d_f/d_s - 1$) could be very large, but this can be relaxed by matching of m planes of the film with n of the substrate. This matching of integral multiples of lattice planes leaves a residual strain of ε_r given by [Narayan 2003]

$$\varepsilon_r = \frac{md_f}{nd_s} - 1 \tag{2.9}$$

or given by [Gevorgian 1998]

$$\varepsilon_r = \frac{2(md_f - nd_s)}{(md_f + nd_s)} \tag{2.10}$$

where m and n are simple integers. In the case of a perfect matching, $md_f = nd_s$, and the residual strain ε_r is zero. If ε_r is finite, then two domains may alternate with a certain frequency to provide for a perfect matching according to

$$(m+\alpha)d_f = (n+\alpha)d_s \tag{2.11}$$

where α is the frequency factor, e.g., if $\alpha = 0.5$, then m/n and $(m+1)/(n+1)$ domains alternate with an equal frequency.

Assuming $d_f > d_s$, we have $n > m$. Therefore,

$$n - m = 1 \quad \text{or} \quad f(m) \tag{2.12}$$

The difference between n and m could be 1 or some function of m.

Based on the above formula, this alignment is thought to be "domain matching epitaxy," in which m of the III–V materials lattice spacing closely matches n of the substrate such as Si spacing. Therefore, the strain can be calculated by following formula:

$$\text{Strain} = \frac{(md_f - nd_s)}{nd_s} \tag{2.13}$$

where

d_f is the lattice constant of the III–V film materials

d_s is the lattice constant of the Si substrate

For example, GaN on Si(111) grows with the GaN c-axis ([001] direction) parallel to the substrate Si [111] direction. Furthermore, in spite of the nearly 20% difference in lattice constant, the GaN aligns azimuthally with the Si substrate very well. Consistent with the literature [Narayan 2003], [Wang 2002], it is found that the GaN a-axis ([100] direction) aligns with the Si [10–1] direction, in which five of the nitride (100) lattice spacing closely matches four of the Si(110) spacing. A strain can be defined based on this matching:

$$\text{Strain} = \frac{\left(5d_N - 4d_{(110)}\right)}{\left(4d_{(110)}\right)} \tag{2.14}$$

In Figure 2.7, this calculated strain, as a function of temperature for the combinations of AlN, GaN, GaAs on Si, $Si_{0.65}Ge_{0.35}$, Ge, SiC, and sapphire as well as GaN on AlN, etc., has been plotted. The positive values are compressive strain, and negative values are tensile strain. From the figure, it is the most compressive or small tensile strain and it decreases with decreasing to lower temperatures. Among them, GaAs on $Si_{0.65}Ge_{0.35}$, GaN on AlN, GaN on $Si_{0.65}Ge_{0.35}$, AlN on Si, AlN on 6H-SiC, GaN on sapphire, AlN on

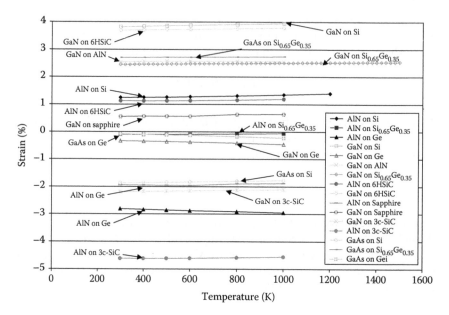

FIGURE 2.7
Strain for "Domain Matching Epitaxy" of various film and substrate combinations as a function of temperature.

$Si_{0.65}Ge_{0.35}$, and GaAs on Ge have the less compressive or tensile strain and almost constant from low-to-high temperatures. Therefore, SiC and sapphire have been widely used as the substrates for III-nitride materials depositions, and AlN, and SiGe as well as Ge have been widely used for the buffer layers for GaN growth on Si, and GaAs growth on Si, respectively. From the figure, the compressive strain is less for GaN on $Si_{0.65}Ge_{0.35}$ than GaN on Si and also is more constant from low-to-high temperatures, which could suggest that SiGe would be a good buffer layer for GaN growth on Si substrate. Furthermore, the strain of GaN on SiGe is very close to that of GaN on AlN, which is known to reduce cracking for GaN growth on Si substrate. Finally, observe that the strain of AlN on $Si_{0.65}Ge_{0.35}$ is very close to zero. On the other hand, the thermal expansion mismatch, which produces the well-known tensile strain and cracking as the film is cooled from the deposition temperature. In Figure 2.6, the thermal expansion mismatch between AlN and Si as well as $Si_{0.65}Ge_{0.35}$ and Si are smaller than that between GaN and Si. So, the combination of $GaN/AlN/Si_{0.65}Ge_{0.35}/Si(111)$ may be a good combination to use for reduced tensile strain and cracking for GaN growth on Si substrate. Based on above calculation and analysis, the thermal mismatch between III–V materials and Si substrate always generate tensile stress on III–V materials films. Therefore, the compress stress generated by controlling the DME to compensate the tensile stress is a very important art for improving the III–V materials film quality on Si substrates.

2.4 Lack of Semi-Insulating Si Substrates

Semi-insulating substrates are desirable for high-speed electronics because they provide circuits and devices with intrinsically low capacitances. Portable electronics, data storage, and fast modems for Internet access, etc., are the fastest growing in the microelectronics industry. Some of the key circuit components for these applications include radio frequency (RF) power amplifiers, intermediate frequency (IF) mixers, and baseband circuitry that have conventionally been in GaAs, Si bipolar, and Si CMOS technologies, respectively. For the III–V materials and devices integrated on a single Si chip, one of the major challenges for single-chip RFIC's is RF cross talk through the Si substrate. Due to lack of semi-insulating Si substrate, the parasitic capacitance effects may occur during high-frequency operation. In other words, noise from switching transient in digital circuits can be transmitted through the conducting Si substrate and degrade the performance of analog circuit elements. Losses as well as substrate-governed cross talk need to be considered. At high frequencies, the electromagnetic field of a signal transmitted along a line extends outside the line. If free carriers are available in the substrate, the transmitted signal loses energy to the carriers. At even higher frequencies, the carriers are no longer in phase with the signal and carrier-caused losses decrease. At high frequencies, a semiconductor, therefore, exhibits dielectric behavior with losses determined by polarization [Gevorgian 1998]. Comparing to CS technology, the lacking of a semi-insulating state in Si material is probably the most challenging technological hurdle in fabricating high-performance RF integrated circuits, especially for passive components such as inductors. Figure 2.8 shows the approximate

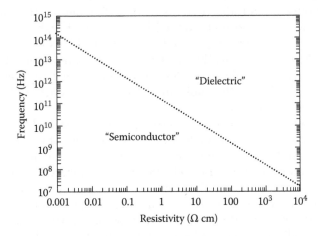

FIGURE 2.8
The dielectric relaxation frequency for silicon of different resistivities. (From Johansson, M. and Bengtsson, S., *J. Appl. Phys.*, 88, 1118, 2000. With permission.)

onset of dielectric behavior for silicon of different resistivities. The straight line marking the division between semiconducting and dielectric behavior is the dielectric relaxation frequency, f_{dr}. It can be determined from [Ashcroft 1976], [Johansson 2000]

$$f_{dr} = \frac{\sigma_0}{2\pi\varepsilon_r\varepsilon_0} \qquad (2.15)$$

where

σ_0 is the DC conductivity of the silicon material
ε_0 is the dielectric constant of vacuum
ε_r is the relative dielectric constant of silicon

From Figure 2.8, it can be concluded that insulators or materials depleted of free carriers can be expected to be good substrates for high-frequency applications.

One big advantage of III–V semiconductors over silicon for high-speed electronics is the availability of semi-insulating substrates such as gallium arsenide, indium phosphor, etc., substrates. In silicon technology, no semi-insulating substrate is available. Previous work related to semi-insulating silicon layers resulted in the formation of semi-insulating polycrystalline silicon (SIPOS) films [Mochizuki 1976]. SIPOS films, which contain oxygen concentrations of a few tens of atomic percent, are used as part of passivation schemes for high-voltage semiconductor devices as well as wide band gap emitters in silicon-based heterojunction bipolar transistors (HBTs). Oxygen-doped silicon epitaxial films (OXSEF) [Takahishi 1987], [Schwartqa 1993] were then introduced as a crystalline substitute for the wide band gap material for silicon-based HBTs. On the other hand, ultrapure highly resistive silicon ~10,000 Ω cm and silicon-on insulator (SOI) materials [Hisamoto 1998] offer advantages over normally doped bulk silicon at high frequencies. In the case of SOI, silicon-on-sapphire (SOS) materials [Johnson 1998] as well as so-called MICROX materials [Hanes 1993] have recently received increased attention for being suitable substrates for high-frequency applications. Both in the case of highly resistive bulk silicon and MICROX material, extremely low-doped wafers are used.

The one of recent approaches is the use of porous silicon. In this material, the lattice has had a large number of its silicon atoms removed by an electrochemical reaction, producing a honeycomb-like structure that exhibits markedly different structural and electrical characteristics compared to conventional silicon. In particular, porous silicon exhibits a resistivity of the order of 10^6 Ω cm, a value approaching the semi-insulating resistivity values achieved in GaAs and InP substrates [Metzger 2001], [Gardes 2008]. The resistivity values of the porous silicon structures have the potential to push the RF operating window for silicon MMICs well into the 20 GHz range, offering real competition in a regime that many consider the exclusive domain of GaAs and InP.

Wafer bonding has also been used to manufacture a silicon material intended as substrate for high-frequency applications. The space charge regions surrounding the bonded silicon/silicon interface deplete the silicon, thereby causing semi-insulating behavior at high frequencies. The material has been characterized electrically for frequencies up to 40 GHz using metal transmission lines on its surface. The results show that transmission lines made on these materials show low losses when signals are transmitted along them. These results have been compared to measurements using similar transmission lines on bulk silicon wafers of different resistivities, on SIMOX wafers, and on quartz [Johansson 2000].

Even though there are many progresses in semi-insulating Si substrate researches and production, developments of new type of semi-insulating Si substrates with high quality and low cost for high-frequency applications are still big challenge in this time.

2.5 Si Diffusion into Compound Semiconductors

During the III–V materials grown, under H_2 flow and high temperature, Si substrate may react with H_2, out–gases of Si, or react and diffuse into III–V materials and this Si will lead to a high n-type background doping, which makes it difficult to achieve p-type doping and poor device performance.

Even though extensive experimental results are available for Si diffusion in GaAs, there is no clear-cut conclusion on the mechanism of diffusion, i.e., the single vacancy and neutral defect pair are some of the possibilities reported. From the theory, there are three kinds of doping in GaAs by Si.

If Si replaces Ga atoms in GaAs, the n-type GaAs is formed. If Si replaces As atoms in GaAs, the p-type GaAs is formed. The third type is Si residual doping in GaAs.

Dutt and Sharma reported the Si diffusion coefficient data in GaAs with various temperatures, as shown in Figures 2.9 and 2.10 [Dutt 1998]. With increasing the temperatures, the Si diffusion coefficient increases. The mechanism for Si diffusion involves the motion of a nearest neighbor [Si–Si] pair silicon atoms through Ga and As vacancies. Paired Si atoms can move substitutionally by exchanging sites with either a Ga or As vacancy. This is expressed by [Greiner 1984]

$$\left(Si_{Ga}^{+} - Si_{As}^{-}\right) + V_{Ga} \leftrightarrow V_{Ga} + \left(Si_{As}^{-} - Si_{Ga}^{+}\right) \tag{2.16}$$

with a similar expression for the exchange with an As vacancy, this mechanism conserves both charge and vacancy type.

In order to understand the exact mode of Si diffusion in GaAs, Murugan, et al. [Murugan 2002] have investigated theoretically all the possible

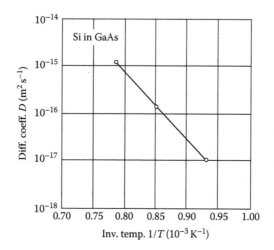

FIGURE 2.9
Diffusion coefficient D of Si in GaAs vs. inverse temperature 1/T. (From Beke, D.L., *Group III Condensed Matter Numerical Data and Functional Relationships in Science and Technology, Diffusion in Semiconductors*, Springer, Berlin, Germany, 1998. With permission.)

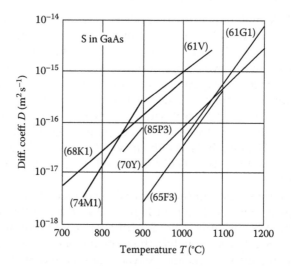

FIGURE 2.10
Diffusion coefficient D of Si in GaAs vs. temperature T. (From Beke, D.L., *Group III Condensed Matter Numerical Data and Functional Relationships in Science and Technology, Diffusion in Semiconductors*, Springer, Berlin, Germany, 1998. With permission.)

mechanisms. By measuring the kinetic energy (ΔK), which is necessary for overcoming the potential barrier so that diffusion is possible, they found that for the single vacancy mechanism, ΔK values are about 87% for all temperatures from 700 to 1300 K, whereas in vacancy-assisted interstitial process, ΔK values are only about 59%. By comparison of ΔK values obtained with the two prominent mechanisms, single vacancy is found to be the most dominant mechanism. Since neutral defect pair mechanism is also reported from experiments, they have worked out this mechanism also theoretically and the ΔK values are found to be about 52%. These results indicate that the most dominant mechanism for Si diffusion in GaAs is through single vacancy compared to the other two mechanisms under normal condition. The other mechanisms may also be possible under specific conditions, either at elevated pressure or temperature or higher doping concentrations. In fact, most possible mechanism can also be understood from the critical value of reaction coordinate.

In order to reduce the Si diffusion during the growth of III–V materials, many researchers have found that the Si diffusion speed is lower when using a Si_3N_4 protection layer than when using an SiO_2 protection layer, and the CS diffusion-limiting layer containing aluminum content such as $Al_xGa_{1-y}As$ ($y \geq x + 0.3$) can also reduce the Si diffusion into III–V materials significantly [Murakami 1992].

Similarly, due to the reaction between the Ga and Si substrate, a GaN film grown directly on the silicon substrate has very poor surface morphology and bad crystal quality. The strong Si diffusion is across the GaN/Si interface when no buffer layer exists. The Si profile penetrates 100–300 nm into the GaN side. Vacancies are known to be vehicles for substitutional atoms. In GaN growth on Si substrates by MBE and MOCVD processes, vacancy clusters are observed, which is evidently stable trace left by diffusion processes during the layer growth. The vacancy clusters are strong interdiffusion across the GaN/Si interface without buffer layer [Manasreh 2000].

Experimental results show that by using NH_3 pretreatment of Si wafer before growth III–V materials layer, the reaction between the Ga and Si can be reduced. Nitridation of Si is self-limiting and the nitridation process with create a thin layer (few nanometers) of Si_xN_y layer and will oppose the Si out-diffusion during growth. However, the SiN_x layer formed on Si substrate results in inhomogeneous growth of GaN. In order to solve this problem, using an Al-contained film as the nucleation layer deposited prior to GaN deposition decreases the Si and Ga interdiffusion, which makes high-quality GaN film on Si be possible. The AlAs nucleation layer is thermally stable even at high temperatures and prevents inhomogeneous growth of GaN due to the formation of SiN_x on the silicon surface at the initial stage of growth [Strittmatter 1999]. Further investigation shows that using Si surface etching and Al-rich AlN buffer layer can reduce Si diffusion into GaN, and high-quality thick epi GaN film on Si substrate can be obtained. The experimental results show that AlN with the highest structural quality can be grown by MOCVD only at very high temperatures of 1100°C and above, which is necessary for the next step of the high-quality epitaxial GaN growth. However, if the AlN buffer layer growth temperature is too high (over 1200°C), the Si diffusion rate increases, and a noncontinuous amorphous SiN layer is formed, which results in a poor GaN layer with an inclined c-axis and also pits where no coalescence occurred [Lin-a 2007], [Schulze 2006], [Lu 2004], [Chen-a 2006].

The typical processes for avoiding Si diffusion into GaN are as follows [Lin-a 2007], [Schulze 2006]. The Si substrate was etched prior to growth using $H_2SO_4:H_2O_2:H_2O$ (3:1:1) and HF (5%). This procedure results in an oxide-free, hydrogen-terminated Si surface. At first, an approximately few nm thin aluminum pre-deposition was grown using TMAl and ammonia as source materials and hydrogen as the carrier gas at growth temperatures of up to about 1150°C and followed by formation high temperature 20–50 nm thick Al-rich AlN layer deposition. The thin aluminum pre-deposition on the clean Si substrate is used to prevent the Si nitridation. Subsequently, an approximately 100 nm thick low-temperature AlN seed layer about 800°C and 100 nm high-temperature AlN buffer layer were deposited at 1100°C. Then, grading $Al_xGa_{1-x}N$ buffer layer or several layers of each an approximately 200 nm thick $Al_xGa_{1-x}N$ with various x buffer layers was deposited at 1100°C. After that, n-GaN layer was grown at 1100°C with an overall thickness of approximately 1–2 μm and optional with interrupted by four thin AlN interlayers grown at 1000°C. The optically active layers consist of a fivefold or above InGaN/GaN multiple quantum well (MQW) grown at 780°C with nitrogen as a carrier gas. A 15 nm AlGaN:Mg electron barrier and a 100 nm and above p-GaN cap layer were grown on top of the MQW. The high-quality blue LED device can be fabricated. The experimental results further confirmed that after Si surface etch, the aluminum pre-deposition followed by high-temperature AlN/low-temperature AlN multilayers, the Si diffusion into GaN can be reduced significantly [Lin-a 2007]. The Si_xN_y layer limitation layer with AlN buffer is also be investigated. The coincidently matched multiple-layer buffer comprises a single-crystal silicon nitride (Si_3N_4) layer formed by the reaction of nitrogen plasma or ammonia with Si substrate at high temperature. Then, an AlN buffer layer or other group III-nitride buffer layer is grown epitaxial on the single-crystal silicon nitride layer. In this way, the Si diffusion into GaN can be reduced, and the high-quality GaN epitaxial layer can be grown on the coincidently matched multiple-layer buffer [Gwo 2006].

2.6 Light Absorption Issue of Optical Device on Si Substrates

Although several groups have demonstrated GaN-based LEDs on Si substrates, there are no reports of volume production of GaN LEDs deposited on Si substrates. A major limitation is the opaque Si substrate that absorbs the light emitted from the active region. Insertion of a high-reflectance distributed Bragg reflector (DBR) between the substrate and the active region would increase light extraction by approximately a factor of 2, thereby doubling luminous efficiency. A survey of the literature reveals the previous highest reflectance for a III-nitride DBR on Si was 78% for a 10× AlN/Al$_{0.2}$Ga$_{0.8}$N DBR grown by MBE [Semond 2001]. Most approaches to DBR growth on Si substrates [Charles 2005] have suffered from the inherent design of the structure. A DBR that does not incorporate the Si substrate as a reflective interface requires approximately 20× quarter-wave stack to achieve 97% reflectance [Ng 2000]. While this approach is feasible for growth on sapphire or SiC [Ng 2000], [Waldrip 2001], the strains generated during growth on Si are prohibitive to such a thick structure. Mastro et al. [Mastro 2005] reported on metalorganic chemical vapor deposition DBRs composed of an AlN/AlGaN superlattice were grown of Si(111) substrates. The first high-reflectance III-nitride DBR on Si was achieved by growing the DBR directly on the Si substrate to enhance the overall reflectance due to the high index of refraction contrast at the Si/AlN interface. A high reflectance (96.8%) of III-nitride 9× DBR structure was grown by MOCVD on a Si(111) substrate, which actually exceeded the theoretical value of 96.1%. The DBR structure is shown in Figure 2.11. The AlN/AlGaN superlattice served the added purpose of compensating for the large tensile strain developed during the growth of a crack-free 500 nm GaN/7× DBR/Si structure. The DBR consisted of alternating layers of AlN and GaN (or AlGaN) that introduced a compressive stress to balance the large tensile stress generated during cool down from growth temperature. Additionally, the dislocation density was found to drop by more than two orders of magnitude through a 500 nm/7× DBR structure. Conceivably, III-nitride LED structures can now be fabricated on low-cost Si without optical absorption loss in the opaque substrate.

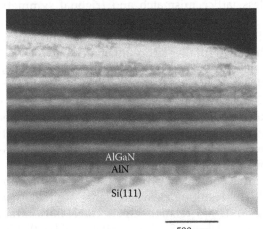

500 nm

FIGURE 2.11
The DBR structure on Si substrate. (From Mastro, M.A. et al., *Appl. Phys. Lett.*, 87, 241103, 2005. With permission.)

Other approaches to solve the light absorption issue of optical device on Si substrates are heterogeneous integration [Gutierrez-Aitken 2008] and wafer transfer and bonding technologies to make vertical LED device [Letertre 2008]. The heterogeneous integration includes selective epitaxial growth, metamorphic growth, and wafer level packaging (WLP) technology. The wafer transfer and bonding technologies deal with recently developed Smart Cut technology. The Smart Cut technology, introduced in the mid-1990s by Bruel [Bruel 1995], is a revolutionary and powerful thin film technology for bringing to industrial maturity engineered substrate solutions. It is a combination of wafer bonding and layer transfer via the use of ion implantation. It allows multiple high-quality transfers of thin layers, from a single-crystal donor wafer onto another substrate of a different nature, allowing the integration of dissimilar materials. As a consequence, it opens the path to the formation of III–V based engineered substrates by integrating, e.g., materials like GaAs, InP, SiC, GaN, germanium, and Si on a silicon, poly SiC, sapphire, ceramic, or metal substrates.

2.7 Approaches of III–V Materials Growth on Si Substrates

2.7.1 Selection of Si Substrates

In order to select suitable Si substrates for III-nitrides and conventional III–V materials epitaxial growth, Si(100), Si(211), and Si(111) have been studied. In the most cases, the Si(111) plane is chosen for III-nitride epitaxial growth because of its less lattice mismatch and trigonal symmetry favoring the GaN(0001) plane, as shown as in Figure 2.12 [Krost 2002], and Si(100) plane is chosen with $Si_{1-x}Ge_x$/Ge buffer layer for GaAs and other III–V materials epitaxial growth due to smaller lattice mismatch between Si(100) plane with GaAs. The most serious problem is the large thermal mismatch between GaN or GaAs and Si. The TEC of GaN, GaAs, and Si are 5.59×10^{-6} K^{-1}, 5.40×10^{-6} K^{-1}, and 3.59×10^{-6} K^{-1} respectively, as shown in Table 2.2, which leads to a large tensile stress during cooling down from the growth temperature to room temperature, and often results in cracks in III–V film layers. The device applications are limited. The tensile stress causes formation of defects, stress, and even cracks. For example, for GaN(0001) epitaxial growth on Si(111) substrate, there are three equivalent primary GaN crack directions along [11$\bar{2}$0], [$\bar{1}$$\bar{2}$0], and [2$\bar{1}$$\bar{1}$0]. The resulting cleavage planes are ($\bar{1}$100), (10$\bar{1}$0), and (01$\bar{1}$0), respectively. The epitaxial relationship is GaN(0001) parallel Si(111) with Si[$\bar{1}$10] parallel GaN[11$\bar{2}$0], and Si[11$\bar{2}$] parallel GaN [$\bar{1}$100]. Since the primary cleavage planes of Si are of type {111} with ⟨110⟩ cleavage directions, GaN and Si always have a common cleavage direction, e.g., GaN[11$\bar{2}$0] and Si[$\bar{1}$10].

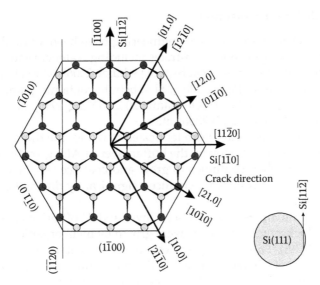

FIGURE 2.12
Epitaxial relationship GaN(0001) on Si(111).

These relations are summarized in Figure 2.12 and the experimental results show the crack formation of thick GaN growth on Si(111) substrate, which are consistent with above analysis, as shown in Figure 2.13. In order to make high-quality and crack-free III–V thick films grown on Si substrates, many kinds of Si substrates engineering have been studied to reduce the thermal tensile stress.

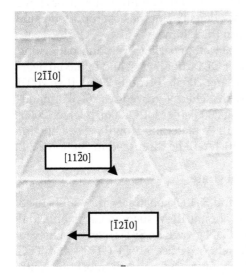

FIGURE 2.13
The crack formation of thick GaN growth on Si(111) substrate.

2.7.2 Si Substrate Engineering

Some novel schemes of substrate engineering have been studied to obtain high-quality GaN layers on Si substrates including wafer curvature [Dadgar 2000], [Feltin 2001], off-oriented substrate [Schulze 2006], N+ ion implantation of an AlN/Si substrate [Jamil 2005], Si substrate containing amorphous or gliding layer, e.g., SIMOX or SOI [Zamir 2002], [Yang 1995], and patterned substrates [Li 2008a], [Chen-b 2006], etc.

The wafer curvatures have been fabricated to reduce the thermal stress. In case of GaN heteroepitaxy on Si, the buffer layer of AlN growth starts with small nuclei, and then continues to grow and coalesce. At the coalescence boundaries, tensile stress can be reduced significantly during GaN growth on Si substrate.

Using a high-temperature AlN seed layer and 4° off-oriented substrates, the 2–3 μm thick, crack-free layers were grown by metalorganic vapor phase epitaxy (MOVPE). This technique allows to grow a flat, fully coalesced, and single-crystalline GaN layer on Si(001). For preventing crack formation, four AlN interlayers were also inserted in the buffer structure.

N+ ion implantation of an AlN/Si substrate is performed to create a defective layer that partially isolates the III-nitride layer and the Si substrate and helps to reduce the strain in the film. Raman spectroscopy shows a substantial decrease in in-plane strain in GaN films grown on nitrogen-implanted substrates. This is confirmed by the enhancement of the E_2 (TO) phonon frequency from 564 to 567 cm^{-1} corresponding to 84% stress reduction and substantial decrease in crack density for a 2 μm thick GaN film.

Si substrate containing amorphous or gliding layer, e.g., SIMOX or SOI has also been studied for GaN growth on Si substrates. The strain management methods such as use of GaN on SOI combined with buffer layers have reported some success in reducing crack density, but the best reported value for crack separation is 120 μm for a 1 μm thick film.

The more efficient works of substrate engineering are nano-patterning combined with other techniques of the SOI, buffer layers, and epitaxial overgrowth [Zubia 1999], [Wang-b 2006], [Zang 2005], [Ferdous 2006], [Lee 2005] such as lateral epitaxial overgrowth with etched V-grove [Lee 2005] or nanopillars, and selective area epitaxy [Zang 2005], [Ferdous 2006], [Hersee 2000] were developed and improvements in III–V materials growth, especially for III-nitrides growth on Si substrates, which will be discussed in Section 2.7.4.

2.7.3 Buffer Layers for III–V on Si Substrate

Based on Section 2.7.1 analysis, Si(111) and Si(100) substrates are widely used for III-nitride such as GaN and conventional III–V such as GaAs epitaxial growth, respectively. The high-quality GaN grown on Si(111) can be achieved only by addressing the significant levels of lattice misfit (~17%) and TEC (~56%) mismatch issues. Comparing sapphire, SiC, and Si substrates with GaN, as shown in Figure 2.14, it is clearly seen that the lattice mismatch and

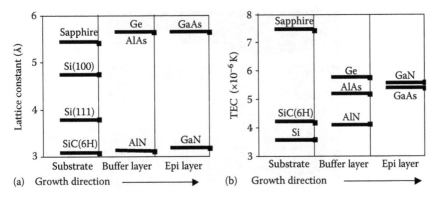

FIGURE 2.14
Schematic of the epitaxial challenges in growing high-quality GaN and GaAs on Si with buffer layers using MOCVD. (a) The lattice mismatch and (b) the TEC mismatch. In both figures, the typical substrates are grouped to the left, buffer layer in the middle, and epitaxial layers to the right in order to preserve a typical growth sequence.

thermal mismatch between GaN and SiC is much smaller than that between GaN and Si or sapphire. The misfit effects will dominate the Si/III-nitride interface, where a high density of misfit dislocations is expected to form during growth. The critical thickness at which cracking occurs depends on the growth conditions such as temperature. In order to achieve crack-free GaN thin film on Si(111), it has been studied using a variety of buffer layers, such as AlN [Zamir 2000], [Raghavan 2005], [Suda 2002], [Chen 2001], AlGaN and AlGaN/GaN superlattices [Jang 2002], [Jang 2003], graded $Al_{1-x}Ga_xN$ [Able 2005], $AlN/Al_{1-x}Ga_xN$ [Lin-a 2007], [Ishikawa 1999], AlInGaN [Wu 2005], SiC [Komiyama 2006], AlAs [Strittmatter 2001], Al_2O_3 [Wakahara 2002], SiN_x [Liu 2000], ZnO [Kim 2003], HfN [Gebauer 2002], $SrTiO_3$ [He 2005], BP [Nishimura 2002], and SiGe [Maa 2007b], etc. Many research groups have attempted, with limited success, to achieve high-quality III-nitrides by using various buffer layers between III-nitrides and silicon, including GaAs, AlAs, ZnO, $LiGaO_2$, SiO_2, Si_3N_4, sputtered AlN, LT-GaN, LT-AlN, and deposited and converted SiC films. Because the Al-rich AlN and AlAs can limit the Si diffusion into GaN, and the TEC of AlN and AlAs is between GaN and Si, it has been demonstrated that using AlN and AlGaN (MOCVD growth) and AlAs (CBE growth) as a nucleation layer on Si(111) results in excellent thin film quality. Recent experimental results show that using high-temperature AlN/low-temperature AlN multilayers with grading $Al_xGa_{1-x}N$ buffer layers can significantly limit the Si diffusion into GaN and release the thermal stress, and 2 μm high-quality crack-free GaN films can be deposited on 6 in. Si(111) wafers [Lin-a 2007]. Before AlN deposition, a few monolayers of Al are deposited over Si substrate, which act as barrier for nitridation of the Si substrate. Low-temperature AlN is for releasing the thermal stress, and high-temperature AlN has a good structural quality, which results in next step

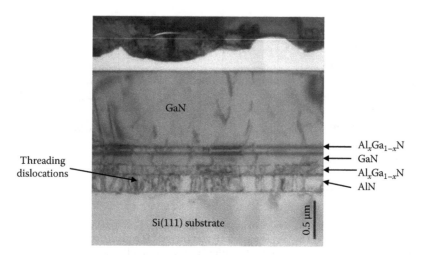

FIGURE 2.15
The multiple buffer structure limits the extension of the threading dislocations.

high-quality epitaxial AlGaN/GaN growth. AlN/$Al_{1-x}Ga_xN$ and $Al_{1-x}Ga_xN$/ GaN heterostructures and superlattices grown at high temperature can serve as effective intermediate layer and is believed that the use of the superlattice layer will reduce the defect density on the Si substrate and improve the heterojunction interface quality. The multiplayer structures show that the extension of treading dislocation and microcracks can be stopped by multiple buffer layer structure, as shown in Figure 2.15.

For GaAs epitaxial growth on Si, because AlAs layer can limit Si into GaAs, and has smaller thermal mismatch with Si, and Si_xGe_{1-x} alloys also have both smaller lattice and thermal mismatch with Si than that between GaAs and Si, the AlAs and Si_xGe_{1-x} alloys buffer layers have been successfully used to make high-quality epitaxial GaAs on Si(100) substrates [Chang 2005], [Lin-b 2007].

2.7.4 Lateral Epitaxial Overgrowth

In order to reduce the dislocation density and release the thermal stress, the lateral epitaxial overgrowth with buffer layer technologies has been developed to grow crack-free and high-quality III–V materials on Si substrates. Lateral epitaxial overgrowth with etched V-grove [Lee 2005] or nanopillars, and selective area lateral overgrowth epitaxy [Yang 1999], [Kawaguchi 1998], [Honda 2002] were reported to improve in III–V materials growth, especially for III-nitrides growth on Si substrates. Recently, using nanosize SiN interlayer and lateral overgrowth epitaxy to reduce dislocation density and release the thermal stress have also been developed. By controlling SiH_4 flow, deposition temperature and time, the high density nano-island SiN layer can be formed. The threading dislocation density of III-nitride films grown on silry the lateral overgrowth epitaxy from the SiN interlayer can be reduced to be smaller than $10^8 cm^{-2}$.

Using the MOCVD selective growth techniques, GaN pyramids selectively grew on Si(111) with a patterned dot structure of a SiO_2 mask and on SiO_2 stripe-patterned Si using an intermediate AlGaN layer, which lead to high-quality GaN grown on Si(111) substrates. It has also been demonstrated using SiO_2-patterned AlN/Si(111) substrates and the lateral epitaxial overgrowth technique. Cracking occurred only when the stripes came into contact and formed a coalesced film. With 60 nm AlN buffer layer thickness, the stripes were bound by the (0001) facet on top, by vertical $\{11\bar{2}0\}$ facets on the edges, and by inclined $\{11\bar{2}2\}$ sidewalls. The distance between cracks on a given stripe was in excess of 300 μm. A threading dislocation density (TDD) $<10^6$ cm^{-2} is reported [Marchand 2001].

Recently, lateral nanoheteroepitaxy overgrowth (LNEO) technologies have been rapidly developed for AlN, GaN, and GaN with buffer layers grown on sapphire, SiC, and Si substrate using MOCVD and hydride vapor phase epitaxy (HVPE) processes and the high-quality III-nitrides with lower defect density such as TDD have been achieved [Zubia 1999], [Zubia 2000], [Zang 2005], [Soh 2007], [Wang-b 2006]. Li et al. [Li 2008b] developed a nano-pattern process using AAO techniques for III–V materials growing on Si substrates. The processes of LNEO of GaN on Si substrates are shown in Figure 2.16 and are described as follows. SiO_2 of 20–70 nm is formed on Si(111) substrates by thermal oxidation. Al film of 1 μm is deposited on SiO_2 by e-beam evaporation using the CHA. The nanoholes are made by an anodized aluminum oxide (AAO) technology and using wet or dry etching to transfer nanoholes from AAO to SiO_2 film. Remove AAO and leave SiO_2 hardmask for GaN with AlN and $Al_{1-x}Ga_xN$ buffer layer LNEO on Si(111) substrates. Finally, chemical-mechanical planarization (CMP) GaN to get a flat surface. Figures 2.17 and 2.18 show the cross section and surface structure of Si substrate with about 50 nm SiO_2 nanoholes made by AAO processes. On the other hand, the

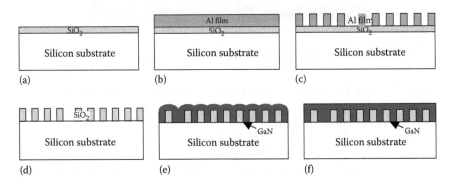

FIGURE 2.16
Processes of LNEO of GaN on Si: (a) thermal oxidation, (b) Al film deposition, (c) anodized aluminum oxide, (d) wet and dry etch processes, (e) GaN LNEO processes, and (f) CMP GaN.

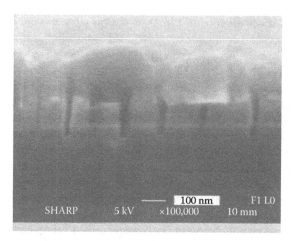

FIGURE 2.17
The cross section of Si substrate with about 50 nm SiO_2 nanoholes.

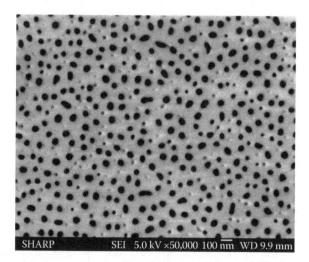

FIGURE 2.18
The surface structure of Si substrate with about 50 nm SiO_2 nanoholes.

porous GaN can be made by optimum etching with an anodization current of 20 mA cm^{-2} for an hour in 2 M NaOH solution, and the porous GaN templates have been used for further GaN LNEO [Soh 2007].

Using these technologies, the GaN LNEO on nanoporous SiO_2 films on the surfaces of GaN/sapphire (0001) using metal organic chemical vapor deposition to realize a continuous and smooth film. The TDD of 10^7 cm^{-2} was achieved at the surface of NLEO GaN [Wang-b 2006]. GaN can also selectively grow by MOCVD into the pores and laterally over the nanoscale-patterned

SiO$_2$ mask on a template of GaN/AlN/Si. Cross-section transmission electron microscopy shows that the threading-dislocation density was largely reduced in this nanoepitaxial lateral overgrowth region [Zang 2005].

Hawker and Russell [Hawker 2005] state that an ideal process would be compatible with existing technological processes and manufacturing techniques; these strategies, together with novel materials, could allow significant advances to be made in meeting both short-term and long-term demands for higher-density, faster devices. While self-assembly alone is sufficient for a number of applications in fabricating advanced microelectronics, directed, self-orienting, self-assembly processes are also required to produce complex devices with the required density and addressability of elements to meet future demands. Both strategies require the design and synthesis of polymers that have well-defined characteristics such that the necessary fine control over the morphology, interfacial properties, and simplicity of processes can be realized. By combining tailored self-assembly processes (a "bottom-up" approach) with microfabrication processes (a "top-down" approach), the ever-present thirst of the consumer for faster, better, and cheaper devices can be met in very simple, yet robust ways.

Li et al. [Li 2008b] also developed the BCP lithography for making a nano-pattern on Si or III–V compounds/buffer/Si such as AlN/Si, GaN/AlN/Si, GaN/AlGaN/AlN/Si or GaAs/SiGe/Si, etc. The self-assembly of BCPs, two polymer chains covalently linked together at one end, provides a robust solution to these challenges. As thin films, immiscible BCPs self-assemble into a range of highly ordered morphologies where the size scale of the features is only limited by the size of the polymer chains and are, therefore, nanoscopic. Figure 2.19 shows the processes for the formation of nano-patterns by the self-assembly of BCPs. This nanosized pattern can be used to make a hardmask for NLEO growth of III–V compounds including GaN on Si(111), GaN/AlN/Si, or III–V compounds/buffers/Si. This will improve the film quality of the III–V compounds including GaN on Si substrates.

Example 2.1: Processes for Making Nano-Patterns on Si Substrates

There are two processes, which are shown in Figure 2.20, and listed in the following steps.

Process A is shown in the left side of Figure 2.20.

1. The starting wafer is silicon substrate. Clean the silicon substrate and form 20–200 nm thermal SiO$_2$ or deposit TEOS on the Si substrate.
2. Make nano-patterns by the self-assembly of BCPs (ozonated sample) on the SiO$_2$ layer.
3. Perform a two-step etch to transfer the nano-pattern from the BCPs to the SiO$_2$ layer.

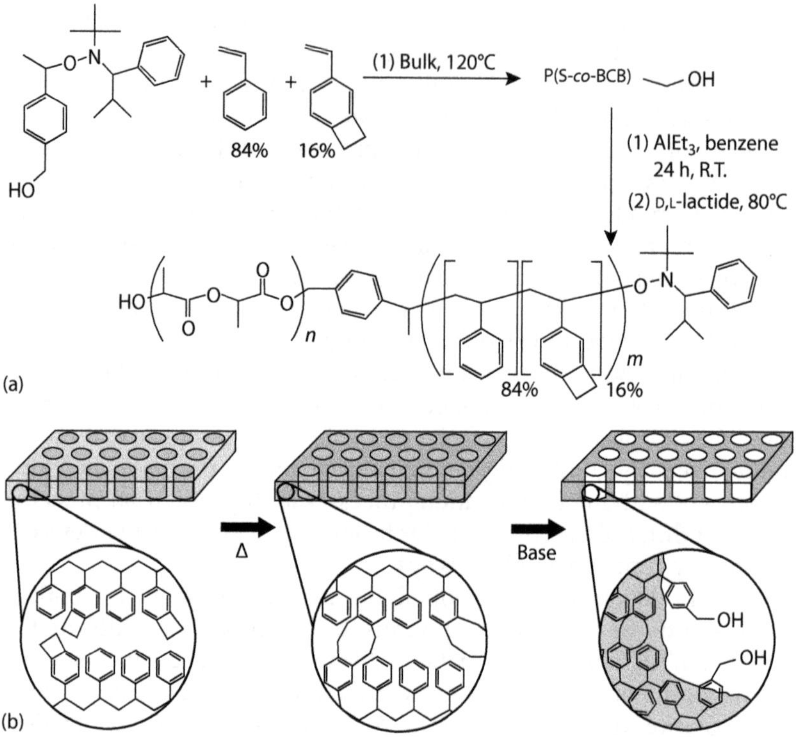

(a)

(b)

FIGURE 2.19
The mechanism and processes of nano-patterning by the self-assembly of BCPs. (From Leiston-Belanger, J.M. et al., *Macromolecules*, 38, 18, 2005. With permission.)

    4. Perform LNEO of III–V compounds, including GaN with buffer layers on the nano-patterned (holes) Si substrate.

    5. The surface of the NLEO GaN may be not flat enough for device fabrication. Therefore, an optional CMP step may be required. After CMP, an additional GaN may be grown to form very smooth GaN film for device fabrication.

Process B is shown in the right side of Figure 2.20.

    1. The starting wafer is silicon substrate. Clean the silicon substrate and form 20–200 nm thermal SiO$_2$ or deposit TEOS on the Si substrate.

    2. Make nano-patterns by the self-assembly of BCPs (stained sample) on the SiO$_2$ layer.

    3. Perform a two-step etch to transfer the nano-pattern from the BCPs to the SiO$_2$ layer.

    4. Perform LNEO of III–V compounds, including GaN with buffer layers on the nano-patterned (dots) Si substrate.

    5. The surface of the NLEO GaN may be not flat enough for device fabrication. Therefore, an optional CMP step may be required. After CMP, an additional GaN may be grown to form very smooth GaN film for device fabrication.

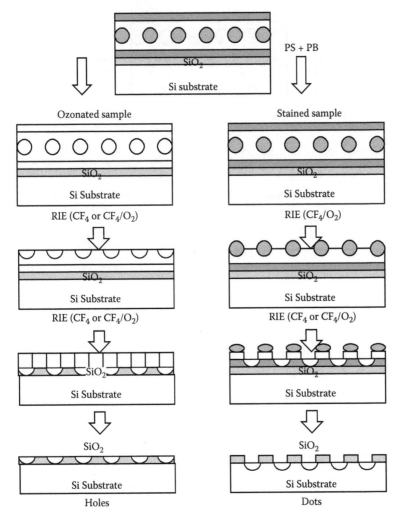

FIGURE 2.20
The processes for making nano-patterns on Si substrates.

Example 2.2: Processes of Making Nano-Patterns on III–V/Buffers/Si Substrate

There are two processes, which are shown in Figure 2.21, and listed in the following steps.

Process A is shown in left side of Figure 2.21.

1. The starting wafer silicon substrate. Clean the silicon substrate, perform reactor conditioning, substrate pretreatment and deposit buffer layers, III–V compounds including GaN, and form 20–100 nm thermal SiO_2 or deposit TEOS on the top III–V compounds including GaN. Make nano-patterns by

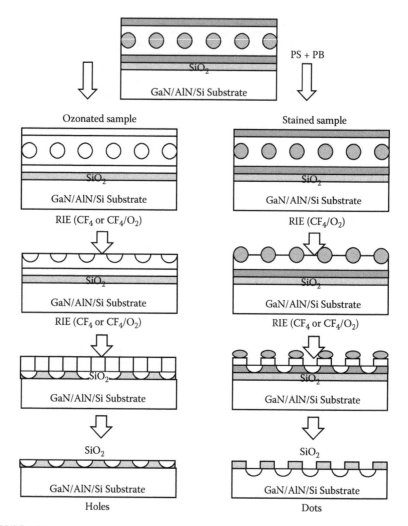

FIGURE 2.21
The processes of making nano-patterns on III–V/buffers/Si substrate.

the self-assembly of BCPs on SiO_2 layer. (Optional: The starting wafer is (111) or (100) oriented silicon substrate. Clean the silicon substrate, perform reactor conditioning, substrate pre-treatment and deposit buffer layers, and form 20–200 nm thermal SiO_2 or deposit TEOS on the top of the buffers.)

2. Make nano-patterns by the self-assembly of BCPs (ozonated sample) on SiO_2 layer.

3. Perform a two-step etch to transfer the nano-pattern from the BCPs to the SiO_2 layer.

4. LNEO of III–V compound including GaN on nano-patterned (holes) buffer layers/Si substrates, or III–V compound including GaN on III–V compound including GaN/buffers/Si substrate.

5. The surface of the NLEO GaN may be not flat enough for device fabrication. Therefore, an optional CMP step may be required. After CMP, additional GaN may be grown to form a very smooth GaN film for device fabrication.

Process B is shown in right side of Figure 2.21.

1. The starting wafer silicon substrate. Clean the silicon substrate, perform reactor conditioning, substrate pretreatment and deposit buffer layers, III–V compounds including GaN, and form 20–100 nm thermal SiO_2 or deposit TEOS on the top III–V compounds including GaN. Make nano-patterns by the self-assembly of BCPs on SiO_2 layer. (Optional: The starting wafer is (111) or (100) oriented silicon substrate. Clean the silicon substrate, perform reactor conditioning, substrate pretreatment and deposit buffer layers, and form 20–200 nm thermal SiO_2 or deposit TEOS on the top of the buffers.)
2. Make nano-patterns by the self-assembly of BCPs (stained sample) on SiO_2 layer.
3. Perform a two-step etch to transfer the nano-pattern from the BCPs to the SiO_2 layer.
4. LNEO of III–V compound including GaN on nano-patterned (dots) buffer layers/Si substrates, or III–V compound including GaN on III–V compound including GaN/buffers/Si substrate.
5. The surface of the NLEO GaN may be not flat enough for device fabrication. Therefore, an optional CMP step may be required. After CMP, additional GaN may be grown to form a very smooth GaN film for device fabrication.

Heteroepitaxial growth of GaAs/Ge/SiGe films on submicrostructured Si substrates is also reported [Vanamu 2006]. One-dimensional, nanometer line width, submicrometer period features were fabricated in Si substrates using interferometric lithography, reactive ion etching, and wet-chemical etching techniques. The defect density of GaAs epilayers grown on submicrostructured Si at 6×10^5 cm^{-2} was two orders of magnitude lower compared with that grown on planar silicon. The optical quality of the GaAs/Ge/SiGe on submicrostructured Si was comparable to that of single-crystal GaAs.

2.7.5 Selective Growth of III–V Materials on Si Substrates

Based on thermal stress calculation in Section 2.3, the thermal stress can be released significantly with decreasing the patterned size. Zamir [Zamir 2001] used reactive ion etching (RIE) method to patterned Si(111) substrates into square mesas of varying sizes L, of 5, 10, 15, 20, 25, 30, 50, and 100 mm. The trenches were 0.5–0.6 mm deep and 2–4 mm wide. GaN was grown on the patterned Si by MOCVD. The growth on prepatterned Si substrates is demonstrated as an efficient way to control the geometrical distribution of the thermal cracks. They achieved maximum crack-free range of GaN on Si as 14.0 ± 0.3 μm only at GaN thickness of ~0.7 μm by lateral confined epitaxy (LCE) growth. Chen [Chen-a 2006] deposited a layer of 150 nm Si_xN_y

on a Si(111) substrate using a plasma-enhanced chemical vapor deposition (PECVD). A mesh pattern of Si_xN_y was produced through a photolithographic process using wet etching. The $80 \times 80\,\mu m^2$ crack-free GaN/AlN multilayers on the mesh-patterned Si(111) has been confirmed.

Tang [Tang 2006] found out that the critical nucleation temperature of GaN on Si(111) surface is as low as 700°C, much lower than that on sapphire or AlN surface. As a result, selective growth of GaN is possible by ammonia molecular beam epitaxy (MBE) on Si(111) substrates using a patterned AlN buffer layer. Based on different nucleation temperatures of GaN on GaN, or on Si_xN_y, or on SiO_2, Li [Li 2010] developed patterning GaN/multilayer $Al_xGa_{1-x}N$/AlN/Si by Si_xN_y and SiO_2 and selective growth technologies to deposit crack-free and high-quality GaN thick films on Si substrate for LED device applications. Jiang [Jiang 2009] grows high-brightness LEDs on patterned Si substrates, which are transferred to a reflective carrier substrate. They grow $4\,\mu m$ of GaN on a strain engineering AlGaN SL structure in patterns up to $1\,mm^2$. The patterning and selective growth technologies have the potential developments for future device applications.

2.7.6 Energetic Neutral Atom Beam Lithography/Epitaxy

Energetic neutral atom beam lithography/epitaxy (ENABLE) is a versatile technique recently developed for patterning nanoscale features into polymer substrates. ENABLE achieves the direct activation of surface chemical reactions by exposing substrates to a beam of energetic neutral atoms. Polymers that form volatile oxidation products may be anisotropically etched using a neutral beam of oxygen atoms at rates exceeding $100\,nm/min$, avoiding problems associated with charged species inherent to other etching techniques. Figure 2.22 shows the schematic of ENABLE used for anisotropic etching.

Akhadov et al. [Akhadov 2005] report on a top-down approach for producing high-aspect-ratio nanoscale structures in polymeric materials using ENABLE. Masking techniques suitable for ENABLE etching are discussed along with applications involving the rapid production of nanoscale features over large areas. Mueller et al. [Mueller 2006] deposited crystalline and polycrystalline GaN films on bare *c*-axis-oriented sapphire at low temperatures (100°C–500°C) using ENABLE. Surface chemistry is activated by exposing substrates to nitrogen atoms with kinetic energies between 0.5 and 5.0 eV and a simultaneous flux of Ga metal, allowing low-temperature growth of GaN thin films. The as-grown GaN films show semiconducting properties, a high degree of crystallinity, and excellent epitaxial alignment. This method of low-temperature nitride film growth opens opportunities for integrating novel substrate materials such as Si with group III nitride technologies.

Reichertz et al. [Reichertz 2008] developed the new thin film growth technique known as ENABLE to grow InGaN on Si(111) substrate with the entire alloy composition range possible without phase separation, which is

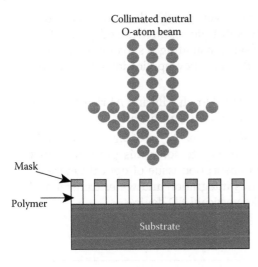

Collimated neutral
O-atom beam

Mask

Polymer

Substrate

FIGURE 2.22
Schematic of ENABLE used for aniso-tropic etching. (From Akhadov, E.A. et al., *Mater. Res. Soc. Symp. Proc.*, 872, J21.3.1, 2005. With permission.)

a significant progress. ENABLE utilizes a collimated beam of ~2 eV nitrogen atoms as the active species, which are reacted with thermally evaporated Ga and In metals. The technique provides a larger N atom flux compared to MBE and reduces the need for high substrate temperatures, making iso-thermal growth over the entire InGaN alloy composition range possible. Electrical characteristics of the junctions between n- and p-type InGaN films and n- and p-type Si substrates were measured and compared with theoreti-cal predictions based on the band edge alignment between those two mate-rials. The predicted existence of a low-resistance tunnel junction between p-type Si and n-type InGaN was experimentally confirmed. The InGaN layers were deposited at 450°C–630°C with the following structure types: (1) with the Ga-rich material on the surface and (2) with the In-rich mate-rial on the surface. The prospect of developing electronic and optoelectronic devices, including solar cells, LEDs, and solid-state lightings, which utilize the wide range of energy gaps of $In_{1-x}Ga_xN$ ($0 \leq x \leq 1$), has led to a considerable research interest in the electronic and optical properties of InN, In-rich gal-lium nitride, Ga-rich gallium nitride, and GaN alloys.

2.8 Summary

In order to make high-quality III–V materials on Si substrates, the following issues have been solved.

1. *Lattice mismatch*: The current approaches are developing substrate engineering and buffer layer technologies.

2. *Thermal mismatch*: The best technologies are buffer layer, domain mis-match epitaxy, nano-lateral epitaxy, and patterning selective growth.

3. *Lack semi-insulating Si substrate*: Even though there are many progresses in semi-insulating Si substrate researches and production,
developments of new type of semi-insulating Si substrates with high
quality and low cost for high-frequency applications are still big
challenge in this time.

4. *Si diffusion into III–V materials*: The Al-rich AlN and Al-rich AlAs can
stop the Si diffusion into III-V materials significantly.

5. *Light absorption issue of optical device on Si substrates*: Using high-
reflectance DBR is the cheapest way to solve this problem. Other
approaches to solve the light absorption issue of optical device on
Si substrates are heterogeneous integration and wafer transfer and
bonding technologies to make vertical LED devices.

References

[Able 2005] A. Able, W. Wegscheider, K. Engl, and J. Zweck, *J. Cryst. Growth*, 276, 415 (2005).

[Adachi 1985] S. Adachi, *J. Appl. Phys.*, 58, R1 (1985).

[Aitken 2008] A. Gutierrez-Aitken, P. Chang-Chien, B. Oyama, K. Tornquist, K. Thai, D. Scott, R. Sandhu, J. Zhou, P. Nam, and W. Phan, *Mater. Res. Soc. Symp. Proc.*, 1068, 209 (2008).

[Akhadov 2005] E. A. Akhadov, A. H. Mueller, and M. A. Hoffbauer, *Mater. Res. Soc. Symp. Proc.*, 872, J21.3.1 (2005).

[Alam 2001] A. Alam, B. Schineller, H. Protzmann, M. Luenenbuerger, M. Heuken, M. D. Bremser, E. Woelk, A. Dadgar, and A. Krost, *Proc. SPIE Int. Soc. Opt. Eng.*, 4278, 158 (2001).

[Amano 1986] H. Amano, N. Sawaki, I. Akasaki, and Y. Toyoda, *Appl. Phys. Lett.*, 48, 353 (1986).

[Ashcroft 1976] N. W. Ashcroft and N. D. Mermin, Crystal structure, Chapter 1. In *Solid State Physics*, Holt-Saunders, Philadelphia, PA (1976).

[Beke 1998] D. L. Beke, *Group III Condensed Matter Numerical Data and Functional Relationships in Science and Technology, Diffusion in Semiconductors*. Springer, Berlin, Germany (1998).

[Bruel 1995] M. Bruel et al., *Electron. Lett.*, 31, 1201 (1995).

[Carlin 2000] J. A. Carlin, S. A. Ringel, E. A. Fitzgerald, M. Bulsara, and B. M. Keyes, *Appl. Phys. Lett.*, 76, 1884 (2000).

[Chakraborty 2004] A. Chakraborty, B. A. Haskell, S. Keller, J. S. Speck, S. P. DenBaars, S. Nakamura, and U. K. Mishra, *Appl. Phys. Lett.*, 85, 5143 (2004).

[Chang 2005] E. Y. Chang, T.-H. Yang, G. Luo, and C.-Y. Chang, *J. Electron. Mater.*, 34, 23 (2005).

[Charles 2005] M. B. Charles, M. J. Kappers, and C. J. Humphreys, *Mater. Res. Soc. Symp. Proc.*, 831, E3.17.1 (2005).

[Chen 2001] P. Chen, R. Zhang, Z. M. Zhao, D. J. Xi, B. Shen, Z. Z. Chen, Y. G. Zhou, S. Y. Xie, W. F. Lu, and Y. D. Zheng, *J. Cryst. Growth*, 225, 150 (2001).

[Chen-a 2006] C. H. Chen, C.-M. Yeh, and J. Hwang, *Appl. Phys. Lett.*, 88, 161912 (2006).

[Chen-b 2006] X. Chen and T. Uesugi, *Appl. Phys. Lett.*, 88, 031916 (2006).

[Chumbes 1999] E. M. Chumbes, A. T. Schremer, J. A. Smart, D. Hogue, J. Komiak, and J. R. Shealy, *IEDM Digest*, 397 (1999).

[Contreras 2005] M. Cervantes-Contreras, C. A. Quezada-Maya, M. López-López, G. González de la Cruz, M. Tamura, and T. Yodo, *J. Cryst. Growth*, 278, 415 (2005).

[Currie 1998] M. T. Currie, S. B. Samavedam, T. A. Langdo, C. W. Leitz, and E. A. Fitzgerald, *Appl. Phys. Lett.*, 72, 1718 (1998).

[Dadgar 2000] A. Dadgar, J. Bläsing, A. Diez, A. Alam, M. Heuken, and A. Krost, *Jpn. J. Appl. Phys.*, 39, L1183 (2000).

[Dalmasso 2000] S. Dalmasso, E. Feltin, P. de Mierry, B. Beaumont, P. Gibart, and M. Leroux, *Electron. Lett.*, 36, 1728 (2000).

[Dobrovinskaya 2009] R. E. Dobrovinskaya, A. L. Lytvynov, and V. Pishchik, *Sapphire Material, Manufacturing, Applications*, Springer, New York (2009).

[Dutt 1998] M. B. Dutt and B. L. Sharma, Diffusion in compound semiconductor, Chapter 3. In *Diffusion in Semiconductors*, ed. D. L. Beke, Springer-Verlag, Berlin, Germany (1998).

[Edwards 1998] N. V. Edwards, M. D. Bremser, R. F. Davis, A. D. Batchelor, S. D. Yoo, C. F. Karan, and D. E. Aspnes, *Appl. Phys. Lett.*, 73, 2808 (1998).

[Egawa 2000] T. Egawa, N. Nakada, H. Ishikawa, and M. Umeno, *Electron. Lett.*, 36, 1816 (2000).

[Eisenbeiser 2002] K. Eisenbeiser. R. Emrick, R. Droopad, Z. Yu, J. Finder, S. Rockwell, J. Holmes, O. Overgaard, and W. Ooms, *IEEE Electron Device Lett.*, 23, 300 (2002).

[Faleev 2005] N. Faleev, H. Temkin, I. Ahmad, M. Holtz, and Y. Melnik, *J. Appl. Phys.*, 98, 123508 (2005).

[Feltin 2001] E. Feltin, B. Beaumont, M. Laugt, P. Mierry, P. Vennegues, M. Leroux, and P. Gibart, *Phys. Status Solidi (A)*, 188, 531 (2001).

[Ferdous 2006] M. S. Ferdous, X. Y. Sun, X. Wang, M. N. Fairchild, and S. D. Hersee, *J. Appl. Phys.*, 99, 096105 (2006).

[Figge 2009] S. Figge, H. Kröncke, D. Hommel, and B. M. Epelbaum, *Appl. Phys. Lett.*, 94, 101915 (2009).

[Follstaedt 1999] D. M. Follstaedt, J. Han, P. Provencio, and J. G. Fleming, MRS Internet, *J. Nitride Semicond. Res.*, 4S1, G3.72 (1999).

[Gardes 2008] P. Gardes and G. Auriel, U.S. patent 7,404,249 B2 (2008).

[Gebauer 2002] J. Gebauer, E. R. Weber, S. Shinkai, and K Sasaki, *Appl. Phys. Lett.*, 81, 1450 (2002).

[Gevorgian 1998] S. Gevorgian, *Int. J. RF Microw. Comput. Aided Eng.*, 8, 433 (1998).

[Glazov 2000] V. M. Glazov and A. S. Pashinkin, *Inorg. Mater.*, 36, 3, 225 (2000).

[Greiner 1984] M. E. Greiner and J. F. Gibbons, *Appl. Phys. Lett.*, 44, 750 (1984).

[Guha 1997] S. Guha and N. A. Bojarczuk, *Electron. Lett.*, 33, 1986 (1997).

[Guha 1998a] S. Guha and N. A. Bojarczuk, *Appl. Phys. Lett.*, 72, 415 (1998).

[Guha 1998b] S. Guha and N. A. Bojarczuk, *Appl. Phys. Lett.*, 73, 1487 (1998).

[Gutierrez-Aitken 2008] A. Gutierrez-Aitken, P. Chang-Chien, B. Oyama, K. Tornquist, K. Thai, D. Scott, R. Sandhu, J. Zhou, P. Nam, and W. Phan, *Mater. Res. Soc. Symp. Proc.*, 1068, 1068-C02-02 (2008).

[Gwo 2006] S. Gwo, U.S. patent 7,012,016 B2 (2006).

[Hanes 1993] M. H. Hanes, A. K. Agarwal, T. W. O'Keeffe, H. M. Hobgood, J. R. Szedon, T. J. Smith, R. R. Siergiej, P. G. McMullin, H. C. Nathanson, M. C. Driver, and R. N. Thomas, *IEEE Electron Device Lett.*, 14, 219 (1993).

[Hawker 2005] C. J. Hawker and T. P. Russell, *MRS Bull.*, 30, 952 (2005).

[He 2005] J. Q. He, C. L. Jia, V. Vaithyanathan, D. G. Schlom, J. Schubert, A. Gerber, H. H. Kohlstedt, and R. H. Wang, *J. Appl. Phys.*, 97, 104921 (2005).

[Heinke 2000] H. Heinke, V. Kirchner, S. Einfeldt, and D. Hommel, *Appl. Phys. Lett.*, 77, 2145 (2000).

[Hersee 2000] S. D. Hersee, *Appl. Phys. Lett.*, 76, 858 (2000).

[Hisamoto 1998] D. Hisamoto, S. Tanaka, T. Tanimoto, and S. Kimura, *IEEE Trans. Electron Devices*, 45, 1039 (1998).

[Honda 2002] Y. Honda, Y. Kuroiwa, M. Yamaguchi, and N. Sawaki, *Appl. Phys. Lett.*, 80, 222 (2002).

[Huang 2002] T. S. Huang, T. B. Joyce, R. T. Murrays, A. J. Papworth, and P. R. Chalker, *J. Phys. D: Appl. Phys.*, 35, 620 (2002).

[Ishikawa 1999] H. Ishikawa, G. Y. Zhao, N. Nakada, T. Egawa, T. Jimbo, and M. Umeno, *Jpn. J. Appl. Phys.*, 38 (Part 2), L492 (1999).

[Jamil 2005] M. Jamil, J. R. Grandusky, V. Jindal, and F. Shahedipour-Sandvika, *Appl. Phys. Lett.*, 87, 082103 (2005).

[Jang 2002] S. H. Jang, S. J. Lee, I. S. Seo, H. K. Ahn, O. Y. Lee, J. Y. Leem, and C. R. Lee, *J. Cryst. Growth*, 241, 289 (2002).

[Jang 2003] S. H. Jang and C. R. Lee, *J. Cryst. Growth*, 253, 64 (2003).

[Jiang 2009] F. Jiang, L. Wang, X. Wang, C. Mo, X. You, C. Zheng, W. Liu, Y. Zhou, C. Xiong, Y. Tang, W. Fang, and B. Lu, H7, *8th International Conference on Nitride Semiconductors*, Jeju, Korea (2009).

[Johansson 2000] M. Johansson and S. Bengtsson, *J. Appl. Phys.*, 88, 1118 (2000).

[Johnson 1998] R. A. Johnson, P. R. de la Houssaye, C. E. Chang, P.-F. Chen, M. E. Wood, G. A. Garcia, I. Lagnado, and P. M. Asbeck, *IEEE Trans. Electron Devices*, 45, 1047 (1998).

[Joshi 2005] P. C. Joshi, A. T. Voutsas, and J. W. Hartzell, *Mater. Res. Soc. Symp. Proc.*, 862, A19.3.1 (2005).

[Kaiser 2000] S. Kaiser, M. Jakob, J. Zweck, W. Gebhardt, O. Ambacher, R. Dimitrov, A. T. Schremer, J. A. Smart, and J. R. Shealy, *J. Vac. Sci. Technol. B*, 18, 733 (2000).

[Kasper 1995] E. Kasper, *Properties of Strained and Relaxed Silicon Germanium*, EMIS Datareviews Series No. 12, INSPEC, London, U.K. (1995).

[Kato 1991] K. Kato, T. Kusunoki, C. Takenaka, T. Takahashi, and K. Nakajima, *J. Cryst. Growth*, 115, 174 (1991).

[Kawaguchi 1998] Y. Kawaguchi, Y. Honda, H. Matsushima, M. Yamaguchi, K. Hiramatsu, and N. Sawaki, *Jpn. J. Appl. Phys.*, 37(Part 2), L966 (1998).

[Kim 2003] K. C. Kim, S. W. Kang, O. Kryliouk, and T. J. Anderson, *Mater. Res. Soc. Symp. Proc.*, 764, c7.7.1 (2003).

[Kim 2008] K.-C. Kim, M. C. Schmidt, F. Wu, M. B. McLaurin, A. Hirai, S. Nakamura, S. P. DenBaars, and J. S. Speck, *Appl. Phys. Lett.*, 93, 142108 (2008).

[Kingery 1976] W. D. Kingery, H. K. Browen, and D. R. Uhlmann, *Introduction to Ceramics*, John Wiley & Sons, Inc., New York (1976).

[Komiyama 2006] J. Komiyama, Y. Abe, S. Suzuki, and H. Nakanishi, *Appl. Phys. Lett.*, 88, 091901 (2006).

[Krost 2002] A. Krost and A. Dadgar, *Mat. Sci. and Eng. B* 93, 77 (2002).

[Lahreche 2000] H. Lahreche, M. Leroux, M. Laugt, M. Vaille, B. Beaumont, and P. Gibart, *J. Appl. Phys.*, 87, 577 (2000).

[Lahreche 2000] H. Lahreche, P. Vennegues, O. Tottereau, M. Laugt, P. Lorenzini, M. Leroux, B. Beaumont, and P. Gibart, *J. Cryst. Growth*, 217, 13 (2000).

[Lee 2005] S. C. Lee, B. Pattada, S. D. Hersee, Y.-B. Jiang, and S. R. J. Brueck, *IEEE J. Quantum Electron.*, 41, 596 (2005).

[LeGoues 1994] F. K. LeGoues, M. C. Reuter, J. Tersoff, M. Hammer, and R. M. Tromp, *Phys. Rev. Lett.*, 73, 300 (1994).

[Leiston-Belanger 2005] J. M. Leiston-Belanger et al., *Macromolecules*, 38, 18 (2005).

[Letertre 2008] F. Letertre, *Mater. Res. Soc. Symp. Proc.*, 1068, 1068-C01-01 (2008).

[Li 2007] T. K. Li, D. Tween, and J. S. Maa, GaN on Si semi-annual report (2007).

[Li 2008a] T. K. Li and S. T. Hsu, U.S. patent 7358160 (2008).

[Li 2008b] T. K. Li, S. T. Hsu, D. J. Tweet, and J. S. Maa, U.S. patent application 20080296616 (2008).

[Li 2010] T. K. Li, T. Gehrke, Patent disclosure (2010).

[Liaw 2000] H. M. Liaw, R. Venugopal, J. Wan, R. Doyle, P. L. Fejes, and M. R. Melloch, *Solid-State Electron.*, 44, 685 (2000).

[Lin-a 2007] K. L. Lin, E. Y. Chang, T. K. Li et al., *Appl. Phys. Lett.*, 91, 222111 (2007).

[Lin-b 2007] Y. C. Lin, H. Yamaguchi, E. Y. Chang, Y. C. Hsieh, M. Ueki, Y. Hirayama, and C. Y. Chang, *Appl. Phys. Lett.*, 90, 023509 (2007).

[Liu 2000] H. Liu, Z. Ye, H. Zhang, and B. Zhao, *Mater. Res. Bull.*, 35, 1837 (2000).

[Lu 2004] Y. Lu, G. Cong, X. Liu, D. Lu, Q. Zhu, X. Wang, J. Wu, and Z. Wang, *J. Appl. Phys.*, 96, 4982 (2004).

[Lueck 2006] M. Lueck, C. L. Andre, A. J. Pitera, M. L. Lee, E. A. Fitzgerald, and S. A. Ringel, *IEEE Electron Devices Lett.*, 27, 142 (2006).

[Maa 2007a] J. Maa, D. Tweet, T. K. Li, J. J. Lee, and S. T. Hsu, U.S. patent 7, 226, 504 (2007).

[Maa 2007b] J. Maa, T. K. Li, and D. Tweet, G. Stecker, and S. T. Hsu, U.S. patent application publication US 2008/0315255 AI (2007).

[Manasreh 2000] O. Manasreh, *III-Nitride Semiconductor: Electrical, Structural and Defect Properties*, Elsevier, Amsterdam, the Netherlands (2000).

[Marchand 2001] H. Marchand, L. Zhao, N. Zhang, B. Moran, R. Coffie, U. K. Mishra, J. S. Speck, S. P. DenBaars, and J. A. Freitas, *J. Appl. Phys.*, 89, 7846 (2001).

[Mastro 2005] M. A. Mastro, R. T. Holm, N. D. Bassim, C. R. Eddy, Jr., D. K. Gaskill, R. L. Henry, and M. E. Twigg, *Appl. Phys. Lett.*, 87, 241103 (2005).

[Metzger 1998] T. Metzger et al., *Philos. Mag. A*, 77, 1013 (1998).

[Metzger 2001] B. Metzger, *Compd. Semiconductor*, 7, 6, News (2001).

[Mochizuki 1976] H. Mochizuki, T. Aoki, H. Yamoto, M. Okayama, and T. Ando, *Suppl. Jpn. I. Appl. Phys.*, 15, 41 (1976).

[Mohammad 1995] S. N. Mohammad, A. A. Salvador, and H. Morkoc, *Proc. IEEE*, 83, 1306 (1995).

[Molina 1999] S. I. Molina, A. M. Sanchez, F. J. Pacheco, R. Garcia, M. A. Sanchez-Garcia, F. J. Sanchez, and E. Calleja, *Appl. Phys. Lett.*, 74, 3362 (1999).

[Mueller 2006] A. H. Mueller, E. A. Akhadov, and M. A. Hoffbauer, *Appl. Phys. Lett.*, 88, 041907 (2006).

[Murakami 1992] T. Murakami, U.S. patent 5,119,150 (1992).

[Murugan 2002] P. Murugan, R. Pothirj, S. D. D. Roy, and K. Ramachandran, *Bull. Mater. Sci.*, 25, 4, 335 (2002).

[Narayan 2003] J. Narayan and B.C. Larson, *J. Appl. Phys.*, 93, 278 (2003).

[NCSR-2007] National compound semiconductor roadmap: Basic materials properties, http://www.onr.navy.mil/sci_tech/31/312/ncsr/properties.asp

[Ng 2000] H. Ng, T. Moustakas, and S. Chu, *Appl. Phys. Lett.*, 76, 2818 (2000).

[Nikishin 1999] S. A. Nikishin, N. N. Faleev, V. G. Antipov, S. Francoeur, L. Grave de Peralta, G. A. Seryogin, H. Temkin, T. I. Prokofyeva, M. Holtz, and S. N. G. Chu, *Appl. Phys. Lett.*, 75, 2073 (1999).

[Nishimura 2002] S. Nishimura, S. Matsumoto, and K. Terashima, *Opt. Mater.*, 19, 223 (2002).

[Ohring 1992] M. Ohring, *The Material Science of Thin Films*, Academic Press, Inc., New York (1992).

[Ohta 2001] J. Ohta, H. Fujioka, H. Takahashi, M. Sumiya, and M. Oshima, *J. Cryst. Growth*, 233, 779–784 (2001).

[Okada 1984)] Y. Okada and Y. Tokumasu, *J. Appl. Phys.*, 56, 314 (1984).

[Osinsky 1998] A. Osinsky, S. Gangopadhyay, J. W. Yang, R. Gaska, D. Kuksenkov, H. Temkin, I. K. Shmagin, Y. C. Chang, J. F. Muth, and R. M. Kolbas, *Appl. Phys. Lett.*, 72, 551 (1998).

[Oye 2007] M. M. Oye, D. Shahrjerdi, I. Ok, J. B. Hurst, S. D. Lewis, S. Dey, D. Q. Kelly, S. Joshi, T. J. Mattord, X. Yu, M. A. Wistey, J. S. Harris, Jr., A. L. Holmes, Jr., J. C. Lee, and S. K. Banerjee, *J. Vac. Sci. Technol. B*, 25, 1098 (2007).

[Pau 2000] J. L. Pau, E. Monroy, F. B. Naranjo, E. Munoz, F. Calle, M. A. Sanchez-Garcia, and E. Calleja, *Appl. Phys. Lett.*, 76, 2785 (2000).

[Ponce 1995] F. A. Ponce, B. S. Krusor, J. S. Major, Jr., W. E. Plano, and D. F. Welch, *Appl. Phys. Lett.*, 67, 410 (1995).

[Raghavan 2005] S. Raghavan, X. Weng, E. Dickey, and J. M. Redwing, *Appl. Phys. Lett.*, 87, 142101 (2005).

[Rao 2000] B. V. Rao, D. Gruznev, T. Tambo, and C. Tatsuyama, *J. Appl. Phys.*, 87, 724 (2000).

[Reichertz 2008] L. A. Reichertz, K. M. Yu, Y. Cui, M. E. Hawkridge, J. W. Beeman, Z. L. Weber, J. W. Ager, W. Walukiewicz, W. J. Schaff, T. L. Williamson, and M. A. Hoffbauer, *Mater. Res. Soc. Symp. Proc.*, 1068, 1068-C06-02 (2008).

[Roder 2005)] C. Roder, S. Einfeldt, S. Figge, and D. Hommel, *Phys. Rev.*, 72, 085218 (2005).

[Sarinanto 1998] M. M. Sarinanto, Y. Yamaguchi, and K. Tsutsui, *Thin Solid Films*, 334, 15 (1998).

[Schremer 2000] A. T. Schremer, J. A. Smart, Y. Wang, O. Ambacher, N. C. MacDonald, and J. R. Schealy, *Appl. Phys. Lett.*, 76, 736 (2000).

[Schulze 2006] F. Schulze, A. Dadgar, J. Bläsing, A. Diez, and A. Krost, *Appl. Phys. Lett.*, 88, 121114 (2006).

[Schwartqa 1993] P. V. Schwartqa, C. W. Liu, and J. C. Sturm, *Appl. Phys. Lett.*, 62, 1102 (1993).

[Semond 2001] F. Semond, N. Antoine-Vincent, N. Schnell, G. Malpuech, M. Leroux, J. Massies, P. Disseix, J. Leymarie, and A. Vasson, *Phys. Stat. Sol. A*, 183, 163 (2001).

[Slack 1975] G. A. Slack and S. F. Bartram, *J. Appl. Phys.*, 46, 89 (1975).

[Soh 2007] C. B. Soh, C. B. Soh, H. Hartono, S. Y. Chow, S. J. Chua, and E. A. Fitzgerald, *Appl. Phys. Lett.*, 90, 053112 (2007).

[Strite 1994] M. S. Strite, G. B. Gao, M.E. Liu, B. Sverdlov, and M. Burns, *J. Appl. Phys.*, 76, 1363 (1994).

[Strittmatter 1999] A. Strittmatter, A. Krost, M. Straßburg, V. Turck, D. Bimberg, J. Blasing, and J. Christen, *Appl. Phys. Lett.*, 74, 1242 (1999).

[Strittmatter, 2001] A. Strittmatter, D. Bimberg, A. Krost, J. Bläsing, and P. Veit, *J. Cryst. Growth*, 221, 293 (2001).

[Suda 2002] J. Suda, K. Miura, M. Honaga, Y. Nishi, N. Onojima, and H. Matsunami, *Appl. Phys. Lett.*, 81, 5141 (2002).

[Sung 2004] M. M. Sung, C. G. Kim, and Y. Kim, *J. Vac. Sci. Technol. A*, 22, 461 (2004).

[Tang 2006] H. Tang, S. Haffouz, and J. A. Bardwell, *Appl. Phys. Lett.*, 88, 172110 (2006).

[Takahishi 1987] M. Takahishi, M. Tabe, and Y. Sakakibara, *IEEE Electron Device Lett.*, EDL-8, 475 (1987).

[Talwar 2002] D. N. Talwar, *Appl. Phys. Lett.*, 80, 1553 (2002).

[Tanaka 1996] S. Tanaka, S. Iwai, and Y. Aoyagi, *J. Cryst. Growth*, 170, 329 (1996).

[Tanaka 2000] S. Tanaka, Y. Kawaguchi, N. Sawaki, M. Hibino, and K. Hiramatsu, *Appl. Phys. Lett.*, 76, 2701 (2000).

[Therrien 2005] R. Therrien, S. Singhal, J. W. Johnson, W. Nagy, R. Borges, A. Chaudhari, A. W. Hanson, A. Edwards, J. Marquart, P. Rajagopal, C. Park, I. C. Kizilyalli, and K. J. Linthicum, *IEEE IEDM Techn. Dig.*, 557 (2005).

[Timoshenko 1925] S. Timoshenko, *J. Opt. Soc. Am.*, 11, 233 (1925).

[Ting 2000] S. M. Ting and E. A. Fitzgerald, *J. Appl. Phys.*, 87, 2618 (2000).

[Tokunaga 2000] H. Tokunaga, H. Tan, Y. Inaishi, T. Arai, A. Yamaguchi, and J. Hidaka, *J. Cryst. Growth*, 221, 616–621 (2000).

[Tran 1999] C. A. Tran, A. Osinski, R. F. Karlicek, Jr., and I. Berishev, *Appl. Phys. Lett.*, 75, 1494 (1999).

[Ujiie 1989] Y. Ujiie and T. Nishinaga, *Jpn. J. Appl. Phys.*, 28 (Part 2), L337 (1989).

[Vanamu 2006] G. Vanamu, A. K. Datye, R. Dawson, and S. H. Zaidi, *Appl. Phys. Lett.*, 88, 251909 (2006).

[Vurgaftman 2001] I. Vurgaftman, J. R. Mayer, and L. R. Ram-Moham, *J. Appl. Phys.*, 89, 5815 (2001).

[Wakahara 2002] A. Wakahara, H. Oishi, H. Okada, A. Yoshida, Y. Koji, and M. Ishida, *J. Cryst. Growth*, 236, 21 (2002).

[Waldrip 2001] K. Waldrip, J. Han, and J. Figiel, *Appl. Phys. Lett.*, 78, 3205 (2001).

[Wang 2002] H. Wang, Process and properties of nitride-based thin film heterostructures, PhD dissertation, North Carolina State University, Raleigh, NC (2002).

[Wang-a 2006)] D. Wang, MOCVD growth of GaN on Si and related high power electronics (2006) [project proposal].

[Wang-b 2006] Y. D. Wang, K. Y. Zang, S. J. Chua, S. Tripathy, H. L. Zhou, and C. G. Fonstad, *Appl. Phys. Lett.*, 88, 211908 (2006).

[Weast 1990] R. C. Weast, D. R. Lide, M. J. Astle, and W. H. Beyer, *CRC Handbook of Chemistry and Physics*, CRC Press, Inc., Boca Raton, FL (1990).

[Weeks 1996] T. W. Weeks, Jr., M. Bremser, K. Ailey, E. Carlson, E. Perry, E. Piner, N. El Masry, and R. F. Davis, *J. Mater. Res.*, 11, 1011 (1996).

[Wu 1998] X. H. Wu, P. Fini, E. J. Tarsa, B. Heying, S. Keller, U. K. Mishra, S. P. DenBaars, and J. S. Speck, *J. Cryst. Growth*, 189–190, 231 (1998).

[Wu 2005] J. Wu, X. Han, J. Li, D. Li, Y. Lu, H. Wei, G. Cong, X. Liu, Q. Zhu, and Z. Wang, *J. Cryst. Growth*, 279, 335 (2005).

[Yang 1995] Z. Yang, F. Guarin, I. W. Tao, W. I. Wang, and S. S. Lyer, *J. Vac. Sci. Technol. B*, 13, 789 (1995).

[Yang 1999] W. Yang, S. A. McPherson, Z. Mao, S. McKernan, and C. B. Carter, *J. Cryst. Growth*, 204, 270 (1999).

[Yang 2000] J. W. Yang, A. Lunev, G. Simin, A. Chitnis, M. Shatalov, M. Asif Khan, J. E. Van Nostrand, and R. Gaska, *Appl. Phys. Lett.*, 76, 273 (2000).

[Yang 2003] V. K. Yang, M. Groenert, C. W. Leitz, A. J. Pitera, M. T. Currie, and E. A. Fitzgerald, *J. Appl. Phys.*, 93, 3859 (2003).

[Zamir 2000] S. Zamir, B. Meyler, E. Zolotoyabko, and J. Salzman, *J. Cryst. Growth*, 218, 181 (2000).

[Zamir 2001] S. Zamir, B. Meyler, and J. Salzman, *Appl. Phys. Lett.*, 78, 288 (2001).

[Zamir 2002] S. Zamir, B. Meyler, and J. Salzman, *J. Cryst. Growth*, 243, 375 (2002).

[Zang 2005] K. Y. Zang, Y. D. Wang, and S. J. Chua, *Appl. Phys. Lett.*, 87, 193106 (2005).

[Zhang 2000] H. Zhang, Z. Ye, and B. Zhao, *J. Appl. Phys.*, 87, 2830 (2000).

[Zubia 1999] D. Zubia and S. D. Hersee, *J. Appl. Phys.*, 85, 6492 (1999).

[Zubia 2000] D. Zubia, S. H. Zaidi, S. R. J. Brueck, and S. D. Hersee, *Appl. Phys. Lett.*, 76, 858 (2000).

Part II

GaN and Related Alloys on Silicon Growth and Integration Techniques

3

III-Nitrides on Si Substrates

Jing Li, Jingyu Lin, Hongxing Jiang, and Nobuhiko Sawaki

CONTENTS

3.1 Introduction

Since the innovative technique of growing high quality GaN on sapphire substrate using a low temperature AlN buffer layer was discovered by Akasaki and coworkers (Amano et al., 1986), the growth of III-nitrides has been developed on different substrates using various buffer layers (Liu and

Edgar, 2002). Nowadays, most III-nitride devices are made on GaN/sapphire templates or on silicon carbide (SiC). However, the fabrication of photonic devices on Si substrates has long been desired for the purpose of integrating photonic devices with Si-based electronic devices. But because of the large lattice mismatch between Si and III–V semiconductors, direct growth and fabrication of photonic materials and devices on Si substrates have been a challenge. Twenty-five years ago, the growth of GaAs on Si substrate was demonstrated to realize the integration of compound semiconductor devices on Si substrates (Soga et al., 1985). However, the quality of the hetero-epitaxial layer was inferior to that obtained on GaAs or InP substrates. One of the issues has been the difference in thermal expansion coefficients, which induces the wafer to bend, preventing fine lithography processes. With the recent rapid progress made in both nitride materials and devices (though mostly on sapphire and SiC substrates), nitride materials and devices on Si substrates naturally attracted much attention and worldwide effort from the nitride community.

Obtaining high quality III-nitrides on Si substrates has been challenging because of the different lattice structures and the large lattice and thermal coefficient mismatches. Some of the difficulties have been overcome by using multiple intermediate layers (Krost and Dadgar, 2002). Another approach is to reduce the thermal effect by reducing the size of the crystal on the wafer (Honda et al., 2002b).

After the successful growth of nitride light-emitting diodes (LEDs) on sapphire substrates, many groups tried to grow nitride LEDs on Si. Guha and Bojarczuk (1998) reported the fabrication and characterization of GaN-based double heterostructure LEDs grown by molecular beam epitaxy (MBE) on Si (111) substrates. They observed electroluminescence (EL) with emission peaks in the ultraviolet (UV) region at 360 and 420 nm. Tran et al. (Tran et al., 2001) grew InGaN/GaN multiple-quantum-well (MQW) blue LED structures on Si (111) using metalorganic chemical vapor deposition (MOCVD) and demonstrated simple LEDs with a turn-on voltage of around 5 V, which emitted a bright EL between 450 and 480 nm.

However, the large lattice mismatch of about 17% $[(a_{Si} - a_{GaN})/a_{Si}; a_{GaN\ (0001)} = 3.189$ Å, $a_{Si\ (111)} = 3.840$ Å$]$, and the thermal mismatch of about 37% between GaN and Si result in the formation of cracks when the GaN epilayer thickness exceeds 1 μm. Several approaches were previously employed to eliminate cracks. The cracking of nitrides on Si can be avoided with substrate patterning (Honda et al., 2002b), stress compensation (Krost and Dadgar, 2002), transition layers (Honda et al., 2002b), or by growing only thin layers (Honda et al., 2002a). However, thin layers often show poor structural and surface quality. Thus, smooth interfaces with a high p-type doping concentration would be difficult to achieve for thin layer LEDs. Nevertheless, a relatively bright LED using only a thin buffer layer was recently presented by Egawa et al. (Edgar et al., 2002). For commercial applications, thicker layers are preferred since dislocation density is reduced and the surface morphology is

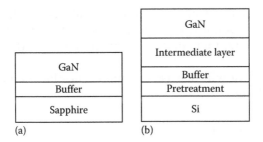

FIGURE 3.1
Typical structures of GaN grown on (a) sapphire and (b) Si substrates.

improved. By using SiN in situ masking on Si (111), Dadgar et al. (2002a) have reported a power output of 0.42 mW at 20 mA in a packaged LED. Now, a few groups have claimed that the performance of LEDs grown on Si substrate is comparable to those grown on sapphire (Mo et al., 2005; Phillips and Zhu, 2009; Fenwick et al., 2009).

For GaN grown on sapphire substrate, the two-step growth method, in which a buffer layer is grown before the GaN epilayer, is used to reduce the effect of lattice mismatch, as shown in Figure 3.1a. For GaN grown on Si substrate, the typical growth procedure is more complicated, as shown in Figure 3.1b. Typically, the growth steps include a pretreatment of Si substrate, a buffer layer, and one or more intermediate layers, which will be reviewed first.

3.2 Approaches for Initial Growth

In the hetero-epitaxial growth of semiconductors, the lattice matching is one of the most important factors in obtaining high crystalline quality. In the case of growing GaN epilayers on sapphire substrates, the nominal lattice mismatch $\Delta a/a$ is as large as 33%. This can be partially reduced by rotating the crystal axis by 30° on the surface (Kaiser et al., 1998). However, even after this procedure, the mismatch is still 13%—far from the value accepted in conventional epitaxial growth. The introduction of an intermediate layer partially overcomes these difficulties (Amano et al., 1986).

For the growth of GaN on Si, the situation is more severe because the crystalline structures are different (Honda et al., 2002a; Liu and Edgar, 2002). The key solution is to introduce an effective intermediate layer between the nitride layer and Si substrate. The first successful work was reported by Koide (1988) who tried to grow a uniform layer of $Al_xGa_{1-x}N$ directly on (111) and (001) Si surfaces by MOCVD. Growth was achieved only at high temperatures. A supply of Al was also essential for obtaining a uniform film. Based on these results, Watanabe and coworkers (Watanabe et al., 1993) achieved a uniform thin GaN film on Si substrate using a high temperature AlN intermediate layer. It was recently found that the insertion of multiple layers is more effective for managing strain in the system (Krost and Dadgar, 2002).

In this section, we will discuss the approaches for the initial growth of GaN-on-Si substrate.

3.2.1 GaN Buffer Layer

A low temperature GaN buffer layer is a well-established approach for the growth of GaN on sapphire. For the growth of GaN on Si, however, early results showed a melt-back etching effect between Ga and Si substrate, which prevented smooth GaN growth on Si substrate (Ishikawa et al., 1998). Figure 3.2 shows surface morphologies of annealed low temperature (LT) GaN on Si (Figure 3.2a and b), and on sapphire (Figure 3.2c and d). Annealing ambient were nitrogen on (a) and (c); and a mixture of nitrogen and ammonia on (b) and (d). The bar indicates a length of 50 μm. While the surfaces of annealed LT-GaN on sapphire were mirrorlike, the surfaces of GaN grown on Si substrate were rough. Flat amorphous-like LT-GaN on Si changed to tetrahedron islands and Ga droplets, and wavy morphologies were also observed under the islands and Ga droplets. This reaction occurs only at

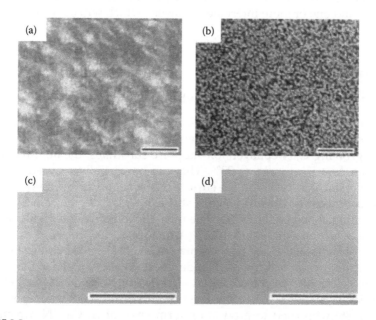

FIGURE 3.2
Surface morphologies of annealed LT-GaN (a), (b) on Si, and (c,d) on sapphire. Annealing ambient are (a) and (c) nitrogen; and (b) and (d) a mixture of nitrogen and ammonia. Bar indicates 50 μm. (After Ishikawa, H. et al., *J. Crystal Growth*, 189/190, 178, 1998. With permission.)

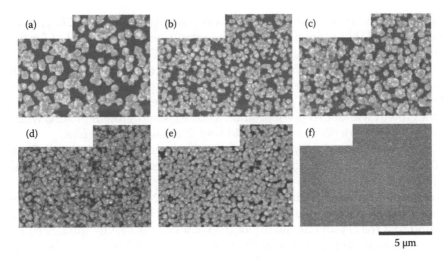

FIGURE 3.3
Surface morphologies of LT-GaN buffer layers grown on Si (111) surfaces with various input mole ratios of H_2 in the carrier gas ($F° = [H_2]/([H_2] + [N_2])$): (a) 1 ($H_2$ carrier gas), (b) 0.1, (c) 0.06, (d) 0.01, (e) 0.004, and (f) 0 (N_2 carrier gas). (After Takemoto, K. et al., *Jpn. J. Appl. Phys.*, 45, L478, 2006. With permission.)

elevated temperatures and is not present during low temperature seed layer growth, even when GaN is grown at low temperatures directly on Si. Si out-diffusion from the substrate through the growing layer is assumed to be the cause of poor surface morphology (Krost and Dadgar, 2002).

Recently, it was found that direct growth of a GaN epilayer on a Si (111) substrate is possible (Takemoto et al., 2006). Figure 3.3 shows the surface morphologies of LT-GaN buffer layers grown on Si (111) surfaces with various input mole ratios of H_2 in the carrier gas. No deterioration in the Si surface as a result of the reaction between Si and Ga vapor was observed. However, when there were Ga droplets on the surface, Ga and Si formed a Ga–Si alloy, which caused the generation of numerous holes on the surface by melt-back etching at high temperatures. In addition, it was revealed that the coverage of the LT-GaN buffer layer on Si was strongly affected by the hydrogen partial pressure in the carrier gas. Using nitrogen carrier gas, a complete coverage of the LT-GaN buffer layer could be achieved directly over the Si surface.

3.2.2 Metal Al Predeposition

Si substrate reacts very easily with ammonia to form amorphous Si_xN, which will passivate the surface and suppress GaN growth. Normally, a few mono-layers of Al deposition before the AlN buffer layer are enough to prevent nitridation (Yang et al., 2000; Dadgar et al., 2003b). It was found that the initial deposition of Al resulted in a very rapid transition to a 2D growth mode of AlN (Nikishin et al., 1999). The rapid transition is essential for the subsequent growth of high quality GaN and AlGaN. This procedure also resulted in complete elimination of cracks in thick (>2 μm) GaN layers. Chen and coworkers (Chen et al., 2001) applied the Al predeposition to HT-AlN layers and found that Al predeposition time is very critical. They showed that growth without Al predeposition leads to island growth of AlN, while excessive Al predeposition time could lead to a rough AlN surface.

Some groups used very thick Al layer for pretreatment (Lu et al., 2002, 2004). Lu and coworkers first exposed the Si surface to a 95 μmol/min TMAl flow for 3 min to deposit an ultrathin layer of liquid Al and followed this treatment with an exposure to 0.05 mol/min of NH_3 for 3 min. This process resulted in the formation of a roughly 30 nm HT-AlN buffer layer. They found that the crystalline quality of the GaN layer is heavily influenced by the Al precovering time. The optimal time is around 3 min. They believed that precovering for too short a time resulted in incomplete Al covering on the Si substrate, while excessive precovering does harm to the crystalline quality of AlN after nitridation.

It was also found that the Al-preseeded layer is a critical factor for accomplishing the AlN and crystalline α-Si_3N_4 layers simultaneously (Chang et al., 2008). The effect of Al layer deposition time on the *c*-axis lattice constants of the subsequent AlN layer and the effect of the crystalline α-Si_3N_4 layer on GaN quality were investigated by x-ray diffraction (XRD) analysis. The 1:2 lattice coincidently matches at the α-Si_3N_4 (0001)/Si (111) interface, and the 5:2 coincident lattice interface at AlN (0001)/α-Si_3N_4 (0001) reduces the lattice mismatch in the AlN/α-Si_3N_4/Si (111) structure. The 5:4 coincident lattice in the AlN/α-Si_3N_4/Si (111) structure is related to the reduction in tensile stress in the AlN epilayers. Figure 3.4 shows the cross-sectional TEM image of GaN on AlN/α-Si_3N_4/Si (111) structure (Chang et al., 2008).

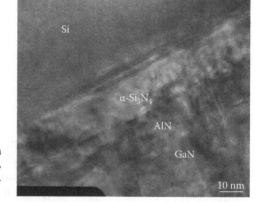

FIGURE 3.4
Cross-sectional TEM image of GaN with AlN/α-Si_3N_4/Si structure. (After Chang, J.R. et al., *J. Appl. Phys.*, 47, 5572, 2008. With permission.)

3.2.3 AlN and AlGaN Buffer Layers

For many reasons, most of the GaN epilayers grown on Si use AlN for buffer layers. One reason is that AlN layers avoid direct contact of Ga with Si. This is important because Ga reacts very easily with Si (Ishikawa et al., 1998). Another reason is that when AlN is used as a buffer, the strain between Si and GaN will change from tenser to compress strain, which will reduce cracks in the GaN epilayers. For the AlN buffer, it is important to find the right order in which to deposit the first layer. Although some groups intentionally grow a thick SiN layer, most groups try to avoid the nitridation of the Si surface. A few monolayers of Al atoms before the AlN buffer seem to obtain good quality GaN epilayers (Zhang et al., 2005b). AlN buffer growth conditions and thicknesses also change the GaN epilayer quality. Some groups found that a HT-AlN buffer improved the GaN epilayer quality, and other groups found that a LT-AlN buffer is very critical for removing cracks.

For growing GaN-on-Si substrate, AlGaN buffers are very similar to AlN buffers. However, Yoshida et al. (2003) found that when the Al composition of the $Al_xGa_{1-x}N$ buffer was above 0.3, the surface morphology of the thick GaN grown on top was rough. It was conjectured that the AlGaN buffer layer had many grains, including polycrystalline grains, which made it difficult to grow GaN on top. When a GaN buffer was used for GaN growth on Si substrates, the thick GaN layer was rough and cracks were generated. Therefore, a GaN buffer layer is not suitable for GaN grown on Si. They confirmed that a suitable Al composition for the $Al_xGa_{1-x}N$ buffer is between 0.05 and 0.2 for x. Sasaki et al. (2007) also confirmed that an optimized AlGaN buffer structure on a Si substrate is very effective for the reduction of pit density.

3.2.4 Other Buffers

Using an AlN buffer is the most successful approach for GaN grown on Si. However, the insertion of an AlN film as thin as 30 nm already restricts the forward current in a GaN/AlN/Si diode structure (Kondo et al., 2006), and this method will not be applicable to high-power vertical devices. Many different buffers were also tried for growing GaN on Si substrate as listed in Table 3.1. A conductive substrate and buffer layer are essential for vertical devices. Using conductive SiC as a buffer layer instead of insulating AlN is preferred for vertical devices (Komiyama et al., 2007, 2009). The (111) oriented 3C-SiC has a lattice structure close to that of GaN (0001). The lattice mismatch between GaN (0001) and 3C-SiC (111) is only 3%—much smaller than the large mismatch between GaN and Si. Komiyama et al. (2009) demonstrated the growth of GaN/SiC/Si heterostructures without using AlN or AlGaN buffer layers by controlling the partial pressure of H_2 in the carrier gas during the low temperature growth of GaN.

Wu et al. (2003) used a stacked buffer structure for heteroepitaxial growth with a large lattice mismatch. The stacked buffer consisted of constituent

TABLE 3.1

List of Other Buffers Used in GaN Grown on Si

Buffer	Structures	Results	Ref.
SiCN (0.1 μm)	Solar cell	Good solar cell efficiency	Liou (2009)
3C-SiC template + AlN	HEMT	Mobility 2050 cm²/V s	Cordier et al. (2008)
3C-SiC/Si + LT GaN	GaN	Good PL	Komiyama et al. (2009)
SiC	GaN	Low quality	Cervantes-Contreras et al. (2001)
3C-SiC (1 μm) + AlN	GaN	Ga polarity in GaN	Komiyama et al. (2007)
SiC (2.5 nm) + GaN:Si	GaN	Reduced dislocation	Wang et al. (2002)
Al₂O₃ (5–20 nm, ALD)	LED	Good LED performance	Fenwick et al. (2009)
Al₂O₃ (4 nm, MBE)	GaN	XRD FWHM 1000 arcsec	Wakahara et al. (2002)
MnS	AlN	(11–20) AlN on Si (100)	Song et al. (2002)
HfN (200 nm)	GaN	Good for GaN-on-Si (100)	Armitage et al. (2002)
ScN + GaN	GaN	GaN islands	Moram et al. (2007)
AlAs	GaN	XRD FWHM 500 arcsec	Strittmatter et al. (2000)
AlAs (20 nm) + GaN	GaN	Prevent SiN formation	Strittmatter et al. (1999)
Nano-rod as buffer		Freestanding GaN	Yang et al. (2009a)
Per Al (6 s) + AlN (55 nm) + N implantation	GaN	XRD FWHM 490 arcsec	Jamil et al. (2007)
InN (300°C) + GaN	GaN	Polycrystalline	Hu et al. (2006)

layers, which formed coincident lattices at layer/layer and layer/substrate interfaces. For the case of GaN-on-Si (111) heteroepitaxy, they utilized the 1:2 and 5:2 coincident lattices formed at the β-Si_3N_4 (0001)/Si (111) and AlN (0001)/β-Si_3N_4 (0001) interfaces, respectively, to facilitate the double-buffer layer for GaN-on-Si heteroepitaxial growth. By using this buffer technique, they resolved the issue of autodoping, which resulted from Si out-diffusion when grown with a single AlN (0001) buffer. As a result, the epitaxial quality of GaN epilayer was improved significantly.

ScN is a semiconducting group IIIB-nitride with an NaCl structure and a lattice mismatch of 0.1% to GaN (in the (111) orientation). It is chemically, thermally, and structurally stable under GaN MOCVD growth conditions. It is expected that the excellent lattice match along with the conductive nature of ScN may prove to be an advantage over currently employed AlN buffer layers for the growth of GaN on Si. Moram et al. (2007) demonstrated the MOCVD deposition of GaN-on-Si (111) substrates, using buffer layers of ScN grown by gas-source MBE. GaN grown directly onto ScN buffer layers at 1020°C displayed island growth with limited wetting of the ScN layer. The reduced ScN/GaN interfacial area resulted in the growth of dislocation-free GaN islands of several microns in diameter. Predeposition of small amounts of GaN on the ScN at either 750°C or 540°C promoted wetting of the ScN by GaN and hence increased coalescence, allowing a continuous film to form; however, the increased ScN/GaN interfacial area resulted in the presence of threading dislocations (TDs).

A reduction of dislocations in the GaN layer on Si substrate was achieved for the film-grown-on-Si (111) substrate, which was engineered to have a polycrystalline defective layer at the AlN/Si interface (Jamil et al., 2005, 2007). The formation of a polycrystalline defective layer at the AlN/Si interface by N+ ion implantation provides substrate conditions that result in a heteroepitaxial GaN film with a much improved surface morphology and better crystalline quality as compared to the film grown directly on AlN/Si.

3.2.5 Intermediate Layer

Nitrides growth on Si with an initial buffer layer still exhibit stress in the epilayers. In order to reduce the stress, many different intermediate layers such as LT-AlN layer, AlN/GaN superlattice, and thin Si_xN in situ mask were used. The full width at half maximum (FWHM) of the XRD (002) rocking curve is typically about two times higher in GaN on Si (varies from 340 to 600 arcsec) than in GaN on sapphire. It indicates that the quality of GaN grown on Si substrates still has room for improvement. The best FWHM result for XRD (002) rocking curve is around 340 arcsec, which was achieved by using an AlN (10 nm) and Ga-rich GaN buffer layer (Mo et al., 2005).

3.3 Epitaxial Lateral Overgrowth of Semipolar and Nonpolar GaN Grown on Si

For growth on Si (111), the <0001> axis of nitride is parallel to the <111> axis of Si substrate. In the case of AlN and/or GaN, growth is more effective toward <0001> than toward <000–1> (Haskell et al., 2003). This fact has been applied to the growth of semipolar and nonpolar nitrides on Si substrates of different planes, such as (100) and (110). This can be done by preparing a {111} facet on a Si substrate. The growth of GaN is achieved on this facet selectively. This is the basic approach used by one of the current authors (Haskell et al., 2003) to achieve the epitaxial lateral overgrowth (ELO) on patterned Si substrates. The formation of the {111} facet is achieved by anisotropy etching of the Si substrate in a KOH solution.

3.3.1 Selective Area Growth

One approach for reducing bending in GaN layer is to make small devices directly on specific areas of Si substrates. By limiting the area of the hetero-epitaxial growth on a wafer, the bending of the wafer is suppressed substantially. If the area of each device is only a fraction of the size of wafer, the effect of bending might be totally neglected. Patterning small devices is achieved by the selective area growth (SAG) method of applying fine lithography and/or micromachining technology not to the grown layer but to the substrate. In the SAG method, the growth of GaN is achieved only on the window area of the patterned substrate. Thus, the device size is determined by the size of the window prepared on the substrate, which can be as small as several tens of nanometers. By reducing the size of individual devices, the problem of stress/strain in the hetero-epitaxy is greatly reduced and the crystalline quality is improved. Moreover, the SAG enables the fabrication of small devices with multiple layer structures. These devices are made on small facets formed in a self-organizing manner during growth (Arai et al., 2001). This method offers a way to directly fabricate nitride devices with high performance on Si wafers.

3.3.2 Formation of Pyramidal GaN on Si

Typical results of the SAG of GaN on Si are shown in Figure 3.5. In this particular example, a mask film on a (111) Si substrate was deposited and followed by the formation of circular windows of 5 μm in diameter. GaN was grown on this substrate and was only grown in the window areas. The *c*-axis of GaN is parallel to the <111> axis of Si, i.e., the growth is organized normal to the substrate surface. Because of the ELO on the mask area, pyramidal GaN structures were grown upon the windows. The size of the pyramids is

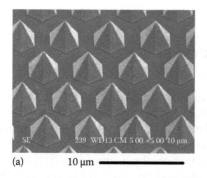
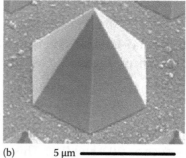

(a)   10 μm ————————   (b)   5 μm ————————

FIGURE 3.5

(a) Array of GaN hexagonal pyramids grown by SAG method on (111) Si substrate. (b) Each pyramid is made of self-organized six {1–101} facets and the top (0001) facet. The height of each pyramid is around 5 μm, which is determined by the diameter of the circular windows (5 μm in diameter) and GaN growth time on the mask film.

essentially determined by the diameter of the windows and the growth time. Pyramidal GaN is formed in a self-organizing manner and the six faces are of atomically flat {1–101} plane. On the top of the pyramid, a narrow (0001) facet as small as several tens of nanometers was occasionally obtained. By adjusting the growth conditions, a flat (0001) platelet on the truncated pyramid was obtained (Honda et al., 2002a). These arrays of GaN pyramids can be used as electron emitters or micro-LED arrays.

3.3.3 Behavior of the Threading Dislocation in GaN Grown on Si

If one makes a stripe mask/window pattern on (111) Si substrate, the growth of GaN is achieved selectively on the window area to form GaN stripes. By increasing the growth time, ELO on the mask will result in the coalescence of the adjacent stripes to form a uniform (0001) GaN surface. To investigate this behavior, 1 μm wide windows separated by 1 μm wide masks were prepared. The stripe was along the <1–10> axis of the Si surface so that the GaN stripes would have self-organized {1–101} facets. Alternatively, the stripe windows can also be prepared along the <11–2> direction of the substrate, where the GaN stripes would have side {11–22} facets. Following the formation of a high temperature AlN IL, GaN stripes inside the window areas were grown. The coalescence of stripes was achieved by increasing the growth time.

High resolution TEM (HRTEM) showed that the AlN/Si interface is made of a direct stack of the wurtzite structure on cubic crystal. Careful inspection of the HRTEM images indicated that the growth of (0001) AlN on (111) Si is managed by a commensurate-like one. That is, a lattice constant mismatch as large as 18% is adjusted by introducing misfit dislocations in every 5.3 lattices in AlN or every 4.3 lattices in Si (Liu et al., 2003). The phenomenon is very similar to one found in the MnAs/GaAs interface (Ploog, 2001).

Figure 3.6 shows typical TEM images of GaN/AlN/(111) Si. The dark areas/ lines represent defects/dislocations. The interface area has a high density of

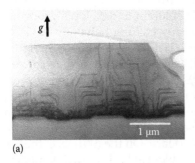
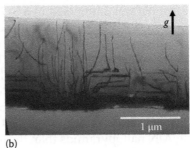

(a)   (b)

FIGURE 3.6
Cross-sectional TEM image of SAG-ELO-GaN grown on (111) Si substrate with $g = (0002)$. Stripe window is along (a) <1–10> axis and (b) <11–2> axis. Because of the stability of {1–101} facets of the GaN, the annihilation of TD is more effective in (a) than in (b). The stripe and mask widths are nominally 1 μm.

dislocations, but at the top of the window area, a lower dislocation density is apparent. It is well known that the TD created at the AlN/Si interface propagates upward along the <0001> axis of the GaN. Some of them turn to the horizontal direction at the {1-101} and/or {11-22} facets (Tanaka et al., 2000). The most remarkable result in Figure 3.6 is the fact that the dark contrast near the hetero-interface persists up to 200 nm and is diminished on the topmost region of GaN epilayer. This is attributed to the presence of small pyramidal GaN near the hetero-interface. In the growth of GaN on Si, the AlN IL is formed at high temperatures to make small AlN crystalline or AlN pyramids. The growth of GaN is then performed on the AlN nanopyramids. Thus, at an early stage of growth, TD turns to the direction perpendicular to the <0001> axis and is annihilated by making a loop. In Figure 3.6, we can see that the annihilation is more effective on a sample with stripes along the <11-20> direction with {1–101} facets than that with stripes along the <1-100> direction with {11-22} facets. This is attributed to the different stabilities of different facets, i.e., under our growth conditions, the {1-101} facets are more stable than the {11-22} facets as shown in Figure 3.5. The strong annihilation, as shown in Figure 3.6a, has not been recognized in samples grown on sapphire substrates, where growth is performed on an LT grown AlN or GaN buffer layer, which is more or less amorphous like.

3.3.4 Selective Growth of GaN on Si (001)

So far, Si integrated circuit (IC) technology has been developed on (001) Si. The growth of GaN on (001) Si substrates is thus particularly important to the integration of Si and nitride technologies and to Si photonics. The growth of GaN, however, is done mostly on Si (111). One of the current authors used the SAG method to successfully grow GaN on Si (001) (Honda et al., 2002a). The {111} facets on Si (001) substrates were prepared first, on which the growth of GaN was performed (Honda et al., 2000, 2002a). A stripe mask pattern was prepared on a (001) Si substrate with the direction along the <–110> axis of Si. The sample was immersed in a KOH solution to develop a groove made of (111) and (–1–11) facets by the anisotropy etching behavior. A GaN epilayer was then grown on these prepared {111} facets as shown in Figure 3.7. Growth was obtained only on the {111} facets as expected. The convergent beam electron diffraction (CBED) analyses on the cross-section of the GaN stripes showed that the growth of GaN is along the <0001> axis on the {111} facet of Si (Tanaka et al., 2002). It was noticed that there was no growth on the bare (001) bottom face. This suggests that the {111} facet is more reactive in forming AlN nanopyramids than the (001) face. On the (001) face, the formation of SiN_x will be effective in the NH_3 ambient at high growth temperature, which might prohibit the nucleation of AlN nanopyramids. It was found that selectivity is sensitive to the growth conditions (Honda et al., 2000). If under a high growth rate, or if the supply of the source precursors are high, polycrystalline was formed on the (001) bottom face.

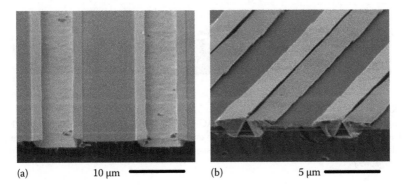

(a)   10 μm ▬▬▬▬   (b)   5 μm ▬▬▬▬

FIGURE 3.7
GaN stripes grown on patterned (001) Si substrate. The growth is achieved on the {111} facets of Si. The window widths are (a) 10 μm and (b) 5 μm, while the depth of the grooves is around 2 μm. The width of the GaN stripe top facets is around 3 μm determined by the depth of the groove.

The most remarkable result from Figure 3.7 is that the surface of the {1–101} GaN is as flat as those found on pyramidal GaN. This is attributed to the stability of the surface, i.e., the surface is produced in a self-organized manner. The atomic force microscopy (AFM) analysis showed that the mean roughness is around 0.2 nm, which is in the order of one monolayer. This indicates the potential for high quality and high performance nitride materials and devices grown on the {1-101} GaN face.

On (001) Si, a groove has two {111} facets. As a result, there are two GaN stripes in a groove as shown in Figure 3.7. Because of the tilting of the <0001> axis on the substrate, the top face of the GaN stripe is on the {1-101} face of GaN. The {1-101} face is at an angle of 62° to the (0001) plane, while the {111} face of Si is at an angle of 54.7° to the (001) face. Thus, the two {1-101} facets of the stripes have opposite angles to the substrate surface as shown in Figure 3.7b.

3.3.5 ELO of (1-101) GaN

As shown in Figure 3.7, the width of the stripes grown by the SAG method on the (1-101) facet of patterned (001) Si substrate is on the same order of the depth as the stripe windows. Thus, in order to increase the stripe width to fabricate a device such as an LED, a mask pattern with a window wider than 300 μm is necessary. Unfortunately, this is not acceptable. To make a uniform (1-101) GaN on (001) Si, the use of an ELO mode would be more practical. To do this, the angles of the {1-101} plane on the substrate must be parallel to each other. For this purpose, a Si (001) substrate is prepared, on which the crystal face is rotated at 7.3° around the <-110> axis so that one of the {1-101} facets is parallel to the substrate surface.

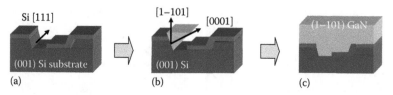

FIGURE 3.8
Schematic diagrams for the process of the SAG-ELO of (1-101) GaN layers grown on (001) Si substrate: (a) etched Si (111) facets with SiO₂ cover areas, (b) growth of GaN (0001) along Si (111) direction, and (c) GaN (1-101) layer after coalescence. (After Sawaki, N. et al., *J. Crystal Growth*, 311, 2867, 2009.)

This process is schematically shown in Figure 3.8 (Sawaki et al., 2009). On a 7.3° off-oriented (001) Si substrate, a stripe mask pattern was prepared as described earlier. The period of the stripe masks was typically 2 µm (1 µm window and 1 µm mask). Si {111} facets can be prepared using KOH anisotropy etching. A thin SiO_2 film was then deposited on the Si (−1–11) and (001) facets so that only the (111) facet is exposed to air. The sample was inserted into the MOCVD chamber to grow GaN stripes by SAG via a high temperature grown AlN IL. By increasing the growth time, the growth of GaN takes place laterally on the mask area (ELO) and the stripes coalesce into each other to make a uniform (1-101) GaN layer. More detail on the growth procedure was reported in reference (Hikosaka et al., 2004).

Figure 3.9 shows a scanning electron microscope (SEM) image of a typical (1-101) GaN grown on 7° off-oriented (001) Si substrate with a uniform and crack-free surface. The difference in the thermal expansion coefficients between GaN and Si is reduced substantially by tilting the crystal axis on the substrate and thus suppressing the cracks. The thermal expansion coefficient of Si is 3.59×10^{-6}/K, which is between that of GaN along (perpendicular to) the c-axis 3.17×10^{-6}/K (5.59×10^{-6}/K) (Sawaki et al., 2002). The XRD analyses showed that there is a strong anisotropy for the in-plane strain while it is biaxial in (0001) GaN grown on Si (111) (Hikosaka et al., 2007).

FIGURE 3.9
SEM image of a SAG-ELO (1-101) GaN grown on (001) Si substrate. (After Sawaki, N. et al., *J. Crystal Growth*, 311, 2867, 2009.)

2 µm

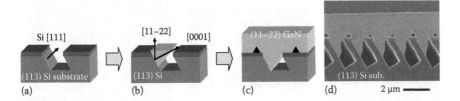

FIGURE 3.10
(a–c) Schematic SAG-ELO processes for a (11-22) GaN on (113) Si substrate and (d) a SEM image of an ELO sample. (After Sawaki, N. et al., *J. Crystal Growth*, 311, 2867, 2009.)

3.3.6 ELO of (11-22) GaN on (113) Si

One of the advantages of growing nitrides on Si substrates is that we can use Si substrates with an arbitrary crystalline orientation (Tanikawa et al., 2009; Yang et al., 2009b). On the substrate, we may develop a {111} facet with arbitrary size on which we can make the SAG of III-nitrides. As an example, we discuss the SAG of nitrides on (113) Si substrates, where the {111} facet is at 58.5° to the (113) plane. The angle is nearly equal to the angle between the {11-22} face and the (0001) face of GaN.

A stripe mask pattern was made along the <21-1> direction on a (113) Si substrate surface. A groove pattern made of {111} facets was obtained on the sides of (−101) or (113) at the bottom after etching with the KOH solution (Tanikawa et al., 2008a). The schematic processes are shown in Figure 3.10a through c. Since (111) facets point upward on the substrate surface while the (−1−1−1) facets are in the opposite direction, efficient growth takes place only on the (111) facets. Additional deposition of a SiO$_2$ film is not required in this case. By the ELO of GaN stripes, a flat (11-22) GaN was obtained and is shown in Figure 3.10d.

3.3.7 ELO of (11-20) GaN on (110) Si

On a (110) Si substrate, the {111} plane is normal to the substrate surface. Therefore, we may grow GaN with the c-axis being parallel to the substrate surface. After coalescence of the stripes, a uniform (11-20) GaN with a nonpolar surface is obtained. To achieve this on a (110) Si substrate, a mask pattern was prepared using SiO$_2$ film followed by etching in a KOH solution to obtain {111} facets. An additional SiO$_2$ film was then deposited on one of the {111} side facets. The SAG took place only on one of the facets in the groove so that the c-axis of the GaN is in a direction parallel to the substrate surface (Tanikawa et al., 2008b). Similar sidewall epitaxy has been demonstrated on an r-plane sapphire substrate (Imer et al., 2006) and on a GaN template (Iida et al., 2008). By the ELO of stripes, a flat (11-20) GaN crystal is achieved. Figure 3.11a and b show the GaN stripes before the coalescence. The cross-section of stripes is not exactly rectangular. It is

(a)   2 µm ━━━━ (b)   2 µm ━━━━ (c)   3 µm ━━━━

FIGURE 3.11
Cross-sectional SEM images of GaN grown on patterned (110) Si substrate. The period of the mask pattern is 3 µm. The cross-sectional shape of stripes depends on the growth conditions: (a) $T_g = 1100°C$ and $P = 100$ Torr, (b) $T_g = 1080°C$ and $P = 500$ Torr, and (c) coalesced after a long growth time. (After Tanikawa, T. et al., *J. Crystal Growth* 310, 4999, 2008b.)

eventually formed with (0001), (000-1), (11-20), and (11-22) facets, depending on the growth conditions. A flat (11-20) surface was obtained after the coalescence and the surface is extremely flat irrespective of the growth conditions.

3.3.8 ELO of *m*-Plane (1-100) GaN on (112) Si

Under normal growth conditions, GaN grown on Si (211) substrate is wurtzite structure with the GaN (0001) plane parallel to the surface (Chen et al., 2002). Ni et al. (2009) demonstrated the concept of nonpolar *m*-plane GaN-on-Si substrates by initiating the growth on the vertical (–1–11) sidewalls of patterned Si (112) substrates. By masking other Si {111} planes using SiO₂, only the vertical Si (–1–11) sidewalls were allowed to participate in GaN growth, which led to *m*-plane GaN films. Figure 3.12 shows a schematic depiction of selective area *m*-plane GaN growth on patterned Si (112) substrates. Since

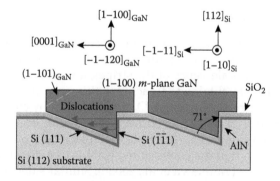

FIGURE 3.12
Schematic depiction of selective area *m*-plane GaN growth on patterned Si (112) substrates. Since growth is initiated on the vertical Si (–1–11) sidewalls, the threading dislocations propagate along the *c*-axis and not toward the surface. (After Ni, X. et al., *Appl. Phys. Lett.*, 95, 111102, 2009. With permission.)

growth is initiated on the vertical Si (–1–11) sidewalls, the TDs propagate along the *c*-axis and not toward the surface. InGaN double heterostructure active layers grown on these *m*-GaN/Si (112) templates exhibited two times higher internal quantum efficiencies (IQE) compared to their *c*-plane counterparts at comparable carrier densities. These results demonstrate a promising method for obtaining high quality nonpolar *m*-GaN films on large area, inexpensive Si substrates.

3.4 Photonic Devices

With successful crack removal and the improved crystalline quality of GaN layers grown on Si substrates, many different nitride device structures have been grown on these GaN/Si templates in the last decade. Below, we briefly summarize nitride photonic devices grown on Si substrates.

3.4.1 LEDs on 2 in. Si Substrates

Dadgar et al. (2002a) demonstrated 2.8 μm thick, entirely crack-free GaN-LEDs on 2 in. Si (111) substrates grown by MOCVD. From *I–V* characteristics, low turn-on voltages and a series resistance of 55 Ω were obtained for a vertically contacted diode. By in situ insertion of a Si_xN_y mask layer, luminescence intensity was significantly enhanced. A light output power of 152 μW at a current of 20 mA and a wavelength of 455 nm was obtained.

Zhang et al. (2005a) reported crack-free thin-film InGaN MQW LEDs, which were successfully transferred from the original Si (111) substrate onto copper carrier by means of metal-to-metal bonding and selective liftoff, using wet-chemical etching. In conjunction with inserting a metal reflector between the LED structure and the copper carrier, the performance of the LEDs fabricated on the substrate removal region was significantly improved compared to LEDs on Si substrate. Operating voltage was set at 20 mA and the series resistances were 3.6 V and 27 Ω, respectively. A 49% increase in optical power compared to LEDs without substrate removal was observed. They also (Zhang et al., 2001, 2003; Egawa et al., 2002) reported crack-free InGaN MQW blue LEDs on 2 in. Si (111) substrate using AlN/GaN multilayers with a thin AlN/AlGaN buffer layer. Operating voltages of 3.7 and 4.2 V and output powers of 34.8 and 34.5 μW at 20 mA were measured for the lateral and vertical configurations, respectively. Nitride LEDs on Si showed a high saturation operating current due to the good thermal conductivity of the Si substrate.

Mo et al. (2005) demonstrated a crack-free InGaN-LED structure on Si using a Ga-rich GaN HT buffer. Operating voltages of 3.8 V, turn-on voltage of about 2.5 V, and a series resistance of 47 Ω were obtained for these

LEDs. These LEDs showed an EL intensity of 20 mcd at an injection current of 20 mA, which is comparable to that of LEDs grown on sapphire. The same group also fabricated freestanding LEDs by removing the Si substrate (Xiong et al., 2006). After transferring LED films to n-type GaN layer, vertical structure LEDs were then fabricated. At 20 mA, the vertical LED had a light output power of 2.8 mW, which was 3.7 times larger than that of the lateral LEDs (Table 3.2).

Fenwick et al. (2009) demonstrated crack-free GaN LEDs grown by MOCVD on Si (111) substrates in 2009 using an atomic layer deposition (ALD) grown Al_2O_3 interlayer. Luminescence intensity versus current density measurements showed higher efficiency for the LEDs on Si substrate compared with those grown on sapphire under high drive currents. These results show comparable performance characteristics for GaN LED on Si and sapphire

TABLE 3.2

List of Nitride LED Structures Grown on 2 in. Si Substrates

Buffer and IL	Power (µW)	Wavelength (nm)	Refs.
LT-AlN + AlN IL + SiN	152	455	Dadgar et al. (2002a)
HT-AlN + AlGaN (30 nm)	34.8	453	Zhang et al. (2003)
MBE AlN + GaN/AlN stack	72	428	Damilano et al. (2008)
AlN (20 nm)		465	Tran et al. (1999)
LT AlN + GaN (200 nm)	20 mcd	460	Mo et al. (2005)
AlN + AlN/GaN SL	6	508	Feltin et al. (2001b)
HT-AlN + AlGaN-on-Si (110)		485	Reiher et al. (2009)
MBE AlN/AlGaN		290	Kipshidze et al. (2002a)
LT-AlN + HT AlN	350	492	Li et al. (2006)
Pattern on 10 nm AlN/Si		465	Yang et al. (2000)
AlN + AlGaN			Fehse et al. (2004)
AlN + AlGaN + SiN	70 lm/W		Phillips and Zhu (2009)
HT-AlN + LT-AlN IL (6 in. SOI substrate)		500 (µ-LED)	Tripathy et al. (2008)
HT-AlN + LT-AlN IL (6 in. SOI substrate)		530	Tripathy et al. (2007)
MBE AlN		360	Guha and Bojarczuk (1998)
HT-AlN + AlN/GaN multilayer	18	478	Egawa et al. (2002)
Al_2O_3 (ALD 5–20 nm)		30% IQE	Fenwick et al. (2009)
Freestanding GaN	2.8 mW	460	Xiong et al. (2006)
AlN + SiN		564	Poschenrieder et al. (2002b)
AlN	420	498	Krost and Dadgar (2002)
			Dadgar et al. (2003b)
LT-AlN + SiN + LT-AlN IL	152	455	Dadgar et al. (2002)
AlN/AlInGaN SL			Kipshidze et al. (2002b)
AlN (32 nm) + GaN		377	Yablonskii et al. (2002)

substrates. The external QE (EQE) was relatively low for devices on both Si (0.08% at 20 mA) and sapphire (0.5% at 20 mA), which is believed to be due to the absence of an $Al_xGa_{1-x}N$ electron blocking layer (Fenwick et al., 2009).

3.4.2 LEDs on Large Size Si Substrates (>5 in. in Diameter)

For the growth of nitride materials on large size Si substrates (>5 in. in diameter), problems associated with cracks and bowing are much more severe compared to those of small size Si wafers because temperature uniformity and mechanical strength over the whole wafer are required. Li et al. (2006) first reported the successful growth of high quality AlN and GaN epilayers and LEDs on 6 in. diameter Si (111) substrates by MOCVD in 2006. By utilizing the combination of a high quality AlN epilayer and a thin $Al_xGa_{1-x}N$ graded layer on Si, they demonstrated that the growth of III-nitride photonic structures on large size Si substrates up to 6 in. in diameter is possible.

One of the keys to their success for obtaining high quality GaN epilayers on large Si substrate was to grow high quality AlN/Si templates. Figure 3.13 compares the room temperature PL spectra of AlN epilayers grown on 2 in. sapphire and 6 in. Si substrates. The PL spectral line shapes from both samples are similar. The PL results show that AlN epilayers grown on Si predominantly exhibit the band-edge PL emission—implying high optical quality.

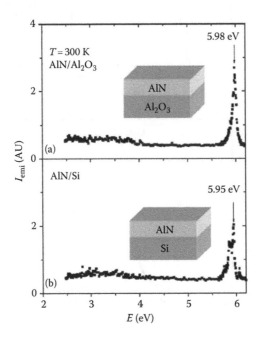

FIGURE 3.13
Room temperature PL spectra of AlN epilayers grown on (a) 2 in. sapphire and (b) 6 in. Si substrates.

Growth begins with deposition of an AlN nucleation layer to ensure that the Si surface does not degrade, followed by a complex buffer structure. Careful control of the composition and thickness of this buffer balances its strain with that induced by the thermal expansion mismatch during cooling. To improve LED performance, GaN and AlGaN layers are added on top of the buffer to reduce dislocation density.

GaN epilayers and LED structures were grown on AlN/Si templates. By inserting a thin $Al_xGa_{1-x}N$ graded layer between the AlN/Si template and GaN epilayer, cracks and wafer bowing were eliminated. Figure 3.14a shows the XRD $\theta/2\theta$ scan curve of a 2 μm GaN epilayer grown on AlN/Si templates, which indicates good crystalline quality in the GaN epilayer. The XRD rocking curve of the (0002) peak of GaN was measured with a typical FWHM of about 565 arcsec, as shown in Figure 3.14b. It is believed that the high quality

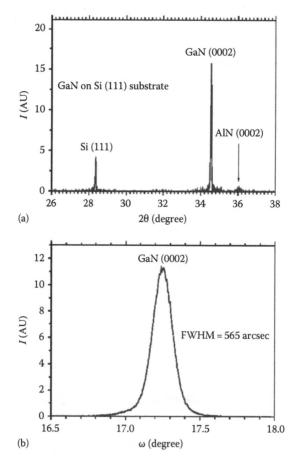

(a)

(b)

FIGURE 3.14
X-ray diffraction spectra of a 2 μm thick GaN epilayer grown on AlN epilayer/Si (111) template: (a) $\theta/2\theta$ scan and (b) rocking curve of the (0002) peak.

AlN epilayer also acted as an effective dislocation filter for the growth of subsequent device layers deposited on Si substrate.

To further investigate the benefits of using AlN epilayer as a template, we have also grown blue LED structures on 6 in. Si substrates. Prior to the growth of the LED active region, a 0.5 μm AlN epilayer was grown on Si at 1050°C as a template. Then, a thin $Al_xGa_{1-x}N$ graded layer was deposited to further minimize cracks and wafer bowing. The subsequent MQW LED structure on AlN/Si templates consisted of a Si doped n-GaN epilayer (2 μm), eight periods of InGaN/GaN MQW active layer, and a Mg doped p-GaN epilayer (0.25 μm).

The characteristics of the fabricated LEDs were also measured. Figure 3.15a shows the *I–V* characteristics of LEDs grown on 6 in. Si. The forward bias voltage V_F is 4.1 V. The reverse leakage current is about 27 μA at a bias voltage of −20 V. The forward differential series resistance is about 25 Ω. Optical microscopy images of a blue LED wafer grown on a 6 in. Si substrate are shown in the inset of Figure 3.15a, which shows that these LEDs have a good surface morphology and are free of cracks. A typical EL spectrum of a 492 nm LED fabricated on Si substrate is shown in the inset of Figure 3.15b. The FWHM of the emission line is about 32 nm. The interference pattern from the EL spectrum can also be observed, again clearly indicating the good surface morphology of the LED wafer. Figure 3.15b shows the *L–I* characteristics (optical power output versus applied current) of LED fabricated on 6 in. Si substrate. The optical power output for an unpackaged LED with dimensions of 0.3 mm × 0.3 mm is about 0.35 mW at 20 mA.

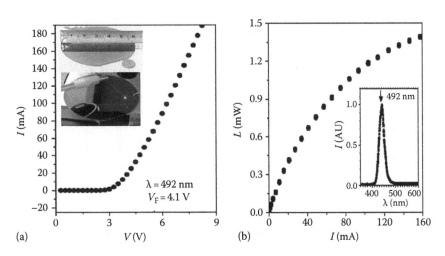

FIGURE 3.15
(a) *I–V* and (b) *L–I* characteristics of fabricated InGaN/GaN MQW LED grown on 6 in. AlN epilayer/Si substrate, measured from the top surface of unpackaged bare chips with a size of 300 × 300 μm². The inset shows (a) the optical microscopy images and (b) EL spectrum of an InGaN/GaN MQW blue LED wafer grown on 6 in. Si substrate. (After Li, J. et al., *Appl. Phys. Lett.*, 88, 171909, 2006. With permission.)

Dadgar et al. (2006) also reported blue LED structures grown on (111) 6 in. Si substrates. They used in situ curvature measurements as shown in Figure 3.16 (for 2 in. wafer) to monitor stress development during growth and the influence of interlayers on strain balancing during cooling. In XRD ω-scans, the GaN (0002) reflection peak is about 380 arcsec and in θ – 2θ measurements, the InGaN/GaN MQW interference peaks are well resolved, indicating the high crystalline quality of the grown structures. However, LED characteristics were not reported.

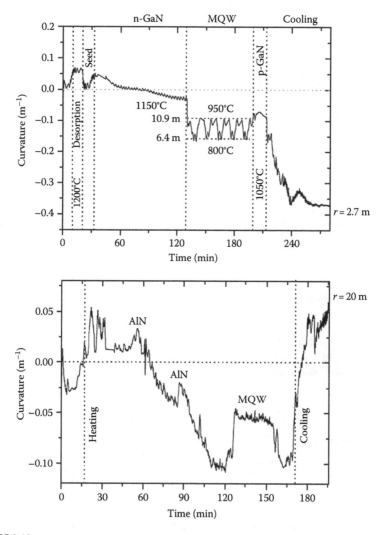

FIGURE 3.16

In situ curvature measurements during the growth of an LED structure on 2 in. sapphire (top) and Si (bottom) substrates. (After Dadgar, A. et al., *J. Crystal Growth*, 297, 279, 2006. With permission.)

Tripathy et al. (2007) reported the fabrication of InGaN/GaN LEDs on nanoscale SOI (111) substrates and Si (111) substrates. Due to a highly reflective Si/SiO$_2$ layer beneath the AlN buffer and high refractive index contrast at the interfaces, they observed multiple interference peaks from LEDs on SOI substrate, which resulted in an increased integrated EL intensity compared with those grown on Si (111). Figure 3.17 shows optical microscopy images and the *I–V* characteristics of blue and green LEDs grown on Si (111) and SOI (111) under current injection. Bright blue and green emissions were clearly observed from LEDs grown on SOI. Figure 3.18 shows the EL spectra

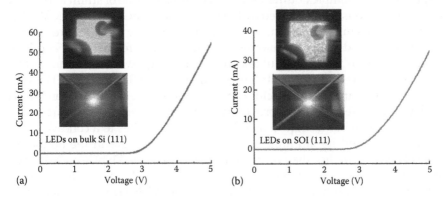

FIGURE 3.17
Optical microscopy images of blue and green LEDs grown on (a) Si (111) and (b) SOI (111) under electrical probing. The *I–V* characteristics of the LEDs are also shown. (After Tripathy, S. et al., *Appl. Phys. Lett.*, 91, 231109, 2007. With permission.)

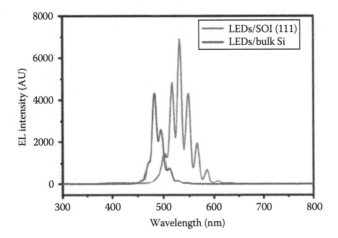

FIGURE 3.18
EL spectra of LEDs grown on Si (111) and SOI (111) under a current injection of 100 mA. Due to the enhanced light extraction and multiple reflections at the SOI substrate, the energetic maxima of the EL interference peak of LEDs grown on SOI (111) are much stronger than those on Si (111). (After Tripathy, S. et al., *Appl. Phys. Lett.*, 91, 231109, 2007. With permission.)

of LEDs grown on Si (111) and SOI (111) under 100 mA current injection. For the case of blue and green LEDs on SOI, very strong multiple interference peaks in the EL spectrum were observed with peaks centered around 530 nm. For the case of LEDs grown on Si (111), the EL intensity modulation is lower and the EL peak is centered around 485 nm. The difference in the EL emission peak wavelengths in these two LED samples is due to different In contents in the MQWs.

Inserting a SiN_x layer, a technique used for nitride film growth on sapphire, further cuts TD density. Growth on 6 in. Si (111) substrates includes the deposition of a complex buffer and IL structures to control strain and wafer curvature. This is then followed by the growth of an InGaN/GaN MQW LED structure that emits at 460 nm, and a Mg-doped p-type GaN (Figure 3.19a).

It was believed that continual monitoring of wafer temperature and bow holds the key to the successful and reproducible growth of flat, uncracked nitride materials and device structures on large size Si substrates. The substrate temperatures in the reactor can be monitored with an Aixtron Argus tool and a LayTec Epicurve, providing real-time wafer-bow measurements. The Si substrate has a slight convex bow, which switches to a concave shape after heating and in situ annealing because temperature in the bottom of the substrate is higher than on the top surface (Figure 3.19b). Concave bowing increases with the addition of an AlN nucleation layer. A convex profile returns with the growth of the buffer and Si-doped GaN layer, which increases compressive stress. The growth of quantum wells and barrier layers produce detectable, small changes in curvature, and the wafer becomes more convex as the reactor temperature is increased for the deposition of Mg-doped GaN. Optimizing the magnitude of this bow during buffer growth can produce a perfectly flat wafer by matching the tensile stress of the film, which results from thermal expansion coefficient differences between GaN and Si substrate.

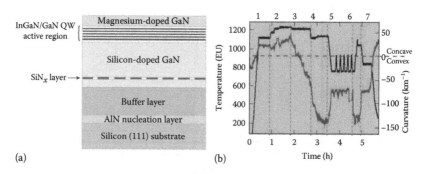

(a)   (b)

FIGURE 3.19
(a) LED structure includes a SiN_x interlayer to reduce dislocations; and (b) wafer bow is monitored by Laytec epicurve and temperatures are measured by the Aixtron Argus tool. Growth can be divided into seven sections: pregrowth heat treatment; growth of the AlN nucleation layer, the buffer, the n-GaN layer, the MQW region, the p-GaN layer; and annealing/cooling. (After Phillips, A. and Zhu, D., *Compound Semiconductor*, 19, 2009. With permission.)

FIGURE 3.20

Weak-beam, dark-field TEM images showing reduction of dislocation density in InGaN/GaN-LED by insertion of SiN_x interlayer. Pure screw and mixed-type dislocations are visible. (After Phillips, A. and Zhu, D., *Compound Semiconductor*, 19, 2009. With permission.)

Higher In content was observed from LEDs grown on Si substrate than those grown on SOI due to higher temperatures on the Si surface, which resulted in better thermal conductivity. Cross-sectional TEM images revealed that the low dislocation density in the device layers was due to a SiN_x layer, which caused defects to bend over and annihilate (Figure 3.20). This also occurs at the AlGaN/GaN interface due to a compressive stress, which results from lattice mismatch between GaN and AlN. The dislocation density in the epiwafers was assessed with plan-view TEM images, which together with AFM images, gives $<10^9$ cm^{-2} in dislocation densities for the best GaN-on-Si material. The IQE of LEDs was evaluated by temperature-dependent PL measurements and gave a room temperature IQE of almost 50%. Similar structures grown on sapphire with a dislocation density of 10^8 cm^{-2} typically have IQE values around 70%.

The best LEDs with dimensions of 0.5 mm × 0.5 mm have very similar *I–V* characteristics to those grown on GaN/sapphire templates (Figure 3.21). The light output from the top side of both devices has been measured using the same optical arrangement. The sapphire-based LEDs produce about twice the power output of LEDs grown on Si. Taking into account the high absorption of Si substrate, the estimated IQE of LEDs on Si is around 37%, a figure based on measurements of the total light emitted in the forward direction.

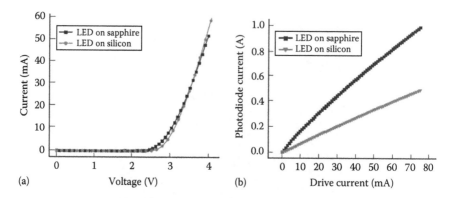

FIGURE 3.21
(a) *I–V* and (b) *L–I* characteristics of LEDs grown on Si and sapphire substrates with a dimension of 0.5 × 0.5 mm². GaN-on-Si LEDs produce very similar *I–V* curves to conventional devices, but light output from GaN-LED on Si is about 50% lower than those on sapphire substrate. (After Phillips, A. and Zhu, D., *Compound Semiconductor*, 19, 2009. With permission.)

3.4.3 LEDs on Semipolar GaN/Si Templates

One of the key issues in Si photonics is the fabrication of a light source on Si. The GaN stripe structures described in Section 3.3.4 offer a natural waveguide/optical cavity (Kim 2006; Hikosaka et al., 2006). In order to demonstrate the feasibility of such structures, optical gain of an AlGaN/GaN heterostructure grown on a stripe was measured. The samples used were three QW AlGaN/GaN structures made on a (1-101) GaN facet grown on a (001) Si substrate. The waveguide is typically 1.5 mm in length along the <11-20> axis of GaN.

The schematic experimental setup is shown in Figure 3.22a (Kim and Lee, 2006; Kim et al., 2006). The sample was illuminated using a N_2 laser (337 nm) and the emission spectrum from the edge of the stripe sample was measured at 77 K. The results are shown in Figure 3.22b for different excitation intensities, I_{exc}. By increasing I_{exc}, narrowing of the spectra was observed. A stimulated emission at 360 nm was observed when $I_{exc} > 4.5$ MW/cm². Optical gain as a function of wavelength in the stripe structure was measured, and a gain as high as 10 cm^{-1} was achieved between 350 and 370 nm under the highest I_{exc} of 4.5 MW/cm². Similar measurements were performed at room temperature. Strong emissions were observed at 370 nm, where the maximum optical gain was 12 cm^{-1}. For InGaN/GaN heterostructures, high optical gain in the dominant emission band was also achieved, which was attributed to the fact that these heterostructures are grown on a self-organized, atomically smooth facet.

In spite of rapid progress in the fabrication of high brightness blue/green LEDs, QE is still low in the green region, which has been attributed partly to the presence of a polarization field in heterostructures grown on (0001) GaN. Many groups have tried to fabricate LED or LD structures on

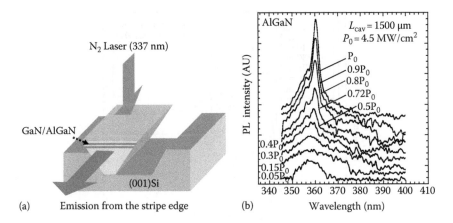

(a) Emission from the stripe edge   (b) Wavelength (nm)

FIGURE 3.22

(a) Experimental setup for optical gain measurements in a AlGaN/GaN stripe grown on Si (001) substrates and (b) the stimulated emission under laser excitation (337 nm).

semipolar or nonpolar GaN (Funato et al., 2006). With the success of GaN epilayers grown on different orientations of Si substrates, LEDs can be grown and fabricated on semipolar GaN such as (1-101) and (11-22) GaN, which are grown on patterned Si substrates (Hikosaka et al., 2008). LED structures were grown on (1-101) GaN prepared on a 7° off-oriented n-type (001) Si substrate by SAG-ELO. Si-doped GaN was prepared on a window of $300 \times 300 \ \mu m^2$ (1-101) GaN using the coalesced GaN stripes as shown in Figure 3.23a. The thickness of the coalescent GaN was typically 700 nm. The LED structure as shown in Figure 3.23b was grown on top of these prepared templates. Figure 3.23c shows the *I–V* characteristics of (1-101) and (11-22) LEDs. The turn-on voltage was around 3–4 V and the forward differential resistance was about 0.04 $\Omega \ cm^2$.

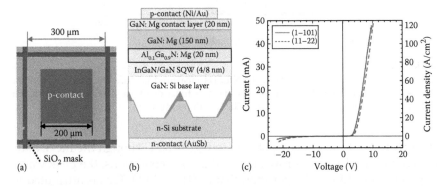

FIGURE 3.23

(a) SEM image and (b) LED structure of (1-101) GaN-LED, and (c) *I–V* characteristics of (1-101) and (11-22) GaN-LEDs. (After Hikosaka, T. et al., *Phys. Stat. Sol. (c)* 5, 2234, 2008. With permission.)

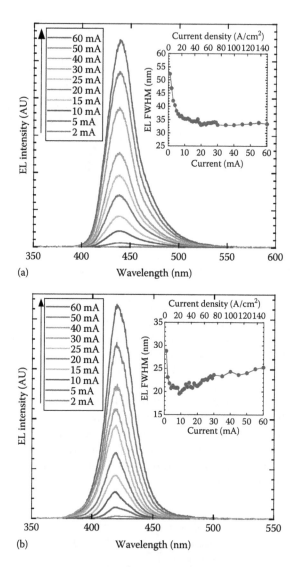

FIGURE 3.24

EL spectra of (a) (1-101) and (b) (11-22) GaN-LEDs measured at different driving currents. The insets show the FWHM of EL emission line as a function of the driving current. (After Hikosaka, T. et al., *Phys. Stat. Sol. (c)* 5, 2234, 2008. With permission.)

EL spectra of these LEDs are shown in Figure 3.24. Though the InGaN layers in both samples were grown under the same conditions, the peak wavelength of (1-101) LED is longer than that of (11-22) LED. This is attributed to different chemical behaviors on the (1-101) and (11-22) surfaces as discussed earlier. In both cases, the EL intensity increased linearly with the driving current while the peak wavelength was slightly blueshifted. The amount of blueshift was nearly equal to the value reported for a (1-101) LED grown on a 4H-SiC substrate (Kamiyama et al., 2005). In the case of the (11-22) LED, the amount of blueshift was extremely small. These results show the advantages of nonpolar GaN-LEDs grown on Si substrates.

Other nitride devices grown on Si substrates, including UV sensors (Chuang et al., 2007; Chang et al., 2009) and solar cells (Liou, 2009), have also been fabricated and characterized. Stevens et al. (1995) fabricated and characterized a GaN:Mg UV photodetector on Si with a cutoff wavelength at 376 nm and a responsivity of 12 A/W at 4 V bias under optical excitation of intensities on the order of 1 W/m^2 and below. The photocurrent increased nearly linearly with optical intensity for up to 10 W/m^2.

3.5 Summary and Future Perspectives

Recent progress and the current status of nitride wide bandgap semiconductors and photonic devices grown on Si substrate, particularly those grown on large size Si substrates and on planes other than (111), have been reviewed. The SAG of GaN on patterned Si substrate has been discussed. Relatively, high crystalline quality nitride materials and the reasonable performance of nitride photonic devices grown on Si wafers up to 6 in. in diameter have been demonstrated.

Though the performance of these LEDs still lags behind those grown on sapphire and SiC substrates, blue/green LEDs grown on Si substrates are already commercially available for niche applications. With the rapid improvements in the crystalline quality of nitride materials and the demonstration of photonic devices grown on Si substrates, further progress in this materials system is anticipated.

One particularly important area, which has not been reported, is the integration of nitride photonic and electrical devices with Si electronic devices. Until this occurs, the full potential of nitride semiconductors on Si will not be realized. GaN/Si integration will open new fields for GaN related technology in photonics and electronics. For example, Si electronic devices could be used to monitor and control nitride photonic devices grown on Si substrates. Another apparent application for GaN/Si integrated devices is the use of nitride emitters on Si for applications in Si photonics.

Acknowledgments

H.X. Jiang and J.Y. Lin would like to acknowledge the support of DOE (DF-FG02-09ER 4655), NSF (DMR-0906879, ECCS-0854619), and ARO. They also gratefully acknowledge the support of the Edward Whitacre and Linda Whitacre endowment chair positions through the AT&T Foundation. Contributions from their group members are also acknowledged. N. Sawaki would like to acknowledge the support by the grant-in-aid from the Japan Society for Promotion of Sciences (JSPS) No. 16106001 and No. 22360009. Collaboration with Dr. Y. Honda, Dr. M. Yamaguchi, Dr. S. Tanaka, and PhD students T. Narita, T. Hikosaka, N. Koide, E. H. Kim, and T. Tanikawa is acknowledged.

References

Amano, H., Sawaki, N., Akasaki, I., and Toyoda, Y. Metalorganic vapor-phase epitaxial-growth of a high quality GaN film using an AlN buffer layer. *Applied Physics Letters* 48 (1986): 353–355.

Arai, M., Nishiyama, N., Azuchi, M., Matsutani, A., Koyama, F., and Iga, K. Monolithic formation of metal organic chemical vapor deposition grown multi-wavelength vertical cavities with highly strained GaInAs/GaAs quantum wells on GaAs (311)B. *Japanese Journal of Applied Physics* 40 (2001): 4056–4057.

Armitage, R., Yang, Q., Feick, H., Gebauer, J., Weber, E.R., Shinkai, S., and Sasaki, K. Lattice-matched HfN buffer layers for epitaxy of GaN on Si. *Applied Physics Letters* 81 (2002): 1450.

Cervantes-Contreras, M., Lopez-Lopez, M., Melendez-Lira, M., Tamura, M., and Hiroyama, Y. Molecular beam epitaxial growth of GaN on (100)- and (111) Si substrates coated with a thin SiC layer. *Journal of Crystal Growth* 227–228 (2001): 425–430.

Chang, C.-A., Lien, S.-T., Liu, C.-H., Shih, C.-F., Chen, N.-C., Chang, P.-H., Peng, H.-C. et al. Effect of buffer layers on electrical, optical and structural properties of AlGaN/GaN heterostructures grown on Si. *Japanese Journal of Applied Physics* 45 (2006): 2516–2518.

Chang, J.R., Yang, T.H., Ku, J.T., Shen, S.G., Chen, Y.C., Wong, Y.Y., and Chang, C.Y. GaN growth on Si(111) using simultaneous AlN/a-Si$_3$N$_4$ buffer structure. *Japanese Journal of Applied Physics* 47 (2008): 5572–5575.

Chang, S.-P., Chang, S.-J., Lu, C.-Y., Chiou, Y.-Z., Chuang, R.W., and Lin, H.-C. Low-frequency noise characteristics of GaN-based UV photodiodes with AlN/GaN buffer layers prepared on Si substrates. *Journal of Crystal Growth* 311 (2009): 3003–3006.

Chen, P., Zhang, R., Zhao, Z.M., Xi, D.J., Shen, B., Chen, Z.Z., Zhou, Y.G., Xie, S.Y., Lu, W.F., and Zheng, Y.D. Growth of high quality GaN layers with AlN buffer on Si(111) substrates. *Journal of Crystal Growth* 225 (2001): 150–154.

Chen, X.-F., Honda, Y., Kato, T., and Sawaki, N. Growth of wurtzite-GaN on Si(211) by metalorganic vapor phase epitaxy. *Journal of Crystal Growth* 237–239 (2002): 1110–1113.

Cheng, K., Leys, M., Degroote, S., Germain, M., and Borghs, G. High quality GaN grown on silicon (111) using a Si$_x$N$_y$ interlayer by metal-organic vapor phase epitaxy. *Applied Physics Letters* 92 (2008): 192111.

Chuang, R.W., Chang, S.P., Chang, S.J., Chiou, Y.Z., Lu, C.Y., Lin, T.K., Lin, Y.C., Kuo, C.F., and Chang, H.M. Gallium nitride metal-semiconductor-metal photodetectors prepared on silicon substrates. *Journal of Applied Physics* 102 (2007): 073110.

Cong, G., Lu, Y., Peng, W., Liu, X., Wang, X., and Wang, Z. Design of the low-temperature AlN interlayer for GaN grown on Si (111) substrate. *Journal of Crystal Growth* 276 (2005): 381–388.

Cordier, Y., Portail, M., Chenot, S., Tottereau, O., Zielinski, M., and Chassagne, T. AlGaN/GaN high electron mobility transistors grown on 3C-SiC/Si(111). *Journal of Crystal Growth* 310 (2008): 4417–4423.

Dadgar, A., Blasing, J., Diez, A., Alam, A., Heuken, M., and Krost, A. Metalorganic chemical vapor phase epitaxy of crack-free GaN on Si (111) exceeding 1 micron in thickness. *Japanese Journal of Applied Physics* 39 (2000): L1183–L1185.

Dadgar, A., Poschenrieder, M., Blasing, J., Fehse, K., Diez, A., and Krost, A. Thick, crack-free blue light-emitting diodes on Si (111) using low-temperature AlN interlayers and in situ Si_xN_y masking. *Applied Physics Letters* 80 (2002a): 3670.

Dadgar, A., Poschenrieder, M., Contreras, O., Christen, J., Fehse, K., Blasing, J., Diez, A., Schulze, F., Riemann, T., Ponce, F.A., and Krost, A. Bright, crack-free InGaN/GaN light emitters on Si (111). *Physica Status Solidi (a)* 192 (2002b): 308–313.

Dadgar, A., Poschenrieder, M., Reiher, A., Blasing, J., Christen, J., Krtschil, A., Finger, T., Hempel, T., Diez, A., and Krost, A. Reduction of stress at the initial stages of GaN growth on Si (111). *Applied Physics Letters* 82 (1) (2003a): 28.

Dadgar, A., Poschenrieder, M., Blasing, J., Contreras, O., Bertram, F., Riemann, T., Reiher, A. et al. MOVPE growth of GaN on Si(111) substrates. *Journal of Crystal Growth* 248 (2003b): 556–562.

Dadgar, A., Hums, C., Diez, A., Blasing, J., and Krosta, A. Growth of blue GaN LED structures on 150-mm Si(111). *Journal of Crystal Growth* 297 (2006): 279–282.

Damilano, B., Natali, F., Brault, J., Huault, T., Lefebvre, D., Tauk, R., Frayssinet, E., Moreno, J.-C., Cordier, Y., Semond, F., Chenot, S., and Massies, J. Blue (Ga,In)N/GaN light emitting diodes on Si(110) substrate. *Applied Physics Express* 1 (2008): 121101.

Egawa, T., Moku, T., Ishikawa, H., Ohtsuka, K., and Jimbo, T. Improved characteristics of blue and green InGaN-based light-emitting diodes on Si grown by metalorganic chemical vapor deposition. *Japanese Journal of Applied Physics* 41 (2002): L663–L664.

Fehse, K., Dadgar, A., Krtschil, A., Riemann, T., Hempel, T., Christen, J., and Krost, A. Impact of thermal annealing on the characteristics of InGaN/GaN LEDs on Si(111). *Journal of Crystal Growth* 272 (2004): 251–256.

Feltin, E., Beaumont, B., Laugt, M., de Mierry, P., Vennegues, P., Lahreche, H., Leroux, M., and Gibart, P. Stress control in GaN grown on silicon (111) by metalorganic vapor phase epitaxy. *Applied Physics Letters* 79 (2001a): 3230.

Feltin, E., Dalmasso, S., de Mierry, P., Beaumont, B., Lahreche, H., Bouille, A., Haas, H., Leroux, M., and Gibart, P. Green InGaN light-emitting diodes grown on silicon (111) by metalorganic vapor phase epitaxy. *Japanese Journal of Applied Physics* 40 (2001b): L738–L740.

Feltin, E., Beaumont, B., Vennegues, P., Vaille, M., Gibart, P., Riemann, T., Christen, J., Dobos, L., and Pecz, B. Epitaxial lateral overgrowth of GaN on Si (111). *Journal of Applied Physics* 93 (2003): 182.

Fenwick, W.E., Melton, A., Xu, T., Li, N., Summers, C., Jamil, M., and Ferguson, I.T. Metal organic chemical vapor deposition of crack-free GaN-based light emitting diodes on Si (111) using a thin Al_2O_3 interlayer. *Applied Physics Letters* 94 (2009): 222105.

Funato, M., Ueda, M., Kawakami, Y., Narukawa, Y., Kosugi, T., Takahashi, M., and Mukai, T. Blue, green, and amber InGaN/GaN light-emitting diodes on semi-polar {11–22} GaN bulk substrates. *Japanese Journal of Applied Physics* 45 (2006): L659–L662.

Guha, S. and Bojarczuk, N.A. Ultraviolet and violet GaN light emitting diodes on silicon. *Applied Physics Letters* 72 (1998): 415.

Haskell, B.A., Wu, F., Craven, M.D., Matsuda, S., Fini, P.T., Fujii, T., Fujito, K., DenBaars, S.P., Speck, J.S., and Nakamura, S. Defect reduction in (110) a-plane gallium nitride via lateral epitaxial overgrowth by hydride vapor-phase epitaxy. *Applied Physics Letters* 83 (2003): 644–646.

Hikosaka, T., Narita, T., Honda, Y., Yamaguchi, M., and Sawaki, N. Optical and electrical properties of (1–101)GaN grown on a 7° off-axis (001)Si substrate. *Applied Physics Letters* 84 (2004): 4717–4719.

Hikosaka, T., Honda, Y., Yamaguchi, M., and Sawaki, N. Al doping in (1–101)GaN films grown on patterned (001)Si substrate. *Journal of Applied Physics* 101 (2007): 103513.

Hikosaka, T., Tanikawa, T., Honda, Y., Yamaguchi, M., and Sawaki, N. Fabrication and properties of semi-polar (1–101) and (11–22)InGaN/GaN MQW light emitting diode on patterned Si substrates. *Physica Status Solidi (c)* 5 (2008): 2234–2237.

Honda, Y., Kawaguchi, Y., Kato, T., Yamaguchi, M., and Sawaki, N. Selective growth of GaN microstructures on (111) facet of a (001)Si substrate by MOVPE. *Proceedings of the International Workshop on Nitride Semiconductors, IPAP Conference Series* 1 (2000): 304–307.

Honda, Y., Kawaguchi, Y., Ohtake, Y., Tanaka, S., Yamaguchi, M., and Sawaki, N. Selective area growth of GaN microstructures on patterned (111) and (001) Si substrate. *Journal of Crystal Growth* 230 (2002a): 346–350.

Honda, Y., Kuroiwa, Y., Yamaguchi, M., and Sawaki, N. Growth of GaN crystal free from cracks on a (111) Si substrate by selective MOVPE. *Applied Physics Letters* 80 (2002b): 222–224.

Hu, F.R., Ochi, K., Zhao, Y., Choi, B.S., and Hane, K. Molecular beam epitaxial growth of GaN thin film on Si substrate with InN as interlayer. *Journal of Crystal Growth* 294 (2006): 197–201.

Iida, D., Kawashima, T., Iwaya, M., Kamiyama, S., Amano, H., and Akasaki, I. Sidewall epitaxial lateral overgrowth of nonpolar a-plane GaN by metalorganic vapor phase epitaxy. *Physica Status Solidi (c)* 5 (2008): 1575–1578.

Imer, B.M., Wu, F., DenBaars, S.P., and Speck, J.S. Improved quality (11–20) a-plane GaN with sidewall lateral epitaxial overgrowth. *Applied Physics Letters* 88 (2006): 061908.

Ishikawa, H., Yamamoto, K., Egawa, T., Soga, T., Jimbo, T., and Umeno, M. Thermal stability of GaN on (111) Si substrate. *Journal of Crystal Growth* 189/190 (1998): 178–182.

Iwakami, S., Yanagihara, M., Machida, O., Chino, E., Kaneko, N., Goto, H., and Ohtsuka, K. AlGaN/GaN heterostructure field-effect transistors (HFETs) on Si substrates for large-current operation. *Japanese Journal of Applied Physics* 43 (2004): L831–L833.

Jamil, M., Grandusky, J.R., Jindal, V., Shahedipour-Sandvik, F., Guha, S., and Arif, M. Development of strain reduced GaN on Si (111) by substrate engineering. *Applied Physics Letters* 87 (2005): 082103.

Jamil, M., Grandusky, J.R., Jindal, V., Tripathi, N., and Shahedipour-Sandvik, F. Mechanism of large area dislocation defect reduction in GaN layers on AlN/Si (111) by substrate engineering. *Journal of Applied Physics* 102 (2007): 023701.

Jang, S.-H., Lee, S.-S., Lee, O.-Y., and Lee, C.-R. The influence of $Al_xGa_{1-x}N$ intermediate buffer layer on the characteristics of GaN/Si(111) epitaxy. *Journal of Crystal Growth* 255 (2003): 220–226.

Johnson, J.W., Piner, E.L., Vescan, A., Therrien, R., Rajagopal, P., Roberts, J.C., Brown, J.D., Singhal, S., and Linthicum, K.J. 12W/mm AlGaN-GaN HFETs on silicon substrates. *IEEE Electron Device Letters* 25 (2004): 459.

Kaiser, S., Preis, H., Gebhardt, W., Ambacher, O., Angerer, H., Stutzmann, M., Rosenauer, A., and Gerthsen, D. Quantitative transmission electron microscopy investigation of the relaxation by misfit dislocations confined at the interface of $GaN/Al_2O_3(0001)$. *Japanese Journal of Applied Physics* 37 (1998): 84–89.

Kamiyama, S., Honshio, A., Kitano, T., Iwaya, M., Amano, H., Akasaki, I., Kinoshita, H., and Shiomi, H. GaN growth on (30–38) 4H-SiC substrate for reduction of internal polarization. *Physica Status Solidi (c)* 2 (2005): 2121–2124.

Kim, D.-W. and Lee, C.-R. N-type doping of GaN/Si(111) using $Al_{0.2}Ga_{0.8}N$/ALN composite buffer layer and $Al_{0.2}Ga_{0.8}N$/GaN superlattice. *Journal of Crystal Growth* 286 (2006): 235–239.

Kim, M.-H., Do, Y.-G., Kang, H.C., Noh, D.Y., and Park, S.-J. Effects of step-graded $Al_xGa_{1-x}N$ interlayer on properties of GaN grown on Si (111) using ultrahigh vacuum chemical vapor deposition. *Applied Physics Letters* 79 (2001): 2713.

Kim, E.H., Hikosaka, T., Narita, T., Honda, Y., and Sawaki, N. Optical spectra of (1–101) InGaN/GaN and GaN/AlGaN MQW structure grown on a 7 degree off axis (001) Si substrate. *Physica Status Solidi (c)* 3 (2006): 1992–1996.

Kipshidze, G., Kuryatkov, V., Borisov, B., Holtz, M., Nikishin, S., and Temkin, H. AlGaInN-based ultraviolet light-emitting diodes grown on Si 111. *Applied Physics Letters* 80 (2002a): 3682.

Kipshidze, G., Kuryatkov, V., Borisov, B., Nikishin, S., Holtz, M., Chu, S.N.G., and Temkin, H. Deep ultraviolet AlGaInN-based light-emitting diodes on Si(111) and sapphire. *Physica Status Solidi (a)* 192 (2002b): 286–291.

Koide, Y. MOVPE growth and optical properties of AlGaN alloys (in Japanese) PhD thesis. Nagoya University, Nagoya, Japan, 1988.

Komiyama, J., Abe, Y., Suzuki, S., and Nakanishi, H. Polarities of GaN films and 3C-SiC intermediate layers grown on (111) Si substrates by MOVPE. *Journal of Crystal Growth* 298 (2007): 223–227.

Komiyama, J., Abe, Y., Suzuki, S., Nakanishi, H., and Koukitu, A. MOVPE of AlN-free hexagonal GaN/cubic SiC/Si heterostructures for vertical devices. *Journal of Crystal Growth* 311 (2009): 2840–2843.

Kondo, H., Koide, N., Honda, Y., Yamaguchi, M., and Sawaki, N. Series resistance in n-GaN/AlN/n-Si heterojunction structure. *Japanese Journal of Applied Physics* 45 (2006): 4015–4017.

Krost, A. and Dadgar, A. GaN based device on Si. *Physica Status Solidi (a)* 194 (2002): 361–375.

Lahreche, H., Nataf, G., Feltin, E., Beaumont, B., and Gibart, P. Growth of GaN on (111) Si: A route towards self-supported GaN. *Journal of Crystal Growth* 231 (2001): 329–334.

Lee, K.J., Shin, E.H., and Lim, K.Y. Reduction of dislocations in GaN epilayers grown on Si (111) substrate using Si_xN_y inserting layer. *Applied Physics Letters* 85 (9) (2004): 1502.

Lee, S.-J., Bak, G.H., Jeon, S.-R., Lee, S.H., Kim, S.-M., Jung, S.H., Lee, C.-R., Lee, I.-H., Leem, S.-J., and Baek, J.H. Epitaxial growth of crack-free GaN on patterned Si(111) substrate. *Japanese Journal of Applied Physics* 47 (2008): 3070–3073.

LeLouarn, A., Vezian, S., Semond, F., and Massies, J. AlN buffer layer growth for GaN epitaxy on (111) Si: Al or N first? *Journal of Crystal Growth* 311 (2009): 3278–3284.

Li, J., Lin, J.Y., and Jiang, H.X. Growth of III-nitride photonic structures on large area silicon substrates. *Applied Physics Letters* 88 (2006): 171909.

Liang, C.-T., Chen, K.Y., Chen, N.C., Chang, P.H., and Chang, C.-A. $Al_{0.15}Ga_{0.85}N/$GaN high electron mobility transistor structures grown on p-type Si substrates. *Applied Physics Letters* 89 (2006): 132107.

Lin, L.-H., Han, S.-S., Chen, K.-M., Zhang, Z.-Y., Chen, K.Y., Huang, J.Z., Sun, Z.-H., Liang, C.-T., Chen, N.C., Chang, P.H., and Chang, C.-A. Correlation field analysis of magnetoresistance of GaN/AlGaN heterostructure grown on Si substrate. *Japanese Journal of Applied Physics* 47 (2008): 4623–4625.

Liou, B.W. High photovoltaic efficiency of $In_xGa_{1-x}N$/GaN-based solar cells with a multiple-quantum-well structure on SiCN/Si(111) substrates. *Japanese Journal of Applied Physics* 48 (2009): 072201.

Liu, L. and Edgar, J.H. Substrate for gallium nitride epitaxy. *Material Science and Engineering* R37 (2002): 61–127.

Liu, R., Ponce, F.A., Dadgar, A., and Krost, A. Atomic arrangement at the AlN/Si (111) interface. *Applied Physics Letters* 83 (2003): 860–862.

Liu, Z., Wang, X., Wang, J., Hu, G., Guo, L., Li, J., Li, J., and Zeng, Y. Effects of buffer layers on the stress and morphology of GaN epilayer grown on Si substrate by MOCVD. *Journal of Crystal Growth* 298 (2007): 281–283.

Lu, Y., Liu, X., Lu, D.-C., Yuan, H., Chen, Z., Fan, T., Li, Y., Han, P., Wang, X., Wang, D., and Wang, Z. Investigation of GaN layer grown on Si(111) substrate using an ultrathin AlN wetting layer. *Journal of Crystal Growth* 236 (2002): 77–84.

Lu, Y., Liu, X., Wang, X., Lu, D.-C., Li, D., Han, X., Cong, G., and Wang, Z. Influence of the growth temperature of the high-temperature AlN buffer on the properties of GaN grown on Si (111) substrate. *Journal of Crystal Growth* 263 (2004): 4–11.

Mastro, M.A., Holm, R.T., Bassim, N.D., Eddy Jr. C.R., Henry, R.L., Twigg, M.E., and Rosenberg, A. Wurtzite III–nitride distributed Bragg reflectors on Si(100) substrates. *Japanese Journal of Applied Physics* 45 (2006): L814–L816.

Mo, C., Fang, W., Pu, Y., Liu, H., and Jiang, F. Growth and characterization of InGaN blue LED structure on Si(111) by MOCVD. *Journal of Crystal Growth* 285 (2005): 312–317.

Moram, M.A., Kappers, M.J., Joyce, T.B., Chalker, P.R., Barber, Z.H., and Humphreys, C.J. Growth of dislocation-free GaN islands on Si(111) using a scandium nitride buffer layer. *Journal of Crystal Growth* 308 (2007): 302–308.

Ni, X., Wu, M., Lee, J., Li, X., Baski, A.A., Özgür, Ü., and Morkoç, H. Nonpolar m-plane GaN on patterned Si (112) substrates by metalorganic chemical vapor deposition. *Applied Physics Letters* 95 (2009): 111102.

Nikishin, S.A., Faleev, N.N., Antipov, V.G., Francoeur, S., Grave de Peralta, L., Seryogin, G.A., Temkin, H., Prokofyeva, T.I., Holtz, M., and Chu, S.N.G. High quality GaN grown on Si(111) by gas source molecular beam epitaxy with ammonia. *Applied Physics Letters* 75 (1999): 2073.

Phillips, A. and Zhu, D. UK cracks GaN-on-silicon LEDs. *Compound Semiconductor*, March 2009: 19.

Ploog, K.H. Materials engineering of advanced device structures by heteroepitaxy of large misfit systems. *Materials Science in Semiconductor Processing* 4 (2001): 451–457.

Poschenrieder, M., Schulze, F., Blasing, J., Dadgar, A., Diez, A., Christen, J., and Krost, A. Bright blue to orange photoluminescence emission from high-quality InGaN/GaN multiple-quantum-wells on Si (111) substrates. *Applied Physics Letters* 81 (2002a): 1591.

Poschenrieder, M., Fehse, K., Schulz, F., Blasing, J., Witte, H., Krtschil, A., Dadgar, A., Diez, A., Christen, J., and Krost, A. MOCVD-grown InGaN/GaN MQW LEDs on Si(111). *Physica Status Solidi (c)* 0 (2002b): 267–271.

Raghavan, S., Weng, X., Dickey, E., and Redwing, J.M. Effect of AlN interlayers on growth stress in GaN layers deposited on (111) Si. *Applied Physics Letters* 87 (2005): 142101.

Reiher, A., Blasing, J., Dadgar, A., Diez, A., and Krost, A. Efficient stress relief in GaN heteroepitaxy on Si(111) using low-temperature AlN interlayers. *Journal of Crystal Growth* 248 (2003): 563–567.

Reiher, F., Dadgar, A., Blasing, J., Wieneke, M., Muller, M., Franke, A., Reimann, L., Christen, J., and Krost, A. InGaN/GaN light-emitting diodes on Si(110) substrates grown by metal–organic vapour phase epitaxy. *Journal of Physics D: Applied Physics* 42 (2009): 055107.

Saengkaew, P., Dadgar, A., Blaesing, J., Hempel, T., Veit, P., Christen, J., and Krost, A. Low-temperature/high-temperature AlN superlattice buffer layers for high-quality $Al_xGa_{1-x}N$ on Si (111). *Journal of Crystal Growth* 311 (2009): 3742–3748.

Sasaki, H., Kato, S., Matsuda, T., Sato, Y., Iwami, M., and Yoshida, S. Investigation of surface defect structure originating in dislocations in AlGaN/GaN epitaxial layer grown on a Si substrate. *Journal of Crystal Growth* 298 (2007): 305–309.

Sawaki, N., Honda, Y., Kameshiro, N., Koide, N., and Tanaka, S. Fabrication of AlGaN/GaN hetero-structures on a (1–101) face of GaN grown on a patterned 7 degree of axis (001) Si substrate. *Institute of Physics Conference Series* 171, CD-ROM Edition D21. 2002.

Sawaki, N., Hikosaka, T., Koide, N., Tanaka, S., Honda, Y., and Yamaguchi, M. Growth and properties of semi-polar GaN on a patterned silicon substrate. *Journal of Crystal Growth* 311 (2009): 2867–2874.

Schenk, H.P.D., Feltin, E., Laugt, M., Tottereau, O., Vennegues, P., and Dogheche, E. Realization of waveguiding epitaxial GaN layers on Si by low-pressure metalorganic vapor phase epitaxy. *Applied Physics Letters* 83 (2003): 5139.

Soga, T., Hattori, S., Sakai, S., Takeyasu, M., and Umeno, M. Characterization of epitaxially grown GaAs on Si substrate with III-V compounds intermediate layers by metalorganic chemical vapor deposition. *Journal of Applied Physics* 57 (1985): 4578–4582.

Song, J.-H., Chikyow, T., Yoo, Y.-Z., Ahmet, P., and Koinuma, H. Epitaxial growth of the wurtzite (11–20) AlN thin films on Si(100) with MnS buffer layer. *Japanese Journal of Applied Physics* 41 (2002): L1291–L1293.

Stevens, K.S., Kinniburgh, M., and Beresford, R. Photoconductive ultraviolet sensor using Mg-doped GaN on Si(111). *Applied Physics Letters* 66 (1995): 3518.

Strittmatter, A., Krost, A., Straburg, M., Turck, V., Bimberg, D., Blasing, J., and Christen, J. Low-pressure metal organic chemical vapor deposition of GaN on silicon (111) substrates using an AlAs nucleation layer. *Applied Physics Letters* 74 (1999): 1242.

Strittmatter, A., Bimberg, D., Krost, A., Blasing, J., and Veit, P. Structural investigation of GaN layers grown on Si(111) substrates using a nitridated AlAs buffer layer. *Journal of Crystal Growth* 221 (2000): 293–296.

Takemoto, K., Murakami, H., Iwamoto, T., Matsuo, Y., Kangawa, Y., Kumagai, Y., and Koukitu, A. Growth of GaN directly on Si(111) substrate by controlling atomic configuration of Si surface by metalorganic vapor phase epitaxy. *Japanese Journal of Applied Physics* 45 (2006): L478–L481.

Tanaka, S., Kawaguchi, Y., Sawaki, N., Hibino, M., and Hiramatsu, K. Defect structure in selective area growth GaN pyramida on (111) Si substrate. *Applied Physics Letters* 76 (2000): 2701–2703.

Tanaka, S., Honda, Y., Kameshiro, N., Iwasaki, R., Sawaki, N., and Tanji, T. Transmission electron microscopy study of the microstructure in selective-area-grown GaN and an AlGaN/GaN heterostructure on a 7-degree off-oriented (001) Si substrate. *Japanese Journal of Applied Physics* 41 (2002): L846–L848.

Tanikawa, T., Hikosaka, T., Honda, Y., Yamaguchi, M., and Sawaki, N. Growth of semi-polar (11–11)GaN on a (113)Si substrate by selective MOVPE. *Physica Status Solidi (c)* 5 (2008a): 2966–2968.

Tanikawa, T., Hikosaka, T., Honda, Y., Yamaguchi, M., and Sawaki, N. Growth of non-polar (11–20)GaN on a patterned (110)Si substrate by selective MOVPE. *Journal of Crystal Growth* 310 (23) (2008b): 4999–5002.

Tanikawa, T., Kagohashi, Y., Honda, Y., Yamaguchi, M., and Sawaki, N. Reduction of dislocations in a (11–22)GaN grown by selective MOVPE on (113) Si. *Journal of Crystal Growth* 311 (2009): 2879–2882.

Tran, C.A., Osinski, A., Karlicek Jr. R.F., and Berishev, I. Growth of InGaN/GaN multiple-quantum-well blue light-emitting diodes on silicon by metalorganic vapor phase epitaxy. *Applied Physics Letters* 75 (1999): 1494.

Tripathy, S., Lin, V.K.X., Teo, S.L., Dadgar, A., Diez, A., Bläsing, J., and Krost, A. InGaN/GaN light emitting diodes on nanoscale silicon on insulator. *Applied Physics Letters* 91 (2007): 231109.

Tripathy, S., Sale, T.E., Dadgar, A., Lin, V.K.X., Zang, K.Y., Teo, S.L., Chua, S.J., Blasing, J., and Krost, A. GaN-based microdisk light emitting diodes on (111)-oriented nanosilicon-on-insulator templates. *Journal of Applied Physics* 104 (2008): 053106.

Ubukata, A., Ikenaga, K., Akutsu, N., Yamaguchi, A., Matsumoto, K., Yamazaki, T., and Egawa, T. GaN growth on 150-mm-diameter (111) Si substrates. *Journal of Crystal Growth* 298 (2007): 198–201.

Uen, W.-Y., Li, Z.-Y., Lan, S.-M., and Liao, S.-M. Epitaxial growth of high-quality GaN on appropriately nitridated Si substrate by metal organic chemical vapor deposition. *Journal of Crystal Growth* 280 (2005): 335–340.

Vezian, S., Le Louarn, A., and Massies, J. Selective epitaxial growth of AlN and GaN nanostructures on Si(111) by using NH_3 as nitrogen source. *Journal of Crystal Growth* 303 (2007): 419–426.

Wakahara, A., Oishi, H., Okada, H., Yoshida, A., Koji, Y., and Ishida, M. Organometallic vapor phase epitaxy of GaN on Si (111) with a g-Al_2O_3 (111) epitaxial intermediate layer. *Journal of Crystal Growth* 236 (2002): 21–25.

Wang, D., Yoshida, S., and Ichikawa, M. Effect of Si doping on the growth and microstructure of GaN grown on Si(111) using SiC as a buffer layer. *Journal of Crystal Growth* 242 (2002): 20–28.

Watanabe, A., Takeuchi, T., Hirosawa, K., Amano, H., Hiramatsu, K., and Akasaki, I. The growth of single crystalline GaN on a Si substrate using AlN as an intermediate layer. *Journal of Crystal Growth* 128 (1993): 391.

Weeks, W. Jr., Piner, E.L., Gehrke, T., and Linthicum, K.J. Gallium nitride materials and methods. US Patent 6,617,060, 2003.

Weng, X., Raghavan, S., Acord, J.D., Jain, A., Dickey, E.C., and Redwing, J.M. Evolution of threading dislocations in MOCVD-grown GaN films on (111) Si substrates. *Journal of Crystal Growth* 300 (2007): 217–222.

Wu, C.-L., Wang, J.-C., Chan, M.-H., Chen, T.T., and Gwo, S. Heteroepitaxy of GaN on Si (111) realized with a coincident-interface AlN/b-Si_3N_4 (0001) double-buffer structure. *Applied Physics Letters* 83 (22) (2003): 4530.

Xiong, C., Jiang, F., Fang, W., Wang, L., Liu, H., and Mo, C. Different properties of GaN-based LED grown on Si(111) and transferred onto new substrate. *Science in China: Series E Technological Sciences* 49 (2006): 313–321.

Yablonskii, G.P., Lutsenko, E.V., Pavlovskii, V.N., Zubialevich, V.Z., Gurskii, A.L., Kalisch, H., Szymakowskii, A., Jansen, R.A., Alam, A., Dikme, Y., Schineller, B., and Heuken, M. Luminescence and stimulated emission from GaN on silicon substrates heterostructures. *Physica Status Solidi (a)* 192 (2002): 54–59.

Yang, J.W., Lunev, A., Simin, G., Chitnis, A., Shatalov, M., Asif Khan, M., Van Nostrand, J.E., and Gaska, R. Selective area deposited blue GaN-InGaN multiple-quantum well light emitting diodes over silicon substrates. *Applied Physics Letters* 76 (2000): 273.

Yang, T.H., Ku, J.T., Chang, J.-R., Shen, S.-G., Chen, Y.-C., Wong, Y.Y., Chou, W.C., Chen, C.-Y., and Chang, C.-Y. Growth of free-standing GaN layer on Si (111) substrate. *Journal of Crystal Growth* 311 (2009a): 1997–2001.

Yang, M., Ahn, H.S., Tanikawa, T., Honda, Y., Yamaguchi, M., and Sawaki, N. Maskless selective growth of semi-polar (11–22) GaN on Si (311) substrate by metal organic vapor phase epitaxy. *Journal of Crystal Growth* 311 (2009b): 2914–2918.

Yoshida, S., Li, J., Takehara, H., and Wada, T. Formation of thin GaN layer on Si (111) for fabrication of high-temperature metal field effect transistors (MESFETs). *Journal of Crystal Growth* 253 (2003): 85–88.

Zhang, B.J., Egawa, T., Ishikawa, H., Nishikawa, N., Jimbo, T., and Umeno, M. InGaN multiple-quantum-well light emitting diodes on Si(111) substrates. *Physica Status Solidi (a)* 188 (2001): 151–154.

Zhang, B., Egawa, T., Ishikawa, H., Liu, Y., and Jimbo, T. High-bright InGaN multiple-quantum-well blue light-emitting diodes on Si (111) using AlN/GaN multilayers with a thin AlN/AlGaN buffer layer. *Japanese Journal of Applied Physics* 42 (2003): L226–L228.

Zhang, B.S., Wu, M., Liu, J.P., Chen, J., Zhu, J.J., Shen, X.M., Feng, G., Zhao, D.G., Wang, Y.T., Yang, H., and Boyd, A.R. Reduction of tensile stress in GaN grown on Si (111) by inserting a low-temperature AlN interlayer. *Journal of Crystal Growth* 270 (2004): 316–321.

Zhang, B., Egawa, T., Ishikawa, H., Liu, Y., and Jimbo, T. Thin-film InGaN multiple-quantum-well light-emitting diodes transferred from Si (111) substrate onto copper carrier by selective lift-off. *Applied Physics Letters* 86 (2005a): 071113.

Zhang, N.H., Wang, X.L., Zeng, Y.P., Xiao, H.L., Wang, J.X., Liu, H.X., and Li, J.M. The effect of the $Al_xGa_{1-x}N/AlN$ buffer layer on the properties of GaN/Si(111) film grown by NH_3-MBE. *Journal of Crystal Growth* 280 (2005b): 346–351.

4

New Technology Approaches

Armin Dadgar

CONTENTS

4.1 Development of GaN Growth Technology on Silicon

The first report of III-nitride growth on silicon dates back to the beginning of metalorganic vapor phase epitaxy in 1971 when Manasevit, Erdmann, and Simpson grew AlN on Si substrates [Manasevit 1971]. After a longer phase of inactivity, GaN growth on silicon started again as a niche topic, mostly motivated by the low cost and scalability of silicon and to a smaller portion by the possible integration with Si electronics. The latter also powered other III–V growth at the end of the 1980s to the early 1990s, which, except for tandem solar cells, was not a big success because of a dislocation density typically above 10^8 cm^{-2}, rarely lower and with single dislocations already destroying laser devices based on InP and GaAs.

At the beginning, GaN on silicon research was focused on suited seeding layers and reasonable good GaN layers grown on top of them and, although motivated by the advantages of a Si substrate, rarely related to device structures. In fact field effect transistor (FET) and light emitting diode (LED) devices on silicon were difficult to reach and the first LED devices presented by Guha and Bojarczuk from IBM [Guha 1998a, Guha 1998b] grown by molecular beam epitaxy (MBE) showed cracking. However, their device ruled out with a belief that blocked the development of GaN on silicon: Si outdiffusion from the substrate and with it the impossibility of p-type doping. Their successful p-type doping demonstrated that this is not a problem and a year later secondary ion mass spectroscopy (SIMS) measurements by Strittmatter et al. demonstrated that also for metalorganic vapor phase epitaxy (MOVPE) grown GaN on silicon outdiffusion is not a topic [Strittmatter 1999]. However, despite a strong increase in GaN on silicon research around 2000, it was soon realized that all layers grown in excess of 0.5–1 µm by MOVPE, and in the best case above 2–3 µm in MBE [Semond 2001] show cracking. Consequently, devices demonstrated, if free of cracks, were usually around or below 1 µm in thickness when grown by MOVPE. Cracking of GaN on silicon originates in tensile stress which is either already generated during growth or, often and to a bigger part, originating in the large mismatch in thermal expansion coefficients between silicon and GaN. While for some devices, a thin-layer sequence may be in principle sufficient for device operation, thicker layers usually show better surface morphology, a lower threading dislocation density, and give more freedom in device design. Therefore, technologies to avoid cracks for GaN on silicon are crucial to achieve device quality layers. The first successful step in this direction was done by Dadgar et al. with the insertion of low temperature (LT) AlN interlayers [Dadgar 2000].

But cracking also depends on material quality. It is usually observed that crack avoiding strain engineering is more efficient when a higher material quality is present [Dadgar 2004, Baron 2009]. This is valid for the degree compressive stress is reduced during growth on top of strain engineering layers and consequently to some part also for the degree of compressive stress that can be applied on such a layer.

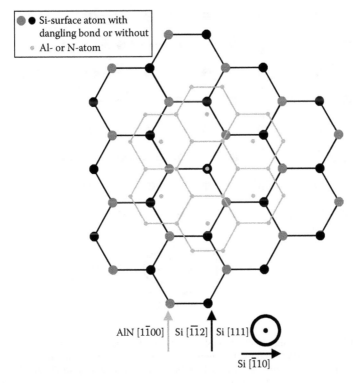

FIGURE 4.1
Possible arrangement of AlN molecules on the Si(111) surface. Only in the inner hexagon a close matching of atoms can be expected.

Most experiments on epitaxial growth on silicon were performed on Si(111) substrates. The reason for that is the matching of symmetries, as the sixfold symmetry of *c*-axis oriented GaN and the threefold symmetry of the silicon (111) surface, enabling only one matching rotational position of GaN on silicon (Figure 4.1) instead of four 90° rotated as for example a Si(001) surface, which will give rise to the growth of two non-coalescing domains (Figure 4.2). As shown in Section 4.3, the search for other substrate orientations can be useful to reduce mismatch and improve material quality.

In this chapter, we do not only present latest approaches for strain engineering and material enhancement but also older concepts, of which some are still used, for comparison.

4.2 Seeding Layers

Seeding layers grown by MOVPE on silicon require either a protecting surface on silicon if they are based on GaN, or a seeding layer different

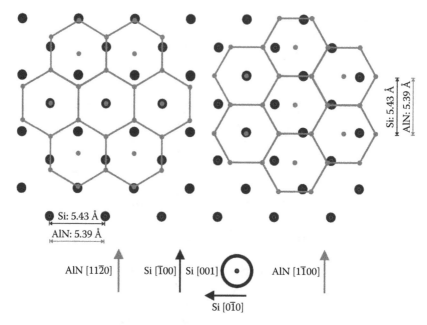

FIGURE 4.2
Theoretically possible alignment of GaN on Si(001). The surface structure with its fourfold symmetry allows two 90° rotated alignments with a relatively good matching (0.7%) for every third layer in AlN matching every fourth in Si for an alignment of AlN[11$\bar{2}$0] ‖ Si[$\bar{1}$00] and perpendicular to it. Although matching is good, this alignment is not observed in experiment (see Figure 4.5).

from GaN. The reason for this is the high reactivity of gallium with silicon under MOVPE growth conditions [Ishikawa 1998, Krost 2002]. The high growth temperatures in GaN MOVPE promote the formation of gallium–silicon–nitrogen alloys [Sunkara 2001]. Likely, the presence of hydrogen and ammonia are driving the subsequent formation of deep holes in the substrate and SiN as well as SiGaN-rich regions (Figure 4.3 bottom) [Krost 2002] also known as meltback etching reaction. Meltback etched surfaces usually look rough in the beginning (Figure 4.3 top left) and, depending on the stage of meltback etching, deep holes are formed in extreme cases. Sometimes, meltback etching results in liquid gallium droplets even at room temperature.

The easiest protection of the silicon surface to date can be given by AlN. In contrast to gallium, Al–Si alloys do not promote a similar reaction: Due to the high bonding energy of the Al–N bond, the formation of metallic aluminum is suppressed during growth. In reality, the degree of protection depends on the amount of craters or holes within the AlN layer, its thickness, and the duration and temperature of subsequent GaN growth. Meltback etching usually occurs if Ga diffuses to the Si surface, mostly this is observed when cracks, craters, etc. are present in the AlN layer and it can be concluded that

FIGURE 4.3
Meltback etching in its different appearances: The typical early meltback etching reaction has a typical appearance with a brighter reaction front at the end of the etched region (top). For heavily etched layers, deep craters are found. The rough surface is likely due to high Si concentrations in the remaining layer (center left). Such reactions can also cause 3D bush or treelike structures (center right). An EDX analysis performed in TEM on a meltback etched region shows that Ga- and Si-rich as well as SiN-rich regions exist within the destroyed layer.

Ga then diffuses through these defects of the AlN layer to the Si surface. Very often, meltback etching is observed in conjunction with small particles.

Alternatives to such a seeding layer directly on Si are protecting layers that are, e.g., SiC [Davis 2000], Al_2O_3, or other crystalline coatings of the Si surface.

When starting with a perfect oxide free surface finish of silicon, it is usually only an optimization task to achieve x-ray diffraction (XRD) full widths at half maximum (FWHMs) for the GaN(0002) and ($10\bar{1}0$) ω-scans below or close to $0.1°–0.2°$. However, surface preparation is often not taken care of. Many groups report on the simple approach from sapphire or SiC for cleaning the substrate: Heating to >1000°C under hydrogen. Under these conditions, hydrogen etches the surface and usually leads to epi-ready surfaces for GaN growth on sapphire and SiC. From our findings, however, heating and desorption is different for silicon. One reason might be that contaminants can absorb on the surface, e.g., from desorbing molecules within the reactor, and destroy the surface structure. This will make epitaxy impossible or at least lead to irregular results when growing. In addition, silicon surface roughening can be significant under hydrogen desorption conditions [Castaldini 2002, Gallois 2005].

Wet chemical substrate preparation prior to epitaxy [Grundmann 1991] usually gives stable growth results and such substrates can be stored for hours in dry nitrogen ambient as, e.g., a well-maintained glove box with low dew point. The HF or buffered HF etching step, however, can lead to surface roughening if it is too long [Adachi 1996, Tomita 2002]. From the manufacturing point of view, chemical preparation is an additional time-consuming pre-epitaxy step which does, however, reduce expensive growth time by removing the desorption step.

For AlN on Si(111), two options exist for the seeding layer, either a LT or a high-temperature (HT) layer. LT seeding layers have been found to be more critical with regard to the growth conditions. The growth temperature around 630°C surface temperature can be only varied a few tens of degrees to higher or lower temperatures. Especially toward higher temperatures, a region where seeding layer growth is usually not successful exists. The reasons for this behavior are presently unknown and difficult to investigate in MOVPE. A possible explanation might be surface reconstruction which undergoes a transition from 7×7 to 1×1 at least under ultra high vacuum (UHV) conditions in MBE at 830°C [Hellman 1998]. For MOVPE growth, Hageman et al. [Hageman 2001] reported that, within error of MOVPE temperature reading, their best LT-AlN seeding layers are grown below the transition temperature [Telieps 1985, Jona 1965]. But MOVPE growth usually takes place under H_2 carrier gas which might influence surface reconstruction and usually a 7×7 surface exposed to hydrogen, even at lower temperatures converts to a 1×1 surface [Owman 1994, Owman 1995]. Therefore, other mechanisms do likely alter the Si surface properties yielding low-quality GaN on Si(111) in a medium temperature regime. It is remarkable that for practically all Si surfaces HT-AlN seeding layers do result in c-axis oriented AlN buffers. Only for the Si(111) orientation high quality c-axis oriented LT-AlN seeding layers and with it GaN layers were reported so far. For other orientations like Si(001) polycrystalline or, in the best case, preferred r-planar growth [Schulze 2004] can be achieved with LT-AlN seeding layers. On all other orientations as Si(110) and other high index surfaces, usually only polycrystalline material can be achieved with LT-AlN seeding layers. A simple explanation might be the good rotational match of AlN on Si(111) even at LTs and the lack of this matching on all other substrates, which, in addition to the large lattice mismatch, leads to polycrystalline growth or the growth of inclined seeds in different rotational orientations instead, or of a single crystalline polarization reduced layer (see Section 4.5.2).

As for LT-AlN seeding layers HT AlN seeding layers can give excellent results. For all AlN seeding layers, mostly an AlN pre-deposition step is performed. In this step, the substrate surface is covered with Al around a monolayer in thickness. This step seems to prevent to some degree uncontrolled surface nitridation. SiN itself is an anti-surfactant and typically does not allow epitaxial growth but only polycrystalline nucleation on it.

However, a patent application by Nitronex [Nitronex 2005] claims that the formation of a thin SiN masking layer by ammonia pre-flow on the silicon substrate can result in a low dislocation density. The masking layer is thought to act as nanomask for nucleation and the subsequent growth of AlN will lead to larger seeds and improved material quality. The validity of the claims is difficult to proof but indeed ammonia pre-flow is possible at elevated temperatures under hydrogen ambient. The formation of SiN, which usually forms a stable layer on all uncovered silicon surfaces is diffusion limited with a final SiN layer thickness around a few nanometers [Ito 1978]. Furtmayr et al. reported on the formation of an amorphous Si_3N_4 layer on Si(111) in MBE, which is limited in thickness and on which c-axis oriented nanorods were grown [Furtmayr 2008]. In MOVPE, the HT hydrogen ambient and the hydrogen radicals from ammonia decomposition influence this process [Morita 2001] and likely delay the formation of a thick SiN layer. This is in agreement with the observation that at LTs ammonia pre-flow was never observed to yield good results, which can be explained by hydrogen etching not being significant at LTs, and a thin SiN layer is more likely to form. The main problem if a SiN interface layer exists is the question where seeding begins and consequently how the seed is oriented, especially how subsequent growth can lead to a well-oriented layer. A high resolution TEM image (Figure 4.4) shows such an amorphous region at the AlN/SiN interface. Being an average image of the whole specimen

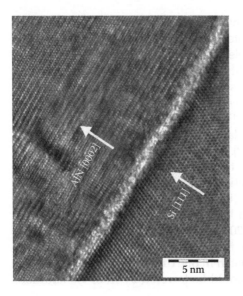

FIGURE 4.4
HR-TEM of an AlN Si(111) interface. The bright partially distorted region is due to the formation of an amorphous or heavily distorted layer at the interface.

any disturbance of the diffracted electron beam does destroy a perfect contrast. Therefore, an amorphous region or strongly misaligned region must exist within the thickness of the sample making the observation of misfit dislocations difficult. But, observing epitaxial growth, also crystalline regions free from amorphous interlayers enabling a well-aligned seeding must exist.

Although AlN is a well-established seeding layer, some authors report the usage of AlGaN [Honda 2007] and also AlInN [Irie 2009] mostly motivated by achieving a low series resistance for vertical contacting. Especially AlInN is in principle well suited as seeding layer due to the absence of a meltback etching reaction. But growth can only be performed at LTs which determine the In-content and the compound has a tendency for spinodal decomposition for In contents between ~15% and 85% [Ferhat 2002, Karpov 2004, Deibuk 2005] depending on thickness [Deibuk 2005, Hums 2007] and most likely also material quality. It has been reported by Irie et al. [Irie 2009] that such a seeding layer is especially suited for achieving a low series resistance across the Si/AlInN interface.

AlGaN growth is more difficult since at high Ga content meltback etching occurs very likely during MOVPE growth, but not in MBE. However, even small additions of Ga and also In might enhance the properties of AlN seeding layers already by enhancing diffusion and the formation of larger initial seeds on the silicon substrate.

4.3 Substrate Orientation

A simple approach for improvements in epitaxial growth is looking for a better lattice match to the substrate than for the commonly used Si(111) plane. However, for silicon, the simple approach of looking for a higher surface atom density, the only possibility for a perfect match, is not possible with Si(111) already having the highest surface atom density. Unfortunately, from all Si orientations the only one well fitting to the hexagonal structure of group-III-nitrides is the threefold Si(111) atom arrangement. Alternatives to that is the growth on coincidence site lattices which can result in significantly reduced lattice mismatch [Krost 2000]. Already for Si(111) an addition of, e.g., indium or small amounts of gallium to the seeding layer might enable this. Comparing possible growth orientations and mismatch on Si(111) (Figure 4.1), Si(001) (Figure 4.5), and Si(110) (Figure 4.6), Si(111) only offers a good local match but no matching over a larger distance even for InN no coincidence lattice is possible [Dadgar 2007a]. The high mismatch results in a high density of misfit dislocations at the Si/AlN interface [Liu 2004]. Si(001) theoretically offers a low 0.7% mismatch in one direction (see Figure 4.2) for every second AlN molecule which could be lowered by In or Ga addition.

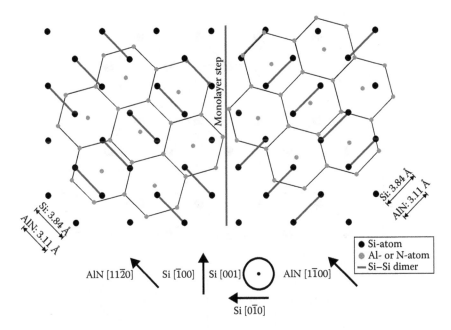

FIGURE 4.5

Alignment of AlN on Si(001) as observed in reality. The Si (2×1) reconstruction leads to the formation of Si dimer rows on the surface. The orientation of these dimers rotates by 90° for every monolayer, leading to two possible alignments of AlN on Si(001).

In the orthogonal direction, 15% are expected. However, this simple view does not work for the real Si(001) surface where surface reconstruction must be taken into account. The formation of a 2×1 surface reconstruction with dimers aligned by 45° to the orthogonal base of the Si substrate is typically observed under standard growth conditions in MBE and MOVPE [Lebedev 2001, Schulze 2006]. XRD measurements show that AlN is oriented with a rotation of 45°, with the m-plane parallel to the dimers, increasing the lattice mismatch to 19% and 30% and leading to an epitaxial relationship in growth direction of [0001] AlN || [001] Si and in plane either $\langle 01\bar{1}0 \rangle$ AlN || [110] Si or $\langle \bar{2}110 \rangle$ || [110]Si [Lebedev 2001]. Adding about 50% indium, a 3:2 coincidence site lattice exists in the AlN[$1\bar{1}00$] || Si[$\bar{1}\,\bar{1}0$] direction or, depending on the dimer orientation, perpendicular to it, however, growing such a high In content seeding layer is very difficult and has not yet been demonstrated. The main disadvantage for the growth on Si[001] is the similarity of two orthogonal growth orientations depending on the Si lattice plane the AlN seeding layer grows on: For [001] oriented surfaces in the zinc-blende lattice, the surface orientation of the Si surface bonds changes every monolayer and with it dimer orientation [Schwarzentruber 1993]. Substrates, even if nominally exactly oriented, usually show a sufficiently high density of steps which will form alternating dimer rows and hinder monocrystalline GaN growth across device relevant distances. For example, the typical

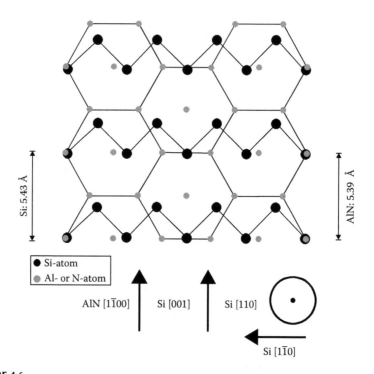

FIGURE 4.6

Alignment of AlN on Si(110). A low relatively mismatch of 0.7% exists for every second plane in the AlN [1$\bar{1}$00] ∥ [001] Si direction. In the perpendicular direction, lattice mismatch is comparable to that with Si(111).

orientation of silicon is only exact with an error of ±0.5°. If the substrate is manufactured with 0.5° misorientation, this can lead to monolayer steps every 31 nm. For a difficult to manufacture accuracy within ±0.1°, the distance is increased for a 0.1° misorientation to about 155 nm. Forcing double-layer steps on the Si surface by misorienting the substrate by a few degrees only one orientation exists due to the formation of double-layer steps. For a 4° misorientation, the monolayer step width is about 3.5 nm, but usually energetically favorable double-layer steps are formed for such large misorientations. The success of this off orientation to achieve single crystalline layers has been first demonstrated in MBE by Lebedev et al. in 2001 [Lebedev 2001]. In MOVPE processes, the high growth temperature does reduce the ordering of these steps and only a slightly elevated density of 62% for one orientation remains [Aumann 1992]. Nevertheless, this is sufficient to grow single crystalline GaN [Joblot 2005a,b, Schulze 2006], although Al-rich seeding and buffer layers usually show both orientations [Schulze 2006]. Figure 4.7 shows the impact of an increasing off orientation on the crystallinity of a 400 nm thick GaN layer determined by electron backscattering diffraction (EBSD) measurements [Schulze 2007b]. The two gray levels denote the

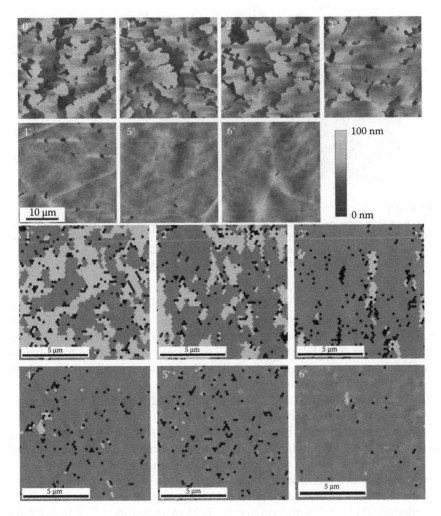

FIGURE 4.7
AFM and EBSD images of GaN on Si(001) samples with increasing off-orientation angle. The gray levels in the EBSD images is correlated to the in-plane orientation of the GaN crystallites showing two by 30° misoriented types of orientation, with increasing off-orientation one in-plane orientation dominates. With further increasing layer thickness, the samples are single crystalline for off-orientations ≥4°. The surface morphology in the AFM image correlates well with the amount of misoriented crystallites. (Adapted from Schulze, F. et al., *J. Cryst. Growth*, 299, 399, 2007. © 2007 by Elsevier.)

two orientations of the crystal at the surface. With increasing off orientation only one of these orientations remains. Due to an imperfect orientation and an elevated twist originating in a high lattice mismatch of 19% and 30% GaN layers grown on Si(001) are still inferior to those grown on Si(111). The high twist can be seen in TEM where the single columns are nearly defect free but twisting leads to a different contrast of those columns (Figure 4.8)

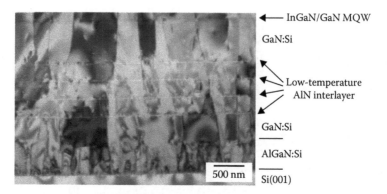

InGaN/GaN MQW

GaN:Si

Low-temperature
AlN interlayer

GaN:Si

AlGaN:Si

500 nm

Si(001)

FIGURE 4.8

TEM cross-section image of an LED structure grown by MOVPE on Si(001). GaN growth columns in the upper part of the structure can be distinguished by their different brightness, originating in a brad variation in twist. (Reprinted from Schulze, F. et al., *Phys. Stat. Sol. (c)*, 4, 41, 2007. With permission from Wiley VCH, © 2007 by Wiley-VCH.)

[Schulze 2007a]. In this image also a higher distortion in the AlGaN buffer is visible. It has been also observed that GaN grows with different inclination angles on the two possible surface orientations [Reiher 2009]. The microscopic origin of this observation is still not clear but similar to this also for GaN/AlN on Si(111) surfaces a ~1° inclination of the grown layer toward the Si surface normal has been observed.

In contrast to Si(001) for Si(110) the situation is different. Here, the atomic view seems to be valid and AlN grows in a 2:1 coincidence site lattice with a relatively small lattice mismatch of 0.7% along the AlN [1$\overline{1}$00] direction being parallel to Si[001]. Perpendicular to that lattice mismatch is equivalent to that of Si(111) but even here a nearly matched 4:1 coincidence site lattice exists. The lower mismatch is indeed already sufficient to reduce twist and tilt. This has been observed for a comparison of Si(111), Si(110), and other high index orientations (Figure 4.9) and proven by TEM investigations [Contreras 2008, Dadgar 2007a, Liu 2004, Reiher 2009].

The successful single crystalline growth on Si(110) also reveals simple rules for suited surfaces to obtain single crystalline GaN layers. Since the hexagonal sixfold symmetry of GaN does not fit to the fourfold symmetry of Si or does result in different growth domains only Si orientations with surface symmetries lower than four are suited for single crystalline growth. Such surfaces are Si(111) (threefold), Si(110) (twofold), and many high index surfaces. Indeed higher index orientations presented in [Reiher 2009, Reiher 2010] show single crystalline GaN. But all of them were grown with a HT AlN seeding layer. LT seeding layers, up to now, could not be successfully applied to any surface different from Si(111) to achieve *c*-axis oriented GaN.

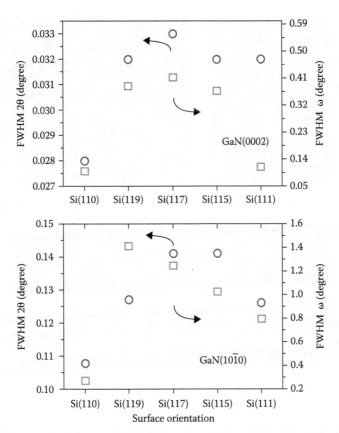

FIGURE 4.9
XRD data of identical Si(*hkl*)/HT-AlN/GaN layer sequences. Best data could be obtained for the growth on Si(110) followed by significantly higher values by Si(111). The good quality on Si(110) is assumed to be originating in the better lattice match of AlN on this surface. (Reprinted from Reiher, F. et al., *J. Cryst. Growth*, 312, 180, 2010. With permission from Elsevier, © 2010 by Elsevier.)

4.4 Si Surface Modification and Texturing

4.4.1 Patterning

A simple surface modification method is surface patterning. By this method, the lateral size of the GaN layer is reduced. The method was first presented for the growth of GaN on silicon for a few micron large dots by Kawaguchi et al. in 1998 [Kawaguchi 1998] and in device relevant size by Yang et al. in 2000 which showed an LED structure grown in that manner [Yang 2000]. But

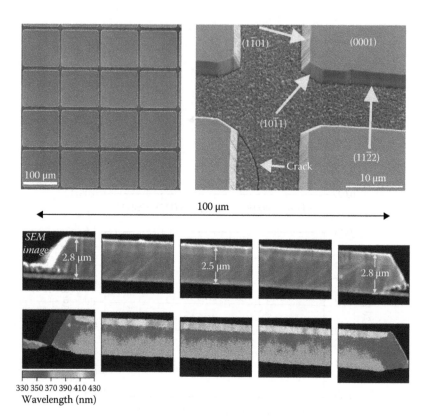

FIGURE 4.10

A LED structure grown on $100 \times 100\,\mu m^2$ patterned Si substrates. Patterning was performed by a SiN mask on a thin GaN buffer layer. The image on the top right shows that cracks are present but propagating solely in the masked region between the LED structure. The cross-sectional SEM image (center) shows that a growth enhancement is observed at the edges due to material transport from the masked region. The lower image documents the enhancement in material quality where defect luminescence (lighter gray) dominates in the lower part and GaN (D^0,X) correlated luminescence in the upper part darker gray before the InGaN MQW luminescence in the top (bright region at the top).

only a reduction in crack density was achieved, cracks were still present. A crack-free LED was presented for $100 \times 100\,\mu m^2$ fields in 2001 [Dadgar 2001a, Dadgar 2001b, Dadgar 2003b] (Figure 4.10). A detailed analysis revealed that cracks were present but only in the Si substrate where the strain fields of the GaN islands induced a high tensile stress. For some of these GaN islands, even a delamination from the silicon substrate was observed.

Although for smaller device sizes patterning offers an alternative when interlayers should be avoided or reduced it leads to problems, especially in processing. There the main difficulty is the growth enhancement at the edges of the patterned field which, depending on the mask width, can give rise to a strong thickness variation (Figure 4.10 bottom). Such a thickness enhancement was also observed by Honda et al. who grew 1.5 μm thick structures in $200 \times 200\,\mu m^2$ fields with good material properties [Honda 2002].

Zamir et al. presented a study on different field sizes [Zamir 2001a, Zamir 2001b, Zamir 2002]. They found that cracks depend on a critical field size. However, they also found, when their etched structures were overgrown with only 700 nm of GaN $14 \times 14\,\mu m^2$ field size is the upper limit to avoid cracking and presented calculations supporting this finding. This is surprising, because such thicknesses can be grown crack-free or nearly crack-free on planar substrates. Cracking on such fields would correspond to a crack distance of only ~7 μm on a planar substrate.

It is most likely that the seeding layer quality was poor causing high tensile stress already during growth and cracks upon cooling. Unfortunately, no information is given on the material quality only an enhancement in PL intensity, which can be already explained by the improved light outcoupling for laterally guided light.

In recent years, patterning was not widely used. Main reasons likely are the requirement of preprocessing and growth artifacts as thickness enhancements at the edges, which can cause difficulties in processing. But the method has a big potential: Latticepower from China announced in 2009 that they grow high brightness LEDs on patterned Si substrates which are transferred to a reflective carrier substrate. They grow 4 μm of GaN on a strain engineering AlGaN SL structure in patterns up to $1\,mm^2$ [Jiang 2009]. For smaller pattern sizes, a crack-free yield of 97% was claimed. Such layer thicknesses of 4 μm without, e.g., a strain engineering LT-AlN layer, were not reported yet. These thicknesses will add additional freedom in device design.

4.4.2 SiC/Si

A protecting layer of silicon to avoid meltback etching and possibly also a better lattice match can be formed by SiC. One motivation to use SiC on Si is the lower cost of such substrates in contrast to SiC and the benefit of a low lattice mismatch and consequently high-quality layers. Takeuchi et al. proposed the use of SiC on Si [Takeuchi 1991] to achieve high-quality GaN layers. The SiC layer can be in principle manufactured by several methods as transformation [Zorman 1995, Steckl 1996, Davis 2001a, Davis 2001b], growth [Liaw 2000, Park 2001, Kang 2001], implantation and annealing [Brink 2004, Sari 2008], or wafer bonding. Main disadvantage is that with most techniques the SiC formed is of the cubic 3C-SiC modification and not of the commonly used 4H or 6H modifications. But the surface structure of 3C-SiC is similar to that of Si(111) on which (0001) oriented GaN growth is preferred. The different structure does, however, also cause defects as stacking faults originating at surface steps when an Al(Ga)N seeding layer is grown on top of the SiC. Many publications on this technique were made around 2000 but the quality of GaN on such layers was not superior to that grown directly on Si, although often of good quality. For such layers cracking is still present, and

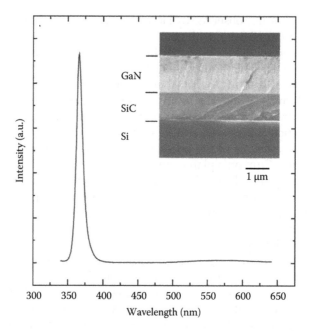

FIGURE 4.11
PL and cross-section SEM image (inset) of a GaN layer grown directly on SiC(111) on Si(111).
(Reprinted from Komiyama, J. et al., *J. Cryst. Growth*, 311, 2840, 2009. With permission from
Elsevier, © 2009 by Elsevier.)

the SiC transformation processes commonly used these days was difficult to
control. Surface roughness is a major difficulty and etching and polishing an
important topic.

Nevertheless, a process enabling high-quality SiC with good structural
and morphological properties can be of use for GaN on Si devices. One can
expect a reduced barrier for vertically contacted devices if the use of AlN
as seeding layer can be avoided. Such a structure has been presented by
Komiyama et al. [Komiyama 2009]. They report the direct growth of GaN on
SiC/Si (Figure 4.11) but have not yet presented electrical or structural data,
except that they obtained a smooth single crystalline GaN layer.

4.4.3 Oxide Layers

Sapphire is a well-established substrate for GaN epitaxy. On silicon, a sapphire quasi-substrate has been employed by several authors. Here, the AlO
layer is fabricated on the Si surface either by oxidation of, e.g., AlAs [Kobayashi
1997] or an AlAs/GaAs structure used for epitaxial lateral overgrowth (ELO)
[Kobayashi 1998] or growth of an Al_2O_3 layer [Fenwick 2009]. The oxidized
layer offers the growth of GaN also with standard GaN seeding layers but it
is reported that the oxide layer is amorphous which is in contradiction to the

results which suggest an epitaxial growth, however, of poor-to-average quality. In the work of Kobayashi et al., the method does yield better results than for a reference sample, but is not comparable to GaN on sapphire.

Fenwick et al. presented GaN grown on 5–20 nm thin Al_2O_3 layers deposited by atomic layer deposition on Si(111) [Fenwick 2009]. Layers grown on such modified substrates with the Al_2O_3 layer likely being an amorphous layer as determined by XRD show an improved crystalline quality with best XRD FWHMs of the (0002) ω-scan of 378 arcsec for a structure with a 5 nm Al_2O_3 layer and 1.5 μm of GaN. The crack density of the layer was significantly reduced for the layers on the Al_2O_3/Si substrates that all have an improved crystalline quality as determined by XRD.

4.4.4 III-As Seeding Layers

The growth of GaAs on top of silicon is already known for 20 years. In the late 1980s, a lot of effort was applied to merge silicon electronics with optoelectronics and the growth of III–Vs on Si was believed to be the solution for this. However, most III–Vs on silicon showed too high dislocation densities for device operation, especially for light emitters as laser devices. For a long time, the only outcome of this research effort has been the growth of GaAs on germanium for high efficiency solar cells. Although the high dislocation density was fatal for most III–V devices on silicon, the dislocation density was lower and overall material quality was much better than for GaN on sapphire or SiC.

Thus, a GaAs or AlAs seeding is in principle interesting for GaN growth. In fact, even more interesting would be a seeding layer based on GaP or AlP which have a closer lattice match to silicon, but the stability of the III-phosphides against nitridation and the low stability of GaP and AlP under GaN growth conditions in MOVPE hinders the application of these layers as seeding layers for GaN growth.

GaAs and AlAs on the other hand can be easily transformed to GaN or AlN, the latter is better suited for GaN growth due to the blocking behavior for meltback etching. Such layers are usually grown after covering the Si surface with arsenic using AsH_3 at LTs. Then AlAs or GaAs is grown in a thickness between 10 and 100 nm. The critical step is heating and transforming AlAs or GaAs to AlN or GaN, respectively, in a reactive nitrogen ambient. Here, the AlAs and GaAs layer should not decompose and lose its epitaxial relationship but arsenic being replaced by nitrogen. If the temperature is chosen too high, a decomposition will take place, however, a sufficiently HT is required to decompose ammonia and force the exchange reaction not only in the surface region of AlAs but in the whole layer.

A first report on such a transformed layer with subsequent GaN growth was by Yang et al. [Yang 1996]. They used a MBE grown GaAs layer transformed to GaN as starting point for AlN layer growth followed by GaN in MOVPE (Figure 4.12). They yielded good PL intensities but the method was

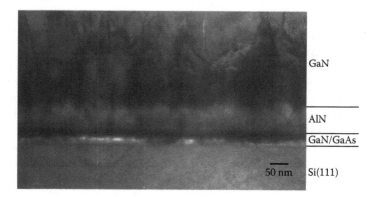

GaN

AlN

GaN/GaAs

50 nm   Si(111)

FIGURE 4.12
Cross-section TEM image of a GaN layer grown on Si with a converted GaAs layer. On this layer, which has been converted to GaN, first an AlN seeding layer was grown. (Reprinted from Yang, J.W. et al., *Appl. Phys. Lett.*, 69, 3566, 1996. With permission from American Institute of Physics, © 1996 by American Institute of Physics.)

not further published by the group. A risk of using GaAs as starting layer is that meltback etching occurs during HT MOVPE growth.

With AlAs, this risk is reduced. Such a layer was applied by Strittmatter et al. [Strittmatter 1999] for GaN growth (Figure 4.13).

A combination of GaAs with GaN was used by Gosh et al. to form a porous GaAs layer underneath the GaN layer [Gosh 2003]. Here, GaAs was grown by MBE followed by a GaN capping layer in MOVPE. Heating to GaN growth temperatures and nitrating leads to a porous GaAs layer underneath

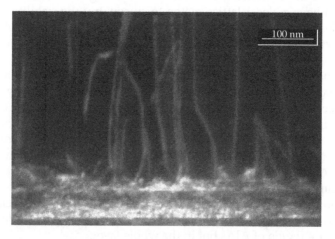

100 nm

FIGURE 4.13
TEM image of the interface region of Si(bottom), nitrided AlAs, and a GaN layer. A sharp AlN/Si interface is observed and most dislocations are annihilated in the seed layer. (Reprinted from Dadgar, A. et al., *Phys. Stat. Sol. (c)*, 0, 1583, 2003. With permission from Wiley VCH, © 2003 by Wiley-VCH.)

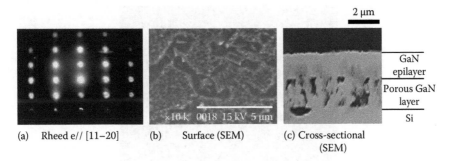

(a) Rheed e// [11–20]   (b) Surface (SEM)   (c) Cross-sectional
                                                      (SEM)

FIGURE 4.14
GaN grown by MOVPE on a MBE grown GaAs layer on Si(111). The overgrowth with GaN was started at LTs. Subsequent high-temperature GaN growth leads to a porous converted (GaAs → GaN) layer. (Reprinted from Ghosh, B.K. et al., *J. Cryst. Growth*, 249, 422, 2003. With permission from Elsevier, © 2003 by Elsevier.)

the GaN. If the GaAs is not capped, the layer is compact and tends to peel off. With an additional improved growth sequence using an additional GaN seeding step on the nitrated GaAs/GaN layer leads to a particle like converted GaAs layer but also a smoother surface (Figure 4.14). All layers on the porous layer, which are in excess of 1 μm in thickness, are reported to be crack-free and not peeling off.

Another approach is using BP which can be grown at GaN MOVPE temperatures thus can be expected to be relatively stable during GaN growth. Nishimura et al. demonstrated the growth of cubic GaN on such buffer layers on Si(001) using MOVPE [Nishimura 2001, Nishimura 2004]. Unfortunately, it was not stated at which temperature the GaN layer was grown to yield these results. BP is in principle, however, an interesting buffer. Being likely stable at high growth temperatures it enables the growth of GaN without an AlN buffer and by this a potentially better vertical contacting of devices. Also using other substrate orientations as, e.g., Si(110) it might be useful for nonpolar GaN layers. Unfortunately, only little data exists on the results, as, e.g., no information is given on XRD FWHMs of the GaN layer and if antiphase domain boundaries, typical for cubic compound semiconductors on exactly oriented Si(001), are present [Morizane 1977]. Cross-sectional SEM images showed a rough morphology [Nishimura 2001] for a GaN layer on BP on Si(001) (Figure 4.15) but a smooth BP layer on Si(001) in a preceding publication (Figure 4.16) [Nishimura 2004]. Whether roughening was due to the growth process in the first case or is related to an instability of BP under GaN growth atmosphere is not clear.

4.4.5 Porous Silicon

Porous silicon can be easily obtained by HF etching, depending on conduction type under illumination and/or an applied current, and is a well-established

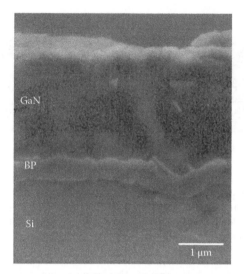

FIGURE 4.15
Cross-section SEM of GaN grown on BP. The interfaces are relatively rough. (Reprinted from Nishimura, S. and Terashima, K., *Mater. Sci. Eng.*, B82, 25, 2001. With permission from Elsevier, © 2001 by Elsevier.)

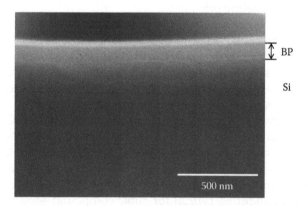

FIGURE 4.16
Cross-sectional SEM of BP grown on Si. The BP itself shows smooth interfaces. (Reprinted from Nishimura, S. et al., *Phys. Stat. Sol. (c)*, 1, 238, 2004. With permission from Wiley VCH, © 2004 by Wiley-VCH.)

process in silicon industry. Main difficulty for the growth on such substrates is the control of the surface structure. For epitaxy the surface should ideally consist of suited planar silicon surfaces for growth. For example, Si(111) when using a Si(111) substrate. Depending on the substrate type (orientation, doping) and etching conditions the formation of porous Si does not only lead to holes but also surface roughening which in turn can lead to difficulties in epitaxial growth. Additionally, such a structure will lead to a high thermal resistivity for devices. On the other hand, a high porosity might have some advantage in local strain relaxation, and when forming the right structure can be used to enhance the reflectivity of the substrate strongly, making it interesting for light emitters.

In the silicon industry, porous Si layers are often used for the fabrication of MEMS structures as, e.g., membranes. By heat treatment, hollows are formed under a thin Si membrane. These processes take place well below the Si melting point, thus it can be expected that also in MOVPE, with its high growth temperatures, the porous Si layer will undergo a transformation. Indeed this is often observed as will be shown in some of the following examples.

Likely due to the difficulties in manufacturing and the growth such porous Si layers suited for epitaxy only few reports exist. Missaoui et al. reported on the early stages of GaN growth on porous Si and reported the observation of polycrystalline GaN [Missaoui 2000, 2001, Boufaden 2004]. They also presented a thicker structure in excess of 1 μm in thickness which roughened significantly. Interestingly, they performed growth at LTs <800°C and started with GaN growth directly on Si without an AlN layer which is usually required at higher temperatures to avoid meltback etching. In XRD, some polycrystalline components were still detected and the surface was rough with an increasing roughness for elevated growth temperatures around 800°C.

Strain reduction using a porous Si(111) substrate was determined by micro Raman measurements by Ishikawa et al. (Figure 4.17), although the sample shows a few cracks in contrast to the growth on planar Si(111) [Ishikawa 2008]. But cracking does occur in contrast to the observation of a lower strain in these layers. An explanation might be the higher pit density of these layers on porous Si which might promote strain relaxation by cracking on these pits acting as predetermined breaking point. The porous Si layer was also reported to undergo structural changes during the MOVPE process (Figure 4.18). It was observed that annealing under hydrogen does not modify the structure but GaN growth. Nevertheless, a continuous layer with reduced tensile stress could be achieved.

A porous silicon layer on Si(111) and Si(001) was overgrown by depositing AlN with an evaporation and nitrating technique and subsequent growth of 500 nm of AlGaN and 1 μm of GaN [Bouzynin 2008]. On Si(001), they observe two differently aligned domains and when grown on porous Si a mixture of hexagonal and cubic GaN. On Si(111), the material is single phase on porous and nonporous surfaces.

Porosified silicon surfaces are certainly worth investigating but difficult to use. In principle, the etching process can be defined in a way where a larger porosity is created underneath a more compact surface region which will ease epitaxy on it. Since the process of porosifying depends on the skills and recipe of the lab and an exact classification does not exist a comparison of the results is even more complicated.

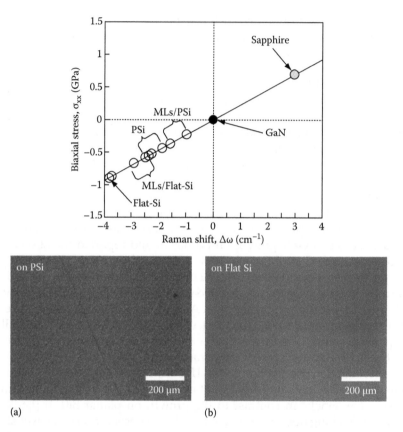

FIGURE 4.17
Stress for GaN layers grown on different substrates with two different buffer layers (top). The growth on porous Si (PSi) leads to a decreased tensile stress after growth in comparison to the growth on flat Si substrates for both cases with and without a strain reducing multilayer. (Reprinted from Ishikawa, H. et al., *J. Cryst. Growth*, 310, 4900, 2008. With permission from Elsevier, © 2008 by Elsevier.)

4.4.6 Nanostructure Growth

Nanostructure growth is one possible route to achieve an improvement in material quality (details on defect reduction are discussed in Section 4.7.5) or for new or enhanced device types. Different growth techniques enable the growth of nanowires that can be used for different applications including, e.g., light emitters, nano-electronics, or field emitters. Nanowires or -rods for LED applications are discussed in Chapter 12.

For the growth of nanowires different methods can be used:

1. Catalyst-based growth
2. Phase separation

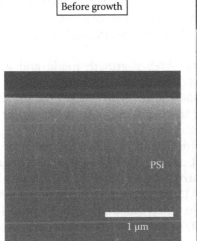
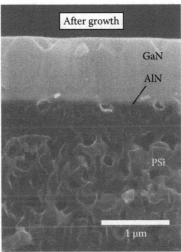

FIGURE 4.18
SEM images showing the change in the microstructure of a porous Si layer before (left) and after (right) growth. (Reprinted from Ishikawa, H. et al., *J. Cryst. Growth*, 310, 4900, 2008. With permission from Elsevier, © 2008 by Elsevier.)

3. Selective (masked) epitaxy
4. Self-organized growth

Catalyst-based growth can be further separated into different techniques based on the growth process as vapor–liquid–solid (VLS) [Wagner 1964, Wagner 1970], solution–liquid–solid (SLS), and vapor–solid (VS) growth. For GaN-based nanowires typically a metal catalyst is evaporated on the substrate or an III-nitride buffer layer. Either self-organization effects or a mask can define the position of the metal catalyst and by the latter the growth of a well-arranged nanowire array is enabled. A control of the nanowire diameter can be achieved by the deposition parameters of the metal catalyst [Simpkins 2007]. Typically, the gaseous precursors get absorbed by the metal catalyst and growth is propagating at the interface of the metal catalyst to the seed, either at the substrate or, e.g., a GaN layer grown prior to metal deposition. The disadvantage of catalyst-based growth is the risk of contaminating the layer or nanowires with the metal catalyst. Such nanowires can then be used for device applications; however, placing them already by growth is difficult.

Phase separation (2) has not yet been reported for the growth of nitride nanowires. But selective epitaxy (3) is one of the most widely used routes to achieve nitride nanowires. It enables a catalyst-free growth, thus it is beneficial to achieve high-purity GaN-based nanowires. The wire shape usually depends on growth conditions, typically having a lower growth rate on the sidewalls. Varying the growth conditions can lead to nanowires with inclined sidewalls or can be used to coalesce the wires to achieve a continuous layer [Kikuchi 2004, Furtmayr 2008], which can yield low dislocation densities if twist and tilt of the nanowires is low. As discussed in Section 4.7.5, nanowires are low if not free of any defects.

Finally, strain can lead to a Volmer–Weber growth mode and a self-organized arrangement of nanowires (4) [Ristic 2008] in MBE.

Core shell growth enables surface passivation [Min 2002] or the growth of radially arranged device types along the nanowires, e.g., a triangular FET nanowire [Vandenbrouck 2009].

For device applications as memory devices [Cha 2006], magnetic devices [Choi 2005], FETs [Vandenbrouck 2009], or LEDs [Lee 2007], it is often required to place nanowires onto a predefined structure. Methods to achieve this have been demonstrated for LEDs (Chapter 13) or by transferring nanowires from the substrate they were grown on onto a new substrate, either a predefined structure [Stern 2005, Vandenbrouck 2009] or the structure was defined afterward [Ham 2006] enabling contacting of the nanowires.

4.4.7 Silicon-on-Insulator

Silicon-on-insulator substrates are based on a carrier substrate and a thin Si overlayer for device manufacturing or epitaxy separated by a thin SiO_2 layer. To manufacture such SOI substrates different techniques can be used. In SIMOX a Si handling wafer is ion implanted with oxygen and by this a buried SiO_2 interfacial layer is formed between an upper and lower Si layer (nitrogen implantation: SIMNI, carbon implantation: SiCOI). Smart Cut® uses hydrogen in an oxidized surface of a Si carrier wafer to break apart a thin layer from a substrate bonded onto this oxide layer.

Silicon-on-insulator has long been thought to solve the cracking issue due to the possibility of an amorphous gliding plane being introduced a few nanometers below a thin silicon layer, which is suited for epitaxial growth. Already in 1995 Yang demonstrated the growth of AlN/GaN on a Si/SiO_2/SiC/Si template [Yang 1995]. The XRD ω-scan FWHM of the (0002) reflection was 1500 arcsec for an AlN/GaN/10x (AlN/GaN SL)/GaN sample with a total thickness of 260 nm. This is a high value, but not if taking into accounts the low total layer thickness.

In another work 750 nm of GaN were grown on a SOI structure [Cao 1998]. The surface was relatively rough, but the overall sample quality better on SOI than on planar Si. Stress was determined to be reduced. This, however, must not necessarily be due to relaxation by the SOI substrate but can also be assigned to the better material quality and with it less intrinsic tensile stress during growth [Dadgar 2004].

With a carbonized SOI substrate resulting in a Si/SiO_2/3C-SiC substrate GaN was grown directly on the SiC layer [Steckl 1996]. It showed a FWHM of the (0002) reflection of 360 arcsec slightly worse than for the growth on 6H-SiC where it yields 310 arcsec. Although these early publications were promising, the activities declined soon. One reason might be cracking, which can be expected to be present for these layers: Assuming a 150 mm SOI substrate and a thermal expansion coefficient of 2.59×10^{-6} K^{-1} in combination with a GaN layer with a thermal expansion coefficient of 5.4×10^{-6} K^{-1}, the difference in length at the edge after cooling from 1000°C to room temperature is 190 μm for Si and 396 μm for the GaN layer, i.e., more than 200 μm. Such a macroscopic movement of the layer on the substrate is usually not observed and it is also difficult to assume a suited mechanism which is not leading to the total ablation of the top layers or cracking of the thin Si top layer and the GaN, since SiO_2 and GaN are quite brittle materials.

Zamir et al. [Zamir 2002] have shown that selective growth on SOI substrates can lead to 100×100 μm^2 fields free of cracks in contrast to the growth on bulk Si. Here, SOI acts as a kind of compliant substrate by partly separating the Si top layer from the Si carrier. The GaN on Si layers, with a GaN thickness that can be estimated to ~1.5 μm, are thus strongly bent (Figure 4.19) and partly separate. A curvature measurement reveals a height difference of 1.6 μm, which is likely overlapped by a thickness enhancement at the edges (see also Figure 4.10), as it can be also concluded from the SEM images.

Although the compliant effect is not as good as expected, SOI substrates are useful for novel devices: First, for GaN MEMS, where SOI enables easy underetching and an easier definition of lateral undercut structures; second, for electronics as transistor devices, although little is published in this field yet. Here, GaN on SOI is interesting with regard to a reduction of parasitic charge carriers which are abundant and difficult to avoid in silicon. Usually, the substrate for such devices is highly resistive float zone Si. But at the GaN growth temperature and from the reactants used it can change its conductivity and negatively influence device performance. At least the latter reason can be reduced when using SOI substrates with a slight disadvantage in the thermal performance of the devices. For high voltage electronics, such a silicon oxide layer, if sufficiently thick can enable a higher breakdown voltage by blocking vertical current flow to the (usually) grounded substrate.

Finally, optoelectronics can gain from such substrates. It facilitates GaN layer separation from the substrate as, e.g., necessary for thin film LEDs. Tripathy et al. have shown that this substrate is useful for other optoelectronic device designs which might also lead to optoelectronic MEMS devices, e.g., for sensing applications, but SOI substrates also act as a mirror thus reduce substrate absorption. This might be useful for medium power LEDs (see Chapter 12).

4.4.8 Layer Transfer

Layer transfer enables the growth on low-cost substrates or high-quality layers on lattice matched substrates. For a separation of the nitride layer from the carrier substrate, several reasons can exist:

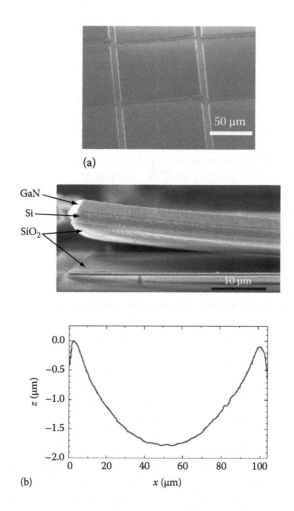

FIGURE 4.19
GaN grown on patterned SOI substrates. Here, the SOI layer leads to a relaxation of tensile stress by delamination at the SiO$_2$ layer (a) in contrast to the samples in Figure 4.10 where some tensile stress is relaxed by cracks in the Si substrate. A line scan of the resulting height profile of the delaminated layer is shown in (b). (Reprinted from Zamir, S. et al., *J. Cryst. Growth*, 243, 375, 2002. With permission from Elsevier, © 2002 by Elsevier.)

1. Use of thermally highly conducting carriers, e.g., for high-power applications
2. Thermally matched layers, e.g., for large area devices which undergo enhanced thermal cycling during operation
3. Hybrid integration of III-nitride devices, e.g., with Si- or GaAs-based devices
4. Designs where the layer has to be turned upside-down for manufacturing and efficient operation as, e.g., for high power LEDs (Chapter 12)

Using sapphire, the common technique for layer transfer is laser lift off [Kelly 1997]: Here a laser with photon energy well above the GaN bandgap energy is focused through the substrate backside onto the seeding layer of the III-nitride structure. Nitrogen evaporates leaving back liquid gallium. Scanning a whole wafer leads to a separated structure which is usually bonded to a carrier substrate prior to this separation process. Liquid gallium on the surface of the separated III-nitride layer is subsequently removed via wet chemical etching. For GaN on silicon, no such process exists or is feasible. In fact it is simpler to remove the substrate in the case of Si. Wet chemical etching comes first to mind but with typical Si surface orientations of (111) etchants with a high etch rate are rare. Simple etchants as KOH have their lowest etch rate in [111] direction. Etchants suited for etching Si(111) are within the system HNO_3–HF–CH_3COOH–H_2O. Typically, HF and HNO_3 are required for etching Si(111) at reasonable etching rates and the most difficult to handle etchants in this system. In addition, this etchant requires a spin etcher unless the bond region and the carrier substrate can be protected from being attacked by this aggressive etch. It is therefore simpler to grind the wafer down to a thickness where it can either be etched with less aggressive etchants which have a sufficiently high etch rate for the layer remaining or to use a spin-etching system. However, also for the latter grinding should be applied to save etching time and etchants.

An alternative to grinding and etching with aggressive etchants could be, as already mentioned in the previous section, SOI substrates where the SiO_2 layer can be etched, e.g., with HF. If a patterned growth process is applied, as is the case for the process by Latticepower, one could also think of a patterned SOI substrate where the patterning is etched deep into the substrate beyond the SiO_2 layer. After LED structure growth is performed, the LED fields grown on the Si on SiO_2 are bonded to a carrier substrate and the SiO_2 is etched away, e.g., with HF or KOH. If the etchand can penetrate into the gap between the bonded layer and patterned substrate, such a process is very fast since only small areas and not one large continuous area are required to be etched. Therefore, such a process can be easily designed not to attack bonding metals or the carrier substrate.

Layer transfer has been demonstrated for LED devices [Zhang 2005, Jiang 2009]. For this device type, it is a requirement to achieve high extraction efficiencies and low thermal resistivity, which can be ideally met by thin film techniques. In addition, an N-face up structure can be easily etched by KOH to form hexagonal pyramidal structures for an improved light extraction at the surface.

In principle, layer transfer is also interesting for hybrid integration of GaN-based devices onto Si electronics (see Chapter 7). This is especially interesting if only a low number of GaN-based devices or device area is required in combination with Si electronics. Then GaN device growth in addition with Si electronics manufacturing on the same substrate is likely too expensive.

4.5 Polarization Reduced GaN on Silicon

Polarization reduced or nonpolar GaN layers are expected to improve LED and LASER performance especially in the longer wavelength regime by the absence or at least a reduction of polarization fields in growth direction. Such layers can be grown easily, e.g., on planar r-plane sapphire [Sano 1976] and a- or m-plane SiC or other suited orientations, but on silicon polarization reduced layers are much more difficult to achieve.

4.5.1 Structured Substrates

Already in 2000 Honda et al. reported the successful growth of GaN($1\bar{1}00$) on 7° off oriented Si(100) [Honda 2000]. With his method, growth still takes place on the Si(111) planes that were defined by photolithography and aniso-tropic etching with KOH (Figure 4.20), but two opposing GaN c-axis orienta-tions occurred hindering the formation of a single domain layer structure. Masking one of two opposing Si(111) facets, this can be avoided and GaN($10\bar{1}1$)

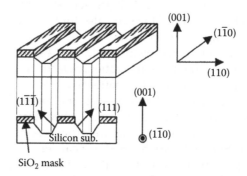

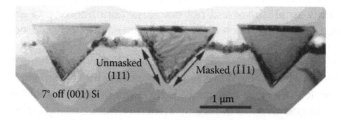

FIGURE 4.20

Si structuring (top) for the growth of ($10\bar{1}1$) GaN by growing c-axis oriented GaN on the Si(111) planes. (Reprinted from Honda, Y. et al., *Proceedings of the International Workshop on Nitride Semiconductors*, IPAP Conference Series 1, 304, 2000. With permission.) The bottom TEM image shows an improved method with one of two opposing Si(111) facets being masked to avoid two domains with opposing c-axis orientation. With the masking of the facets, the triangular GaN bars can coalesce and form a continuous layer. (Reprinted from Tanaka, S. et al., *Jpn. J. Appl. Phys.*, 41, L 846, 2002. With permission from Japan Society of Applied Physics, © 2000 by Japan Society of Applied Physics.)

layers can be grown on 7° off oriented Si(001) [Tanaka 2002], or GaN(11$\bar{2}$2) on Si(311) [Tanikawa 2008a] and nonpolar GaN(11$\bar{2}$0) on Si(110) [Tanikawa 2008b]. The most remarkable point is that with this method, where GaN grows dominantly in c-axis orientation on the Si(111) surface until coalescence, stacking faults are extremely low in density. In addition, the threading dislocation density after coalescence is typically below 1×10^8 cm^{-3}, which depends on the growth parameters following the well-known behavior of faceted growth [Hiramatsu 2000]. It is therefore a very interesting technique to obtain high-quality nonpolar GaN. More can be found in Chapter 3.

4.5.2 Planar Substrates

The growth of polarization reduced material on planar Si is much more complicated and yet only little work has been performed on this topic.

With a 50 nm MnS layer on a Si(001) substrate, Song et al. achieved already in 2002 single crystalline a-planar AlN layers by pulsed laser deposition (PLD) [Song 2002]. Without the MnS layer, the AlN was always c-axis oriented. For the a-planar AlN, the epitaxial relationship found was AlN(11$\bar{2}$0) || MnS(100) || Si(100) with an in-plane alignment of AlN[0001] || MnS[010] || Si[010] and AlN[1$\bar{1}$00] || MnS[001] || Si[001].

Using hydride vapor phase epitaxy (HVPE) and an Al anodic mask, Polyakov et al. were also able to grow a-planar GaN on Si(001) substrates without an AlN seeding layer [Polyakov 2009]. The crystalline material quality achieved is high with a XRD (11$\bar{2}$0) rocking curve FWHM of 450 arcsec and a comparatively low stacking fault density, with regions visible in microcathodoluminescence free of stacking fault luminescence (Figure 4.21). However, the layers are reported to be rough, even after 200 µm of growth. The mechanism of a-planar growth is believed not to be epitaxial growth on Si but that the channels in the oxidized Al layer are seeding a-planar growth (Figure 4.21 bottom). With the quality achieved for the a-planar GaN layer, this usually requires a high crystalline orientation of the Al-oxide layer which is reported to be amorphous.

Nishimura et al. reported on the growth of cubic GaN on a BP buffer on Si(001) grown by MOVPE (see Section 4.4.4) [Nishimura 2001, Nishimura 2004].

Schulze et al. reported that applying a LT-AlN seeding layer, it is possible to achieve the growth of r-planar GaN directly on Si(001) [Schulze 2004b]. On this surface, however, the fourfold surface symmetry together with the sixfold GaN symmetry leads to the growth of four orientations of GaN. An increasing off orientation to 4° did help in increasing the amount of one of these four orientations, however, did not solve the problem to achieve a continuous layer.

Interestingly only on Si(111), LT seeding layers lead to the growth of c-axis oriented GaN layers. On all other Si directions, GaN grows polycrystalline. In contrast to that, HT seeding layers lead to c-axis oriented GaN layers practically on all Si surfaces [Reiher 2010].

Thus, for growing non-c-axis oriented layers on silicon LT seeding layers are advantageous. With such layers, Ravash et al. [Ravash 2009] have

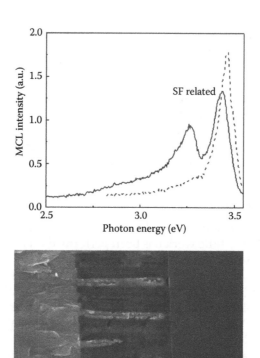

FIGURE 4.21

Microcathodoluminescence (top) and SEM image (bottom) of a-planar GaN grown on an Al-masked Si(001) substrate. The SEM image shows the substrate (left), the masked region (middle), and the AlN buffer. The AlN protrudes into hollows in the Al mask but is assumed not to be in contact with the Si substrate. (Reprinted from Polyakov, A.Y. et al., *Appl. Phys. Lett.*, 94, 022114, 2009. With permission from American Institute of Physics, © 2009 by American Institute of Physics.)

achieved GaN layers on Si(211) with a tilted *c*-axis of 18°. It is assumed that on the Si(211) surface, which typically consists of alternating (111) and (001) surfaces [Chadi 1984]. GaN only nucleates with a preferred GaN(0001) orientation on the (111) planes which are declined by 18° toward the surface normal (Figure 4.22).

One can conclude that offering Si(111) planes will lead to a preferred growth of *c*-axis oriented GaN on them. However, with increasing angle, the width of (111) planes and (001) planes usually changes, e.g., (001) planes get broader while (111) planes get narrower when going from (211) to (911) surfaces. Due to the preferential polycrystalline growth of LT AlN seeding layers on Si(001) [Schulze 2004, Schulze 2004b] successful single-crystalline growth on the (001) planes can be only achieved with HT-AlN seeding layers.

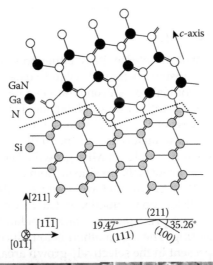

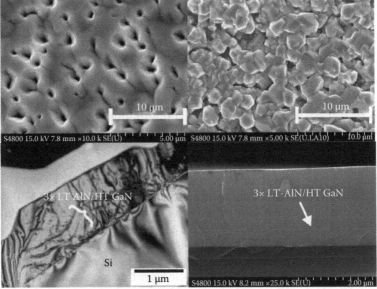

FIGURE 4.22

Model of the Si(211)/AlN(0001) interface (top). The growth of the seeding layer perpendicular to the Si(111) surface planes leads to a ~18° inclination of the GaN (0001) surface. The SEM image (middle) shows the comparison of a GaN layer on an optimized LT-AlN seeding layer in comparison to a standard LT-AlN seeding layer leading to polycrystalline growth as also HT-AlN seeding layers do. The TEM image (bottom left) shows the evolution of growth by the three strain engineering LT-AlN interlayers acting as markers. First only islands are grown, which then lead to a more and more continuous layer by a high lateral growth rate. A cross-sectional SEM image (bottom right) of a ~2 μm thick structure shows that the layer can be compact if grown sufficiently long enough. (Reprinted from Ravash, R. et al., *Appl. Phys. Lett.*, 95, 242101, 2009. With permission from American Institute of Physics, © 2009 by American Institute of Physics.)

4.6 Strain Engineering

Strain is always present in heteroepitaxy either from differences in lattice constant, thermal expansion coefficient, island coalescence, or dislocation climb [Dadgar 2004]. For GaN on silicon strain can be related to all of these sources and is, except for InGaN on GaN, tensile. For the growth of crack-free GaN layers in excess of 1 µm on silicon the main difficulty is the large thermal mismatch of these materials and the high growth temperatures in MOVPE growth, thus it is simpler in MBE growth. The cracking problem is the main reason not to grow GaN on silicon. Some of the methods applied to avoid cracks in GaN on Si growth have been used already in the 1990s in other context as, e.g., dislocation reduction in GaN [Amano 1998].

There are two different principles for crack avoidance of which only one is true strain engineering. The first method is to reduce the growth area so that cracks are not present in the selectively grown area, which then enables the controlled manufacturing of devices, as already mentioned in Section 4.3. The second real strain engineering method is, although performed in different ways, always based on the induction of compressive stress during growth to compensate tensile thermal stress upon cooling.

All other methods, especially methods as compliant substrates have not proven to yield an improved material quality, especially to fully avoid cracks.

In principle, when using two differently strained materials, depending on relaxation, four different strain configurations can exist as shown in Figure 4.23. Assuming two materials with a larger (A) and a smaller (B) lattice constant, the case described in (a) where material B grows relaxed on material A and after this A compressively strained on B is described in Section 4.6.3. If pseudomorphic growth of B occurs, either by choosing a layer thickness below the critical thickness or by the growth conditions A on B will not be influenced and grows with the same lattice parameter as the underlying layer A shown in (d). Case (e) occurs either for a freestanding layer sequence where pseudomorphic growth leads to an average lattice distance or, if the layer stack partially relaxes as it is likely the case for the layers described in Section 4.6.2. Growing such a layer sequence with single layer thicknesses above the critical thickness the stack will consist of relaxed layers and usually an inferior material quality as shown in (f). A compositional step grading a buffer with smaller lattice constant is shown in (c). This is a method described in Section 4.6.1.

All strain engineering methods are of little effect if the material quality is low. Strain fields inherent with these methods can lead to dislocation climb reducing the compressive (and also increasing tensile) stress. Such a dislocation climb has to date mostly been reported for silicon doped layers [Cantu 2003, Romanov 2003, Dadgar 2004, Cantu 2005], but is likely also the driving force when high strain fields are present as reported by Wang et al. [Wang 2006b] (see also Chapter 5). Therefore, the effectiveness of such methods is also a direct measure of the material quality.

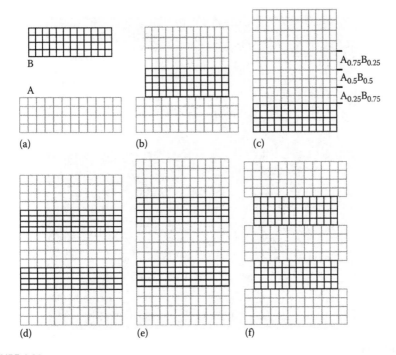

FIGURE 4.23
Different possibilities of lattice mismatched growth. (a) Shows two relaxed materials. In (b) material B grows relaxed on layer A but layer A will grow pseudomorphic on B. This is assumed to be the case for LT-AlN layers or thick HT-AlN layers. A (step) graded transition from material B to A does lead to a compressive strain in layer A, if it is grown pseudomorphic or only partially relaxed. In superlattice layers either on one layer (buffer A) layer B grows pseudomorphically and no strain engineering effect will be observed for a subsequently grown layer A (d) or both grow pseudomorphic with an equilibrium lattice constant, which will enable subsequent compressive growth of layer A (e). If both grow fully relaxed (f) on each other, no compressive strain can be induced on a subsequently grown layer A, as in the case of example (d).

4.6.1 AlGaN Intermediate Layers and AlGaN Grading

An AlGaN intermediate or (step) graded layer between the AlN seeding and GaN buffer layer is probably the oldest method to obtain crack-free GaN on silicon. First results were published in the late 1990s with a crack-free layer thickness around and above $1\,\mu m$ [Marchand 2001, Ishikawa 1999a, Ishikawa 1999b, Kim 2001] difficult to achieve crack-free without any strain engineering applied in MOVPE. Starting from an AlN seeding layer the composition is graded from AlN to GaN or a single AlGaN layer is used. If material quality, especially seeding layer quality, is sufficiently high, such a transition can induce a compressive stress on the subsequently grown GaN layer. For low layer qualities with high dislocation densities, the probability for stress relaxation via dislocation climb and annihilation processes is significant (see, e.g., Figure 4.24), and it is usually observed that for such cases the method will not enable the growth of

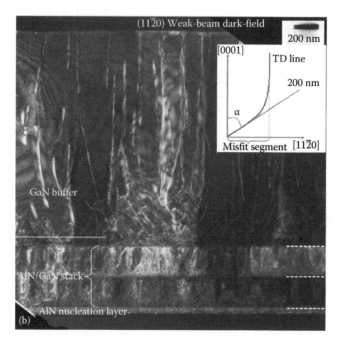

FIGURE 4.24
GaN/AlN layer stack on a AlN seeding layer on Si. The subsequently grown GaN layer shows a strong bending of the edge dislocations after the AlN/GaN stack, which likely originates in a strong compressive strain induced onto the GaN layer. This compressive strain is reduced by the dislocation movement. This leads to a decrease of the dislocation bending (inset). (Reprinted from Baron, N. et al., *J. Appl. Phys.*, 105, 033701, 2009. With permission from American Institute of Physics, © 2009 by American Institute of Physics.)

thick GaN layers. Many different types of gradients are possible with regard to composition, step width and composition change, and total thickness.

The compositional and with it strain changes are also in favor for a reduction in dislocation density because the compressive strain fields can force dislocation bending and by this enhance the chance of dislocation annihilation (Figure 4.25) [Raghavan 2006]. However, edge dislocation climb does also reduce compressive strain, thus a high edge dislocation density is lowering the effectiveness of the transition layer. Nitronex is using the method of a step-graded buffer for the growth of FET device layers but is limited to a total thickness of the III-nitride layers around 2.5 μm most likely limited by edge dislocation climb as already described. By optimizing the buffer structure, an overgrown crack-free thickness of 2.6 μm of GaN was achieved by Arslan et al. using an AlN/LT-AlN/AlN/and in five steps graded AlGaN buffer layer [Arslan 2008]. Able et al. were only able to overgrow an AlGaN transition layer crack-free with 800 nm of GaN [Able 2005] and Leys et al. only achieved overgrowth of 1.3 μm GaN over a 200 nm AlN/400 nm AlGaN ([Al] ~40%) buffer structure [Leys 2008].

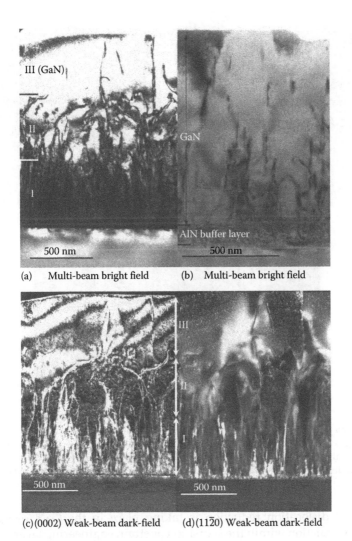

FIGURE 4.25

TEM images of a Si(111)/AlN/AlGaN grading/GaN layer sequence. The dislocation density is strongly reduced by the composition graded AlGaN layer. (Reprinted from Raghavan, S. et al., *Appl. Phys. Lett.*, 88, 041904, 2006. With permission from American Institute of Physics, © 2006 by American Institute of Physics.)

These results demonstrate that a thick overgrowth depends a lot on a successful optimization of the AlGaN intermediate or transition layers and the seeding layer.

A comparative study of different strain engineering buffer layers with each structure optimized was performed by Yoshida et al. [Yoshida 2006]. They investigated an AlN (18 nm)/GaN (5 nm) SL an AlGaN grading and a threefold AlN (50 nm)/GaN (200 nm) buffer layer (Figure 4.26). For all layers,

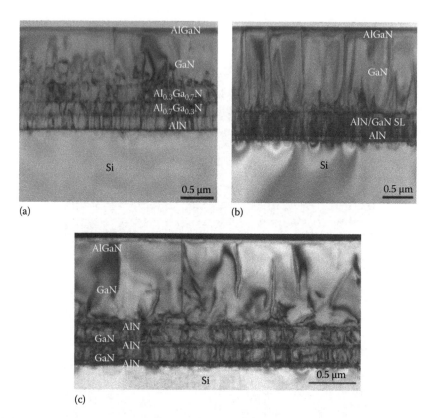

(a)   (b)

(c)

FIGURE 4.26
TEM images showing a comparison of three different strain engineering methods. A step-graded AlGaN buffer (a), a AlN/GaN superlattice (b), and a GaN/AlN multilayer (c). (Reprinted from Yoshida, S. et al., *Phys. Stat. Sol. (a)*, 203, 1739, 2006. With permission from Wiley VCH, © 2006 by Wiley-VCH.)

significant dislocation reduction is taking place in the GaN overgrowth layer. Annihilation within the transition or strain engineering layers is much weaker pronounced. In this study, the best material quality was obtained with the threefold AlN (50 nm)/GaN (200 nm) buffer layer.

4.6.2 GaN/AlN Multilayers

AlN having a ~2.5% smaller lattice constant compared to GaN can induce compression onto an (Al)GaN layer grown on top of it as it has been described in the previous section for a graded AlGaN buffer. Prerequisite when grown on a GaN buffer is at least a partial relaxation of the AlN layers. When grown directly on silicon without a buffer layer to improve the material quality, a similar case as in the previous section for AlGaN grading is likely. Simple AlN seeding layers usually are of too low quality to induce compression, causing GaN growth to be fully relaxed on them.

After the growth of a GaN layer on an AlN seeding layer, the material quality usually improves significantly. AlN grown on such a layer either grows fully strained or, if thicker, (partially) relaxed. GaN grown on such a fully or partially relaxed AlN surface will start growth with the new in-plane lattice parameter given by the AlN. Growth will therefore be under compressive stress for GaN. The strained growth can lead to 3D growth as such systems were also described for the growth of quantum dots formed by a Stranski–Krastanov growth mode [Daudin 1997, Chama 2003].

Depending on the amount of tension induced by AlN/GaN multilayer growth and subsequent compression onto the GaN layer during growth this compression can be sufficient to compensate tensile thermal stress upon cooling. In MBE, Nikishin et al. demonstrated the benefit of such an AlN/GaN SL to improve the material quality [Nikishin 1999]. Feltin et al. [Feltin 2001a] demonstrated this in 2001 for a ~2.4 μm thick green LED structure. They introduced three stacks of 10-fold superlattice layers with AlN and GaN thicknesses of 3 and 4 nm, respectively separated by 200 nm of GaN. This is shown in Figure 4.27 in TEM for the first SL. Dislocation annihilation is high and seems to take mostly place in the second GaN layer directly after the SL stack. This indicates the presence of high strain fields induced by the SL stack. Obviously, although 3 nm of AlN can be grown with little relaxation [Bykhovski 1997], the multilayer has caused some relaxation and can therefore apply compressive stress on the subsequently grown GaN layer. Finally, they could overgrow this buffer crack-free with 1.5 μm of a GaN LED structure. However, the samples showed some cracks after cooling and an extreme bow of 1 m⁻¹. This strong bow likely originates in plastic deformation of the substrate, often occurring with such thick layers on thin 2 in. substrates.

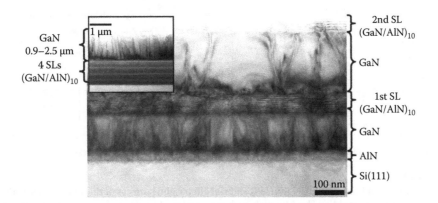

FIGURE 4.27
TEM of the lower part of a multifold (AlN/GaN) SL/GaN structure enabling the growth of a >2 μm thick LED structure. Dislocation reduction is preferably observed after the AlN/GaN SL stack. (Reprinted from Feltin, E. et al., *Appl. Phys. Lett.*, 79, 3230, 2001. With permission from American Institute of Physics, © 2001 by American Institute of Physics.)

Further work has been published, e.g., by Sugahara, Lee, and Ohtsuka where a similar method is applied by growing 20-, 40-, or 80-fold superlattice with 5 nm AlN/followed by 15, 20, or 40 nm of GaN [Sugahara 2004]. Most effective was reported a structure with 40 nm of GaN. But in their publication further experiments were then performed with a 20 nm thick GaN layer. After an AlN/GaN SL layer up to 600 nm GaN could be grown crack-free. Their SEM image (Figures 4.28 and 4.33) of an AlN surface nicely shows a

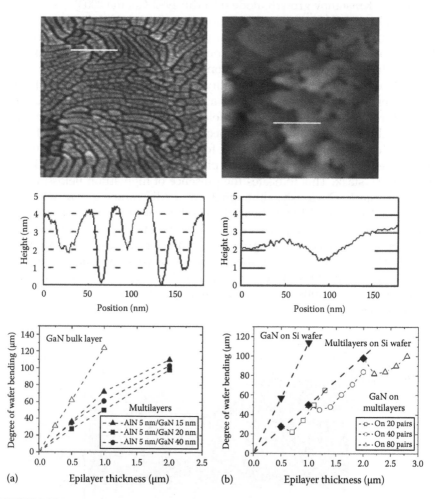

FIGURE 4.28

AFM and AFM linescan of AlN (top left) and GaN (top right) surfaces of the 40-fold AlN/GaN SL stack (top and middle). The AlN surface shows an island type of growth or cracks, the latter are however unlikely the dominating source of the surface structure since cracks are usually propagating along preferred crystallographic directions. The bottom images show the wafer bow after growth of different layer sequences. (Reprinted from Sugahara, T. et al., *Jpn. J. Appl. Phys.*, 43, L1595, 2004. With permission from Japan Society of Applied Physics, © 2004 by Japan Society of Applied Physics.)

surface structure with deep trenches, indicating at least partial relaxation via micro-crack generation or a 3d growth transition, which can be often observed for AlN layers in high resolution AFM or SEM images.

4.6.3 AlN Interlayers

AlN interlayers grown at LTs were first described by Amano (see Section 4.7.1) [Amano 1998]. On silicon, such layers were applied to demonstrate a method for crack-free GaN on silicon in 2000 [Dadgar 2000]. The mechanism originates in the growth of AlN which does lead to a partial or full relaxation for temperatures below 1000°C thermocouple reading, which is usually about 150°C below the GaN growth temperature. But the effectiveness of the layer and the behavior depends a lot on the growth conditions as can be seen in Figure 4.29 (top): here the temperature and doping does have a significant impact on the effectiveness of the layer [Blaesing 2002, Reiher 2003]. The thickness determines the amount of stress that can be applied, most likely by an increasing relaxation of the layer with increasing thickness (Figure 4.29 bottom), which saturates around 10–12 nm layer thickness. Alternatively to such a LT layer, the AlN layer could, in principle, be grown with a thickness sufficient to relax it to a larger part. But then cracks propagating into the GaN layer below are common and this method has to date not been widely used except for MBE growth and it can lead to an increase in dislocation density (see below). The main advantage of LT-AlN layers is that these thin LT-AlN interlayers can be introduced again and again. Usually, with increasing layer thickness and improving material quality, these layers can be applied in larger distances presently up to 2 μm for highly Si-doped layers and even higher for undoped layers. Consequently, the top layer of a structure can be 2 μm for highly Si-doped GaN and even up to 3 μm for undoped GaN, sufficient for LED and FET devices. The reason for this thickness difference is the increase in dislocation climb by Si doping which reduces compressive stress and can even lead to tensile stress during growth.

With this method ~7 μm thick GaN layers were already presented in 2003 [Riemann 2003, Krost 2003]. Cross-sectional cathodoluminescence and Raman measurements on such layers show the impact of the strain engineering layers on the strain state. Especially for the first interlayer, it was possible to resolve the change in strain state by micro-Raman measurements (Figure 4.30) [Riemann 2003]. This is also validated by CL measurements which show a shift of the (D^0,X) excitonic emission which is influenced by the strain of the layer.

In MBE also a sequence of 40 nm AlN seeding layer followed by a 250 nm GaN/250 nm AlN layer allows the subsequent growth of 2 μm of GaN [Cordier 2005a] or 1.5 μm of AlGaN with Al content up to 20% [Cordier 2005b]. Dislocation densities are around $5-7 \times 10^9$ cm^{-3} which are mostly generated by the 250 nm thick AlN strain engineering layer after a dislocation density of 5×10^8 cm^{-3} after the first GaN layer was reached.

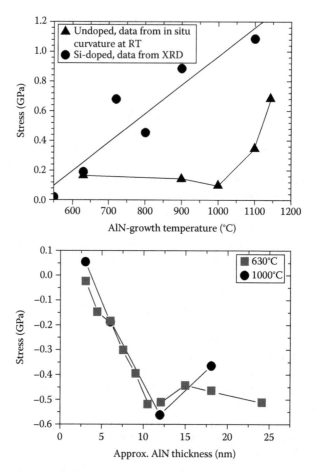

FIGURE 4.29

Impact of growth temperature and growth conditions (here doping) on the effectiveness of AlN interlayers as strain engineering layers (top). The efficiency depends on the thickness of the layer until saturation is reached for layer exceeding approximately 12 nm in thickness, independent of the growth temperature (bottom).

4.7 Dislocation Reduction

Dislocation reduction is prerequisite for high-quality nitride devices, on all heterosubstrates, due to the, compared to other III–Vs extremely high threading dislocation density originating in poor lattice match and (misoriented) island growth at the beginning of the epitaxial process. Therefore, many techniques were developed to reduce the dislocation density. Most methods which locally result in quasi-perfect material with dislocation densities

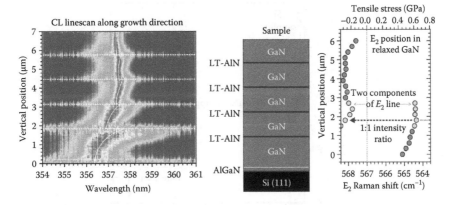

FIGURE 4.30
Cathodoluminescence (left) and Raman measurements (right) on a 7 μm thick GaN structure on Si(111) with strain reducing LT-AlN interlayers [Riemann 2003]. The shift of the (D^0,X) emission in CL can be attributed to a change in strain state. The broadening reflects the GaN quality. In μ-Raman measurements, a drastic change in the strain state is found for the first LT-AlN interlayer; for the other interlayers the resolution is too weak to resolve the change.

well below 10^8 cm^2 are based on lateral overgrowth which can be performed dislocation free, except for the starting and coalescence region.

Edge and screw type dislocations in c-planar GaN usually propagate vertically in growth direction. Reducing the dislocation density either means that at least two dislocations react with each other and, in the case the Burger vectors are of opposite sign they annihilate; if this is not the case they can react and form one new dislocation. Alternatively, dislocations can end at surfaces as, e.g., masking layers used in ELOG processes. Because dislocations propagate vertically the probability for them to react is rather low and with decreasing density this gets even lower (Figure 4.31). To enable lateral propagation and increase interaction strain fields are useful. They can be applied by growing a lattice mismatched heterostructure but also intrinsic sources, e.g., from island growth and coalescence are present. Additionally, dislocations themselves cause local strain fields and, if closely spaced, are likely to interact.

4.7.1 AlN Interlayers

Amano et al. applied LT-AlN interlayers for GaN on sapphire to improve the material quality [Amano 1998] and later to grow crack-free AlGaN on top of a GaN buffer [Iwaya 2000]. They demonstrated that an improvement of the layer quality can be achieved with such layers with a thickness around 10 nm. This has been also observed for the growth of GaN on silicon [Dadgar 2000, Blaesing 2002, Reiher 2003]. In Figure 4.32, cross-sectional TEM images show the impact of the interlayers and SiN layers on the propagation of screw- and edge-type dislocations. All dislocations undergo an enhanced recombination at or shortly after the interlayers. But this reduction also depends on

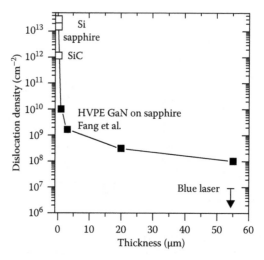

FIGURE 4.31
Development of dislocation density with grown layer thickness. The initial values are theoretical values for the misfit dislocation density at the substrate/seeding layer interfaces which do not necessarily reflect the threading dislocation density. (HVPE data is taken from Fang, Z.Q. et al., *Appl. Phys. Lett.*, 78, 332, 2001.)

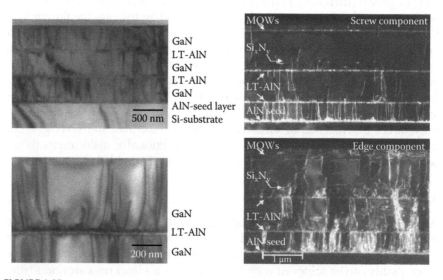

FIGURE 4.32
Impact of differently grown LT-AlN interlayers on the dislocation density. The example on the left shows a steady number of dislocations before and after the AlN layer while for the sample on the right dislocations are reduced at or in the vicinity of this interface. (Reprinted from Dadgar, A. et al., *Phys. Stat. Sol. (c)* 0, 1583, 2003. With permission from Wiley-VCH, © 2003 by Wiley-VCH.)

the growth temperature and thickness in the case of the AlN interlayers. Too thick layers or too low growth temperatures will even deteriorate the material quality. In the left image of Figure 4.32, growth parameters were chosen less favorable than at right. Directly after the LT-AlN layer, the dislocation density increases and the overall material quality is not improving. In the right image, not taking into account the two SiN layers the dislocation density is always improved about 100 nm after the interlayer. The only differences here were the growth conditions that have to be optimized to achieve a reduction in dislocation density and inducing compressive stress.

4.7.2 Superlattices

Superlattices are well known from conventional III–Vs to enhance dislocation reaction by strain field–induced dislocation bending. The same can be also achieved for III-nitrides where AlN/GaN are typically applied for strain engineering as, e.g., presented by Feltin et al. [Feltin 2003]. It is also used with different thicknesses of the SL layers by Sanken Electric for growing thick high-voltage FET layers (Figures 4.28 and 4.33) [Iwakami 2004, Sugahara 2004]. In the work of Feltin et al., the threading dislocation density could be reduced about a factor of 5 to 2×10^9 cm^{-2}, which is well visible in TEM cross-sectional images, where the first SL stack already significantly

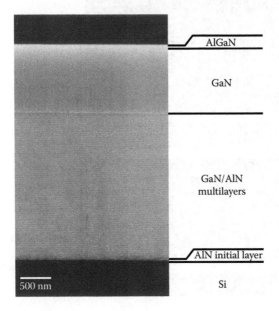

FIGURE 4.33
SEM image of a 40-fold GaN/AlN superlattice for the growth of a AlGaN/GaN HV-FET. For wafer bow after growth see also Figure 4.28. (Reprinted from Iwakami, S. et al., *Jpn. J. Appl. Phys.*, 43, L831, 2004. With permission from Japan Society of Applied Physics, © 2004 by Japan Society of Applied Physics.)

reduces the dislocation density (Figure 4.27) and an increasing number of SL stacks improves the XRD ω-scan FWHM. In the work by Sanken, a correlation between SL thickness and residual bow is given (Figure 4.28). The thickness above the SL that can be overgrown is limited to 600 nm, less than in the work of Feltin, however, sufficient for lateral FET devices. A SL was also applied by Ubukata et al. for the growth of 400 nm GaN on a 150 mm Si substrate. Here, they used a ~500 nm thick AlN/GaN SL for strain compensation which also reduced the dislocation density significantly. A drastic reduction in dislocation density is observed at the beginning of the GaN top layer (Figure 4.34). It seems that during AlN/GaN SL growth, the reduction is less effective than after the last AlN layer which seems also to be valid for other cases (see Figures 4.26b and 4.27).

Lin have used a HT-AlN/LT-AlN/HT-AlN multilayer [Lin 2007]. Based on this Saengkaew et al. have used an extended HT-AlN/LT-AlN/HT-AlN growth scheme [Saengkaew 2009a, Saengkaew 2009b], namely, a LT/HT AlN superlattice to improve the material quality of subsequently grown AlGaN layers on silicon [Saengkaew 2009a, Saengkaew 2009b]. In contrast to the growth of GaN dislocation reduction is very difficult by SiN masking when growing AlGaN due to the poor lateral growth and difficulties in coalescing

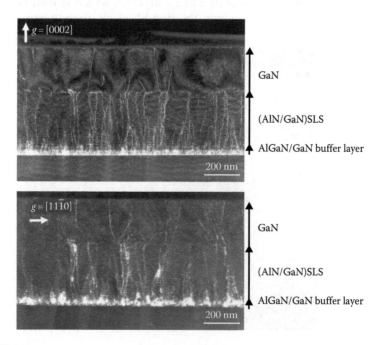

FIGURE 4.34
TEM of a strain engineering and dislocation reducing AlN/GaN SL structure. The dislocation density is strongly reduced after the SL stack in agreement with other publications (see also Figures 4.26 and 4.27). (Reprinted from Ubukata, A. et al., *J. Cryst. Growth*, 298, 198, 2007. With permission from Elsevier, © 2007 by Elsevier.)

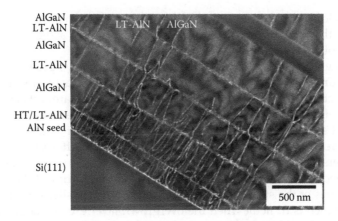

AlGaN
LT-AlN
AlGaN
LT-AlN
AlGaN
HT/LT-AlN
AlN seed
Si(111)

LT-AlN   AlGaN

500 nm

FIGURE 4.35
TEM image of an AlGaN layer sequence with strain engineering LT-AlN interlayers on a HT/ LT-AlN superlattice on Si(111).

such a layer. The AlN was grown at high (1100°C–1200°C) and LT (700°C) directly on the HT-AlN seeding layer. With an optimized structure with a 3 nm thick HT-AlN layer and a 7 nm thick LT-AlN layer, the best material quality with a XRD FWHM of the (0002) reflection of 540 arcsec and 1400 for the ($10\bar{1}0$) ω-scan was obtained. Further optimized samples were grown with 0.5 nm of LT-AlN and 10.5 nm of HT-AlN.

In TEM, one cannot distinguish the HT and LT layers even for thicker LT-AlN layers (Figure 4.35). However, the material improvement is well visible in XRD where the FWHM of the (0002) ω-scan is improved for a thinner AlGaN layer from an initial value of 1220–1000 arcsec, and from 2330 to 1450 arcsec for the ($10\bar{1}0$) reflection. Although the thin LT-AlN layers can be thought to be unimportant the layers are of inferior quality if grown without. This also if only a temperature ramping step to LT-AlN growth temperature not followed by growing LT-AlN but subsequent heating to HT-AlN growth temperature and growing HT-AlN is applied. A possible explanation for the material improvement is the different strain state of the AlN layer at high and LTs enhancing dislocation reactions. Indeed Amano et al. used single LT-GaN layers to improve the GaN material quality [Amano 1998].

4.7.3 ELOG and Pendeo-Epitaxy

A dislocation reduction method which was for a long time the method to locally achieve layers with a quality sufficient for laser operation is lateral overgrowth over a mask or a gap region [Usui 1997, Kung 1999, Linthicum 1999]. Prerequisite for this method, which can be applied either directly on a substrate or on a buffer layer, are structuring and masking or etching structures, which can then be overgrown. For GaN-on-silicon with its comparable high dislocation density, this would be an ideal method to improve the layer quality. However, the main difficulty of tensile thermal stress cannot be solved with this method; thus, the main difficulty is to achieve coalescence and to avoid cracking of this coalesced area.

Several groups have attempted to grow such lateral overgrowth structures with the first report by Kung et al. [Kung 1999]. In the last years, it got quiet since defect densities achievable without such techniques are sufficient for LED or FET operation and the technique difficult to control. The first problem occurs for etched structures and is related to proper masking of Si areas not or only thin covered with the AlN seeding and buffer layer. After several hundred nanometers of growth a too-thin protecting layer, but also stress at the edges of such etched structures, can lead to meltback etching starting at the edges and irregularities of the structures. The second problem on silicon is strain engineering which is difficult when the overgrown layer exceeds about a micron in thickness. In addition to that typical ELO problems as wing tilt can occur.

Kawaguchi et al. [Kawaguchi 1998] presented the growth of GaN in small few micron and submicron large openings on Si(111). They showed that from these openings pyramidal structures increase in diameter which also enables an ELOG. Unfortunately, this concept was not published for fully coalesced layers.

The group of R.F. Davis was the first to present Pendeo-epitaxy (from Latin *pendeo*—hanging on, suspended from) on Si(111). In fact the growth was not performed directly on Si(111) but on a 2 µm thick 3C-SiC layer [Davis 2000, Gehrke 2000, Linthicum 2000, Davis 2001a, Davis 2001b]. On this substrate, an AlN seeding and a GaN buffer layer were grown, which were subsequently structured into trenches. When resuming GaN growth the masked top area allows GaN only to grow on the sidewalls of the trenches. For longer growth times, this layer also coalesces above the masked region and a continuous layer with low threading dislocation densities is obtained. X-ray reveals the usual problem of wing tilt with GaN(0002) x-ray FWHMs of 860 arcsec parallel to the trenches and 2124 arcsec perpendicular to them. An advantage of this method is that SiC did help in protecting the substrate from meltback etching. Cracking is expected and indeed it still persists. To solve this problem, the Si top layer of a SOI substrate was completely converted to SiC and reported to avoid cracking [Linthicum 2000]. The origin of this mechanism is not understood especially since SOI was never observed to act as efficient compliant substrate.

Strittmatter et al. used a pendeo or cantilever epitaxy process with less steps and especially only one growth step [Strittmatter 2001, Dadgar 2003b]. They structured a substrate surface into trenches and grew directly on this structure. Since no masking was present AlN (on an AlAs seeding layer) and GaN growth takes place on all surfaces. But for sufficiently deep trenches, the lateral growth from the top surface is fast enough to close the openings and lead to an undisturbed growth of GaN with a low threading dislocation density (Figure 4.36). A similar approach was applied by Katona et al. (Figure 4.37) [Katona 2001]. For these approaches, cracking was observed for coalesced layers.

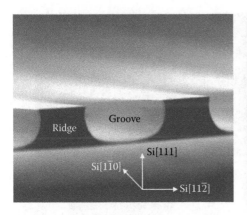

FIGURE 4.36
SEM of a Si substrate with etched grooves before (left) and after (right) GaN growth, see also
[Strittmatter 2001].

Tanaka also used a single growth process by patterning the silicon substrate
with a SiO_2 mask and starting growth with AlGaN [Tanaka 2001]. With a
$1\,\mu m$ wide $2\,\mu m$ period stripe pattern, they obtained a threading dislocation
density of 2×10^9 cm^{-2} for a $2\,\mu m$ thick layer, but also cracks were observed.
Today, such a threading dislocation density can be already obtained without
an ELO pattern for $2\,\mu m$ thick GaN layers; an explanation for this might be
found in a previous report from the same group where small AlGaN nuclei
were observed on the mask. Such additional likely misaligned seeds might
deteriorate GaN overgrowth [Tanaka 2000b]. Cracking was avoided by an
approach of Naoi et al. [Naoi 2003] who used a combination of ELO and patterning "selected-ELO." ELO was applied for improving the material quality and patterning was performed to avoid cracks. The type of patterning,
ridges, or SiO_2 masking did not result in a significant difference in material
quality which was improved by a factor of three with regard to the threading dislocation density (Figure 4.38). They used a $5\,\mu m$ ridge/uncovered and
$5\,\mu m$ grooves/SiO_2 covered structure in a $100 \times 100\,\mu m^2$ pattern and a FACELO
[Hiramatsu 2000] growth process resulting in $2\,\mu m$ thick layers free of cracks
in the case of ELO masking (Figure 4.39). For the growth on the ridges, the
structure also was nitrided at 800°C for 10 min in an ammonia atmosphere.

Low dislocation densities of 10^6 cm^{-2} were reported by Marchand et al.
[Marchand 1999] in laterally overgrown regions. As mentioned above, protecting the Si regions but also the mask edges is critical. Marchand observed
a degradation of the layers before coalescence when a too thin AlN buffer was used. This is likely due to meltback etching. Also, coalesced layers
showed cracks and wing tilt.

The combination of a stress reducing buffer with ELO, which should be an
enabler for thicker layers, was presented by Feltin et al. [Feltin 2003]. They
used a stress reducing AlN/GaN superlattice buffer layer as described in

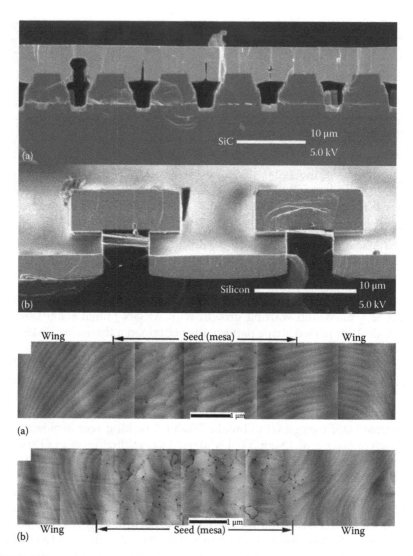

FIGURE 4.37
SEM (top) and AFM (bottom) images of GaN grown on grooves etched into SiC or Si substrates.
The upper image shows samples on SiC, the lower on Si(111). In both substrates, the dislocation
in the wing region is low; however, in the post region it is higher for GaN on Si(111). (Reprinted
from Katona, T.M. et al., *Appl. Phys. Lett.*, 79, 2907, 2001. With permission from American
Institute of Physics, © 2001 by American Institute of Physics.)

Section 4.7.2 for the lateral overgrowth of GaN on Si(111) [Feltin 2001b, Feltin
2001c]. A reduction in threading dislocation density from an initial 8×10^9
cm^{-2} to 5×10^7 cm^{-2} in the wing regions and 8×10^8 cm^{-2} in the mask opening
was determined. But although strain engineering layers were applied cracks
were also present.

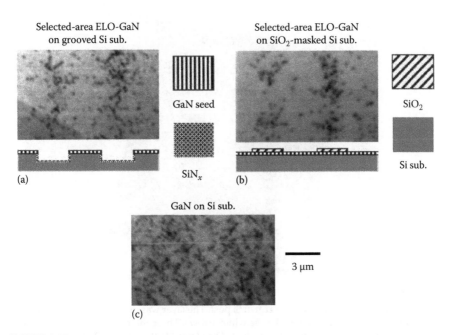

Selected-area ELO-GaN on grooved Si sub. (a)

Selected-area ELO-GaN on SiO₂-masked Si sub. (b)

GaN seed

SiN$_x$

SiO₂

Si sub.

GaN on Si sub. (c)

3 µm

FIGURE 4.38
CL showing defect related darkspots of GaN ELO layers grown on different Si substrate preparations: A grooved substrate with SiN masks (a). On such substrates, the GaN layer usually has no contact to the substrate in the overgrown region. A SiO₂ masked substrate (b), and on plain Si (c). (Reprinted from Naoi, H. et al., *J. Cryst. Growth*, 248, 573, 2003. With permission from Elsevier, © 2003 by Elsevier.)

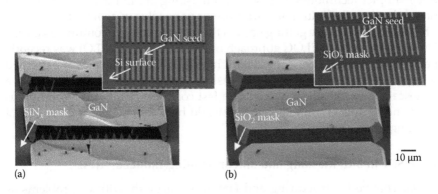

GaN seed

Si surface

SiN$_x$ mask   GaN

(a)

GaN seed

SiO₂ mask

GaN

SiO₂ mask

10 µm

(b)

FIGURE 4.39
SEM images showing GaN grown by a combination of ELO and patterning. By using ELO, the material quality is enhanced, patterning inhibits crack formation, both enabling the growth of thick crack-free GaN layers. (Reprinted from Naoi, H. et al., *J. Cryst. Growth*, 248, 573, 2003. With permission from Elsevier, © 2003 Elsevier.)

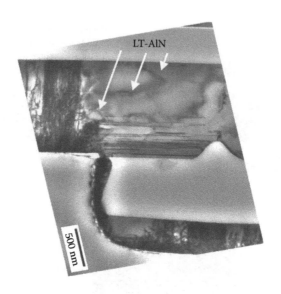

FIGURE 4.40
Cross-section TEM of a laterally overgrown Si post. The layer was grown with strain reducing LT-AlN interlayers that are marked by the white arrows. These interlayers although reducing stress, do not introduce threading dislocations (Data from Strittmatter, A., 2009.)

Strittmatter et al. also applied strain engineering to their etched trench structures when overgrowing. Three thin ~8 nm LT-AlN layers were used and efficient [Strittmatter 2009]. Using LT-AlN might introduce new dislocations but here the practically threading dislocation free wing region is not negatively affected by the interlayers, indeed no dislocation generation is observed (Figure 4.40).

Today classical ELO and Pendeo on silicon are rather insignificant, and other simpler techniques as SiN in situ masking are in the focus of dislocation reduction. But with nanostructure or nanocolumn growth (see Section 4.8.5) similar topics are getting back in the focus to achieve continuous layers. On sapphire, e.g., an ELOG approach on the nanoscale (NELO) was recently performed [Wang 2006a, Zang 2008]. There a GaN buffer layer was patterned with a SiO₂ mask in nanoscale dimensions and overgrown. The benefit of a nanoscale mask on silicon would be a fast coalescence to form a continuous layer and by this strain engineering would be simpler, lowering the risk of cracking upon cooling.

4.7.4 SiN In Situ Masking and Silicon-Doping-Induced Stress

An alternative to structuring and lateral overgrowth, with its problems as wing tilt and meltback etching, for a reduction in dislocation density is the application of a SiN nano-mask, simply by a growth interruption under ammonia and the flow of silane which will form a thin SiN mask on the surface [Tanaka 2000a]. This mask has an arbitrary shape mostly depending

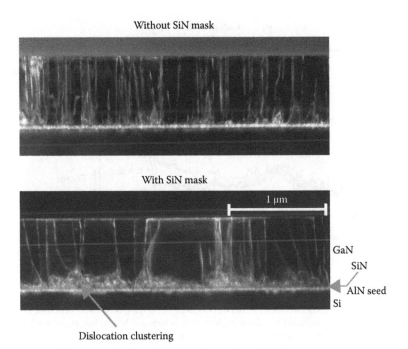

Without SiN mask

With SiN mask

1 μm

GaN

SiN

AlN seed

Si

Dislocation clustering

FIGURE 4.41
Cross-section TEM of an AlN/GaN top and an AlN/SiN/GaN layer sequence on Si(111). The usage of a SiN masking layer leads to an enhanced dislocation reaction above the mask and reduces the threading dislocation density.

on the surface structure and depending on the deposition time. It masks part of the substrate surface and, acting as anti-surfactant, growth will be suppressed on these SiN islands. This masking method has been applied on GaN on sapphire and is helpful on silicon. The challenge is to coalesce within a thickness of around 1 μm to avoid high tensile stresses after cooling and to efficiently reduce dislocations, e.g., by dislocation bending and annihilation. For the latter, faceted growth is helpful [Hiramatsu 2000]. Growth parameters are best adjusted to have a preferred growth of inclined facets and enhanced lateral growth to reduce dislocations. A study on this coalescence shows that a GaN growth forcing vertical growth followed by an enhanced lateral growth increasing ammonia flow in MOVPE improved the (0002) and (10$\bar{1}$2) ω-scans to FWHMs of 450 and 624 arcsec compared to 528 and 1008 arcsec for a sample with modulated flow in comparison with a sample without enhanced ammonia flow rate, respectively [Naisen 2009].

The material quality is usually observed to improve upon masking with SiN. Figure 4.41 shows a TEM comparison of two layers where a SiN mask has been deposited directly on the AlN seeding layer. With a SiN masking layer material quality improves with increasing thickness until saturation is reached [Dadgar 2003a]. The improvement in material quality is also reflected by the strain observed in XRD and PL measurements (Figure 4.42). A thicker

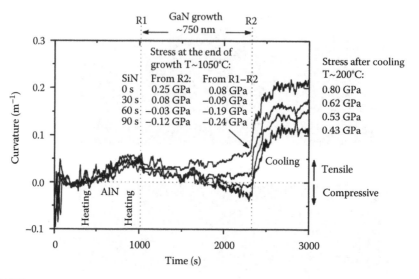

FIGURE 4.42

In situ curvature measurement and stress development of a AlN/SiN/GaN structure on Si(111). Tensile growth stress, induced by island coalescence, is reduced by the insertion of a SiN masking layer. (Reprinted from Dadgar, A. et al., in *Advances in Solid State Physics* 44, Kramer, B. (ed.), Springer, Berlin, Germany, 2004. With permission from Springer, © 2004 Springer.)

SiN mask leads to lower tensile stress after cooling. For this observation, two mechanisms likely apply. First, the increase in island size leads to a reduction in crystallite boundaries. These boundaries, even when the two crystallites are perfectly aligned with regard to twist and tilt, do induce tensile stress. This originates in the independently grown islands which are unlikely to match perfectly on the lateral scale if grown on a lattice mismatched substrate or layer. Thus, the remaining gap is closed over a distance typically larger than the atomic distance and local tensile stress is induced by this. Second, the reduction in dislocation density as shown in Figure 4.41 reduces the amount of dislocations available for dislocation climb and the probability for the introduction of tensile stress.

For material with SiN masking layers, it has been also shown that they are helpful in reducing tensile stress which can be introduced by high silicon doping levels. Silicon does promote the inclination of edge type dislocations [Cantu 2003, Dadgar 2004, Cantu 2005, Dadgar 2007a,b] of which the mechanism, surface roughening [Romanov 2003] or dislocation core masking [Dadgar 2004, Dadgar 2007a] or other effects [Romano 2000, Follstaedt 2005, Raghavan 2005, Wang 2006b, Follsteadt 2009, Manning 2009], is still under debate. Some of them are only valid for compressively strained layers thus cannot be explained in the cases of increasing tensile stress. Figure 4.43 shows the impact of a differently thick SiN masking layer on the in situ measured curvature of a heavily Si-doped GaN layer. The curvature during growth shows a stronger tensile stress for thinner SiN masking layers.

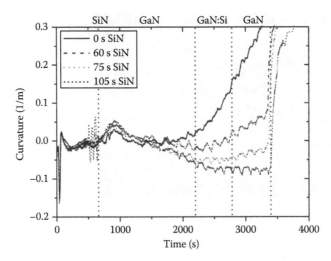

FIGURE 4.43
In situ curvature during MOVPE growth of an AlN/SiN/GaN/GaN.Si/GaN structure with differently thick SiN masking. (Reprinted from Dadgar, A. et al., in *Advances in Solid State Physics* 44, Kramer, B. (ed.), Springer, Berlin, Germany, 2004. With permission from Springer, © 2004 Springer.)

This also demonstrates that low dislocation densities are a prerequisite for highly Si-doped devices on silicon as, e.g., LEDs and that the material quality can be already monitored in situ when the structure involves a highly Si-doped layer where the increase in tensile stress is a direct measure of material quality.

But not for all device types SiN masking is well suited. Even for a perfect SiN layer, the boundary to GaN includes Si–N bonds where Si can act as n-type dopand. In SEM images, often a contrast, which can be attributed to an enhanced carrier concentration, is visible at the position of SiN nanomasks (Figure 4.44). Consequently, the operation of FETs might be negatively influenced by introducing such a masking layer. On the other hand for LEDs such a δ-doped layer might improve current spreading.

4.7.5 Nanorods and Nanowires

To reduce dislocations and also to create novel light-emitting structures, epitaxial structures called nanorods, nanowires, or nanocolumns are well suited. A single nanorod with its large surface does contain no or only few dislocations and if they are present they are likely to eliminate, e.g., simply by bending to a surface. Additionally, their large surface is helpful for light outcoupling making them good candidates for efficient light emitters. But such a large surface also is a high risk with regard to surface states and surface degradation.

Devices based on nanorods are, however, much more complicated to fabricate than for planar epitaxial growth. A simple approach for nanorod

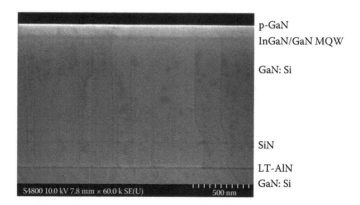

p-GaN
InGaN/GaN MQW

GaN: Si

SiN
LT-AlN
GaN: Si

S4800 10.0 kV 7.8 mm × 60.0 k SE(U)   500 nm

FIGURE 4.44
FE-SEM cross-section image of the upper part of an LED structure on Si. The position of the
SiN masking layer is visible as dark horizontal line. Such a contrast can be correlated to an
enhanced carrier concentration.

growth is by MBE where a low III–V ratio does lead to the growth of 3D
nanorod structures on silicon (Figure 4.45) [Park 2004, Yamashita 2005, Kim
2005, Chen 2006, Calleja 2007]. It has been found that for ~25 nm wide nanorods misfit dislocations are rare and do annihilate at the free surface of the
nanorod [Calleja 2007]. Thus, except for their large surface and some misfit
dislocations at the interface to Si, these single nanorods are perfect crystals
and strain can be easily accommodated in the vertical direction, possibly
enabling the growth of highly strained materials without strain relaxation.
This has been demonstrated by the growth of a nanorod cavity with AlN/
GaN Bragg-reflectors (Figure 4.46) [Ristic 2005].

Another approach for nanorod growth has been demonstrated by the
growth on a GaN buffer layer where the growth conditions of the buffer
layer determine the density of the subsequently grown nanorods at high
V–III ratio [Hsiao 2006] (Figure 4.47).

Being stable against meltback etching under MBE growth conditions as
starting material often GaN is used in MBE on silicon. But to avoid SiN formation at the interface, and by this a possible loss of epitaxial orientation on
the substrate, AlN or a thin Al predeposition is preferential [Calleja 2007].

The typical growth orientation of nanorods on Si(111) [Chen 2006] follows
the same epitaxial relationship as for GaN layers with GaN[0001] ∥ Si[111]
and GaN[11$\bar{2}$0] ∥ Si[$\bar{1}$10] [Morimoto 1973, Lei 1993, Krost 2002]. HVPE grown
nanorods were also observed with a 30° in-plane rotation with GaN[10$\bar{1}$0] ∥
Si[$\bar{1}$00] [Lee 2008]. Growing nanorods on Si(001), as for the growth on planar
substrates [Schulze 2006] where the in-plane orientation has at least two possible alignments, the same two alignments are present. However, also other
not well-oriented nanorods are observed [Cerutti 2006]. Nonpolar a-plane
InN nanorods were reported by Gandal et al. [Grandal 2009]. Here, InN was
grown on a a-plane GaN layer by MBE.

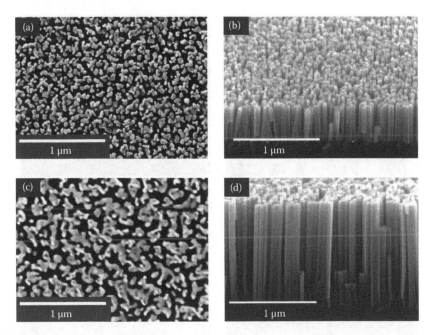

FIGURE 4.45
FE-SEM images of vertically aligned GaN nanorods grown on Si(111) by MBE with 0.4 μm (a and b) and 1 μm (c and d) length in plan view and cross section. (Reprinted from Chen, H.-Y. et al., *Appl. Phys. Lett.*, 89, 243105, 2006. With permission from American Institute of Physics, © 2006 American Institute of Physics.)

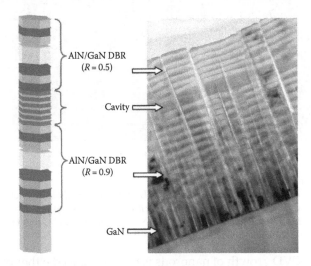

FIGURE 4.46
GaN nanorods with grown in AlN/GaN DBR. Such AlN/GaN layers usually exhibit cracking due to a large lattice mismatch. Here, they can release their stress due to the low diameter of the structure. (Reprinted from Ristić, J. et al., *Phys. Rev. Lett.*, 94, 146102, 2005. With permission from American Physical Society, © 2005 American Physical Society.)

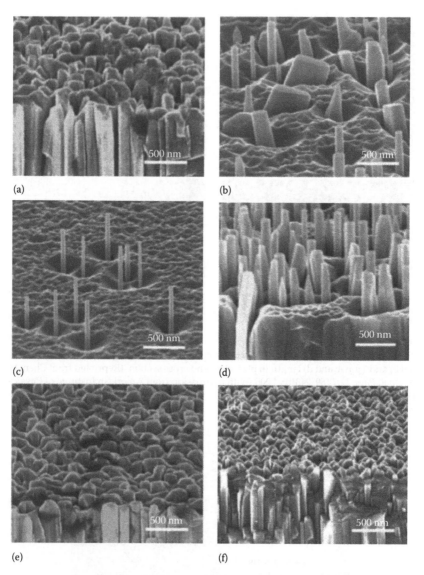

FIGURE 4.47
FE-SEM images of GaN nanorods grown by MBE at different GaN buffer temperatures: (a) 460°C, (b) 510°C, (c) 600°C, (d) 620°C, (e) 690°C, (f) without buffer. Two types of buffer structures are observed: either a dense array or single nanorods in a matrix. (Data from Hsiao, C.L. et al., *J. Vac. Sci. Technol.*, B, 24, 845, 2006.)

MOVPE or CVD growth of nanorods is typically based either on a process involving a metal catalyst [Cai 2006, Simpkins 2007, Dinh 2009, Narukawa 2009] or on structuring and growth on such a structured surface [Fündling 2008] with the latter usually resulting in well-oriented columns (Figure 4.48). For the growth of planar surfaces or enabling simple contacting,

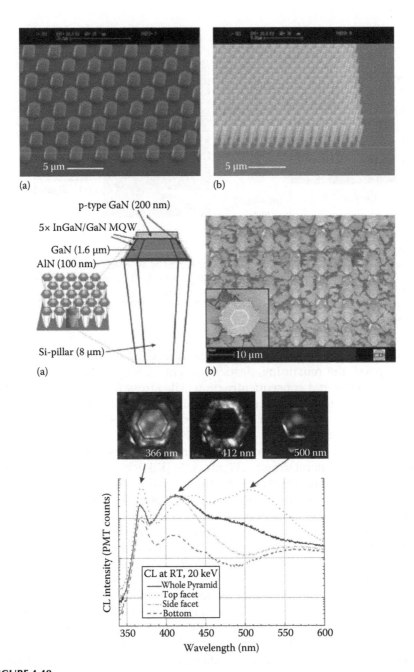

FIGURE 4.48
MOVPE grown nanorods with InGaN/GaN MQWs on a patterned substrate. The benefit of patterning is a regular array of the nanorods. PL is shown in the lower image. For these faceted structures, the emission energy depends on the crystalline facet the MQW was grown on. (Reprinted from Fündling, S. et al., *Phys. Stat. Sol. (a)* 206, 1194, 2009. With permission from Wiley VCH, © 2009 Wiley VCH.)

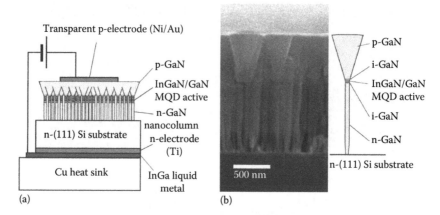

FIGURE 4.49

MBE-grown nanorod LED structures that were coalesced by an Mg-doped p-GaN layer on top. Such a coalesced layer is helpful in contacting the structures which is much more difficult for an uncoalesced structure. (Reprinted from Kikuchi, A. et al., *Jpn. J. Appl. Phys.*, 43, L1524, 2004. With permission from Japan Society of Applied Physics, © 2004 by Japan Society of Applied Physics.)

nanorods need to coalesce to form a continuous layer. If the columns are well aligned, the remaining elasticity might be helpful to achieve a gap-less coalescence and consequently no tensile stress from island (nanorod) coalescence. In sum such a process is similar to an ELO or Pendeo approach but on a smaller scale with a different scaling factor of the diameter of the post and its length.

In MBE, such a planarization has been already demonstrated by adding Mg which promotes lateral growth (Figure 4.49) [Kikuchi 2004, Furtmayr 2008]. An LED has been demonstrated by this growth method (see Chapter 12). However, it has not been applied to the growth of thicker layers yet where the nanorods can act as compliant substrate to accommodate tensile stress occurring during coalescence and also, but only to a small portion, upon cooling. Chen et al. did also report on the formation of dislocations at the coalescence boundary [Chen 2006]. This likely originates in the remaining twist and tilt of the nanorods and the insufficient height vs. diameter ratio for the stiff group-III-nitrides: The low elasticity does most likely inhibit the necessary rotation of coalescing columns. Consequently, using nanorods as templates for the growth of high-quality epilayers is difficult because of the formation of dislocations at coalescence boundaries. A lower density of nanorods would be helpful in reducing this problem, with the disadvantage of a longer coalescence time.

Also for light emitters, the density of the nanorods should not be too high to reduce reabsorption of light. Therefore, techniques based on patterning well-defined positions and density might be better suited than the self-assembled process often employed. By this, photonic bandgap structures can be integrated into the light-emitting device by properly choosing the nano-rod distance, increasing, e.g., vertical light emission and overall efficiency possibly with integrated Bragg mirrors.

4.8 Future Developments

It is never very successful predicting the future of growth techniques and the question is if there will be any revolutionary technique to solve all existing problems or if it is just a question of combining the best-suited growth techniques for the device structure and optimizing them. One should keep in mind that the driving force behind GaN on silicon, except for some niche applications, has always been the application as light emitter and for electronics. Consequently, GaN growth on silicon is highly cost driven and all techniques for material enhancement will therefore be weighted with regard to their cost.

4.8.1 Bow and Homogeneity

But one important point has been left out until now, mostly because nothing relevant has yet been published and the solution is simple: bowing and homogeneity when increasing substrate size. As for silicon but also for, e.g., sapphire or SiC or other substrates (even GaN) bow occurs during MOVPE nitride epitaxy. The first reason for bow is heating (Figure 4.51) [Dadgar 2006b]. Typical MOVPE systems have a heated substrate carrier and a cooled ceiling or showerhead. This leads to a vertical temperature gradient and with it to concave bowing (Figure 4.50). For semi-standard 675 µm thick 150 mm c-plane sapphire substrates, the inhomogeneities correlated with strong bowing usually lead to the destruction of them upon heating to growth temperature. Using a-plane sapphire avoids cracking but bow is still extreme and in the mm range (Figure 4.51). Thicker substrates help to avoid this problem but thinning thick sapphire is not a simple task and costly. For silicon, curvatures are much lower and this cracking and extreme bowing problem does not exist.

In addition to bowing by heating tensile stresses (sapphire) or, for crack-free GaN on silicon, compressive stresses (strain engineered Si) are typical. Temperature ramping, e.g., for InGaN growth will always lead to a change in curvature. Consequently, it is unavoidable that wafers in heteroepitaxy are curved during growth. For InGaN MQWs, this is already a major issue when increasing the substrate diameter from 50 to only 100 mm. The larger wafer diameter leads to a much stronger bow and with it a strong inhomogeneous cooling of the wafer surface [Dadgar 2006a, Krost 2005]. Nevertheless, homogeneous wafers are obtainable on 150 mm Si substrates, at least in single

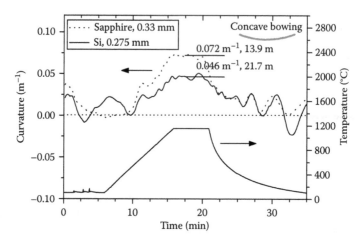

FIGURE 4.50
Curvature measurements showing the curvature (upper lines) for a Si and for a sapphire substrate upon heating to GaN growth temperature. The concave bowing, although relatively small, leads to a significant problem when using large diameter substrates. (Reprinted from Dadgar, A. et al., *Proc. SPIE*, 6355, 63550R, 2006. With permission.)

FIGURE 4.51
a-Plane sapphire substrate at ~650°C wafer carrier temperature after heating to GaN growth temperature. The bow is extreme and lifts the substrate some millimeters at the edge. (Reprinted from Dadgar, A. et al., *Proc. SPIE*, 6355, 63550R, 2006. With permission.)

wafer growth systems with differently heated radial arranged heater zones [Dadgar 2006a, Li 2006, Cheng 2008, Schulze 2008]. Looking at the expected substrate diameters of future GaN on silicon technology, this is a major problem which is only partially compensated by the increasing substrate thickness with increasing substrate diameter. For epitaxy on 100 mm substrates, curved substrate holders are available to achieve a homogenous temperature during the growth of the active layers but this will get increasingly difficult for larger substrate diameters. The best solution for this problem is to use hot wall or at least warm wall reactors with a temperature of the opposite walls identical or only slightly below the substrate temperature. By this, homogenous layers can be easily achieved. The main challenge is to improve gas flow design to avoid the excessive loss of precursors on the heated reactor ceilings.

4.8.2 Strain Management—Cracking

For the most specific problem of GaN on silicon, cracking, many different techniques are nowadays available for the growth of crack-free high-quality GaN on silicon layers. Device design limitations given by strain-engineering techniques can be reduced mostly by optimizing the material quality [Nikishin 1999, Semond 2001]. This has two positive effects: dislocations are reduced and device performance enhanced and strain engineering is more efficient. In return, the latter enables the growth of thicker and higher Si-doped layers without suffering from cracking. Such layers are required for the economically most important topic to date, low-cost high-efficiency light emitters for general illumination.

In general, the techniques that will be used for strain engineering and material improvement will always depend on the application: thick AlN/GaN multilayers are currently of little use for LEDs, but advantageous for high-voltage devices.

But also other device types are advantageous to be grown on silicon: high-voltage devices, which gained big interest in recent years, and surely also MEMS. All such devices will profit from an enhancement in material quality.

In the end, the most important layer to be optimized is always the seeding layer. When growing the seeding layer on a perfect silicon crystal, the crystalline quality of the following seeding and buffer layer is strongly deteriorated in comparison to silicon. Improving poor material quality is a huge effort and requires additional layers and thicker buffers typically ending up only with average results. Thus, growing a high-quality seeding layer reduces the effort to improve the material quality and increases the freedom in device design.

Acknowledgment

Part of this work has been or is still funded by several BMBF and DFG within the framework of SPP "Group III-Nitrides and Their Heterostructures: Growth, Characterization and Applications" and FOR957 which we gratefully acknowledge.

References

[Able 2005] A. Able, W. Wegscheider, K. Engl, and J. Zweck, *Journal of Crystal Growth* **276**, 415 (2005).

[Adachi 1996] S. Adachi, T. Arai, and K. Kobayashi, *Journal of Applied Physics* **80**, 5422 (1996).

[Amano 1998] H. Amano, M. Iwaya, T. Kashima, M. Katsuragawa, I. Akasaki, J. Han, S. Hearne, J.A. Floro, E. Chason, and J. Figiel, *Japanese Journal of Applied Physics* **37**, L1540 (1998).

[Arslan 2008] E. Arslan, M.K. Ozturk, A. Teke, S. Ozcelik, and E. Ozbay, *Journal of Physics D: Applied Physics* **41**, 155317 (2008).

[Aumann 1992] C. Aumann, J.J. de Miguel, R. Kariotis, and M.G. Lagally, *Surface Science* **275**, 1 (1992).

[Baron 2009] N. Baron, Y. Cordier, S. Chenot, P. Vennéguès, O. Tottereau, M. Leroux, F. Semond, and J. Massies, *Journal of Applied Physics* **105**, 033701 (2009).

[Blaesing 2002] J. Bläsing, A. Reiher, A. Dadgar, A. Diez, and A. Krost, *Applied Physics Letters* **81**, 2722 (2002).

[Boufaden 2004] T. Boufaden, A. Matoussi, S. Guermazi, S. Juillaguet, A. Toureille, and Y. Mlik, *Physics Status Solidi* (a) **201**, 582 (2004).

[Bouzynin 2008] Yu.N. Buzynin, Yu.N. Drozdov, M.N. Drozdov, A.Yu. Luk'yanov, O.I. Khrykin, A.N. Buzynin, A.E. Luk'yanov, E.I. Rau, and F.A. Luk'yanov, *Bulletin of the Russian Academy of Sciences: Physics* **72**, 1499 (2008).

[Brink 2004] D.J. Brink, J. Camassel, and J.B. Malherbe, *Thin Solid Films* **449**, 73 (2004).

[Bykhovski 1997] A.D. Bykhovski, B.L. Gelmont, and M.S. Shur, *Journal of Applied Physics* **81**, 6332 (1997).

[Cai 2006] X.M. Cai, A.B. Djurišić, and M.H. Xie, *Thin Solid Films* **515**, 984 (2006).

[Calleja 2007] E. Calleja, J. Ristić, S. Fernández-Garrido, L. Cerutti, M.A. Sánchez-García, J. Grandal, A. Trampert, U. Jahn, G. Sánchez, A. Griol, and B. Sánchez, *Physics Status Solidi* (b) **244**, 2816 (2007).

[Cantu 2003] P. Cantu, F. Wu, P. Waltereit, S. Keller, A.E. Romanov, U.K. Mishra, S.P. DenBaars, and J.S. Speck, *Applied Physics Letters* **83**, 674 (2003).

[Cantu 2005] P. Cantu, F. Wu, P. Waltereit, S. Keller, A.E. Romanov, S.P. DenBaars, and J.S. Speck, *Journal of Applied Physics* **97**, 103534 (2005).

[Cao 1998] J. Cao, D. Pavlidis, Y. Park, J. Singh, and A. Eisenbach, *Journal of Applied Physics* **83**, 3829 (1998).

[Castaldini 2002] A. Castaldini, D. Cavalcoli, A. Cavallini, D. Jones, V. Palermo, and E. Susi, *Journal of the Electrochemical Society* **149**, G633 (2002).

[Cerutti 2006] L. Cerutti, J. Ristić, S. Fernández-Garrido, E. Calleja, A. Trampert, K.H. Ploog, S. Lazic, and J.M. Calleja, *Applied Physics Letters* **88**, 213114 (2006).

[Cha 2006] H.-Y. Cha, H. Wu, S. Chae, and M.G. Spencer, *Journal of Applied Physics* **100**, 024307 (2006).

[Chadi 1984] D.J. Chadi, *Physical Review B* **29**, 785 (1984).

[Chama 2003] V. Chamard, T.H. Metzger, M. Sztucki, V. Holý, and M. Tolan, *Europhysics Letters* **63**, 268 (2003).

[Chen 2006] H.-Y. Chen, H.-W. Lin, C.-H. Shen, and S. Gwo, *Applied Physics Letters* **89**, 243105 (2006).

[Cheng 2008] K. Cheng, M. Leys, S. Degroote, J. Derluyn, B. Sijmus, P. Favia, O. Richard, H. Bender, M. Germain, and G. Borghs, *Japanese Journal of Applied Physics* **47**, 1553 (2008).

[Choi 2005] H.-J. Choi, H.-K. Seong, J. Chang, K.-I. Lee, Y.-J. Park, J.-J. Kim, S.-K. Lee, R. He, T. Kuykendall, and P. Yang, *Advanced Materials* **17**, 1351 (2005).

[Contreras 2008] O.E. Contreras, F. Ruiz-Zepeda, A. Dadgar, A. Krost, and F.A. Ponce, *Applied Physics Express* **1**, 061104 (2008).

[Cordier 2005a] Y. Cordier, M. Hugues, F. Semond, F. Natali, P. Lorenzini, Z. Bougrioua, J. Massies, E. Frayssinet, B. Beaumont, P. Gibart, and J.-P. Faurie, *Journal of Crystal Growth* **278**, 383 (2005).

[Cordier 2005b] Y. Cordier, F. Semond, M. Hugues, F. Natali, P. Lorenzini, H. Haas, S. Chenot, M. Laügt, O. Tottereau, P. Vennegues, and J. Massies, *Journal of Crystal Growth* **278**, 393 (2005).

[Dadgar 2000] A. Dadgar, J. Bläsing, A. Diez, A. Alam, M. Heuken, and A. Krost, *Japanese Journal of Applied Physics* **39**, L1183 (2000).

[Dadgar 2001a] A. Dadgar, M. Poschenrieder, J. Bläsing, K. Fehse, T. Riemann, A. Diez, J. Christen, and A. Krost, *MRS Fall Meeting*, Boston, MA, I4.7, 2001.

[Dadgar 2001b] A. Dadgar, A. Alam, T. Riemann, J. Bläsing, A. Diez, M. Poschenrieder, M. Straßburg, M. Heuken, J. Christen, and A. Krost, *Physica Status Solidi (a)* **188**, 155 (2001).

[Dadgar 2003a] A. Dadgar, M. Poschenrieder, A. Reiher, J. Bläsing, J. Christen, A. Krtschil, T. Finger, T. Hempel, A. Diez, and A. Krost, *Applied Physics Letters* **82**, 28 (2003).

[Dadgar 2003b] A. Dadgar, A. Strittmatter, J. Bläsing, M. Poschenrieder, O. Contreras, P. Veit, T. Riemann, F. Bertram, A. Reiher, A. Krtschil, A. Diez, T. Hempel, T. Finger, A. Kasic, M. Schubert, D. Bimberg, F.A. Ponce, J. Christen, and A. Krost, *Physica Status Solidi (c)* **0**, 1583 (2003).

[Dadgar 2004] A. Dadgar, R. Clos, G. Strassburger, F. Schulze, P. Veit, T. Hempel, J. Bläsing, A. Krtschil, I. Daumiller, M. Kunze, A. Kaluza, A. Modlich, M. Kamp, A. Diez, J. Christen, and A. Krost, in *Advances in Solid State Physics* 44, Kramer, B. (ed.), Springer, Berlin, Germany, 2004.

[Dadgar 2006a] A. Dadgar, C. Hums, A. Diez, J. Bläsing, and A. Krost, *Journal of Crystal Growth* **297**, 279 (2006).

[Dadgar 2006b] A. Dadgar, C. Hums, A. Diez, F. Schulze, J. Bläsing, and A. Krost, *Proceedings of SPIE* **6355**, 63550R (2006).

[Dadgar 2007a] A. Dadgar, F. Schulze, M. Wienecke, A. Gadanecz, J. Bläsing, P. Veit, T. Hempel, A. Diez, J. Christen, and A. Krost, *New Journal of Physics* **9**, 389 (2007).

[Dadgar 2007b] A. Dadgar, P. Veit, F. Schulze, J. Bläsing, A. Krtschil, H. Witte, A. Diez, T. Hempel, J. Christen, R. Clos, and A. Krost, *Thin Solid Films* **515**, 4356 (2007).

[Daudin 1997] B. Daudin, F. Widmann, G. Feuillet, Y. Samson, M. Arlery, and J.L. Rouvière, *Physical Review B* **56**, R7069 (1997).

[Davis 2000] R.F. Davis, T. Gehrke, K.J. Linthicum, T.S. Zheleva, P. Rajagopal, C.A. Zorman, and M. Mehregany, *MRS Internet Journal of Nitride Semiconductor Research* **5S1**, W2.1 (2000).

[Davis 2001a] R.F. Davis, T. Gehrke, K.J. Linthicum, P. Rajagopal, A.M. Roskowski, T. Zheleva, E.A. Preble, C.A. Zorman, M. Mehregany, U. Schwarz, J. Schuck, and R. Grober, *MRS Internet Journal of Nitride Semiconductor Research* **6**, 14 (2001).

[Davis 2001b] R.F. Davis, T. Gehrke, K.J. Linthicum, E. Preble, P. Rajagopal, C. Ronning, C. Zorman, and M. Mehregany, *Journal of Crystal Growth* **231**, 335 (2001).

[Deibuk 2005] V.G. Deìbuk and A.V. Voznyì, *Semiconductors*, **39**, 623 (2005).

[Dinh 2009] D.V. Dinh, J.H. Yang, S.M. Kang, S.W. Kim, and D.H. Yoon, *Journal of the Korean Physical Society* **55**, 202 (2009).

[Fang 2001] Z.-Q. Fang, D.C. Look, J. Jasinski, M. Benamara, Z. Liliental-Weber, and R.J. Molnar, *Applied Physics Letters* **78**, 332 (2001).

[Feltin 2001a] E. Feltin, B. Beaumont, M. Laügt, P. de Mierry, P. Vennéguès, H. Lahrèche, M. Leroux, and P. Gibart, *Applied Physics Letters* **79**, 3230 (2001).

[Feltin 2001b] E. Feltin, B. Beaumont, M. Laügt, P. de Mierry, P. Vennéguès, M. Leroux, and P. Gibart, *Physics Status Solidi* (a) **188**, 531 (2001).

[Feltin 2001c] E. Feltin, B. Beaumont, P. Vennéguès, T. Riemann, J. Christen, J.P. Faurie, and P. Gibart, *Physics Status Solidi* (a) **188**, 733 (2001).

[Feltin 2003] E. Feltin, B. Beaumont, P. Vennéguès, M. Vaille, P. Gibart, T. Riemann, J. Christen, A. Dobos, and B. Pécz, *Journal of Applied Physics* **93**, 182 (2003).

[Fenwick 2009] W.E. Fenwick, N. Li, T. Xu, A. Melton, S. Wang, H. Yu, C. Summers, M. Jamil, and I.T. Ferguson, *Journal of Crystal Growth* **311**, 4306 (2009).

[Ferhat 2002] M. Ferhat and F. Bechstedt, *Physical Review B* **65**, 075213 (2002).

[Follstaedt 2005] D.M. Follstaedt, S.R. Lee, P.P. Provencio, A.A. Allerman, J.A. Floro, and M.H. Crawford, *Applied Physics Letters* **87**, 121112 (2005).

[Follstaedt 2009] D.M. Follstaedt, S.R. Lee, A.A. Allerman, and J.A. Floro, *Journal of Applied Physics* **105**, 083507 (2009).

[Fündling 2008] S. Fündling, Ü. Sökmen, E. Peiner, T. Weimann, P. Hinze, U. Jahn, A. Trampert, H. Riechert, A. Bakin, H.-H. Wehmann, and A. Waag, *Nanotechnology* **19**, 405301 (2008).

[Fündling 2009] S. Fündling, S. Li, Ü. Sökmen, S. Merzsch, P. Hinze, T. Weimann, U. Jahn, A. Trampert, H. Riechert, E. Peiner, H.-H. Wehmann, and A. Waag, *Physica Status Solidi (a)* **206**, 1194 (2009).

[Furtmayr 2008] F. Furtmayr, M. Vielemeyer, M. Stutzmann, J. Arbiol, S. Estradé, F. Peirò, J.R. Morante, and M. Eickhoff, *Journal of Applied Physics* **104**, 034309 (2008).

[Gallois 2005] B.M. Gallois, T.M. Besmann, and M.W. Stott, *Journal of the American Ceramic Society* 77, 2949 (2005).

[Gehrke 2000] T. Gehrke, K.J. Linthicum, P. Rajagopal, E.A. Preble, and R.F. Davis, *MRS Internet Journal of Nitride Semiconductor Research* **5S1**, W2.4 (2000).

[Gosh 2003] B.K. Ghosh, T. Tanikawa, A. Hashimoto, A. Yamamoto, and Y. Ito, *Journal of Crystal Growth* **249**, 422 (2003).

[Grandal 2009] J. Grandal, M.A. Sánchez-García, E. Calleja, E. Gallardo, J.M. Calleja, E. Luna, A. Trampert, and A. Jahn, *Applied Physics Letters* **94**, 221908 (2009).

[Grundmann 1991] M. Grundmann, A. Krost, and D. Bimberg, *Applied Physics Letters* **58**, 284 (1991).

[Guha 1998a] S. Guha and N.A. Bojarczuk, *Applied Physics Letters* **72**, 415 (1998).

[Guha 1998b] S. Guha and N.A. Bojarczuk, *Applied Physics Letters* **73**, 1487 (1998).

[Ham 2006] M.-H. Ham, J.-H. Choi, W. Hwang, C. Park, W.-Y. Lee, and J.-M. Myoung, *Nanotechnology* **17**, 2203 (2006).

[Hageman 2001] P.R. Hageman, S. Haffouz, V. Kirilyuk, A. Grzegorczyk, and P.K. Larsen, *Physica Status Solidi (a)* **188**, 523 (2001).

[Hellman 1998] E.S. Hellman, D.N.E. Buchanan, and C.H. Chen, *MRS Internet Journal of Nitride Semiconductor Research* **3**, 43 (1998).

[Hiramatsu 2000] K. Hiramatsu, K. Nishiyama, M. Onishi, H. Mizutani, M. Narukawa, A. Motogaito, H. Miyake, Y. Iyechika, and T. Maeda, *Journal of Crystal Growth* **221**, 316 (2000).

[Honda 2000] Y. Honda, Y. Kawaguchi, T. Kato, M. Yamaguchi, and N. Sawaki, in *Proceedings of the International Workshop on Nitride Semiconductors, Nagoya, Japan, IPAP Conference Series* **1**, 304 (2000).

[Honda 2002] Y. Honda, Y. Kuroiwa, M. Yamaguchi, and N. Sawaki, *Applied Physics Letters* **80**, 222 (2002).

[Honda 2007] Y. Honda, S. Kato, M. Yamaguchi, and N. Sawaki, *Physica Status Solidi (c)* **4**, 2740 (2007).

[Hsiao 2006] C.L. Hsiao, L.W. Tu, T.W. Chi, H.W. Seo, Q.Y. Chen, and W.K. Chu, *Journal of Vacuum Science and Technology* B **24**, 845 (2006).

[Hums 2007] C. Hums, J. Bläsing, A. Dadgar, A. Diez, T. Hempel, J. Christen, A. Krost, K. Lorenz, and E. Alves, *Applied Physics Letters* **90**, 022105 (2007).

[Irie 2009] M. Irie, N. Koide, Y. Honda, M. Yamaguchi, and N. Sawaki, *Journal of Crystal Growth* **311**, 2891 (2009).

[Ishikawa 1998] H. Ishikawa, K. Yamamoto, T. Egawa, T. Soga, T. Jimbo, and M. Umeno, *Journal of Crystal Growth* **189/190**, 178 (1998).

[Ishikawa 1999a] H. Ishigawa, G.Y. Zhao, N. Nakada, T. Egawa, T. Soga, T. Jimbo, and M. Umeno, *Physica Status Solidi (a)* **176**, 599 (1999).

[Ishikawa 1999b] H. Ishigawa, G.-Y. Zhao, N. Nakada, T. Egawa, T. Jimbo, and M. Umeno, *Japanese Journal of Applied Physics* **38**, L492 (1999).

[Ishikawa 2008] H. Ishikawa, K. Shimanaka, F. Tokura, Y. Hayashi, Y. Hara, and M. Nakanishi, *Journal of Crystal Growth* **310**, 4900 (2008).

[Ito 1978] T. Ito, T. Nozaki, H. Arakawa, and M. Shinoda, *Applied Physics Letters* **32**, 330 (1978).

[Iwakami 2004] S. Iwakami, M. Yanagihara, O. Machide, E. Chino, N. Kaneko, H. Goto, and K. Ohtsuka, *Japanese Journal of Applied Physics* **43**, L831 (2004).

[Iwaya 2000] M. Iwaya, S. Terao, N. Hayashi, T. Kashima, H. Amano, and I. Akasaki, *Applied Surface Science* **159–160**, 405 (2000).

[Jiang 2009] F. Jiang, L. Wang, X. Wang, C. Mo, X. You, C. Zheng, W. Liu, Y. Zhou, C. Xiong, Y. Tang, W. Fang, B. Lu, in *High Power InGaN-Based Blue LEDs Grown on Si Substrate by MOCVD*, H7, ICNS-8, Jeju, Korea, 2009.

[Joblot 2005a] S. Joblot, E. Feltin, E. Beraudo, P. Vennéguès, M. Leroux, F. Omnès, M. Laügt, and Y. Cordier, *Journal of Crystal Growth* **280**, 44 (2005).

[Joblot 2005b] S. Joblot, F. Semond, F. Natali, P. Vennéguès, M. Laügt, Y. Cordier, and J. Massies, *Physica Status Solidi (c)* **2**, 2187 (2005).

[Jona 1965] F. Jona, *Applied Physics Letters* **6**, 205 (1965).

[Kang 2001] H. Kang, M.K. Kwon, J.I. Rho, J.W. Yang, K.Y. Lim, and K.S. Nahm, *Physica Status Solidi (a)* **188**, 527 (2001).

[Karpov 2004] S.Y. Karpov, N. Podolskaya, I.A. Zhmakin, and A.I. Zhmakin, *Physical Review B* **70**, 235203 (2004).

[Katona 2001] T.M. Katona, M.D. Craven, P.T. Fini, J.S. Speck, and S.P. DenBaars, *Applied Physics Letters* **79**, 2907 (2001).

[Kawaguchi 1998] Y. Kawaguchi, Y. Honda, H. Matsushima, M. Yamaguchi, K. Hiramatsu, and N. Sawaki, *Japanese Journal of Applied Physics* **37**, L966 (1998).

[Kelly 1997] M.K. Kelly, O. Ambacher, R. Dimitrov, R. Handschuh, and M. Stutzmann, *Physica Status Solidi (a)* **159**, R3 (1997).

[Kikuchi 2004] A. Kikuchi, M. Kawai, M. Tada, and K. Kishino, *Japanese Journal of Applied Physics* **43**, L 1524 (2004).

[Kim 2001] M.-H. Kim, Y-G. Do, H.C. Kang, D.Y. Noh, and S.-J. Park, *Applied Physics Letters* **79**, 2713 (2001).

[Kim 2005] Y.H. Kim, J.Y. Lee, S.-H. Lee, J.-E. Oh, and H.S. Lee, *Applied Physics A* **80**, 1635–1639 (2005).

[Kobayashi 1997] N.P. Kobayashi, J.T. Kobayashi, P.D. Dapkus, W.-J. Choi, and A.E. Bond, *Applied Physics Letters* **71**, 3569 (1997).

[Kobayashi 1998] N.P. Kobayashi, J.T. Kobayashi, X. Zhang, P.D. Dapkus, and D.H. Rich, *Applied Physics Letters* **74**, 2836 (1999).

[Komiyama 2009] J. Komiyama, Y. Abe, S. Suzuki, H. Nakanishi, and A. Koukitu, *Journal of Crystal Growth* **311**, 2840 (2009).

[Krost 2000] A. Krost, J. Bläsing, F. Schulze, O. Schön, A. Alam, and M. Heuken, *Journal of Crystal Growth* **221**, 251 (2000).

[Krost 2002] A. Krost and A. Dadgar, *Material Science and Engineering* **B93**, 77 (2002).

[Krost 2003] A. Krost, A. Dadgar, G. Strassburger, and R. Clos, *Physica Status Solidi (a)* **200**, 26 (2003).

[Krost 2005] A. Krost, F. Schulze, A. Dadgar, G. Straßburger, K. Haberland, and T. Zettler, *Physica Status Solidi (b)* **242**, 2570 (2005).

[Kung 1999] P. Kung, D. Walker, M. Hamilton, J. Diaz, and M. Razeghi, *Applied Physics Letters* **74**, 570 (1999).

[Lebedev 2001] V. Lebedev, J. Jinschek, J. Kräusslich, U. Kaiser, B. Schröter, and W. Richter, *Journal of Crystal Growth* **230**, 426 (2001).

[Lee 2007] S.-K. Lee, T.-H. Kim, S.-Y. Lee, K.-C. Choi, and P. Yang, *Philosophical Magazine* **87**, 2105 (2007).

[Lee 2008] K.H. Lee, Y.H. Kwon, S.Y. Ryu, T.W. Kang, J.H. Jung, D.U. Lee, and T.W. Kim, *Journal of Crystal Growth* **310**, 2977 (2008).

[Lei 1993] T. Lei, K.F. Ludwig Jr., and T.D. Moustakas, *Journal of Applied Physics* **74**, 4430 (1993).

[Leys 2008] M. Leys, K. Cheng, J. Derluyn, S. Degroote, M. Germain, G. Borghs, C.A. Taylor, and P. Dawson, *Journal of Crystal Growth* **310**, 4888 (2008).

[Li 2006] J. Li, J.Y. Lin, and H.X. Jiang, *Applied Physics Letters* **88**, 171909 (2006).

[Liaw 2000] H.M. Liaw, R. Doyle, P.L. Fejes, S. Zollner, A. Konkar, K.J. Linthicum, T. Gehrke, and R.F. Davis, *Solid State Electronics* **44**, 747 (2000).

[Lin 2007] K.-L. Lin, E.-Y. Chang, Y.-L. Hsiao, W.-C. Huang, T. Li, D. Tweet, J.-S. Maa, S.-T. Hsu, and C.-T. Lee, *Applied Physics Letters* **91**, 222111 (2007).

[Linthicum 1999] K. Linthicum, T. Gehrke, D. Thomson, E. Carlson, P. Rajagopal, T. Smith, D. Batchelor, and R. Davis, *Applied Physics Letters* **75**, 196 (1999).

[Linthicum 2000] K.J. Linthicum, T. Gehrke, R.F. Davis, D.B. Thomson, and K.M. Tracy, US 6,255,198.

[Liu 2004] R. Liu, F.A. Ponce, A. Dadgar, and A. Krost, *Applied Physics Letters* **83**, 860 (2003).

[Manasevit 1971] H.M. Manasevit, F.M. Erdmann, and W.J. Simpson, *Journal of the Electrochemical Society* **118**, 1864 (1971).

[Manning 2009] I.C. Manning, X. Weng, J.D. Acord, M.A. Fanton, D.W. Snyder, and J.M. Redwing, *Journal of Applied Physics* **106**, 023506 (2009).

[Marchand 1999] H. Marchand, N. Zhang, L. Zhao, Y. Golan, S.J. Rosner, G. Girolami, P.T. Fini, J.P. Ibbetson, S.P. DenBaars, J.S. Speck, and U.K. Mishra, *MRS Internet Journal of Nitride Semiconductor Research* **4**, 2 (1999).

[Marchand 2001] H. Marchand, L. Zhao, N. Zhang, B. Moran, R. Coffie, U.K. Mishra, J.S. Speck, S.P. DenBaars, and J.A. Freitas, *Journal of Applied Physics* **89**, 7846 (2001).

[Min 2002] B. Min, J.S. Lee, K. Cho, J.W. Hwang, H. Kim, M.Y. Sung, S. Kim, J. Park, H.W. Seo, S.Y. Bae, M.-S. Lee, S.O. Park, and J.-T. Moon, *Journal of Electronic Materials* **32**, 1344 (2003).

[Missaoui 2000] A. Missaoui, M. Saadoun, H. Ezzaouia, B. Bessais, T. Boufaden, A. Rebey, and B. El Jani, *Physica Status Solidi (a)* **182**, 189 (2000).

[Missaoui 2001] A. Missaoui, M. Saadoun, T. Boufaden, B. Bessaïs, A. Rebey, H. Ezzaouia, and B. El Jani, *Materials Science and Engineering* **B82**, 98 (2001).

[Morimoto1973] Y. Morimoto, K. Uchiho, and S. Ushio, *Journal of the Electrochemical Society: Solid-State Science and Technology* **120**, 1783 (1973).

[Morita 2001] Y. Morita, T. Ishida, and H. Tokumoto, *Surface Science* **486**, L524 (2001).

[Morizane 1977] K. Morizane, *Journal of Crystal Growth* **38**, 249 (1977).

[Naisen 2009] Y. NaiSen, W. Yong, W. Hui, N. KaiWei, and L. KeiMay, *Science in China Series E: Technological Sciences* **52**, 2758 (2009).

[Naoi 2003] H. Naoi, M. Narukawa, H. Miyake, and K. Hiramatsu, *Journal of Crystal Growth* **248**, 573 (2003).

[Narukawa 2009] M. Narukawa, S. Koide, H. Miyake, and K. Hiramatsu, *Journal of Crystal Growth* **311**, 2970 (2009).

[Nikishin 1999] S.A. Nikishin, N.N. Faleev, V.G. Antipov, S. Francoeur, L. Grave de Peralta, G.A. Seryogin, H. Temkin, T.I. Prokofyeva, M. Holtz, and S.N.G. Chu, *Applied Physics Letters* **75**, 2073 (1999).

[Nishimura 2001] S. Nishimura and K. Terashima, *Materials Science and Engineering* **B82**, 25 (2001).

[Nishimura 2004] S. Nishimura, S. Matsumoto, and K. Terashima, *Physica Status Solidi (c)* **1**, 238 (2004).

[Nitronex 2005] US2005/0285142.

[Owman1994] F. Owman and P. Martensson, *Surface Science* **303**, L367 (1994).

[Owman 1995] F. Owman and P. Martensson, *Surface Science* **324**, 211 (1995).

[Park 2001] C.I. Park, J.H. Kang, K.C. Kim, E.-K. Suh, K.Y. Lim, and K.S. Nahm, *Journal of Crystal Growth* **224**, 190 (2001).

[Park 2004] Y.S. Park, C.M. Park, D.J. Fu, T.W. Kang, and J.E. Oh, *Applied Physics Letters* **85**, 5718 (2004).

[Polyakov 2009] A.Y. Polyakov, A.V. Markov, M.V. Mezhennyi, A.V. Govorkov, V.F. Pavlov, N.B. Smirnov, A.A. Donskov, L.I. D'yakonov, Y.P. Kozlova, S.S. Malakhov, T.G. Yugova, V.I. Osinsky, G.G. Gorokh, N.N. Lyahova, V.B. Mityukhlyaev, and S.J. Pearton, *Applied Physics Letters* **94**, 022114 (2009).

[Raghavan 2005] S. Raghavan, X. Weng, E. Dickey, and J.M. Redwing, *Applied Physics Letters* **87**, 121112 (2005).

[Raghavan 2006] S. Raghavan, X. Weng, E. Dickey, and J.M. Redwing, *Applied Physics Letters* **88**, 041904 (2006).

[Ravash 2009] R. Ravash, J. Bläsing, T. Hempel, M. Noltemeyer, A. Dadgar, J. Christen, and A. Krost, *Applied Physics Letters* **95**, 242101 (2009).

[Reiher 2003] A. Reiher, J. Bläsing, A. Dadgar, A. Diez, and A. Krost, *Journal of Crystal Growth* **248**, 563 (2003).

[Reiher 2009] F. Reiher, Thesis, Otto-von-Guericke-Universität Magdeburg, Germany 2009.

[Reiher 2010] F. Reiher, A. Dadgar, J. Bläsing, M. Wieneke, and A. Krost, *Journal of Crystal Growth* **312**, 180 (2010).

[Riemann 2003] T. Riemann, J. Christen, U. Haboeck, A. Hoffmann, C. Thomsen, P. Veit, R. Clos, A. Dadgar, and A. Krost, *ICNS5*, Nara, Japan, 2003.

[Ristic 2005] J. Ristić, E. Calleja, A. Trampert, S. Fernández-Garrido, C. Rivera, U. Jahn, and K.H. Ploog, *Physical Review Letters* **94**, 146102 (2005).

[Ristic 2008] J. Ristic, E. Calleja, S. Fernández-Garrido, L. Cerutti, A. Trampert, U. Jahn, and K.H. Ploog, *Journal of Crystal Growth* **310**, 4035 (2008).

[Romano 2000] L.T. Romano, C.G. Van de Walle, J.W. Ager III, W. Götz, and R.S. Kern, *Journal of Applied Physics* **87**, 7745 (2000).

[Romanov 2003] A.E. Romanov and J.S. Speck, *Applied Physics Letters* **83**, 2569 (2003).

[Saengkaew 2009a] P. Saengkaew, A. Dadgar, J. Blaesing, T. Hempel, P. Veit, J. Christen, and A. Krost, *Journal of Crystal Growth* **311**, 3742 (2009).

[Saengkaew 2009b] P. Saengkaew, A. Dadgar, J. Blaesing, B. Bastek, F. Bertram, F. Reiher, C. Hums, M. Noltemeyer, T. Hempel, P. Veit, J. Christen, and A. Krost, *Physica Status Solidi (c)* **6**, S455 (2009).

[Sano 1976] M. Sano and M. Aoki, *Japanese Journal of Applied Physics* **15**, 1943 (1976).

[Sari 2008] A.H. Sari, S. Ghorbani, D. Dorranian, P. Azadfar, A.R. Hojabri, and M. Ghoranneviss, *Applied Surface Science* **255**, 2180 (2008).

[Schulze 2004] F. Schulze, A. Dadgar, J. Bläsing, and A. Krost, *Journal of Crystal Growth* **272**, 496 (2004).

[Schulze 2004b] F. Schulze, J. Bläsing, A. Dadgar, and A. Krost, *Applied Physics Letters* **84**, 4747 (2004).

[Schulze 2006] F. Schulze, A. Dadgar, J. Blaäsing, T. Hempel, A. Diez, J. Christen, and A. Krost, *Journal of Crystal Growth* **289**, 485 (2006).

[Schulze 2007a] F. Schulze, A. Dadgar, F. Bertram, J. Bläsing, A. Diez, P. Veit, R. Clos, J. Christen, and A. Krost, *Physica Status Solidi (c)* **4**, 41 (2007).

[Schulze 2007b] F. Schulze, O. Kisel, A. Dadgar, A. Krtschil, J. Bläsing, M. Kunze, I. Daumiller, T. Hempel, A. Diez, R. Clos, J. Christen, and A. Krost, *Journal of Crystal Growth* **299**, 399 (2007).

[Schulze 2008] F. Schulze, A. Dadgar, A. Krtschil, C. Hums, L. Reissmann, A. Diez, J. Christen, and A. Krost, *Physica Status Solidi (c)* **5**, 2238 (2008).

[Schwarzentruber 1993] B.S. Schwartzentruber, N. Kitamura, M.G. Lagally, and M.B. Webb, *Physical Review B* **47**, 13432 (1993).

[Semond 2001] F. Semond, P. Lorenzini, N. Grandjean, and J. Massies, *Applied Physics Letters* **76**, 1842 (2000).

[Simpkins 2007] B.S. Simpkins, P.E. Pehrsson, M.L. Taheri, and R.M. Stroud, *Journal of Applied Physics* **101**, 094305 (2007).

[Song 2002] J.-H. Song, T. Chikyow, Y.-Z. Yoo, P. Ahmet, and H. Koinuma, *Japanese Journal of Applied Physics* **41**, L1291 (2002).

[Steckl 1996] A.J. Steckl, J. Devrajan, C. Tran, and R.A. Stall, *Applied Physics Letters* **69**, 2264 (1996).

[Stern 2005] E. Stern, G. Cheng, E. Cimpoiasu, R. Klie, S. Guthrie, J. Klemic, I. Kretzschmar, E. Steinlauf, D. Turner-Evans, E. Broomfield, J. Hyland, R. Koudelka, T. Boone, M. Young, A. Sanders, R. Munden, T. Lee, D. Routenberg, and M.A. Reed, *Nanotechnology* **16**, 2941 (2005).

[Strittmatter 1999] A. Strittmatter, A. Krost, M. Straßburg, V. Türck, D. Bimberg, J. Bläsing, and J. Christen, *Applied Physics Letters* **74**, 1242 (1999).

[Strittmatter 2001] A. Strittmatter, S. Rodt, L. Reißmann, D. Bimberg, H. Schröder, E. Obermeier, T. Riemann, J. Christen, and A. Krost, *Applied Physics Letters* **78**, 727 (2001).

[Strittmatter 2009] A. Strittmatter, TU-Berlin.

[Sugahara 2004] T. Sugahara, J.-S. Lee, and K. Ohtsuka, *Japanese Journal of Applied Physics* **43**, L 1595 (2004).

[Sunkara 2001] M.K. Sunkara, S. Sharma, R. Miranda, G. Lian, and E.C. Dickey, *Applied Physics Letters* **79**, 1546 (2001).

[Takeuchi 1991] T. Takeuchi, H. Amano, K. Hiramatsu, N. Sawaki, and I. Akasaki, *J. Cryst. Growth* **115**, 634 (1991).

[Tanaka 2000a] S. Tanaka, M. Takeuchi, and Y. Aoyagi, *Japanese Journal of Applied Physics* **39**, L831 (2000).

[Tanaka 2000b] S. Tanaka, Y. Kawaguchi, K. Yamada, N. Sawaki, M. Hibino, and K. Hiramatsu, *IPAP Conference Series* **1**, 300 (2000).

[Tanaka 2001] S. Tanaka, Y. Honda, N. Sawaki, and M. Hibino, *Applied Physics Letters* **79**, 955 (2001).

[Tanaka 2002] S. Tanaka, Y. Honda, N. Kameshiro, R. Iwasaki, N. Sawaki, and T. Tanji, *Japanese Journal of Applied Physics* **41**, L 846 (2002).

[Tanikawa 2008a] T. Tanikawa, T. Hikosaka, Y. Honda, M. Yamaguchi, and N. Sawaki, *Physica Status Solidi (c)* **5**, 2966 (2008).

[Tanikawa 2008b] T. Tanikawa, D. Rudolph, T. Hikosaka, Y. Honda, M. Yamaguchi, and N. Sawaki, *Journal of Crystal Growth* **310**, 4999 (2008).

[Telieps 1985] W. Telieps and E. Bauer, *Surface Science* **162**, 163 (1985).

[Tomita 2002] N. Tomita and S. Adachi, *Journal of the Electrochemical Society* **149**, G245 (2002).

[Ubukata 2007] A. Ubukataa, K. Ikenagaa, N. Akutsua, A. Yamaguchi, K. Matsumotoa, T. Yamazakia, and T. Egawa, *Journal of Crystal Growth* **298**, 198 (2007).

[Usui 1997] A. Usui, H. Sunakawa, A. Sakai, and A.A. Yamaguchi, *Japanese Journal of Applied Physics* **36**, L899 (1997).

[Vandenbrouck 2009] S. Vandenbrouck, K. Madjour, D. Théron, Y. Dong, Y. Li, C.M. Lieber, and C. Gaquiere, *IEEE Electron Device Letters* **30**, 322 (2009).

[Wagner 1964] R.S. Wagner and W.C. Ellis, *Applied Physics Letters* **4**, 89 (1964).

[Wagner 1970] R.S. Wagner, *Whisker Technology*, A.P. Levitt (ed.), Wiley, New York, 1970.

[Wang 2006a] Y.D. Wang, K.Y. Zang, S.J. Chua, S. Tripathy, H.L. Zhou, and C.G. Fonstad, *Applied Physics Letters* **88**, 211908 (2006).

[Wang 2006b] J.F. Wang, D.Z. Yao, J. Chen, J.J. Zhu, D.G. Zhao, D.S. Jiang, H. Yang, and J.W. Liang, *Applied Physics Letters* **89**, 152105 (2006).

[Yamashita 2005] T. Yamashita, S. Hasegawa, S. Nishida, M. Ishimaru, Y. Hirotsu, and H. Asahi, *Applied Physics Letters* **86**, 082109 (2005).

[Yang 1996] J.W. Yang, C.J. Sun, Q. Chen, M.Z. Anwar, M. Asif Khan, S.A. Nikishin, G.A. Seryogin, A.V. Osinsky, L. Chernyak, H. Temkin, C. Hu, and S. Mahajan, *Applied Physics Letters* **69**, 3566 (1996).

[Yang 2000] J.W. Yang, A. Lunev, G. Simin, A. Chitnis, M. Shatalov, M.A. Kahn, J.E. Van Nostrand, and R. Gaska, *Applied Physics Letters* **76**, 273 (2000).

[Yoshida 2006] S. Yoshida, S. Katoh, H. Takehara, Y. Satoh, J. Li, N. Ikeda, K. Hataya, and H. Sasaki, *Physica Status Solidi (a)* **203**, 1739 (2006).

[Zamir 2001a] S. Zamir, B. Meyler, and J. Salzman, *Journal of Crystal Growth* **230**, 341 (2001).

[Zamir 2001b] S. Zamir, B. Meyler, and J. Salzman, *Applied Physics Letters* **78**, 288 (2001).

[Zamir 2002] S. Zamir, B. Meyler, and J. Salzman, *Journal of Crystal Growth* **243**, 375 (2002).

[Zang 2008] K.Y. Zang, S.J. Chua, J.H. Teng, N.S.S. Ang, A.M. Yong, and S.Y. Chow, *Applied Physics Letters* **92**, 243126 (2008).

[Zhang 2005] B. Zhang, T. Egawa, H. Ishikawa, Y. Liu, and T. Jimbo, *Applied Physics Letters* **86**, 071113 (2005).

[Zorman 1995] C.A. Zorman, A.J. Fleischman, A.S. Dawa, M. Mehregany, C. Jakob, S. Nishino, and P. Pirouz, *Journal of Applied Physics* **78**, 5136 (1995).

Part III

III–V Materials and Device Integration Processes with Si Microelectronics

5

Group III-A Nitrides on Si: Stress and Microstructural Evolution

Srinivasan Raghavan and Joan M. Redwing

CONTENTS

5.1 Introduction, Historical Perspective, and Current Status

A direct band gap that covers the entire optical spectrum from <1 to >6 eV, a high saturation velocity, and high breakdown field strength makes GaN and its alloys with AlN and InN, hereafter referred to as III-nitrides, of importance for optoelectronic and high-power and/or high-frequency electronic applications. Following some earlier work and as a continuation of research on group III-A (Ga, Al, and In) phosphides and arsenides, GaN was first shown to have a direct band gap of 3.39 eV in 1969 by Maruska and Tietjen at RCA laboratories (Maruska and Tietjen 1969). However, lack of native substrates for epitaxial growth, such as for GaAs and Si, hampered further progress until 1986, when Amano and Akasaki reported growth of uniform layers on c-plane sapphire using low-temperature (LT) (~600°C) AlN buffer layers by organometallic vapor phase epitaxy (OMVPE) (Amano et al. 1986). Further research by Amano and Akasaki and by Nakamura between 1989 and 1995 firmly established sapphire as the substrate and OMVPE as the method of choice for fabrication of group III-A nitride-based optoelectronic devices (Nakamura et al. 1991; Nakamura and Fasol 1997).

On sapphire, the large threading dislocation density (TDD) in GaN epilayers, $>10^8$ cm^{-2}, unless some form of lateral growth technique is used to reduce densities to 10^6 cm^{-2} as in laser diodes, is a hindrance to further improving device performance. GaAs-based devices have TDDs of less than 10^4 cm^{-2}. The low thermal conductivity of sapphire also makes it less desirable for high-power devices. Currently, apart from sapphire, SiC, GaN templates, bulk AlN, bulk GaN, and Si are the other substrates most commonly being pursued (Liu and Edgar 2002). While SiC is the most suitable of all the heteroepitaxial substrates—it has the least epitaxial and thermal coefficient mismatch and the highest thermal conductivity—the lack of cheap, large, and defect-free substrates is the main impediment to further progress. Further, dislocation densities on SiC are not significantly lower from that on sapphire (Davis et al. 2003). Free-standing GaN substrates (www.kyma.com and www.lumilog.com) up to 500 µm thick and GaN templates up to 200 µm thick, grown by halide vapor phase epitaxy (HVPE) on SiC or sapphire, among other substrates, are currently available, but like SiC substrates, they are expensive as well. While thermal mismatch stress and hence curvature remain a problem, they can however be fabricated with dislocation densities of about 10^6–10^7 cm^{-2} or lower (Albrecht et al. 1999). Large area, $>1''$, bulk (produced by a substrate-free method) GaN and AlN substrates are yet commercially unavailable (Hashimoto et al. 2008).

In comparison to sapphire and SiC, the two most popular substrates, Si substrates are the most unfavorable for growth of group III-A nitrides. As is evident from Table 5.1, lattice mismatch and thermal coefficient mismatch between the III-nitrides and the growth substrate is highest for (111) Si substrates. Epilayers on Si are more defective than on the other substrates, devices lag far behind and the films are stressed in tension, see Table 5.1, thereby causing cracking. However, Si is the cheapest of all the substrates and it is available in large sizes up to 12″ in diameter. Si is also appealing because of the possibility of integrating GaN electronic and optoelectronics with the well-established Si CMOS technology. Si also lends itself easily to surface engineering (Chen et al. 2006), a feature not that easily possible on sapphire or SiC. Given these incentives, there has been a major research effort to integrate GaN on Si. The cracking problem has more or less been solved through the use of suitable transition layers to manage stress (Rajagopal et al. 2003). Dislocation densities in Si layers now routinely approach that on sapphire. Through substrate engineering, GaN has been deposited on Si by lateral epitaxial overgrowth (LEO) (Gibart 2004), laterally confined epitaxy (Zamir 2001), and by pendeoepitaxial methods (Davis et al. 2001). Commercially available GaN-RF power transistors on Si from Nitronex Corporation, Durham, NC (www.nitronex.com), possible commercialization

TABLE 5.1

Interfacial Mismatch between III-Nitrides and Commonly Used Substrates[a]

Material	Lattice Constant (Å) (Basal Plane)	Epitaxial Mismatch with GaN (%)	CTE ($\times 10^{-6}$ K^{-1}) (‖Basal Plane) at 298 K	Calculated % Thermal Strain in GaN Films on Cooling from 1000°C $\int_{298}^{1273} \Delta\alpha \cdot dT$
GaN (0001)	3.189		3.9469[b]	
AlN (0001)	3.110	−2.4	3.0420[c]	
Al$_2$O$_3$ (0001)	4.765	**−14.4**[d]	5.0	−0.14[e]
SiC (0001), 6-H	3.0806	−3.4	4.46	
Si (111)	3.840[f]	+20.4	3.6[g]	+0.13

[a] All data, unless indicated or calculated, from Liu and Edgar (2002) a "−ve" sign implies a compressive stress and a "+ve" sign a tensile stress.
[b] $\alpha = 10^{-6} \times (1.6761 + 0.01036T - 9.535 \times 10^{-6}T^2 + 2.9537 \times 10^{-9}T^3)$; obtained from fitting a polynomial to data from Reeber and Wang (2000).
[c] $\alpha = 10^{-6} \times (-0.9214 + 0.01683T - 1.2921 \times 10^{-5}T^2 + 3.5139 \times 10^{-9}T^3)$; obtained from fitting a polynomial to data from Wang and Reeber (1998).
[d] Calculated from data in Kosicki and Kahng (1969).
[e] Calculated from data in Hearne et al. (1999) using 470 GPa as the biaxial modulus of GaN.
[f] Calculated from $a = 5.4310$ Å along [100].
[g] $\alpha = 10^{-6} \times (0.1547 + 0.01122T - 1.0412 \times 10^{-5}T^2 + 3.3118 \times 10^{-9}T^3)$; obtained from fitting a polynomial to data from Okada and Tokumaru (1984).

of GaN-LEDs on Si by Azzurro Semiconductor in Germany (http://www.azzurro-semiconductors.com/), and recent reports (Dadgar et al. 2006; Li et al. 2006; Zhu et al. 2009) of growth of GaN light-emitting diodes (LEDs) on 6″ Si can be looked upon as testimony to the success of all these research efforts.

This chapter is a summary of the methods used to grow GaN and AlN epilayers on (111) Si, such as the ones used in LEDs or high electron mobility transistors (HEMT). Growth on (100) Si will be dealt with briefly. The main emphasis will be on the management of stress and defects, particularly TDDs, and the interdependence between them during growth. A description of the sources of stress is provided at appropriate locations. While the discussion here will be limited to growth of oriented films grown by OMVPE, it is also applicable to growth by molecular beam epitaxy (MBE), the other most common technique used to deposit nitride epilayers. The main difference between OMVPE and MBE is that MBE growth occurs at lower temperatures. Thus, while MBE grown layers will have lower thermal stresses, stress and defect evolution during growth are not expected to be very dissimilar unless growth rates are so vastly different that time-dependent relaxation processes start influencing growth behavior. Growth of GaN by other methods such as sputtering and pulsed laser deposition that yield a polycrystalline film will not be discussed. Growth stress behavior in such polycrystalline or polydomain films are expected to be similar to that in other polycrystalline films of high-melting-point materials. Growth of InN will also be neglected as InN-rich alloys are not directly deposited on any of the substrates mentioned but only on a set of transition layers that is some combination of AlN and GaN. Thus, the growth of InN is expected to be influenced more by this transition layer scheme than by the substrate.

5.2 Interfaces and Epitaxial Relationships

Both stress and microstructural evolution, the subject matter of this chapter, begin at the interface. Hence, a brief description of the interfaces involved is provided below.

For growth of GaN on Si, one generally starts with a 40–100 nm thick, high-temperature (>1000°C) AlN buffer layer and hence an AlN/Si interface. Direct growth of GaN on Si is prevented by the formation of low-melting-point Ga–Si compounds and significant diffusion of Si to the film-vapor interface resulting in the formation of Si_xN_y therein (Dadgar et al. 2003). The former results in the so-called melt-back etching, whereas the latter results in a rough interface. The presence of a defective Si_xN_y layer has been reported at the AlN/Si interface as well. However, it does not prevent epitaxy and it is still not quite clear if it forms prior to deposition of the AlN buffer layer or during subsequent exposure of the AlN/Si interface to high temperatures during growth.

The (111) plane is the most commonly used surface of Si for growth of the group III-A nitrides, the epitaxial orientation relationship in general being $\{111\}_{Si} < 1\bar{1}0 >_{Si} // \{0001\}_{GaN} < 11\bar{2}0 >_{GaN}$. The sixfold symmetry* of arrangement of atoms on the unreconstructed {111} surface (filled circles in the schematic crystal structure of Si shown in Figure 5.1), favors the growth of the wurtzite phase with hexagonal symmetry of Ga-polarity (Ga face up). However, as shown in Table 5.1, the mismatch, defined as $(a_{substrate} - a_{film})/a_{film}$, a being the lattice parameter, is greater than 20% for both GaN and AlN and of a nature that would result in a tensile (or positive) stress in the film when grown directly on (111) Si. It should be noted that for (111) Si, $a_{substrate}$, given the

---

* The unreconstructed (111) surface of Si is made up of a bilayer of atoms with threefold symmetry, but each of the two atomic layers constituting this bilayer has sixfold symmetry. In the image in Figure 5.1, if the (111) surface were to be created by breaking the bond represented by the dotted line, then the bilayer at the (111) surface would consist of one layer made up by the hatched circles and another layer made up by the solid circles. In this case, the (111) surface would have one dangling bond per atom.

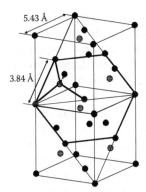

5.43 Å

3.84 Å

FIGURE 5.1

Two unit cells of Si show-ing the location of atoms and the hexagonal symme-try of the (111) plane. Solid atoms belong to one FCC lattice while atoms with a grid pattern belong to the second FCC lattice.

orientation relationship, should be the interatomic distance along [1$\bar{1}$0] and not the often quoted lattice parameter of Si, which is the interatomic distance along [100].

For such highly mismatched systems, strain is very quickly relaxed by either the formation of mis-fit dislocations or by formation of islands. The criti-cal thickness for nucleation of a misfit dislocation can be estimated using the Matthews and Blakeslee relation (see Tsao 1993, p. 166 for instance).

$$h_c = \left[\left(\frac{b}{8\pi f \cos \lambda}\right)\right]\left[\frac{(1-\nu \cos^2\beta)}{(1+\nu)}\right]\ln\left(\frac{4h_c}{b}\right) \quad (5.1)$$

For a misfit $f=0.02$ or 2%, $\lambda=0$, and $\beta=90$ for a pure edge dislocation line lying on the interface, $\nu=0.3$ and $b=3.1$ Å one obtains a critical thick-ness of 1 Å. More detailed calculations yield slightly higher numbers in the range of 10–25 Å for the AlN/GaN interface with a mismatch of 2.5% (Bykhovski et al. 1995; Holec et al. 2008). Thus, at misfits of 20% for the AlN/Si interface, dislocations become energetically favorable at calculated thicknesses smaller than a monolayer. Experimentally, it has been observed by Bourret et al. (1998) using transmission electron microscopy (TEM) that strains are relaxed to below 0.2% within one bilayer at the AlN/Si interface grown by MBE (Bourret et al. 1998). Geometrical analysis of the interfaces also suggest that the mismatches in excess of 20% can be reduced to −1.2% (negative or compressive strain) for AlN/Si and −3.6% for GaN/Si through a 4–5 relationship ($4a_{Si} \approx 5a_{AlN\ or\ GaN}$). While TEM evidence for such a relation-ship through the periodic presence of misfit dislocations has been presented by Schenk et al. (1999) and more recently by Liu et al. (2003), the analysis of Bourret et al. points to the lack of any long-range order over areas larger than about 15 Å at the interface.

The large mismatch, discussed above, between the nitrides and the substrates used for growth currently, tends to distract from the fact that lat-tice mismatch between the nitrides themselves is significant. For growth on Si, this factor becomes important at the GaN/AlN buffer layer interface and the AlN interlayer/GaN interface. AlN interlayers, about 10 nm thick and grown at both low (~600°C) and high temperatures (~1000°C), are inserted to introduce a compressive stress and thus enable the growth of crack-free thick GaN layers as will be discussed later. Given the lattice parameters in Table 5.1 and using 470 and 510 GPa as the biaxial modulus of GaN and AlN, respectively, the mismatch translates to a stress of −11.3 and +12.6 GPa in GaN on AlN and AlN on GaN, respectively. Thus, the discussion in the previous paragraph applies equally well to interfaces between the nitrides themselves as well. TEM observations have revealed a triangular grid of misfit dislocations with a spacing of 15 nm in GaN islands growing on an AlN buffer layer on SiC substrates (Moran et al. 2004). Calculations show that the critical thickness for dislocation formation in AlN or InN films on GaN is ≤10 Å and that for $Al_{0.2}Ga_{0.8}N$ or $In_{0.2}Ga_{0.8}N$ films on GaN is ≤100 Å (Holec et al. 2008). It is emphasized here that these calculations based on just energetic considerations are subject to kinetic modifications. The actual thickness at which misfit dislocation-induced relaxation is observed in Si for instance, is often greater than what is predicted thermodynamically (Nix 1989). This would be even more so in the case of the III-nitrides, where the stress state is equibiaxial and hence resolved shear stresses on the basal plane, the one most conducive to dislocation glide is zero. The behavior of films grown by metal organic chemical vapor deposition (MOCVD) at relatively high rates would be even more prone to deviate from thermodynamic predicted ones. As will be discussed later in connection with AlGaN films on GaN, films eventually relax by cracking at critical thicknesses exceeding that predicted above for relaxation by formation of misfit dislocations. Relaxation could also happen by surface roughening leading to formation of islands. Nitrides in tension, such as AlN on GaN, seem to prefer relaxation by nucleation of misfit dislocations or cracking, but nitrides in compression seem to prefer formation of islands, strained or relaxed by dislocations (Feuillet et al. 1998).

While (111) Si is the most commonly used substrate, research on growth of III-nitrides on (100) Si is currently on the rise (Dadgar et al. 2007). The possibility of integrating GaN with the well-established (100) Si CMOS technology and the possibility of growing III-nitrides in directions other than the polar [0001] direction are the main incentives. The presence of an electric field across active device regions due to spontaneous and piezoelectric polarization along the [0001] axis is undesirable for many device applications. It is interesting that in spite of the fourfold symmetry of the (100) Si face and the sixfold symmetry of the (0001) III-nitride face, the orientation relationship observed is $(100)_{Si}[110]_{Si}//(0001)_{GaN}(10\bar{1}0)_{GaN}$ when grown directly on (100) Si (Dadgar et al. 2007). Since, there are two orthogonal <110> type directions on the (100) face, there are two crystallographic variants, two hexagons separated by a 90° rotation. By using a suitable transition layer scheme, a different orientation relationship $(100)_{Si}[110]_{Si}//(10\bar{1}2)_{GaN}(10\bar{1}0)_{GaN}$ with four variants can be obtained (Dadgar et al. 2007). The important difference when compared to growth on (111) Si or (0001) sapphire or SiC is that stress will now not be isotropic in the growth plane.

In summary, the interfaces between the nitrides and substrates have a large density of misfit dislocations at sub-monolayer thicknesses and residual unrelaxed stress. Interfaces between the nitrides themselves could either be just stressed or relaxed to various extents depending on the film thickness and variation in composition across the substrate. A quick analysis of the various energies involved then provides an insight into the possible growth mode. To begin with the Si–N bond energy, $\Phi_{N/Si}$, of 4.5 eV (Wang et al. 1996) is lower than the Ga–N and Al–N bond energies of 9.12 and 11.7 eV, respectively (Morkoc 2008). Thus, there is no great incentive from the bonding view point to spread out along the substrate or, in other words, for the two-dimensional (2D) growth mode. In addition, as discussed previously, III-nitride layers have both strain and defects, which would raise the chemical potential with respect to the bulk even further. Thus, the chemical potential of the nth layer from the interface given by (Markov 1998)

$$\mu(n) = \mu_{bulk} + \Delta\Phi_{desorption}(n) + \Phi_{strain}(n) + \Phi_{dislocations}(n) \qquad (5.2)$$

would decrease with distance from the interface and tend to μ_{bulk} with increasing n. This is so because, given the bonding energies above and assuming similar bonding configuration across interfaces, $\Delta\Phi_{desorption}(n) \approx \Phi_{N/Ga} - \Phi_{N/Si}$ is either positive (first layer) or close to zero (subsequent layers) and the other two terms decrease with thickness. Since, the chemical potential of the first monolayer is higher than that of the subsequent layers, thermodynamics indicates that III-nitrides would prefer to form multilayer islands rather than spread out as a single layer on the substrate and for that matter even on other III-nitride layers. The resultant growth mode, based on observations at thickness greater than a bilayer, is thus the Volmer-Weber mode, involving nucleation of three-dimensional (3D) islands that then grow laterally and coalesce. It is this mode of beginning to the growth of the III-nitrides on Si, which results in all the subsequent evolutionary processes to be discussed in the following sections. It is emphasized that this discussion has been on purely thermodynamic grounds and the growth mode actually observed is therefore subject to kinetic modifications.

5.3 Defects in III-Nitride Layers

III-nitride films on Si are host to all the major defect types including point defects, line defects or dislocations, plane defects such as stacking faults, inversion domains and cracks, and 3D defects such as V-shaped pits and pipes (Morkoc 2008). Even though III-nitride devices have been commercialized, it is nevertheless very well documented that defects affect both device performance and lifetimes. Extensive research has thus been devoted to understanding the origin, structure, and evolution of defects during growth. Given the discussion in the previous section and as will become amply clear later on, stress plays a major role in the origin and evolution of these defects and also in their interactions.

Of the defects mentioned above, III-nitride devices stand apart because of their high TDDs. In contrast to the misfit dislocations discussed in the previous section where the dislocation line lies entirely at the interface, threading dislocations have lines that propagate through the thickness of the film. The dislocation line can either be straight and perpendicular or inclined to the interface. They could also bend and form loops. Dislocations are particularly important as the evolution of their configuration during growth is related to evolution of stresses in the film.

The TDD, henceforth only called dislocation density, in 1 μm thick GaN layers grown on (111) Si is typically higher than 10^9 cm^{-2} at the surface of the film, the corresponding number for films on sapphire and SiC being 10^8 cm^{-2}. It is important to note the mention of film thickness, as dislocation density is not a constant but varies across the film. At the beginning of growth of the GaN layer on (111) Si for instance, the dislocation density can be in excess of 10^{14} cm^{-2} close to the GaN/AlN interface and its reduction with thickness depends on the buffer layer scheme used. Dislocation density can be brought down to about 10^6 cm^{-2} by using lateral growth techniques (Gibart 2004) and this is still about two orders of magnitude higher than that found in GaAs-based devices.

The dislocations observed in GaN films are of three types, a, c, and a + c as shown in Figure 5.2a. The a-type dislocations, extra $(11\bar{2}0)$ planes are pure edge dislocations, with Burger's vector **a** $(1/3\langle11\bar{2}0\rangle)$ and dislocation line along the c-axis. The c-type dislocations are pure screw dislocations with a Burger's vector **c** $(\langle0001\rangle)$ and line direction along the c-axis. The a + c type dislocations are mixed dislocations with Burgers vector **a + c** $(1/3\langle11\bar{2}3\rangle)$ with line direction inclined at 12.2° to the c-axis (Mathis et al. 2001). The magnitudes of the Burgers vectors can be calculated by using the lattice parameters of the corresponding III-nitride. Since a < c, dislocations with an a-component to their Burgers vector have lower energies. While reported fractions vary greatly, the a-type constitutes about 70%–90% of the population, the remainder being made up of the c and a + c type (Follstaedt et al. 1999).

III-nitride films on all substrates have domains that are twisted or tilted with respect to the substrate film interface to varying extents. The difference in twist between domains can be accommodated by the pure edge dislocations, whereas difference in tilt can be accommodated by the pure screw dislocations. When a combination of both difference in twist and tilt between domains exist, then all three types of dislocations would be required. It is pointed out that pure twist between domains with respect to the substrate which is normally measured using asymmetric scans, Figure 5.2b, results in pure tilt grain boundaries between them in the film. Pure tilt with respect to the substrate Figure 5.2c, could result in tilt boundaries, twist boundaries, or a combination. Only the twist boundary formation is shown in Figure 5.2c.

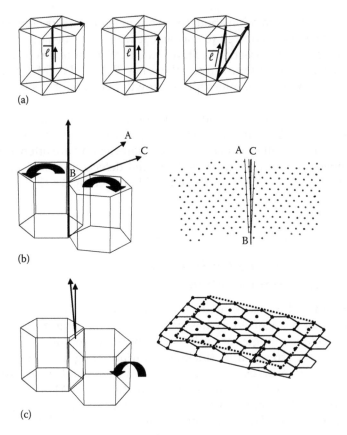

FIGURE 5.2
(a) Common dislocation types observed in III-nitrides. $\bar{\ell}$ denotes line direction. The dark arrows represent the Burgers vectors, **a, c, and a + c** from left to right. (b) Twist with respect to the substrate results in a pure tilt grain boundary represented by angle ABC. Schematic on the right shows the top view of that on the left in a simple hexagonal lattice, tilt angle ABC, and its accommodation by edge dislocations of type a represented by lines. Such twist is studied by asymmetric x-ray scans. (c) Tilt with respect to the substrate results in a pure twist boundary (left) and is accommodated by a screw dislocation as shown on the right of type c. Dotted lines show a Burgers circuit that can be completed by a vector $\bar{c}$.

The boundaries shown are symmetric tilt and twist boundaries. They could, however, also be asymmetric. X-ray diffraction (XRD) using a four-circle diffractometer (Metzger et al. 1998), TEM (Metzger et al. 1998), and atomic force microscopy (AFM) (Oliver et al. 2006b) are the three most commonly used techniques to characterize these dislocations. XRD only gives tilt (symmetric 0002 scans) and twist values (ideally from {hki0} planes (see Chapter on: Stress and strain evaluation and calculations by Prof. Rainer Clos and Prof. Alois Krost) or if not, from planes inclined with respect to the c-axis) in the epitaxial layer, which are then used to back calculate the dislocation densities. Direct information about the evolution of the dislocation structure across the thickness, required to correlate it with stress evolution, can only be obtained from cross-sectional TEM. Quantitative estimates of actual dislocation densities are obtained from plan view TEM and AFM. Quantitative evolutionary data require interrupting growth at different thicknesses and studying the surface.

Apart from dislocations, the other main defect type that interact with dislocations and possibly stress are pipes and pin holes, both of which, at their origin, have the so-called V-shaped defect character and are often associated with inversion domain boundaries. V-shaped defects are "conical-shaped" hollows with a hexagonal cross section and bounded by the $(10\bar{1}0)$ planes.

5.4 In Situ Monitoring for Stress and Microstructural Evolution

Given the difficulties associated with growth of the III-nitrides, it is not surprising that a lot of effort has gone into monitoring growth in situ. Optical reflectance, single- or multi-wavelength, for monitoring curvature (and hence film stress), film thickness, surface roughness, composition (through refractive index measurements), and temperature is one of the most commonly used techniques in conjunction with MOCVD (Floro et al. 1997). With MBE, apart from optical reflectance, reflection high-energy electron diffraction (RHEED) is a more commonly used technique. RHEED has been used to study strain relaxation (by calculating lattice parameter changes) and strain-induced changes to surface morphology during growth (Feuillet et al. 1998). Apart from these two commonly used methods, as summarized in Table 5.2, spectroscopic ellipsometry, spectroscopic reflectance, mass spectroscopy, Raman spectroscopy, cathodoluminescence,

TABLE 5.2

In Situ Monitoring Methods and References

Method	References
1. In situ curvature measurement using optical reflectance	Floro et al. (1997), Griffiths et al. (2007)
2. RHEED	Feuillet et al. (1998)
3. Spectroscopic ellipsometry	Bonanni et al. (2003)
4. Spectroscopic reflectance	Liu et al. (2006)
5. Mass spectroscopy	Carreno et al. (2003)
6. Raman spectroscopy	Zahn (1998)
7. Cathodoluminescence	Hove et al. (1997)
8. X-ray diffraction	Kharchenko et al. (2005)

and XRD have been used to monitor growth in situ. The addition of an x-ray diffractometer is interesting in particular as it allows one to monitor the variation in crystal quality during growth. A combination of a stress measurement method and an x-ray diffractometer would thus instantly provide one with information on the interrelation between the two. Mass spectroscopy has been used to determine the polarity of the growing film in situ.

With the introduction of a multiple beam method, optical reflectance using radiation of a single wavelength, typically from a 650 nm diode laser, has become a very handy tool to install and monitor III-nitride growth, especially stress, in situ. Given the problems associated with stress during growth on Si, in situ analysis aids the development of growth methods considerably. Optical reflectance tools from multiple sources are now routinely available as a standard addition with most MOCVD reactors, though optical access to the growth area can vary considerably from one reactor to another. Hence, the choice of the reactor is often dictated by its ability to accommodate the optical geometry of such tools. In brief, an array of laser spots, obtained by splitting the beam from a diode laser is reflected off the growth surface into a camera coupled to a personal computer. Both spot spacing and spot intensity are monitored in real time during growth. The change in spot intensity can be used to monitor surface roughness, thickness, and composition. The change in spot spacing is used to calculate the change in substrate curvature and hence stress through the Stoney's formula, Equation 5.3a. Strictly speaking, the Stoney's formula is only valid under certain assumptions and errors in quantitative data will arise from its improper use (Suresh and Freund 2003). For instance, for growth of continuous, uniform AlN layers on (111) Si substrates, the ratio of biaxial modulus of film to substrate, $M_{film}/M_{substrate} = 510/290 = 2.2$. For this ratio, the error in the estimation of curvature using Stoney's equation crosses 10%, i.e., $\kappa/\kappa_{Stoney} = 0.9$, when the ratio of film to substrate thickness, $h_{film}/h_{substrate}$, becomes greater than about 0.02 in the absence of edge effects. On a 500 µm (111) Si substrate, this translates to an AlN film thickness of 10 µm and increases for sapphire and SiC substrates with higher moduli. The spot spacing data is typically plotted as a stress-thickness vs. thickness plot. For systems such as the nitrides in which there is apparently no stress relaxation during growth such as due to diffusion or dislocation glide, and there is no other source of contribution to the change in curvature other than the addition of a certain thickness of stressed film material, the slope of the stress thickness vs. thickness plot, also called the incremental stress, gives the stress in the film at the growth surface, $\sigma(z)$. Thus, if one sectioned the film and measured stress as a function of thickness, z, by an ex situ technique such as TEM from variation in lattice parameter, it would be the same as that obtained from measuring the slope of the aforementioned plots at different thickness values. The average film stress, $\langle \sigma_f \rangle$, on the other hand, is obtained by the net change in curvature induced due to film growth.

The relation between the mean stress and incremental stress is given by Equation 5.3b.

$$\langle \sigma_f \rangle h_f = \int_0^{h_f} \sigma(z)dz = \Delta\kappa \left(\frac{M_s h_s^2}{6} \right) \tag{5.3a}$$

$$\sigma(h) = \frac{d\left[\langle \sigma_f \rangle h_f \right]}{dh_f} \tag{5.3b}$$

When stress in the growing film is zero, $\sigma(z) = 0$, then no change in curvature will be observed from in situ measurements. When the stress in the growing film is a constant, $\sigma(z) = $ constant, the stress thickness vs. thickness plot would be a straight line with slope equal to the value of stress. A positive slope corresponds to a tensile stress, whereas a negative slope corresponds to a compressive one. When the stress in the film changes with thickness due to microstructural evolution, then the slope of the stress-thickness vs. thickness plot will change accordingly. Hence, when such data on change in stress-thickness with thickness is viewed in combination with data on microstructural evolution from *ex situ* characterization, the interrelationship between the two can be deduced. Thus, the in situ sensor is not only a stress monitor, but also an in situ growth monitor that provides valuable insights into the growth mechanism. In addition, commonly used *ex situ* tools such as XRD cannot easily provide information on the variation of stress across the thickness of the film, as is readily obtained by an in situ measurement, but only a volume-averaged value.

As is evident from the Stoney's equation, the sensitivity of the multiple beam optical reflectance method to such microstructural and stress evolution depends on its ability to resolve curvature changes, the biaxial modulus of the substrate and, especially, its thickness. During growth by MOCVD on 500 μm thick (111) Si substrates, a stress of 0.3 GPa in films as thin as 50 nm, i.e., a stress-thickness product of 0.02 GPa μm, was found to be easily measurable (Raghavan and Redwing 2004a,b, 2005a,b). With a 250 μm wafer, the corresponding value would be 0.005 GPa μm. In systems with less vibration than MOCVD, such as MBE, the resolution and sensitivity are expected to be even better. Often nonspherical or inhomogenous substrate bow also limits the resolution of these measurements. In such cases, measuring curvature at the same location and wafer orientation helps reduce noise. This can be done for instance by coupling the shutter of the camera to the rotation of the substrate through a suitable position-sensitive trigger.

5.5 Growth of AlN on (111) Si

In spite of the highly unfavorable relation across the interface discussed in Section 5.2, oriented AlN nitride films with typical XRD full width at half maximum (FWHMs) for the (0002) omega scan of 0.3° or better have been synthesized by sputtering (Ivanov et al. 1995), MBE (Sheldon et al. 2005), and MOCVD (Raghavan and Redwing 2004a,b). While differences are present in the nature of microstructural evolution, these are mostly confined to a few monolayers of growth. In oriented crystalline films that are about 50 nm or thicker, layers grown by all the above techniques have a lateral domain size of about 50–100 nm, Figure 5.3a, and this domain size does not appreciably

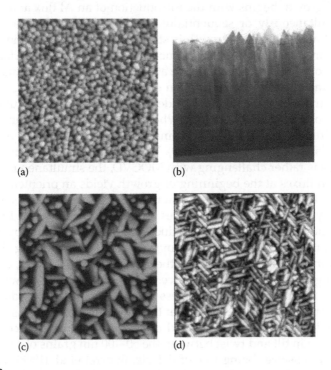

FIGURE 5.3
Microstructure of AlN films on (111) Si deposited by MOCVD at a growth rate of 2 Å/s. (a) 1 μm × 1 μm × 50 nm AFM scan of 200 nm film surface deposited at 1100°C (From Raghavan, S. and Redwing, J.M., *J. Cryst. Growth*, 261 (2–3), 294, 2004a; Raghavan, S. and Redwing, J.M., *J. Appl. Phys.*, 96 (5), 2995, 2004b. With permission.); (b) cross-sectional TEM image. (Courtesy of Xiaojun Weng, The Pennsylvania State University, University Park, PA) of 1 μm thick film consisting of ~100 nm wide columns deposited at 1100°C; (c) 1 μm × 1 μm × 200 nm AFM scan of the 1 μm thick film surface in (b) deposited at 1100°C; (d) 1 μm × 1 μm × 50 nm AFM scan of 200 nm film surface deposited at 800°C. (From Raghavan, S. and Redwing, J.M., *J. Cryst. Growth*, 261 (2–3), 294, 2004a; Raghavan, S. and Redwing, J.M., *J. Appl. Phys.*, 96 (5), 2995, 2004b. With permission.)

change with thickness, at least at growth rates >1 Å/s. Thus, thick AlN films on (111) Si are essentially a stack of 50–100 nm wide columns separated by low-angle grain boundaries and domain boundaries as shown in Figure 5.3b (Sheldon et al. 2007). Another essential feature evident from these images is that there is considerable increase in surface roughness in going from the 100 nm film, Figure 5.3a to the 1 μm thick AlN film, Figure 5.3c AFM measured average surface roughness/maximum surface roughness data for the 100 nm and 1 μm film grown by MOCVD were 6.4/61.2 and 31.6/240 nm, respectively. This appears to be a common feature of all nitride films growing under tension. However, significant differences in the nature of stress evolution are observed and these are a result of the different kinetics of various processes involved and not just thermodynamics.

All AlN growth begins with the introduction of an Al flux and an N flux either simultaneously or sequentially into the growth chamber. In MBE (Louran et al. 2009), when growth is started with the introduction of an N flux with or without an Al flux, some nitridation of the Si surface resulting in the formation of silicon nitride at the monolayer scale is unavoidable. However, intentional deep nitridation of the Si surface resulting in the formation of an amorphous silicon nitride is generally considered undesirable from the point of view of epitaxy. When the Al flux is introduced following the initial N flux in MBE, Al atoms react with the nitrided Si zones to form AlN islands (Louran et al. 2009). While such fine control over growth parameters is rather challenging with MOCVD, the simultaneous introduction of both fluxes at the beginning of growth yields an oriented AlN film, thus suggesting that any nitridation at this point is not deep enough to prevent epitaxy. Both the above techniques yield metal-polar nitride. In contrast, a recent study has shown that introduction of an Al flux first by MBE can result in nitrogen-polar films (Dasgupta et al. 2009). Following the minor difference in growth mode details at the level of few monolayers, the islands so formed grow laterally and coalesce. In AlN films on Si growing under tension, the microstructure more or less remains frozen-in beyond this stage. The popular view so far has been that the threading dislocations discussed in the previous section are generated during coalescence to accommodate the difference in tilt and twist between the 50–100 nm grains observed, with the grains themselves being free of defects. Bourret et al. (1998) have however suggested that threading dislocations could be generated at the even smaller length scales of 15 Å, mentioned before, to accommodate orientation differences between such "epitaxial patches." More recent views (Oliver et al. 2006a) on the origin of threading dislocations in the GaN layer, which suggest that coalescence is not the only source, will be discussed in the next section.

The discussion in the previous paragraphs pertained to the growth of oriented crystalline films. Such growth is typically done at relatively higher temperatures, greater than 800°C by MOCVD. It is emphasized here that depending on the method of temperature measurement in different reactors, the actual value at the growth surface will be significantly different from the reported value. Thus, in comparing results from multiple sources, trends in data with change in temperature are more meaningful than those based on absolute values. With a reduction in growth temperature, there is a general deterioration in crystal structure, an increase in the FWHM of x-ray omega scans. Films deposited at 600°C by MOCVD, for instance, are for all practical purposes almost amorphous (Raghavan and Redwing 2004a,b). X-ray data representing this effect of change in growth temperature on film structure is summarized in Figure 5.4. In addition, while films deposited at higher temperatures have a more equiaxed island structure when examined in an AFM, Figure 5.3a, at lower temperatures, the islands get elongated in the [11$\bar{2}$0] directions, Figure 5.3d resulting in a more anisotropic structure. TEM analysis shows that such elongated islands are bound by {1$\bar{1}$01} planes, the anisotropy resulting from slow growth is the direction perpendicular to these facets (Rehder et al. 1999).

The stress evolution that accompanies the microstructural evolution is relatively much more diverse. Figure 5.5 summarizes the evolution of growth stresses in oriented epitaxial AlN films deposited by sputtering (Meng et al. 1994), MBE (Sheldon et al. 2005), and MOCVD (Raghavan and Redwing 2004a,b) at relatively higher temperatures. As mentioned before, due to lack of stress relaxation in AlN films at the deposition temperatures involved, the slope of these plots gives the actual value of stress in the films at that thickness, z, location. The following features are noteworthy:

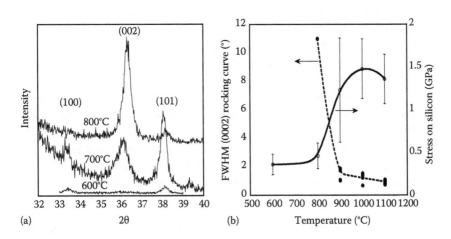

(a)   (b)

FIGURE 5.4

(a) θ–2θ scans showing the general deterioration in orientation with reduction in growth temperature of AlN on (111) Si. (b) The corresponding change in FWHM of (0002) rocking scans showing a sharp increase in tilt with respect to the substrate between 800°C and 900°C and accompanied with a drop in growth stress. (From Raghavan, S. and Redwing, J.M., *J. Cryst. Growth*, 261 (2–3), 294, 2004a; Raghavan, S. and Redwing, J.M., *J. Appl. Phys.*, 96 (5), 2995, 2004b. With permission.)

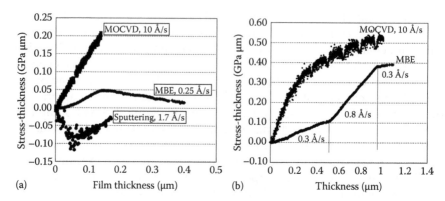

FIGURE 5.5
Stress-thickness vs. thickness plots during the growth of AlN films on Si by sputtering (Meng et al. 1994), MBE (Sheldon et al. 2005), and MOCVD (Raghavan and Redwing 2004). A negative slope corresponds to a compressive stress whereas a positive slope corresponds to a tensile stress. (From Meng, W.J. et al., *J. Appl. Phys.*, 75, 3446, 1994; Sheldon, B.W. et al., *J. Appl. Phys.*, 98, 043509, 2005; Raghavan, S. and Redwing, J.M., *J. Cryst. Growth*, 261 (2–3), 294, 2004a; Raghavan, S. and Redwing, J.M., *J. Appl. Phys.*, 96 (5), 2995, 2004b. With permission.)

1. For films deposited by sputtering (Meng et al. 1994) and by MBE (Sheldon et al. 2005) at low growth rates, the growth stress is initially compressive (stage I), and then transitions to a tensile stress (stage II).

2. For films deposited by MBE (see Fig. 3(c) in Ref. Sheldon et al. 2005) and MOCVD, Figure 5.5, at higher growth rates >0.5 Å/s, the initial compressive stress (stage I), is not observed.

3. For films grown by MBE, a third compressive stage, Figure 5.5a is observed at low growth rates, 0.25 Å/s, but not at higher ones, following the tensile stress in stage II.

4. In AlN films grown by MOCVD, stress relaxation sets in at about 400 nm, Figure 5.5b, but no compressive stress as in the MBE films is observed.

5. The tensile stress in stage II is growth rate dependent, notice different slopes in Figure 5.5b, at 0.3 and 0.8 Å/s, in regimes accessible by MBE at <1 Å/s. It is independent of growth rate in films deposited by MOCVD at rates >1 Å/s (Raghavan and Redwing 2004a,b).

In films that grow by the Volmer-Weber mode, nucleation and coalescence of islands as in the present case, any initial compressive stress, stage I, is typically attributed to lattice mismatch stress and surface stress, the transition to a tensile stress in stage II to island coalescence, and the compressive stress in stage III to injection of adatoms into grain boundaries (Freund and Suresh 1993; Chason et al. 2002; Floro et al. 2002). Each of these stages is now dealt with in greater detail.

In nitride films as can be seen from the stress-thickness vs. thickness plots, a stage I compressive stress is not always observed. The reader is reminded that in stage I, island coalescence is incomplete, meaning that substrate curvature is caused by isolated islands. For such a mechanism to be operative, the areal coverage of islands must be dense enough for stress fields in the substrate generated by transfer of stress across the interface to interact. In the absence of any coalescence-related tensile stresses, the lattice mismatch stress in an isolated island, across a (111) Si/AlN interface is expected to be −6.1 GPa, given the 4–5 relationship discussed before. For compound semi-conductors, the sign of surface stress, f, changes with the terminating plane. For a Ga-terminated surface, it is expected to be compressive (implying a tensile strain inside the island) and tensile (implying a compressive strain inside the island) for a N-terminated surface. Thus, for typical nitride films growing with Ga polarity, if it is assumed that the surfaces are on average covered with Ga, surface stress, from the top surface will counter the lattice mismatch stress. Typical calculated values of surface stress for Ga- and As-terminated GaAs surfaces are −1 and 0.5 N/m, respectively (Cammarata 1994). This would translate to an internal stress of $2/R$ GPa and $-1/R$ GPa in Ga- and N-terminated isolated islands not attached to any substrate where R is the island size in nm. To determine the effect of surface stresses on internal stresses in nitride islands attached to a substrate, one needs to have a measure of the size at which the islands get locked to the substrate and are, therefore, prevented from relaxing further. The upper bound on such stresses is $-2f/R_{locked}$ or $2/R_{locked}$ GPa for metal-terminated films. To summarize, if the "lock-down" mechanism of surface stress is indeed accurate (see Freund and Suresh 1993 for a critical evaluation), the biaxial stress in isolated islands of Ga- and N-terminated nitride islands are estimated to be $-6.1+2/R_{locked}$ and $-6.1-1/R_{locked}$ GPa, respectively based on current understanding. From the stress-thickness vs. thickness plots, Figure 5.5, the measured initial growth stresses range from almost 0 to −2.5 GPa in metal-terminated nitrides. It is assumed here that since growth of nitrogen-terminated nitrides is relatively more difficult on (111) Si substrates, unless otherwise specifically mentioned in the references, the films will be assumed to be metal terminated. Part of the difference is obviously due to the conversion of curvature data to stress thickness data using the Stoney's equation, which is only valid for continuous films (for an estimate of this error, see Freund and Suresh 1993, pp. 68–69). In such cases, the Stoney's equation underestimates the magnitude of stress thickness and thus could partially account for the observed difference. Part of the remaining difference needs a better experimental understanding of the sources of stress discussed. However, none of the above can explain the complete absence of stage I in films grown at relatively larger rates of >1 Å/s.

The tensile stresses observed in stage II or right from the beginning of growth are typically attributed to island coalescence. Equation 5.4 represents a simple thermodynamic analysis of grain coalescence as given by Nix and Clemens (1999).

$$\sigma_{max} = \frac{M\Delta_{max}}{L} = \left[\frac{2(2\gamma_{sv} - \gamma_{gb})M}{L}\right] \qquad (5.4)$$

Grains of size L are said to "stretch" out across a gap of size Δ to form a grain boundary of energy γ_{gb} and in the process get stressed. Such a mechanism is energetically feasible only if the reduction in surface energy, $2\gamma_{sv} - \gamma_{gb}$, can balance the increase in mechanical energy of the film. M is the biaxial modulus of the film and γ_{sv} is the solid vapor interface energy. By its very definition, island coalescence implies that there should be a pre-coalescence stage in which growth occurs at a lower value of tensile stress and at coalescence, there is a positive change to the slope of the stress-thickness vs. thickness plot. Changes in slope are, however, less obvious for the so-called low-mobility films such as the III-nitrides compared to the high-mobility films (Thurner and Abermann 1990; Schneeweiss and Abermann 1992; Floro et al. 2002). Also, the change in slope could be gradual, if the ratio of grain boundary height to film thickness height evolves with film thickness and is not established at a single contact and coalescence event (Sheldon et al. 2001). Nevertheless, the absence of any observation of a change in slope raises questions about the origin of this tensile stress. The change in this constant tensile stress observed right from the beginning of growth with temperature is plotted in Figure 5.4b. It is observed that there is a sharp drop in the value of the tensile growth stress as the growth temperature is decreased from 900°C to 800°C. The corresponding change in FWHM of the (0002) AlN rocking curves shows a sharp rise in the same temperature range. The sharp rise in FWHM of the omega scans and the change in the θ–2θ scans shown in Figure 5.4a represent a general decrease in the degree of orientation in the film. This in turn implies boundaries with larger levels of tilt and twist between them and therefore higher energies, γ_{gb}. The energetics of grain coalescence as represented by Equation 5.4 shows that with an increase in γ_{gb}, coalescence-related tensile stresses should decrease as is the observed behavior. Also, expected from the same equation is the small rise in the tensile stress from 1100°C to 900°C, which is accompanied by a reduction in grain size. Such trends in data, combined with observations that films whose growth otherwise includes a pre-coalescence stage I that disappears on increasing the growth rate (MBE AlN films on Si(111) in Figure 5.5) or increasing the defect level (MOCVD AlN films on sapphire in Raghavan and Redwing 2004a,b), point to grain coalescence as the most obvious source of this tensile stress right from the beginning of growth.

The constant tensile stress following stage I or right from the start of growth is limited to films that are about 200 nm thick or lower. In thicker films a compressive stress is observed at higher thicknesses Figure 5.5a, when growth rates are lower than 0.3 Å/s. At the higher growth rates, Figure 5.5b while a compressive stress is not observed, the films tend to relax, by microcracking along grain boundaries as shown in the cross-sectional TEM as seen in Figure 5.5f. This stage III compressive stress is typically attributed to injection of adatoms into the grain boundaries in the film, though other mechanisms such as a change in surface stress with supersaturation could be operative as well (Chason et al. 2002; Sheldon et al. 2003). The driving force for such a mechanism is the difference in chemical potential between an adatom on a supersaturated surface and an atom in the grain boundary. The driving force increases with film tensile stresses. The absence of the compressive stage in epitaxial films devoid of grain boundaries and their presence in polycrystalline films of the same material system lends support to the possible role of grain boundaries, though there have also been reports of a compressive stress in films devoid of boundaries. In the particular case of nitrides, it is noted again that since significant long-range diffusion-related relaxation is negligible, the processes described above happen at the intersection of the grain boundary with the growth surface.

The discussion so far has been purely based on thermodynamics. As growth rate increases, the stage III compressive stress regime disappears at 0.3 Å/s, the tensile stress in stage II increases till 1 Å/s and then it too becomes independent of growth rate at higher temperatures. This dependence of the observed stress behavior on growth rate points to a role of kinetics in stress evolution and not just thermodynamics. In brief, the observed steady-state stress, i.e., no change in stress with thickness, which is tensile in stage II and compressive in stage III, is modeled as being due to a balance between the net rates of a compressive stress generating mechanism such as atom injection and a tensile stress generating mechanism due to the formation of bonds across the grain boundary during the addition of each new layer (Sheldon et al. 2003, 2005). If one envisioned film growth as the addition of layers of atoms, then a compressive stress would be generated if the number of atoms in the $(n+1)$th layer is greater than the number of atoms in the nth layer. The opposite would lead to a tensile stress. When there is no change in the number of atoms from the nth to $(n+1)$th layer, the stress in the $(n+1)$th layer would be the same as that in the nth layer and film would grow under a steady-state compressive or tensile stress.

As pointed out previously, for the compressive stress generation by atom incorporation into the grain boundary to be active, a boundary is required in the first place. It is formed at coalescence and is accompanied by the generation of a tensile stress. Hence, it is typical to see a tensile stress preceding a compressive one. Atom incorporation into these grain boundaries would result in an increase in the number of atoms and the generation of a compressive stress. Hence, the tensile stress would decrease and eventually transition to a compressive stress. Note, that the transition to a compressive stress itself decreases the driving force for the mechanism, thus makes itself limiting. As growth rate increases, the relative contribution of the compressive stress–generating mechanism, which requires diffusion of adatoms, decreases and eventually becomes negligible. The net effect observed is an increasing trend toward a tensile stress with increase in growth rate. While the mechanism that leads to generation of a compressive stress has supporting experimental evidence, its counterpart, the mechanism that leads to higher tensile stress at larger growth rates is yet unclear for the AlN films in question. Within the framework described, for larger tensile stresses, a mechanism that would result in a reduction in the number of atoms with successive layers needs to be identified. Sheldon et al. have proposed a possible mechanism involving bond formation at grain boundaries that involves a loss of free volume but experimental evidence for the same is yet unavailable (Sheldon et al. 2003).

Finally, at very large growth rates such as in the MOCVD films, the maximum tensile stress at which a film could grow would be limited by the cohesive stress of the boundary (Sheldon et al. 2007), which when exceeded would result in cracking as seen from Figure 5.3b. Stresses as high as 1.5 GPa have been observed in sub-200 nm thick AlN films deposited on (111) Si at growth rates of 1–10 Å/s (Raghavan and Redwing 2004a,b). Stresses in MBE grown films at 0.85 and 0.3 Å/h are lower at 0.85 and 0.23 GPa, respectively (Sheldon et al. 2005).

Growth of oriented AlN films is typically done at temperatures in excess of 1100°C by MOCVD and in excess of 700°C by MBE and sputtering. Given the CTE mismatch listed in Table 5.1 (CTE data for AlN is theoretical and not experimental), an additional calculated tensile stress of about 0.5 GPa is imposed on the film during cooling. The critical thickness for the observation of channeling cracks based on the studies in Raghavan and Redwing (2004a,b) was determined to 200 nm. If one assumed a growth stress of 1.2 ± 0.4 GPa at 1100°C, then Equation 5.5a yields a thermal mismatch stress of 0.3–0.7 GPa.

$$h_{\text{critical}} = \frac{\Gamma E}{(1-\nu)^2 Z \sigma^2} \qquad (5.5a)$$

Equation 5.5a gives the critical thickness for channeling of cracks that extend to the film–substrate interface (Beuth 1992). Γ is fracture resistance of the material and equal to 2γ for pure brittle fracture, γ is the specific surface energy and for III-nitrides, the energy of the ($10\bar{1}0$) plane along which cracking is observed, ν is Poisson's ratio, σ is the stress under which cracking occurs, E is the Young's modulus and $Z = g(\alpha,\beta)^*\pi/2$ can be calculated (Beuth 1992) from data listed in Table 5.3.

The combined effect of tensile growth stresses and thermal expansion coefficient mismatch stresses leads to cracking in films that are thicker than 200–300 nm when cooled to room temperature by MOCVD. As already mentioned, films thicker than 400 nm relax almost completely during growth itself. However, thicker crack-free AlN films on Si are desirable for many applications. By inserting a LT, 700°C, 30 nm thick AlN interlayer at 30 nm

TABLE 5.3

List of Physical Constants

	M (GPa) at 25°C	ν	$\gamma_{(10\bar{1}0)}$ (J/m²)	$g(\alpha, \beta) \approx g(0.4,0)$
GaN	470[a]	0.18[b]	1.97[c]	1.737[d]
AlN	510[a]	0.24[a]	2.43[e]	1.737[d]
Si	229[f]	0.3[g]		

Source: Raghavan, S. and Redwing, J.M., *J. Appl. Phys.* 98, 023514, 2005. With permission.

[a] Wright (1997).
[b] Perry et al. (1997).
[c] Northrup and Neugebauer (1996).
[d] Beuth (1992).
[e] Felice and Northrup (1997); Pandey et al. (1997).
[f] Hall (1967).
[g] Franca and Blouin (2004).

from the Si/AlN interface Mastro et al. have increased the crack-free thickness of AlN deposited at 1100°C to 600 nm (Mastro et al. 2006). As will be discussed in the following sections, by optimizing the growth temperature of the interlayer, a "disconnect" in the transmission of strain but not as much in orientation can be introduced resulting in similarly oriented films but with lower growth stresses, thus mitigating cracking.

5.6 Growth of GaN on (111) Si Using AlN Buffer Layers

As mentioned before, growth of group III-nitrides on (111) Si invariably begins first with growth of AlN. Even in the case when AlN is not the first layer to be deposited on (111) Si, given the higher reactivity of GaN, an AlN layer or interlayer typically precedes the growth of GaN on other buffer layers as well. Thereafter, it may be followed by various other transition strategies before the growth of a GaN layer. A list of the transition layer schemes used to integrate GaN with (111) Si is provided in Table 5.4. A graded AlGaN layer (Rajagopal et al. 2003) seems to be one of the most popular transition layers for both GaN photonics (Zhu et al. 2009) as well as electronics (Johnson et al. 2004) currently. Hence, the following discussion will be limited to AlGaN transition chemistries. It is instructive to begin with the growth of GaN on a simple 100 nm thick AlN layer to understand the problems involved, which then necessitate the more complicated solutions such as the graded buffer layer, LT interlayers, etc.

The III-nitride multilayer structures used are typically about 5 μm thick and are deposited on Si substrates that are about 250–500 μm in thickness. As discussed previously, this system can be approximated without

TABLE 5.4

Various Transition Layer Schemes Used and References

	Buffer Layer or Transition Layer for Growth of GaN	References (Not Meant to Be a Historical Record)
1.	200 nm 3-C SiC layer	Takeuchi et al. (1991)
2.	40–100 nm AlN layer	Watanabe et al. (1993), Raghavan et al. (2005)
3.	Linearly graded AlGaN buffer layer	Rajagopal et al. (2003), Raghavan et al. (2005)
4.	Step-graded buffer AlGaN layer	Kim et al. (2001)
5.	AlN buffer layer on 19 nm Sc_2O_3 on (111) Si	Lee et al. (2009)
6.	Porous GaN converted from GaAs	Ghosh et al. (2003)
7.	AlAs	Strittmatter et al. (1999)
8.	Al-oxide obtained by oxidizing AlAs on Si	Kobayashi et al. (1997)
9.	AlN on GaAs on (111) Si	Yang et al. (1996)
10.	AlInN on AlN layer on (111) Si	Irie et al. (2009)
11.	Porous (111) Si	Ishikawa et al. (2008)
12.	AlN on in situ formed Si_3N_4 layer on Si	Chang et al. (2008)
13.	HfN	Armitage et al. (2002)
14.	ScN	Moram et al. (2007a)

significant error, <10%, by a thin film that is uniformly stressed without any significant stress gradients across the thickness. Given this platform, as in the case of AlN, the first problem in integrating GaN with Si arises from the thermal expansion mismatch stress. Experimental data on thermal expansion coefficients of the III-nitrides are currently unavailable. Given the theoretical data listed in Table 5.1, the thermal expansion mismatch stress in GaN films deposited on (111) Si substrates at 1100°C and cooled to 25°C is estimated to be about 0.6 GPa. Experimental estimates of thermal stress based on cracking of GaN on Si is about 1.1 GPa (Raghavan and Redwing 2005a). Data from Table 5.3 in combination with Equation 5.5a yields Equation 5.5b for GaN on Si. Equation 5.5b predicts that a stress of 1.1 GPa will limit the maximum crack-free thickness that can be deposited on (111) Si to about 500 nm. Therefore, in order to increase the critical thickness to cracking, one would need a transition layer scheme that could introduce compressive stresses during growth so as to reduce the net tensile film stress following cooling to room temperatures. An alternative is to use confined layer epitaxy wherein growth is limited to areas smaller than $14 \times 14 \ \mu m^2$ (Zamir et al. 2001).

$$h_{\text{critical}} = \frac{576}{\sigma^2} \qquad\qquad (5.5b)$$

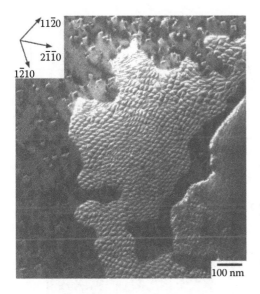

FIGURE 5.6
Misfit dislocation in GaN island on AlN on SiC substrate. (From Moran, B. et al., *J. Cryst. Growth*, 273, 38, 2004. With permission from Elsevier.)

The lattice mismatch between AlN and GaN is 2.4% and translates to a mismatch stress of −11.3 GPa. However, the AlN buffer layers on Si are themselves under a tensile stress of 1.5 GPa and hence this mismatch stress reduces to −10 GPa. The introduction of misfit dislocations at GaN layer thicknesses calculated to be less than 10 Å (Holec et al. 2008) and shown in Figure 5.6 further significantly reduces the magnitude of compressive stresses from the mismatch source. Surface stresses also appear to contribute a hydrostatic compressive stress (Krost et al. 2004).

The net effect of all the sources mentioned is that GaN films on the AlN buffer layer start growth under a compressive stress whose measured in-plane value is about −1 GPa ± 0.5 GPa as seen from Figure 5.7 (Krost et al. 2004; Raghavan and Redwing 2005a). This is obviously much lower than what might have been expected from the mismatch between GaN and AlN. Nevertheless, even if this compressive stress itself were retained, it would counter a significant portion of the tensile thermal mismatch stresses and increase the critical thickness to cracking. However, as seen from Figure 5.7, this initial stress quickly transforms to a tensile growth stress within 88 ± 24 nm of growth.

The remainder of the film then grows under a tensile stress of 0.25 ± 0.1 GPa. Hence, on cooling down to room temperature GaN films crack under the combined 1.35 GPa stress (1.1 CTE mismatch stress + 0.25 growth stress) at thicknesses in excess of about 400 nm as shown in Figure 5.7d. All the AlGaN buffer layer schemes and low-temperature interlayers schemes essentially revolve around trying to incorporate as much of the AlN-GaN mismatch compressive stress in the GaN layer during growth. As a result,

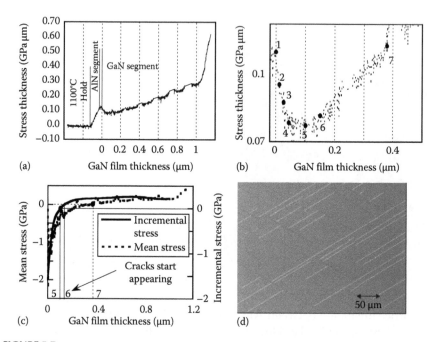

FIGURE 5.7

(a) Stress-thickness vs. film thickness during growth of GaN films on an AlN buffer on (111) Si. The discontinuities seen in the GaN segment correspond to thicknesses at which complete laser spot extinction occurs due to destructive interference between reflections from the surface of the film and the interface between the film and the substrate. (b) Initial portion of the curve corresponding to the GaN segment in (a) showing the points corresponding to the AFM scans in Figure 5.8. (c) Mean film stress and incremental stress in the GaN film. Points 5, 6, and 7 from (b) are as indicated. Cracks start appearing in the film after point 6 by which time the mean stress in the film is close to zero. (d) A heavily cracked 500 μm GaN film on an AlN buffer layer. (From Raghavan, S. and Redwing, J.M., *J. Appl. Phys.*, 98, 023514, 2005. With permission.)

the net tensile stress on cool down is reduced and this in turn helps to prevent cracking, the first step required for integrating GaN with Si.

In order to design an efficient growth scheme, it is first necessary to understand the reasons for the transition from the initial compressive to a final tensile stress. AFM images of the evolution in microstructure during MOCVD growth that accompanies the stress evolution discussed in the previous paragraph is shown in Figure 5.8. The numbers on the images correspond to those in Figure 5.7b. GaN nuclei, Figure 5.8b, are formed on the AlN buffer layer, Figure 5.8a first. These then grow laterally and coalesce. Given all the prior discussion on the role of coalescence in the generation of tensile stresses during the growth of AlN, it might at first appear that the same also holds true for GaN growth. However, in addition to coalescence, the transition in stress from tensile to compressive is also accompanied by a sharp increase in lateral grain size close to the interface. The lateral island

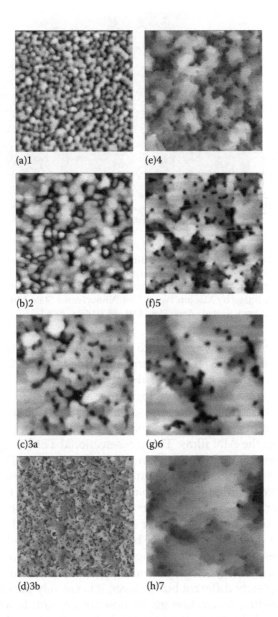

FIGURE 5.8

Sequential AFM scans showing evolution of the surface morphology of the film. Numbers next to the scans correspond to those in Figure 5.1b. Image scan sizes are as follows. (a)–(c)—1 μm × 1 μm × 50 nm; image (d)—5 μm × 5 μm × 50 nm; image (e)—5 μm × 5 μm × 100 nm; images (f)–(h)—5 μm × 5 μm × 200 nm. The GaN film thickness in the images correspond to 0, 13, 40, 40, 70, 100, 132, and 400 nm. The AFM roughness values (R_a) for images (d)–(h) are 3.2, 11.8, 19.9, 20.7, and 9.3 nm, respectively. (From Raghavan, S. and Redwing, J.M., *J. Appl. Phys.*, 98, 023514, 2005. With permission.)

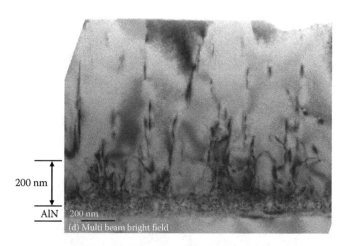

200 nm

AlN   200 nm
(d) Multi beam bright field

FIGURE 5.9
Cross-sectional TEM image of GaN deposited direction on an AlN buffer layer on (111) Si. Most microstructural evolution in the form of a reduction in dislocation density is complete within about 200 nm from the GaN/AlN interface in agreement with AFM data that shows an increase in lateral grain size. Image by Xiaojun Weng, The Pennsylvania State University, University Park, PA. (From Raghavan, S. et al., *Appl. Phys. Lett.*, 88, 041904, 2006.)

size in GaN increases from 50 nm to 2.5 μm over a thickness of 100 nm (compare images Figure 5.8b and g corresponding to points 2 and 6, respectively). This is in sharp contrast to the absence of any such evolution in lateral grain size in AlN films growing under a tensile stress. However, following the transition to a tensile stress, point 6 to point 7 in Figure 5.7b, further evolution in microstructure in relatively nonexistent, compare Figure 5.8g and h, as in the AlN films. The cross-sectional TEM image, Figure 5.9 shows the corresponding evolution in dislocation density (Raghavan et al. 2006). Thus, the transition from the initial compressive stress to a final tensile stress is accompanied by both coalescence and a reduction in TDD or increase in lateral grain size. For the case of GaN on Si, it is most likely that the dislocations in the GaN layer are due to the tilt and twist between the GaN nuclei dictated by the AlN buffer layer below and are therefore formed at coalescence. However, in light of recent debate on the origins of threading dislocations in GaN layers on sapphire (Oliver et al. 2006a), which employs a completely different buffer layer, it is entirely possible that other sources of threading dislocation generation are present. In any event, it is not the source of dislocations, but their evolution—more specifically the formation of misfit segments with an edge component that is parallel to the interface as discussed later—which would affect stress evolution. Hence, this debate while noted is not discussed further. Data from MBE growth of GaN on Si seems to indicate that the microstructural evolution involved there is not very different from the one discussed above in MOCVD growth (Adikimenakis et al. 2009).

In situ stress evolution data on growth of GaN on Si by MBE is unavailable to compare the effect of growth rates. However, as for AlN, there is very little effect of growth rate of GaN when varied from 0.4 to 1 nm/s on the evolution described above. Growth rate of the buffer or its thickness, as would be expected, also do not have any effect (Raghavan and Redwing 2005).

The interrelationship between stress and microstructural evolution will be discussed after the next section on observations during growth of graded AlGaN buffer layers. It will be seen that while the transition layer schemes have been successful in solving the first problem in integrating GaN with Si, i.e., cracking, the eventual TDD in thick, >1 μm, layers is still about 10^9 cm^{-2} or higher. As with other substrates, lateral growth methods can be employed on Si substrates as well to reduce TDDs to below 10^8 cm^{-2} (Gibart 2004).

5.7 Growth of AlGaN Buffer Layers

The GaN on AlN system discussed in the previous section is a special case of GaN on graded AlGaN on AlN. It is a drastic step grade. The processes happening across this step grade can be better understood if that interface is drawn out over a graded AlGaN layer. This grading can be done in various ways with minor differences in microstructural evolution, but they have not yet been able to significantly reduce the eventual TDD. Hence, discussion will be limited to a linearly graded buffer layer to bring out the essential interrelated features between stress and microstructural evolution.

The stress-thickness vs. thickness plots representing growth of a 1 μm thick linearly graded AlGaN buffer layers is shown in Figure 5.10a (Raghavan and Redwing 2005b). It is seen that films begin to grow under a tensile stress (positive slope) of about 1.5 GPa, which eventually transitions to a compressive stress midway during growth at a composition of $Al_{0.5}Ga_{0.5}N$. The source of this initial tensile stress is exactly the same as that discussed previously for growth of AlN on Si. The change in lattice parameter due to grading of composition from AlN to GaN imposes a compressive stress on this initial tensile stress. Hence, the eventual transition to a net compressive stress. For reasons that have not yet been completely explained, as seen from Figure 5.10a, the transition to a compressive stress always occurs midway corresponding to the composition $Al_{0.5}Ga_{0.5}N$. Thus, 1 and 0.5 μm films transition at 0.5 and 0.25 μm thickness, respectively.

If the initial tensile stress and the lattice mismatch were the only two sources of stress, then this evolution could be represented by an equation of the form

$$\sigma(h) = \sigma_{\circ} - 122.5 \ln\left(\frac{1 + 0.021h}{t}\right) \tag{5.6}$$

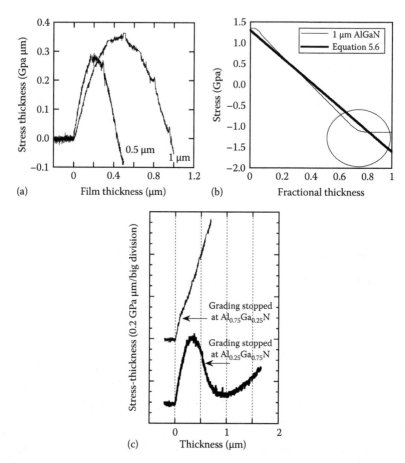

FIGURE 5.10
(a) Stress-thickness vs. thickness plots during growth of a 0.5 and 1 µm linearly graded AlGaN (AlN-GaN) layer on (111) Si. (b) Incremental stress (slope of stress-thickness vs. thickness plots), $\sigma(h)$, vs. fractional thickness (h/film thickness) for a 1 µm graded AlGaN layer and its comparison with Equation 5.6. (c) Stress-thickness vs. thickness plots showing the effect of interrupting the change in composition during the tensile portion and the compressive portion of the graded AlGaN stress-thickness vs. thickness plots and growing a layer of the same composition as at the interruption. (From Raghavan, S. and Redwing, J.M., *J. Appl. Phys.*, 98, 023515, 2005. With permission.)

where σ_o is the initial tensile stress of about 1.5 GPa coming from the AlN layer and t is the total layer thickness. The second term summarizes the incremental compressive stress contribution coming from the changing lattice parameter. The slope of the stress-thickness vs. thickness curve in Figure 5.10a, which yields stress in the film, is compared with Equation 5.6 above in Figure 5.10b. They differ in the encircled region. It is seen that while Equation 5.6 predicts that stress will continue to become increasingly compressive, actual stress evolution tends toward an apparently constant value.

This difference implies that there is a source of tensile stress that compensates the compressive stress arising from the changing lattice parameter, the only nonconstant source accounted for by Equation 5.6. Indeed, as would then be expected, when grading is stopped at some composition, to take away the source of compressive stress, and followed by growth of a film of the same composition, a transition to a tensile stress should be observed. As shown in Figure 5.10c, when grading is stopped at $Al_{0.25}Ga_{0.75}N$ in the compressive segment, followed by growth of the same composition, a transition to a tensile stress is indeed observed. However, when grading is stopped at $Al_{0.75}Ga_{0.25}N$, the tensile portion, the film continues to grow at the same value of tensile stress prior to the interruption.

Since the $Al_{0.25}Ga_{0.75}N$ films in the discussion above are considerably past the coalescence stage, which occurs at <100 nm, the data above clearly demonstrate that there is a source of tensile stress in these films apart from island coalescence. Such delayed transitions to a tensile stress have also been observed during the growth of GaN on AlN interlayers to be discussed in the next section by multiple groups in films that are as thick as 2 μm.

The microstructural evolution that accompanies the stress evolution above in the linearly graded 1 μm AlGaN film is shown in Figure 5.11. It is very clearly seen that all evolution in the dislocation structure, namely a reduction in dislocation density occurs only when the film is growing under a compressive stress. Given the prior discussions on AlN films and GaN films, the first common observation is that microstructural evolution in the form of a reduction in TDD or an increase in lateral grain size occurs only under a compressive growth stress. In GaN films on AlN buffer layers, this reduction

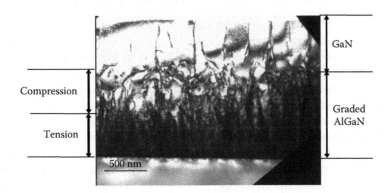

FIGURE 5.11

Cross-sectional TEM images of the graded 1 μm AlGaN buffer layer. When viewed in conjunction with the stress data in Figure 5.10a it is seen that there is very little microstructural evolution in the first 500 nm when the film is growing under a tensile stress as in AlN films on Si. Most change, in the form of a reduction in dislocation density, occurs in the next 500 nm when the film is growing under a compressive stress as in GaN films on AlN buffer layers. Image by Xiaojun Weng, The Pennsylvania State University, University Park, PA. (From Raghavan, S. et al., *Appl. Phys. Lett.*, 88, 041904, 2006.)

happens very close to the GaN/AlN interface and so does the transition from a compressive stress to a tensile stress. In graded AlGaN films, it happens far from the interface and is again accompanied by a similar transition. Thus, the second common observation is that it is the reduction in dislocation density that is primarily responsible for the transition of the compressive stress arising from the lattice mismatch to an eventual tensile one.

It is important to note that it is not just relaxation of stresses, compressive or tensile, such as, e.g., by the formation of misfit dislocations. In such cases of relaxation, the stress level in the film tends to zero. Rather in the case of the III-nitrides, there is an active source of tensile stress, as will be discussed later, related to the ubiquitous threading dislocation lines that not only relaxes any compressive stresses present to zero but eventually strains the film in tension. Films growing under tension such as the AlN films in Figure 5.5b or the $Al_xGa_{1-x}N$ films in Figure 5.10 with $x < 0.5$ seem to be incapable of relaxing by such a mechanism. These experimental observations imply that if dislocation structure evolution is prevented in films growing under a compressive stress, then the transition to a tensile stress can be prevented as well. A net compressive growth stress can be incorporated in such GaN films to offset the tensile mismatch stress and thus prevent or mitigate the cracking problem. As will be seen in the next section, this indeed is the mechanism used to prevent cracking in GaN film on Si using the graded AlGaN buffer layer just discussed.

5.8 Growth of GaN on Graded AlGaN Buffer Layers

Stress-thickness vs. thickness and stress vs. thickness plots obtained from data collected during growth of the GaN epilayer only on graded AlGaN buffer layers of varying thicknesses are shown in Figure 5.12a and b. The topmost curve is a reproduction of the plot in Figure 5.7a and b for growth of GaN directly on an AlN buffer layer. All dislocation density evolution in this case occurs in the GaN layer and hence the quick transition to a tensile stress. With increase in the thickness of the graded AlGaN buffer layer to 1 μm, the thickness of the subsequent GaN layer that grows under compression before undergoing the transition to a tensile stress increases as well. As can be seen from Figure 5.12a, in GaN layers deposited on 1 μm thick graded AlGaN buffer layers, the transition is just about to begin at 1 μm thickness. In contrast, in GaN layers deposited on AlN buffer layers, it happens at 100 nm. Finally, in GaN layers deposited on thicker AlGaN buffer layers, the transition is not observed even after 2 μm of growth. As a result, a net compressive growth stress is incorporated in the GaN film. Given the discussion in the previous sections, it is obvious that with an increase in thickness of the graded AlGaN buffer layer, the thickness of the portion that grows under a compressive stress increases. Thus, greater fractions of dislocation density

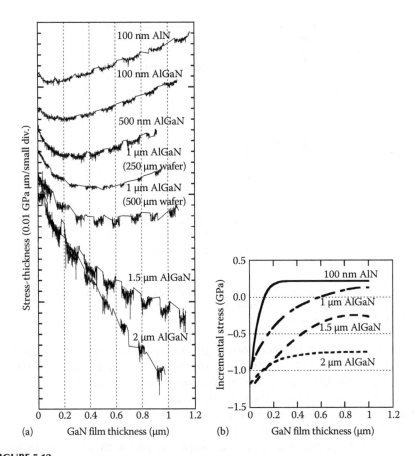

FIGURE 5.12

(a) Stress-thickness vs. thickness plots showing effect of thickness of the graded AlGaN layer on stress evolution during growth of the GaN layer. (b) Slopes of the stress-thickness vs. thickness plots showing effect of AlGaN layer thickness on incremental stress evolution in the GaN layer. (From Raghavan, S. and Redwing, J.M., *J. Appl. Phys.*, 98, 023515, 2005. With permission.)

evolution occur in the AlGaN buffer layer itself. Hence, with an increase in the thickness of the graded AlGaN buffer layer, the ability of such evolution to engender a transition to a tensile stress in the subsequent GaN layer diminishes and eventually becomes nonexistent. The GaN layer then grows under compression.

The microstructural evolution in such GaN layers is summarized in the x-ray data in Figure 5.13. With the increase in thickness of the AlGaN layer, the FWHM of both symmetric and asymmetric scans from the GaN decrease, thereby indicating a lowering of tilt and twist with respect to the substrate, a lowering of the corresponding dislocation density, and an overall improvement in crystal structure. This improvement is again due to increasing reduction in TDD occurring in the AlGaN buffer layer itself. From plan view TEM measurements, the dislocation density, a and a+c type, on top of

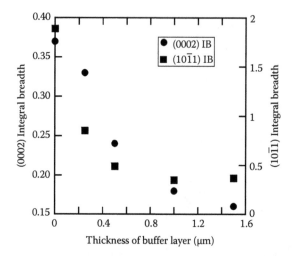

FIGURE 5.13

Effect of thickness of the graded AlGaN buffer layer on the integral breadth of (0002) and ($10\bar{1}1$) peaks of a 0.5 μm thick GaN layer. (From Raghavan, S. and Redwing, J.M., *J. Appl. Phys.*, 98, 023515, 2005. With permission.)

the 0.5 μm GaN layer deposited on a 1 μm AlGaN buffer layer was estimated to be 10^{10} cm^{-2} (Weng et al. 2006). In comparison, the corresponding density for a 1 μm GaN film on an AlN buffer layer was also about 1.5×10^{10} cm^{-2}. Thus, as mentioned previously, the major contribution of the graded AlGaN buffer layer is in preventing cracking and not as much in reducing the eventual TDD in the subsequent GaN layer.

The effect of having incorporated a compressive growth stress on cracking following cooling at room temperature is shown in Figure 5.14a. It is seen that the critical thickness to cracking has been increased from about 400 nm on thin AlN buffer layers to greater than 2 μm on a graded AlGaN buffer layer. It is important to note here as shown in Figure 5.14b, a 1 μm thick GaN film deposited on an 1 μm thick AlGaN buffer layer, that cracks are not observed even after subjecting the wafer to dicing, thus implying that the crack-free system is a stable one. The importance of this observation will become apparent after the next section. A comparison of Figures 5.14b and 5.7d shows the effect of the AlGaN buffer layer in reducing the cracking problem.

5.9 Growth of GaN Using AlN Interlayers

As mentioned before, the only source of compressive stress for GaN films deposited on AlGaN films is from the lattice mismatch across the GaN/AlN interface. As also discussed, the compressive stress from this mismatch is

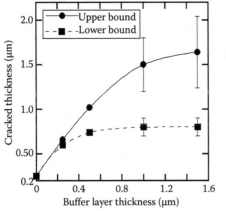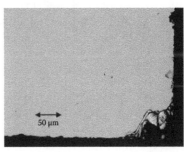

FIGURE 5.14

(a) Effect of thickness of the graded AlGaN buffer layer on the "thickness for cracking" in GaN layers. The lower and upper bound, see ref. Raghavan and Redwing 2005b for more details, are due to different ways in obtaining the crack-free GaN layer because of growth-related thickness variations. The actual behavior is expected to follow the upper bound as seen in image (b). (b) Optical microscope image of a corner obtained by dicing a 1 μm GaN on a 1 μm AlGaN buffer layer structure deposited on a 2" wafer. (From Raghavan, S. and Redwing, J.M., *J. Appl. Phys.*, 98, 023515, 2005. With permission.)

offset to varying extents by the evolution in density of defects that are also generated at or very close to the interface requiring the need for graded AlGaN layers. Thus, a relaxed AlN layer deposited on a GaN layer in which most defect evolution has been completed would be expected to serve as a new source of compressive stress and one that would not cause the transition to a tensile stress in the subsequent GaN layer. The use of an AlN interlayer (or AlGaN interlayer) falls under this category of solutions and involves the deposition of thin (~10 nm) AlN interlayers at different locations within the GaN layer. The stress-thickness vs. thickness plot from in situ data during the growth of GaN layers that incorporate AlN interlayers is shown in Figure 5.15 (Krost et al. 2003). As seen while a compressive stress is introduced in the GaN layer immediately following the AlN interlayer, these stresses relax quickly and eventually transition to a tensile stress as shown in Figure 5.16 (Raghavan et al. 2005).

The stress evolution and the microstructural evolution following the introduction of an interlayer are dependent on many parameters.

1. *Thickness and composition of the interlayer:* Interlayers that are too thin and thus not relaxed with respect to the GaN layer below will transmit both the stress and the microstructure and hence will not serve any purpose. The interlayer needs to be at least thick enough for some mechanism such as misfit dislocation formation or cracking to cause relaxation. Interlayers that have a thickness much beyond one at which

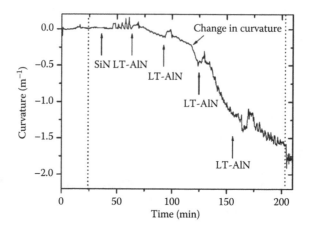

FIGURE 5.15

Effect of AlN interlayer insertion on the stress evolution during growth of thick GaN layers on Si. (From Krost, A. et al., *Phys. Stat. Sol. (a)*, 200, 26, 2003. With permission.)

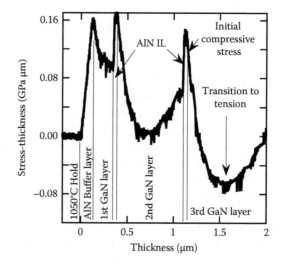

FIGURE 5.16

Growth of GaN beyond the point of zero stress in Figure 5.15 eventually leads to the transition to a net tensile growth stress. (From Raghavan, S. et al., *Appl. Phys. Lett.*, 87, 142101, 2005. With permission.)

they relax completely will also be of no further use. Studies by Lee et al. (2004) show that $Al_xGa_{(1-x)}N$ layers on GaN, $x=1$ for AlN interlayers, relax by surface cracking and formation of misfit dislocations involving glide on non-basal systems, though the former mechanism appears to precede the latter. The critical thickness for relaxation by cracking, both experimentally measured and theoretically calculated,

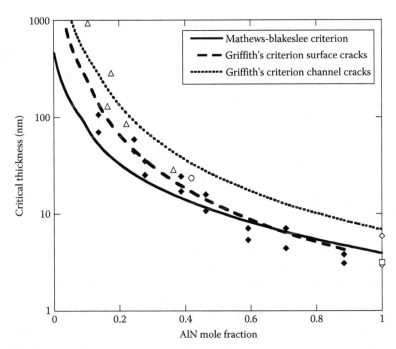

FIGURE 5.17
Critical thickness for relaxation of AlGaN layers on GaN. Data points are from various literature sources. For details see ref. Lee et al. (2004). (From Lee, S.R. et al., *Appl. Phys. Lett.*, 87, 142101, 2005. With permission.)

is plotted in Figure 5.17 for different Si-doped AlGaN/GaN systems grown at the same temperature. The behavior in the absence of Si doping could change but is not expected to be very different as data from other sources in the literature were in agreement with the observations of Lee et al. For AlN interlayers, the critical thickness for cracking is about 4 nm. The typical 10 nm AlN interlayer used would thus be expected to be completely relaxed by cracking.

2. *Temperature of growth of the interlayer*: The discussion above was for interlayers deposited at the same temperature of growth as the GaN layer. For Si-doped layers, Reiher et al. (2003) observed a reduction in average tensile stress in the GaN layer as measured by *ex situ* XRD, Figure 5.18, with a reduction in the temperature of growth of the AlN interlayer between 1100°C and 550°C. Such an effect was not observed for undoped layers (Krost et al. 2003). The decrease in tensile stress was attributed to increasing incoherency between the AlN interlayer and the GaN layer below with a reduction in growth temperature. This results in an increasingly relaxed AlN interlayer and, therefore, greater compressive mismatch between it and the subsequent GaN or AlGaN layer.

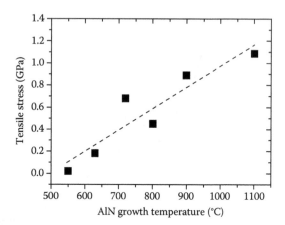

FIGURE 5.18
Effect of temperature of deposition of AlN interlayer on stress
in GaN. (From Reiher, A. et al., *J.
Cryst. Growth*, 248, 563, 2003. With
permission.)

3. *Thickness of the first GaN layer preceding the AlN interlayer*: As mentioned
   before, most microstructural evolution involving a reduction in dislocation density is completed within about 150 nm from the first AlN
   buffer layer/GaN interface. Thus, the structure of the AlN interlayer
   and the one it transmits to the subsequent second GaN layer would be
   expected to vary with increase in thickness of the first GaN layer up to
   150 nm and cease beyond that. Figure 5.19 shows the effect of a change
   in thickness of the first GaN layer on the stress evolution in the second
   GaN layer. As expected, with an increase in the thickness of the first
   GaN layer to 150 nm, the thickness for transition to a tensile stress in
   the second GaN layer increases, with very little change thereafter.

Both, the observation in point 3 above and the transition to a tensile stress
seen in GaN layers deposited on multiple interlayers at cumulative thicknesses greater than 2 μm again strongly suggest that grain coalescence alone
cannot be the only source of tensile stress. Microstructural evolution in the
form of evolution of the dislocation structure is the most likely source as
discussed in greater detail in the next section. In spite of the transition following each interlayer, however, as seen in Figure 5.15, the net change in
curvature observed on judicious insertion of the interlayer at the point of
transition implies that the average compressive stress in the multilayer stack
has increased. The stress in the GaN film at each transition thickness, $\sigma(h)$
would, however, be at a lower level of compressive stress. In terms of cracking, therefore, care needs to be exercised as a crack-free system on cool down
to room temperature need not necessarily be a stable system. To illustrate
this point, Figure 5.20 compares cracking in GaN films grown on (111) Si
without AlN interlayers (a) and with two interlayers prior to and after dicing, (b) and (c), respectively. Growth in the sample with two interlayers was
interrupted only after the GaN had undergone the transition to a tensile
stress. The increase in crack density due to dicing in this heterostructure

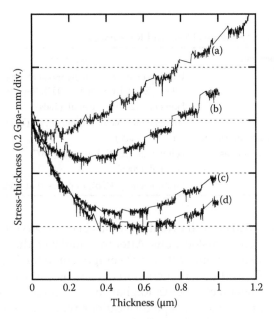

FIGURE 5.19
Effect of thickness of the first GaN layer preceding the interlayer on the stress evolution in the GaN layer following the Al N interlayer. (a) GaN on AlN buffer layer. (b), (c), and (d) GaN on AlN interlayer. Thickness of the first GaN layer in (b), (c), and (d) are 50, 150, and 300 nm, respectively. (From Raghavan, S. et al., *Appl. Phys. Lett.*, 87, 142101, 2005. With permission.)

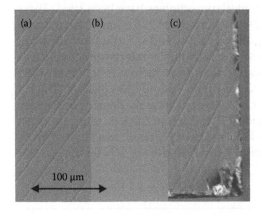

FIGURE 5.20
Effect of dicing on cracking of GaN layers on Si using AlN interlayers.

as compared to the absence of cracking, see Figure 5.14b, in GaN on graded AlGaN is an indicator of the instability of this structure. Interruption of growth when the GaN is under a compressive stress would help in preventing such cracking. Such ability highlights the utility of in situ stress monitoring.

TABLE 5.5

Various Interlayer Schemes Used and References

Interlayer	References
1. AlN	Waldrip et al. (2001), Kuwano et al. (2002), Dadgar et al. (2000), Krost et al. (2003), Raghavan et al. (2005)
2. AlGaN	Han et al. (2001), Lee et al. (2004)
3. Si_xN_y	Arslan et al. (2009), Kappers et al. (2007)
4. In-doped AlGaN interlayer	Wu et al. (2008)
5. AlN-GaN superlattices	Feltin et al. (2001), Liu et al. (2007)
6. InN	Hu et al. (2006)
7. TiN, ScN	Moram et al. (2007b), Kappers et al. (2007)

In terms of microstructural evolution, it has been observed that the interlayer itself is a source of dislocations. After the interlayer, there is an increase in the density of dislocations with an "a" component to the Burgers vector in the GaN layer, followed by a reduction in density as at the GaN/AlN buffer layer interface (Kuwano et al. 2002; Raghavan et al. 2005). In addition, apart from a reduction in density, a-type dislocations with a line direction along the *c*-axis are seen to get inclined (Follstaedt et al. 2009). Both these microstructural features are believed to contribute to the observed stress evolution as discussed in the next section. As with the linearly graded AlGaN buffer layers, it is seen that while AlN interlayers are effective in mitigating cracking, they do not have a significant effect on the dislocation density on the surface (Weng et al. 2006).

As with the buffer layer, various interlayer schemes have been tried. These are summarized in Table 5.5. It has been reported that TDDs have been reduced to $9 \times 10^7\,cm^{-2}$ and $3 \times 10^7\,cm^{-2}$ using silicon nitride and scandium nitride interlayers, respectively. In the case of silicon nitride, the interlayer actually acts as an in situ random mask. Dislocation density reduction is achieved when islands grow up through the mask and then laterally to coalesce. Thus, a larger size at coalescence translates to a lower dislocation density. In the case of ScN interlayers obtained by nitriding 5 and 20 nm thick Sc films, the exact opposite is observed. Thinner ScN layers apparently result in a larger density of GaN nuclei and lower dislocation density than thicker ScN layers.

5.10 Dislocations and Compressive Stress Reduction in GaN Layers

The common thread running through all the previous sections on stress and microstructural evolution is the relaxation of compressive stresses with a corresponding evolution in the dislocation structure in the film. There are

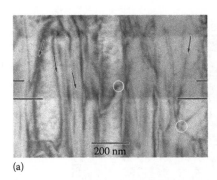 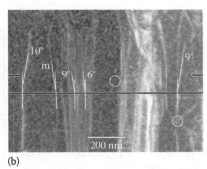

(a)   (b)

FIGURE 5.21
TEM images showing "a" type dislocations with line direction along the c-axis inclining at the AlGaN/AlN interface. (a) Bright field. (b) Weak-beam image. (From Follstaedt, D.M. et al., *J. Appl. Phys.*, 105, 083507, 2009. With permission.)

two components to the evolution in dislocation structure. One is a reduction in the density of threading dislocation segments. Density reduction is seen, e.g., in Figures 5.9 and 5.11. The second is bending and inclination of otherwise straight dislocation lines at interfaces across which there is a compressive mismatch such as when GaN (or $Al_xGa_{(1-x)}N$) is deposited on AlN (or $Al_yGa_{(1-y)}N$, $y < x$) as shown in Figure 5.21 (Follstaedt et al. 2009). Inclination without significant reaction and subsequent reduction in density is typically observed during growth of AlGaN layers on fairly thick GaN templates, >1 μm, with much lower defect densities than that present at the AlN buffer layer/GaN layer interface (Wang et al. 2006; Follstaedt et al. 2009).

In terms of observed microstructural evolution density reduction is most likely the dominant reason for compressive stress relaxation in GaN and graded AlGaN films deposited on AlN buffer layers on (111) Si while dislocation density reduction and inclination is most likely the reason for the observed stress evolution in GaN layers deposited on the AlN interlayer. Dislocation inclination is expected to be the more dominant mechanism with increase in thickness of both the GaN layer preceding the interlayer and after the interlayer. However, it is entirely possible that the two are different manifestations of one and the same phenomenon, namely the change in number of atoms with the addition of new layers. Stress and microstructural evolution then would not be entirely of thermodynamic origin, but would also be influenced by kinetic considerations as for the AlN films. As discussed then, a net compressive stress will be generated when the number of atoms in the nth layer is greater than the number of atoms in the $(n-1)$th layer. The opposite will result in a tensile stress.

Following the observation of inclination of lines of the a-type dislocations in Si-doped AlGaN layers by Cantu et al. (2003), Romanov and Speck proposed that the bending of dislocations with a Burgers vector **a** is equivalent to relaxation by the deposition of a misfit segment (Romanov

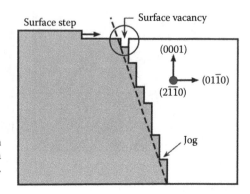

FIGURE 5.22
Schematic of how dislocation inclination can give rise to a tensile stress. (From Follstaedt, D.M. et al., *J. Appl. Phys.*, 105, 083507, 2009. With permission.)

and Speck 2003). Since the angle of inclination following bending remains constant, the effective length of the misfit segment increases with thickness and hence a linear relation is observed between film thickness and strain relaxation. A schematic of the above mechanism as presented by Follstaedt et al. is shown in Figure 5.22 (Follstaedt et al. 2009). An inclined a-type edge dislocation made up by the extra ($11\bar{2}0$) can be looked upon as an array of vertical segments separated by jogs. The projected cumulative length of the jogs is the length of the misfit segment. The average biaxial relaxation in strain due to these segments at the top of the layer can then be given as

$$\Delta\varepsilon = \left(\frac{3}{2}\right)bL\left(\frac{\rho}{3}\right) = \left(\frac{3}{2}\right)bh\tan\theta\left(\frac{\rho}{3}\right) \tag{5.7}$$

where
  b is the Burgers vector
  L is the projected length of the dislocation
  ρ is the dislocation density of the entire family made up of three sets 120° apart
  the factor of 3/2 comes from the contribution of three linear strain fields 120° apart corresponding to each of the three sets to an equi-biaxial field
  h is the film thickness and θ is the angle of inclination

This linear relationship obtained by Romanov and Speck (2003) has since been verified by multiple other groups (Wang et al. 2006; Acord et al. 2008; Follstaedt et al. 2009).

However, the reason for the inclination remains open to debate. Romanov and Speck used a purely thermodynamic analysis, balancing strain energy decrease due to relaxation and line energy increase due to inclination, to come up with a critical thickness for inclination much like that for formation of misfit dislocations. Follstaedt et al. (2009) have however observed bending at thicknesses much smaller than that predicted by the Romanov and Speck model, thus bringing into question the use of pure thermodynamics to predict bending. The fact remains that most crystal growth happens at large supersaturations, far from thermodynamic equilibrium, and at rates much faster than those required for the surface to attain equilibrium. In the representation in Figure 5.22, for a compressive stress to relax, the extra half plane needs to retract with increasing thickness. Such retraction happens when the new step approaching the point of intersection of the dislocation with the growth surface forms a bond across the surface vacancy shown, creating a single atom jog. This essentially results in a reduction in the number of atoms in the new step and hence a tensile stress. Thus, this description of events is now very similar to the one presented by Sheldon et al. (2005) in their kinetic theory to explain tensile stresses in AlN. Essentially, there is loss of free volume associated with the atoms at the dislocation cores. Since, these dislocation arrays essentially form low-angle grain boundaries, this volume loss is equivalent to that lost on "grain growth," the model used by Raghavan et al. (Raghavan and Redwing 2005) to explain lateral increase in grain size as a means of tensile stress generation. In addition, if the retracting plane were to intersect a surface, another retracting plane, or a boundary, it would form a loop and thus cease to thread further or depending on the reaction continue as a single dislocation. Such reactions and reduction in dislocation density are expected to happen more rapidly in areas of higher dislocation density such as the GaN-AlN buffer interface (Romanov et al. 1996). Any reduction in number of planes due to such interactions would thus result in both a reduction in density and the generation of a tensile stress. Thus, while the formulations have been very different, they seem to be converging to the basic idea of reduction in number of atoms or planes being responsible for the generation of a tensile stress. Observations of bending being dependent on surface roughness of the films and the fact that the inclined lines are "frozen in" seem to further indicate that the processes responsible for these mechanisms happen on the growth surface and at the "grain boundaries" or points of intersections of dislocation lines with the growth surface. The eventual transition of a film growing under compression to a net tensile stress and its cracking due to continued inclination of dislocation, however, cannot be answered by the above model in Figure 5.22 and remains open.

Acknowledgments

Srinivasan Raghavan would like to acknowledge support from the Government of India, Ministry of Defense, under project number DECS/08/08147/2436/D (R&D)/TD2008/SPL147.

References

Acord, J. D., I. C. Manning, X. Wang, D. W. Snyder, and J. M. Redwing. 2008. In-situ measurement of stress generation arising from dislocation inclination in AlGaN: Si thin films. *Appl. Phys. Lett.* 93:111910.

Adikimenakis, A., S.-L. Sahonta, G. P. Dimitrakopulos, J. Domagala, Th. Kehagias, Ph. Komninou, E. Iliopoulos, and A. Georgakilas. 2009. Effect of AlN interlayers in the structure of GaN-on-Si grown by plasma assisted MBE. *J. Cryst. Growth* 311:2010.

Albrecht, M., I. P. Nikitina, A. E. Nikolaev, Yu. V. Melnik, V. A. Dmitriev, and H. P. Strunk. 1999. Dislocation reduction in AlN and GaN bulk crystals grown by HVPE. *Phys. Stat. Sol. (a)* 176:453.

Amano, H., N. Sawaki, I. Akasaki, and Y. Toyoda. 1986. Metalorganic vapor phase epitaxial growth of a high quality GaN film using an AlN buffer layer. *Appl. Phys. Lett.* 48 (5):353.

Armitage, R., Q. Yang, H. Feick, J. Gebauer, and E. R. Weber. 2002. Lattice matched HfN buffer layers for epitaxy of GaN on Si. *Appl. Phys. Lett.* 81 (8):1450.

Arslan, E., M. K. Ozturk, S. Ozcelik, and E. Ozbay 2009. The effect of Si_xN_y interlayer on the quality of GaN epitaxial layers grown on Si(111) substrates by MOCVD. *Curr. Appl. Phys.* 9:472.

Beuth, J. L. 1992. Cracking of thin bonded films in residual tension. *Int. J. Solids Structures* 29 (13):1657–1675.

Bonanni, A., D. Stiffer, A. Montaigne-Ramil, K. Schmidegg, K. Hingerl, and H. Sitter. 2003. In Situ spectroscopid ellipsometry of MOCVD-grown GaN compounds for online composition determination and growth control. *J. Cryst. Growth* 248:211.

Bourret, A., A. Barski, J. L. Rouviere, G. Renaud, and A. Barbier. 1998. Growth of aluminum nitride on (111) Si: Microstructure and interface structure. *J. Appl. Phys.* 83 (4):2003–2009.

Bykhovski, A. D., B. L. Gelmont, and M. S. Shur. 1995. Elastic strain relaxation in GaN-AlN-GaN semiconductor-insulator-semiconductor structures. *J. Appl. Phys.* 78 (6):3961.

Cammarata, R. C. 1994. Surface and interface stress effects in thin films. *Prog. Surf. Sci.* 46 (1):1.

Cantu, P., F. Wu, P. Waltereit, S. Keller, A. E. Romanov, U. K. Mishra, S. P. DenBaars, and J. S. Speck. 2003. Si doping effect on strain reduction in compressively strained $Al_{0.5}Ga_{0.51}N$ thin films. *Appl. Phys. Lett.* 83:674.

Carreño, L. A., C. Boney, and A. Bensaoula. 2003. In situ determination of surface composition, polarity, crystallographic relationship, and periodicity of GaN films by mass spectroscopy of recoiled ions and direct recoiled spectroscopy. *J. Appl. Phys.* 94 (12):7883.

Chang, J. R., T. H. Yang, J. T. Ku, S. G. Shen, Y. C. Chen, Y. Y. Wong, and C. Y. Chang. 2008. GaN growth on Si(111) using simultaneous AlN/α-Si_3N_4 buffer structure. *Jpn. J. Appl. Phys.* 47:5572.

Chason, E., B. W. Sheldon, L. B. Freund, J. A. Floro, and S. J. Hearne. 2002. Origin of compressive residual stress in polycrystalline thin films. *Phys. Rev. Lett.* 88 (15):156103.

Chen, C.-H., C.-M. Yeh, J. Hwang, T.-L. Tsai, C.-H. Chiang, C.-S. Chang, and T.-P. Chen. 2006. Band gap shift in the GaN/AlN multilayers on the mesh-patterned Si(111). *Appl. Phys. Lett.* 88:161912.

Dadgar, A., J. Blasing, A. Diez, A. Alam, M. Heuken, and A. Krost. 2000. Metalorganic chemical vapor phase epitaxy of crack-free GaN on Si(111) exceeding 1 micron in thickness. *Jpn. J. Appl. Phys.* 39:L1183–1185.

Dadgar, A., A. Strittmatter, J. Blasing, M. Poschenrieder, O. Contreras, P. Veit, T. Riemann, F. Bertram, A. Reiher, A. Krtschil, A. Diez, T. Hempel, T. Finger, A. Kasic, M. Schubert, D. Bimberg, F. A. Ponce, J. Christen, and A. Krost. 2003. Metalorganic chemical vapor phase epitaxy of GaN on silicon. *Phys. Stat. Sol.* (c) 6:1583.

Dadgar, A., C. Hums, A. Diez, J. Bläsing, and A. Krost. 2006. Growth of blue GaN LED structures on 150-mm Si(111). *J. Cryst. Growth* 297:279.

Dadgar, A., P. Veit, F. Schulze, J. Blasing, A. Krtschil, H. Witte, A. Diez, T. Hempel, J. Christen, R. Clos, and A. Krost. 2007. MOVPE growth of GaN on Si-substrates and strain. *Thin Solid Films* 515:4356.

Dasgupta, S., F. Wu, J. S. Speck, and U. Mishra. 2009. Growth of high quality N-polar AlN(00011) on Si(111) by plasma assisted MBE. *Appl. Phys. Lett.* 94:151906.

Davis, R. F., T. Gehrke, K. J. Linthicum, T. S. Zheleva, E. A. Preble, P. Rajagopal, C. A. Zorman, and M. Mehregany. 2001. Pendeo-epitaxial growth of thin films of gallium nitride and related materials and their characterization. *J. Cryst. Growth* 225:134.

Davis, R. F., S. Einfeldt, E. A. Preble, A. M. Roskowski, Z. J. Reitmeier, and P. Q. Miraglia. 2003. Gallium nitride and related materials: Challenges in materials processing. *Acta Mater.* 51:5961.

Felice, R. Di, and J. E. Northrup. 1997. Theory of the AlN/SiC (1010) Interface. *Phys. Rev. B* 56 (15):9213–9216.

Feltin, E., B. Beaumont, M. Laugt, P. de Mierry, P. Vennegues, and H. Lahreche. 2001. Stress control in GaN grown on Si(111) by MOVPE. *Appl. Phys. Lett.* 79 (20):3230.

Feuillet, G., B. Daudin, F. Widmann, J. L. Rouviere, and M. Arlery. 1998. Plastic versus elastic misfit relaxation in III-nitrides grown by molecular beam epitaxy. *J. Cryst. Growth* 189/190:142.

Floro, J. A., E. Chason, S. R. Lee, R. D. Twesten, R. Q. Hwang, and L. B. Freund. 1997. Real time stress evolution during $Si_{1-x}Ge_x$ heteroepitaxy: Dislocations, islanding and segregation. *J. Electron. Mater.* 26 (9):969.

Floro, J. A., E. Chason, R. C. Cammarata, and D. J. Srolovitz. 2002. Physical origins of intrinsic stresses in Volmer-Weber thin films. *MRS Bull.* 27 (1):19–25.

Follstaedt, D. M., J. Han, P. Provencio, and J. G. Fleming. 1999. Microstructure of GaN grown on (111) Si by MOCVD. *MRS Internet J. Nitride Semicond. Res.* 4S1, G3.72.

Follstaedt, D. M., S. R. Lee, A. A. Allerman, and J. A. Floro. 2009. Strain relaxation in AlGaN multilayer structures by inclined dislocations. *J. Appl. Phys.* 105:083507.

Franca, D. R. and A. Blouin. 2004. All-optical measurement of in-plane and out of plane Young's modulus and Poisson's ratio in silicon wafers by means of vibration modes. *Meas. Sci. Technol.* 15:859–868.

Freund, L. B. and S. Suresh. 1993. *Thin Film Materials: Stress, Defect Formation and Surface Evolution.* Cambridge, U.K.: Cambridge University Press.

Ghosh, B. K., T. Tanikawa, A. Hashimoto, A. Yamamoto, and Y. Ito. 2003. Reduced stress GaN epitaxial layers grown on Si(111) by using a porous GaN interlayer converted from GaAs. *J. Cryst. Growth* 249:422.

Gibart, P. 2004. MOVPE of GaN and lateral overgrowth. *Rep. Prog. Phys.* 67:667.

Griffiths, C. L., and K. J. Weeks. 2007. Optical monitoring of MBE growth of AlN/ GaN using single wavelength laser interferometry: A simple method of tracking real-time changes in growth rate. *J. Vac. Sci. Technol* 25 (3):1066.

Hall, J. J. 1967. Electronic effects in the elastic constants of n-type silicon. *Phys. Rev.* 161:756–761.

Han, J., K. E. Waldrip, S. R. Lee, J. J. Figiel, S. J. Hearne, G. A. Peterson, and S. M. Myers. 2001. Control and elimination of cracking of AlGaN using low temperature AlGaN interlayers. *Appl. Phys. Lett.* 78 (1):67.

Hashimoto, T., F. Wu, M. Saito, K. Fujito, J. S. Speck, and S. Nakamura. 2008. Status and perspectives on the ammonothermal growth of GaN substrates. *J. Cryst. Growth* 310:876.

Hearne, S., E. Chason, J. Han, J. A. Floro, J. Figiel, J. Hunter, H. Amano, and I. S. T. Song. 1999. Stress evolution during metalorganic chemical vapor deposition of GaN. *Appl. Phys. Lett.* 74 (3):356–358.

Holec, D., Y. Zhang, D. V. Sridhara Rao, M. J. Kappers, C. McAleese, and C. J. Humphreys. 2008. Equilibrium critical thickness for misfit dislocations in III-nitrides. *J. Appl. Phys.* 104:123514.

Hove, J. M. Van, P. P. Chow, A. M. Wowchak, J. J. Klaassen, M. F. Rosamond, and D. R. Croswell. 1997. Optimization of AlGaN films grown by RF atomic nitrogen plasma using in-situ cathodoluminiscence. *J. Cryst. Growth* 175/176:79.

Hu, F. R., K. Ochi, Y. Zhao, B. S. Choi, and K. Hane. 2006. MBE growth of GaN thin film on Si substrate with InN as interlayer. *J. Cryst. Growth* 294:197.

Irie, M., N. Koide, Y. Honda, M. Yamaguchi, and N. Sawaki. 2009. MOVPE growth and properties of GaN on (111) Si using an AlInN intermediate layer. *J. Cryst. Growth* 311:2891.

Ishikawa, H., K. Shimanaka, F. Tokura, Y. Hayashi, Y. Hara, and M. Nakanishi. 2008. MOCVD growth of GaN on porous Si substrates. *J. Cryst. Growth* 310:4900.

Ivanov, I., L. Hultman, K. Jarrendahl, P. Martensson, J.-E. Sendgren, B. Hjovarsson, and J. E. Greene. 1995. Growth of epitaxial AlN(0001) on Si (111) by reactive magnetron sputter deposition. *J. Appl. Phys.* 78 (9):5721.

Johnson, J. W., E. L. Piner, A. Vescan, R. Therrien, P. Rajagopal, J. C. Roberts, J. D. Brown, S. Singhal, and K. J. Linthicum. 2004. 12 W/mm AlGaN-GaN HFETs on silicon substrates. *IEEE Electron Device Lett.* 25 (7):459.

Kappers, M. J., M. A. Moram, Y. Zhang, M. E. Vickers, Z. H. Barber, and C. J. Humphreys. 2007. Interlayer methods for reducing dislocation density in GaN. *Physica B* 401–402:296.

Kharchenko, A., K. Lischka, K. Schmidegg, H. Sitter, J. Bethke, and J. Woitok. 2005. In situ and real time characterization of MOCVD growth by high resolution x-ray diffraction. *Rev. Sci. Inst.* 76:033101.

Kim, M., Y. Do, H. C. Kang, D. Y. Noh, and S. Park. 2001. Effects of step graded AlGaN interlayer on properties of GaN grown on Si(111) using UHV CVD. *Appl. Phys. Lett.* 79 (17):2713.

Kobayashi, N. P., J. T. Kobayashi, P. D. Dapkus, W.-J. Choi, A. E. Bond, X. Zhang, and D. H. Rich. 1997. GaN growth of Si(111) substrate using oxidized AlAs as an intermediate layer. *Appl. Phys. Lett* 71:3569.

Kosicki, B. B. and D. Kahng. 1969. Preparation and structural properties of GaN thin films. *J. Vac. Sci. Technol.* 6:593.

Krost, A., A. Dadgar, G. Strassburger, and R. Clos. 2003. GaN-based epitaxy on silicon: Stress measurements. *Phys. Stat. Sol. (a)* 200 (1):26.

Krost, A., A. Dadgar, J. Blasing, A. Diez, T. Hempel, S. Petzold, J. Christen, and R. Clos. 2004. Evolution of stress in GaN heteroepitaxy on AlN/Si(111): From hydrostatic compressive to biaxial tensile. *Appl. Phys. Lett.* 85 (6):3441.

Kuwano, N., T. Tsuruda, Y. Adachi, S. Terao, S. Kamiyama, H. Amano, and I. Akasaki. 2002. Annihilation of threading dislocations in GaN/AlGaN. *Phys. Stat. Sol. (a)* 192 (2):366.

Lee, S. R., D. D. Koleske, K. C. Cross, J. A. Floro, K. E. Waldrip, A. T. Wise, and S. Mahajan. 2004. *In situ* measurements of the critical thickness for strain relaxation in AlGaN/GaN heterostructures. *Appl. Phys. Lett.* 85 (26):6164.

Lee, W. C., Y. J. Lee, J. Kwo, C. H. Hsu, C. H. Lee, S. Y. Wu, H. M. Ng, and M. Hong. 2009. GaN on Si with nm-thick single crystal Sc_2O_3 as a template using MBE. *J. Cryst. Growth* 311:2006.

Li, J., J. Y. Lin, and H. X. Jiang. 2006. Growth of III-nitride photonic structures on large area silicon substrates. *Appl. Phys. Lett.* 88:171909.

Liu, L. and J. H. Edgar. 2002. Substrates for gallium nitride epitaxy. *Mater. Sci. Eng. R* 37:61–127.

Liu, R., F. A. Ponce, A. Dadgar, and A. Krost. 2003. Atomic arrangement at the AlN/Si(111) interface. *Appl. Phys. Lett.* 83 (3):860.

Liu, C., S. Stepanov, P. A. Shields, A. Gott, W. N. Wang, E. Steimetz, and J. T. Zettler. 2006. In situ monitoring of GaN epitaxial lateral overgrowth by spectroscopic reflectometry. *Appl. Phys. Lett* 88:101103.

Liu, Z., X. Wang, J. Wang, G. Hu, L. Guo, J. Li, J. Li, and Y. Zeng. 2007. Effects of buffer layers on the stress and morphology of GaN epilayer grown on Si substrate by MOCVD. *J. Cryst. Growth* 298:281.

Louran, A. L., S. Vezian, F. Semond, and J. Massies. 2009. AlN buffer layer growth for GaN epitaxy on (111) Si: Al or N first? *J. Cryst. Growth* 311:3278.

Markov, I. V. 1998. *Crystal Growth for Beginners.* Singapore: World Scientific.

Maruska, H. P. and J. J. Tietjen. 1969. The preparation and properties of vapor deposited single crystalline GaN. *Appl. Phys. Lett.* 10 (5):327.

Mastro, M. A., C. R. Eddy, D. K. Gaskill, N. D. Bassim, J. Casey, A. Rosenberg, R. T. Holm, R. L. Henry, and M. E. Twigg. 2006. MOCVD growth of thick AlN and AlGaN superlattice structures on Si substrates. *J. Cryst. Growth* 287:610.

Mathis, S. K., A. E. Romanov, L. F. Chen, G. E. Beltz, W. Pompe, and J. S. Speck. 2001. Modeling of threading dislocation reduction in growing GaN layers. *J. Cryst. Growth* 231:371.

Meng, W. J., J. A. Snell, T. A. Perry, L. E. Rehn, and P. M. Baldo. 1994. Growth of aluminum nitride thin films on Si(111) and Si(001): Structural characteristics and development of intrinsic stresses. *J. Appl. Phys.* 75 (7):3446–3455.

Metzger, T., R. Hopler, E. Born, O. Ambacher, and M. Stutzmann. 1998. Defect structure of epitaxial GaN films determined by transmission electron microscopy and triple-axis x-ray diffractometry. *Phil. Mag. A* 77 (4):1013.

Moram, M. A., M. J. Kappers, Z. H. Barber, and C. J. Humphreys. 2007. Growth of low dislocation density GaN using transition metal nitride masking layers. *J. Cryst. Growth* 298:268.

Moram, M. A., M. J. Kappers, T. B. Joyce, P. R. Chalker, Z. H. Barber, and C. J. Humphreys. 2007. Growth of dislocation free GaN islands on Si(111) using a ScN buffer layer. *J. Cryst. Growth* 308 (2):302.

Moran, B., F. Wu, A. E. Romanov, U. K. Mishra, S. P. Denbaars, and J. S. Speck. 2004. Structural and morphological evolution of GaN grown by metalorganic chemical vapor deposition on SiC substrates using an AlN initial layer. *J. Cryst. Growth* 273:38.

Morkoc, H. 2008. *Handbook of Nitride Semiconductors and Devices* (Volume 1: Material Properties, Physics and Growth). Weinheim, Germany: Wiley-VCH.

Nakamura, S. and G. Fasol. 1997. *The Blue Laser Diode.* Berlin, Germany: Springer-Verlag.

Nakamura, S., Y. Harada, and M. Seno. 1991. Novel metalorganic chemical vapor deposition system for GaN growth. *Appl. Phys. Lett.* 58 (18):2021.

Nix, W. D. 1989. Mechanical properties of thin films. *Metal. Trans. A* 20A:2217–2245.

Nix, W. D. and B. M. Clemens. 1999. Crystallite coalescence: A mechanism for intrinsic tensile stresses in thin films. *J. Mater. Res.* 14 (8):3467–3473.

Northrup, J. E. and J. Neugebauer. 1996. Theory of GaN(1010) and (1120) surfaces. *Phys. Rev. B* 53 (16):R10477.

Okada, Y. and Y. Tokumaru. 1984. Precise determination of lattice parameter and thermal expansion coefficient of silicon between 300 and 1500 K. *J. Appl. Phys.* 56 (2):314–320.

Oliver, R. A., M. J. Kappers, and C. J. Humphreys. 2006a. Insights into the origin of threading dislocations in GaN/Al₂O₃ from atomic force microscopy. *Appl. Phys. Lett.* 89:011914.

Oliver, R. A., M. J. Kappers, J. Sumner, R. Datta, and C. J. Humphreys. 2006b. Highlighting threading dislocations in MOVPE grown GaN using an in-situ treatment with silane and ammonia. *J. Cryst. Growth* 289:506.

Pandey, R., P. Zapol, and M. Causa. 1997. Theoretical study of non-polar surfaces of AlN: Zinc blende (110) and wurtzite (1010). *Phys. Rev. B* 55 (24):R16009–R16012.

Perry, W. G., T. Zheleva, K. J. Linthicum, M. D. Bremser, R. F. Davis, W. Shan, and J. J. Song. 1997. Bound exciton energies, biaxial strains, and defect microstructures in GaN/AlN/6H-SiC(0001) heterostructures. *MRS Symp. Proc. III–V Nitrides,* Ed. F. A. Ponce, T. D. Moustakas, I. Akasaki, and B. A. Monemar, 415:3–14.

Raghavan, S. and J. M. Redwing. 2004a. In-situ stress measurements during the MOCVD growth of AlN buffer layers on (111) Si substrates. *J. Cryst. Growth* 261 (2–3):294–300.

Raghavan, S. and J. M. Redwing. 2004b. Intrinsic stresses in AlN layers grown by metal organic chemical vapor deposition on (0001) sapphire and (111) Si substrates. *J. Appl. Phys.* 96 (5):2995.

Raghavan, S. and J. M. Redwing. 2005a. Growth stresses and cracking in metal-organic vapor phase deposited GaN on (111) Si: Part 1, AlN buffer layers. *J. Appl. Phys.* 98:023514.

Raghavan, S. and J. M. Redwing. 2005b. Growth stresses and cracking in metal-organic vapor phase deposited GaN on (III) Si: Part 2, AlN buffer layers. *J. Appl. Phys.* 98:023515.

Raghavan, S., X. Weng, E. Dickey, and J. M. Redwing. 2006. Correlation of growth stress and structural evolution during metalorganic chemical vapor deposition of GaN on (111) Si. *Appl. Phys. Lett.* 88:041904.

Rajagopal, P., T. Gehrke, J. C. Roberts, J. D. Brown, T. W. Weeks, E. L. Piner, and K. J. Linthicum. 2003. Large-area device quality GaN on Si using a novel transition layer scheme. *Mat. Res. Soc. Symp. Proc.*, Ed. C. Wetzel, E. T. Yu, J. S. Speck, A. Rizzi, and Y. Arakawa. 743:L.1.2.

Reeber, R. R. and K. Wang. 2000. Lattice parameters and thermal expansion of GaN. *J. Mater. Res.* 15 (1):40–44.

Rehder, E., M. Zhou, L. Zhang, N. R. Perkins, S. E. Babcock, and T. F. Kuech. 1999. Structure of AlN on Si(111) deposited with metal organic vapor phase epitaxy. *MRS Internet J. Nitride Semicond. Res.* 4S1:G3.56.

Reiher, A., J. Blasing, A. Dadgar, A. Diez, and A. Krost. 2003. Efficient stress relief in GaN heteroepitaxy on Si(111) using low temperature AlN interlayers. *J. Cryst. Growth* 248:563–567.

Romanov, A. E. and J. S. Speck. 2003. Stress relaxation in mismatched layers due to threading dislocation inclination. *Appl. Phys. Lett.* 83:2569.

Romanov, A. E., W. Pompe, G. E. Beltz, and J. S. Speck. 1996. An approach to threading dislocation reaction kinetics. *Appl. Phys. Lett.* 69 (22):3342.

Schenk, H. P. D., U. Kaiser, G. D. Kipshidze, A. Fissel, J. Krausslich, H. Hobart, J. Schulze, and Wo. Richter. 1999. Growth of atomically smooth AlN film with a 5:4 coincidence interface on Si(111) by MBE. *Mater. Sci. Eng. (B)* 59:84.

Schneeweiss, H. J. and R. Abermann. 1992. Ultra high vacuum measurements of the internal stress of PVD titanium films as a function of thickness and its dependence on substrate temperature. *Vacuum* 43 (5–7):463–465.

Sheldon, B. W., K. H. A. Lau, and A. Rajamani. 2001. Intrinsic stress, island coalescence and surface roughness during the growth of polycrystalline films. *J. Appl. Phys.* 90 (10):5097.

Sheldon, B. W., A. Ditkowski, R. Beresford, E. Chason, and J. Rankin. 2003. Intrinsic compressive stress in polycrystalline films with negligible grain boundary diffusion. *J. Appl. Phys.* 94 (2):948.

Sheldon, B. W., A. Rajamani, A. Bhandari, E. Chason, S. K. Hong, and R. Beresford. 2005. Competition between tensile and compressive stress mechanisms during Volmer-Weber growth of aluminum nitride films. *J. Appl. Phys.* 98:043509.

Sheldon, B. W., A. Bhandari, A. F. Bower, S. Raghavan, X. Weng, and J. M. Redwing. 2007. Steady state tensile stresses during the growth of polycrystalline films. *Acta Mater.* 55:4973.

Strittmatter, A., A. Krost, M. Strassburg, V. Türck, D. Bimberg, J. Bläsing, and J. Christen. 1999. Low pressure metal organic chemical vapor deposition of GaN on Si(111) substrates using an AlAs nucleation layer. *Appl. Phys. Lett.* 74 (9):1242.

Suresh, S. and L. B. Freund. 2003. *Thin Film Materials: Stress, Defect Formation and Surface Evolution.* Cambridge, U.K.: Cambridge University Press.

Takeuchi, T., H. Amano, K. Hiramatsu, N. Sawaki, and I. Akasaki. 1991. Growth of single crystalline GaN film on Si substrate using 3C-SiC as an intermediate layer. *J. Cryst. Growth* 115:634.

Thurner, G. and R. Abermann. 1990. Internal stress and structure of ultra high vacuum evaporated chromium and iron films and their dependence on substrate temperature and oxygen partial pressure during deposition. *Thin Solid Films* 192:277–285.

Tsao, J. Y. 1993. *Material Fundamentals of Molecular Beam Epitaxy.* San Diego, CA: Academic Press.

Waldrip, K. E., J. Han, J. J. Figiel, H. Zhou, E. Makarona, and A. V. Nurmikko. 2001. Stress engineering during MOCVD of AlGaN/GaN distributed Bragg reflectors. *Appl. Phys. Lett* 78 (21):3205.

Wang, K. and R. R. Reeber. 1998. Thermal expansion of GaN and AlN. *MRS Symp. Proc.* Ed. by F. A. Ponce, S. P. DenBaars, B. K. Meyer, S. Nakamura and S. Strite. 482:863.

Wang, C.-M., X. Pan, and M. Ruhle. 1996. Silicon nitride crystal structure and observations of lattice defects. *J. Mater. Sci.* 31:5281.

Wang, J. F., D. Z. Yao, J. Chen, J. J. Zhu, D. G. Zhao, D. S. Jian, H. Yang, and J. W. Liang. 2006. Strain evolution in GaN layers grown on high-temperature AlN interlayers. *Appl. Phys. Lett.* 89:152105.

Watanabe, A., T. Takeuchi, K. Hirosawa, H. Amano, K. Hiramatsu, and I. Akasaki. 1993. The growth of single crystalline GaN on a Si substrate using AlN as an intermediate layer. *J. Cryst. Growth* 128:391.

Weng, X., S. Raghavan, J. D. Acord, A. Jain, E. C. Dickey, and J. M. Redwing. 2006. Evolution of threading dislocations in MOCVD-grown GaN films on (111) Si substrates. *J. Cryst. Growth* 300:217.

Wright, A. F. 1997. Elastic properties of zinc-blend and wurtzite AlN, GaN and InN. *J. Appl. Phys.* 82 (6):2833–2839.

Wu, J., L. Zhao, G. Zhang, X.Liu, Q. Zhu, Z. Wang, Q. Jia, L. Guo, and T. Hu. 2008. Effect of In-doped interlayer on the strain relief in GaN films grown on Si(111). *Phys. Stat. Sol. (a)* 205 (2):294.

Yang, J. W., C. J. Sun, Q. Chen, M. Z. Anwar, M. Asif Khan, S. A. Nishikin, G. A. Seryogin, A. V. Osinsky, L. Chernyak, H. Temkin, C. Hu, and S. Mahajan. 1996. High quality GaN-InGaN heterostructures grown on (111) Si substrates. *Appl. Phys. Lett.* 69 (23):366.

Zahn, D. R. T. 1998. Raman monitoring of wide bandgap MBE growth. *Appl. Surf. Sci.* 123/124:276.

Zamir, S., B. Meyer, and J. Salzman. 2001. Lateral confined epitaxy of GaN layers on Si substrates. *J. Cryst. Growth* 230:341.

Zhu, D., C. McAleese, K. K. Mclaughlin, M. Haberlen, C. O. Salcianu, E. J. Thrush, M. J. Kappers, W. A. Phillips, P. Lane, D. J. Wallis, T. Martin, M. Astles, S. Thomas, A. Pakes, M. Heuken, and C. J. Humphreys. 2009. GaN based LEDs grown on 6-inch diameter Si(111) substrates by MOVPE. *Proc. SPIE*, Ed. K. P. Streubel, H. Jeon and L. W. Tu. 7231:723118–1.

6

Direct Growth of III–V Devices on Silicon

Thomas Kazior, Katherine J. Herrick, and Jeffrey LaRoche

CONTENTS

6.1 Introduction

As silicon technology has scaled to shorter dimensions and higher frequency operation, silicon integrated circuits (ICs) are beginning to be used in applications traditionally served by III–V compound semiconductors (e.g., microwave/millimeter wave amplifiers and high-frequency mixed signal and data converter ICs). However, while the silicon-based ICs clearly provide more cost-effective solutions and increased integration density, they exhibit significant performance limitations when compared to III–V-based ICs. Due to the superior transport properties of III–V materials, III–V devices offer higher gain, efficiency, bandwidth, and dynamic range/linearity, and lower noise characteristics and RF loss at relaxed geometries than silicon-based devices. In addition, the direct band gap III–V devices are better suited as optical sources and receivers in optoelectronics.

As a result, the future of ICs will include the integration of high-performance III–V electronic and/or optoelectronic devices with standard

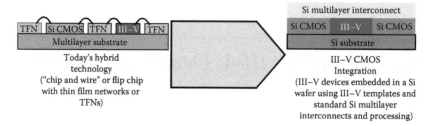

FIGURE 6.1
Traditional hybrid assembly (left) and the direct monolithic integration of III–V and CMOS on a SOLES substrate (right). (From Kazior, T.E. et al., A high performance differential amplifier through the direct monolithic integration of InP HBTs and Si CMOS on silicon substrates, in: *Microwave Symposium Digest, 2009. MTT'09. IEEE MTT-S International Microwave Symposium*, 7–12 June 2009, Boston, MA, 1113–1116, 2009a. With permission.)

Si complementary metal oxide semiconductor (CMOS). While traditional hybrid approaches, such as wire bonded or flip chip multichip assemblies (Figure 6.1, left), may provide short-term solutions, the variability and losses of the interconnects and the limitation in the placement of III–V devices relative to CMOS transistors will limit the performance and utility of these approaches. Recently, investigators successfully demonstrated the heterogeneous integration of InP heterojunction bipolar transistors (HBTs) and silicon CMOS using variations on wafer-bonding techniques (Li et al. 2008, Rosker et al. 2008), where the III–V epitaxial layers or the completed devices are bonded to the surface of a completed Si CMOS wafer. A more attractive approach is the direct integration of CMOS and III–V devices on a common silicon substrate (Figure 6.1, right). In this way, circuit performance can be optimized by the strategic placement of III–V devices adjacent to CMOS transistors and cells. While the direct growth and fabrication of III–V devices on silicon substrates has been pursued for over 30 years (Fang et al. 1990), recent advances in strain and lattice-engineered materials and epitaxial growth techniques have enabled the direct growth of high-quality III–V device layers on silicon substrates (Lubyshev et al. 2004, Herrick et al. 2008a–c, Liu et al. 2009).

In this chapter, the challenges and the recent progress on the direct heterogeneous integration of InP HBTs and Si CMOS on a silicon substrate are presented. The process is analogous to SiGe Bipolar CMOS and is essentially a III–V BiCMOS process. As a demonstration vehicle, a high-speed, low-power-dissipation differential amplifier, which serves as the basic building block for high-performance mixed signal circuits, such as analog to digital converters (ADCs) and digital to analog converters (DACs), was designed and fabricated. Results are also presented for the integration of GaN high electron mobility transistors (HEMTs) and Si CMOS. Finally, the performance advantages and the potential insertion opportunities for the III–V BiCMOS process in data converters, transceiver ICs, and power conditioning circuitry are discussed.

6.2 Substrate Engineering

The direct integration approach presented in this chapter is based on a unique "engineered" silicon substrate that is similar to a standard silicon on insulator (SOI) wafer. The silicon-on-lattice engineered substrate (SOLES), invented at MIT (Dohrman et al. 2006, Chilukuri et al. 2007) and manufactured by SOITEC using their Smart-Cut™ process (Maleville et al. 2004), contains a buried III–V template layer that enables the direct growth of high-quality III–V epitaxial material in windows directly on the silicon substrate (Figure 6.2). At present, the buried III–V template layer is Ge, although the substrate fabrication process is compatible with GaAs or InP template layers as well. Ge was chosen as the III–V template layer due to that fact that (1) Ge is readily accepted into today's silicon wafer fabs and (2) GaAs is closely lattice matched to Ge ($\Delta a/a \sim 0.07\%$), thus facilitating the growth of GaAs and InP-based III–V devices. The crystal orientation of the Ge layer is 6° off (001) toward <111>, which promotes III–V epilayer growth with a minimal anti-phase domain (APD) formation (Sieg et al. 1998, Ting and Fitzgerald 2000). A cross-sectional transmission electron micrograph (TEM) of a SOLES wafer is shown in Figure 6.3. SOLES have been successfully scaled to 200 mm diameter wafers (see Figure 6.4) and are compatible with and can be readily inserted into a standard silicon CMOS foundry.

6.3 III–V CMOS Integration Process Flow

While the direct integration approach described below is for the integration of InP HBTs on SOLES, effectively creating a high-performance InP BiCMOS process, the approach is equally applicable to other III–V electronic (filed

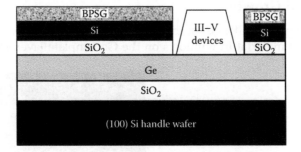

FIGURE 6.2
Schematic cross section of SOLES wafers showing the placement of the III–V device in windows. (From Kazior, T.E. et al., A high performance differential amplifier through the direct monolithic integration of InP HBTs and Si CMOS on silicon substrates, in: *Microwave Symposium Digest, 2009. MTT'09. IEEE MTT-S International Microwave Symposium*, 7–12 June 2009, Boston, MA, 1113–1116, 2009a. With permission.)

effect transistors (FETs), HEMTs) and optoelectronic (photodiodes, vertical cavity surface emitting lasers (VCSELS)) devices. In fact, the process flow is similar to a SiGe BiCMOS process flow (see Figure 6.5): (1) starting substrate, (2) Si CMOS device fabrication, (3) window formation (placement of III–V devices relative to CMOS devices), (4) III–V epitaxial growth in windows, (5) III–V device fabrication, and (6) multilayer heterogeneous interconnect fabrication. After the completion of the CMOS device fabrication, the CMOS transistors are coated with borophosphosilicate glass (BPSG) to protect them during the III–V growth process, and windows are lithography defined and etched into the SOLES wafer to reveal the III–V template layer. Since the III–V growth

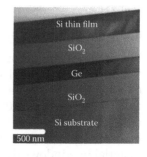

FIGURE 6.3
Cross-sectional TEM of a SOLES wafer showing the buried III–V (Ge) template layer.

windows are defined as part of the CMOS fabrication process, the III–V epitaxial material can be grown selectively and arbitrarily across the substrate as required for the particular circuit or applications. Figure 6.6 shows an example of a scanning electron microscope (SEM) image of a completed InP HBT in close proximity to a CMOS transistor prior to interconnect formation. To facilitate the interconnecting of the III–V devices and CMOS transistors, the thickness of the III–V epitaxial layers and the depth of the windows are optimized, such that the III–V devices and the CMOS transistors are planar. Figure 6.7 shows an example of a daisy chain test structure interconnecting InP HBTs and Si CMOS. With this truly planar approach, interconnect lengths (III–V – CMOS separation) as small as 2.5 µm have been demonstrated.

6.4 Material Growth

One of the biggest challenges of this approach is the growth of high-quality III–V epitaxial material in windows on the Ge template layer. (*Note*: all the

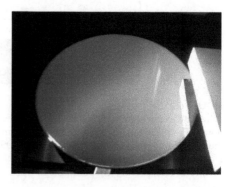

FIGURE 6.4
Optical image of a 200 mm diameter SOLES wafer.

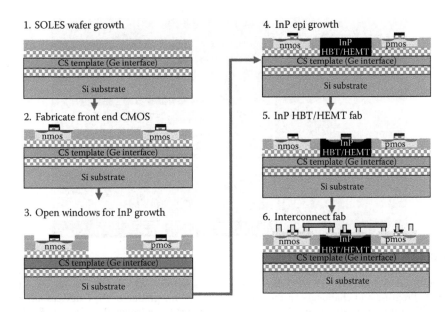

FIGURE 6.5
III–V CMOS integration process flow.

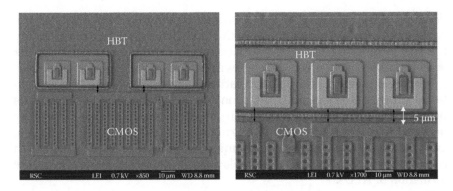

FIGURE 6.6
SEM image of a completed InP HBT in close proximity to a Si CMOS transistor prior to heterogeneous interconnect formation. (From Kazior, T.E. et al., A high performance differential amplifier through the direct monolithic integration of InP HBTs and Si CMOS on silicon substrates, in: *Microwave Symposium Digest, 2009. MTT'09. IEEE MTT-S International Microwave Symposium*, 7–12 June 2009, Boston, MA, 1113–1116, 2009a. With permission.)

III–V epitaxial material reported in this work is grown by molecular beam epitaxy (MBE)). For InP HBTs, a GaAs nucleation layer on the Ge template layer is grown, followed by a metamorphic buffer layer, and then the InP HBT epilayers. Challenges include windows patterning, surface preparation, and growth methods that support the nucleation of low defect density, anti-phase

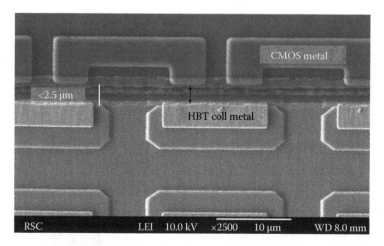

FIGURE 6.7

SEM image of a heterogeneous interconnect daisy chain test prior to final interconnect metallization. InP HBT–Si CMOS, the interconnect spacing is <2.5 μm. (From Kazior, T.E. et al., Progress and challenges in the direct monolithic integration of III–V devices and Si CMOS on silicon substrates, in: *IEEE International Conference on Indium Phosphide and Related Materials (IPRM'09)*, *Digest*, 10–14 May 2009, Newport Beach, CA, 100–104, 2009b. With permission.)

domain (APD) free GaAs on the Ge surface in windows in SOLES; the prevention of Ge out-diffusion; the growth of low defect density, device-quality InP HBT structures on the GaAs surface using a suitable metamorphic buffer layer (M-buffer); and minimizing growth defects on the edges of windows.

While GaAs and Ge are closely lattice matched ($\Delta a/a \sim 0.07\%$), the epitaxial growth of GaAs on Ge is prone to APD formation because of its polar-on-nonpolar nature. APDs can be suppressed in MBE growth by using off-cut substrates, by optimizing the initial nucleation condition, and by using thermal annealing techniques (Sieg et al. 1998). Thus, growth conditions for the GaAs nucleation layer are optimized to minimize the formation of APDs.

The basic requirement of an M-buffer is to accommodate the lattice mismatch between GaAs and the InP-based alloys of the device structure and to absorb this strain while minimizing the nucleation of dislocations. The resulting surface should also have low roughness and minimal warp for device reliability and processing requirements. Therefore, the M-buffer is designed for transition from the GaAs lattice constant to the InP lattice constant while minimizing propagation of threading dislocations associated with strain relaxation into the device layers. In this example, an InAlAs-based linearly graded metamorphic buffer layer with an inverse step is grown to transition to the lattice constant of InP. Details of the growth of the graded InAlAs M-buffer can be found elsewhere (Lubyshev et al. 2004). Using the graded InAlAs M-buffer, a large area HBT figure-of-merit (DC current gain/base

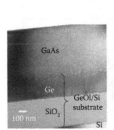
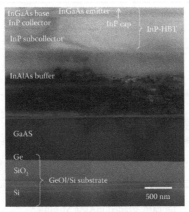

FIGURE 6.8

Cross-sectional TEM (left, center) and schematic cross section (right) of InP HBT grown in windows on SOLES. For optimized growth conditions, there is no evidence of APDs at the Ge–GaAs interface. Dislocations are confined to the metamorphic buffer layer.

sheet resistance) and a breakdown voltage comparable to control structures on InP substrates have been achieved.

Figure 6.8 shows a cross-sectional TEM of the InP HBT grown on Ge in windows in SOLES. The Ge–GaAs interface shows no evidence of APDs. Dislocations are confined to the M-buffer, in fact, the dislocations are confined (are annihilated) in the linearly graded region before the inverse step. The inverse step region is virtually dislocation free due to strain relaxation. The InP-HBT, layers above the M-buffer have well-defined interfaces. The device active layers exhibit low dislocation density as measured by plan view TEM ($(3.5 \pm 0.6) \times 10^7$ cm^{-2}). X-ray analysis of the III–V device layers show well-defined x-ray peaks and line widths (see Figure 6.9). The full width half maximum (FWHM) for the GaAs and the HBT layers are close to the values measured on bare substrates, indicating that the structural quality of the material grown on patterned substrates is close to the material grown on bare substrates. Atomic force microscope (AFM) images of the III–V surface show good surface morphology (surface roughness <1 nm, comparable to epilayers grown on native substrates) (Figure 6.10). The AFM images inside the windows exhibit cross-hatch patterns characteristic of high-quality metamorphic growth with low root mean squared (RMS) roughness of 1.7 nm for a 20×20 μm^2 AFM scan.

Due to the high sticking coefficients of group III atoms, the selective growth of InGaAs-based materials on masked areas pose an additional challenge, as growth proceeds across the entire wafer, resulting in a polycrystalline growth on the mask material (BPSG protective layer over the CMOS transistors) that requires a patterned dry etch process for its removal. The epitaxial challenge is to optimize the growth condition for a high filling factor within the

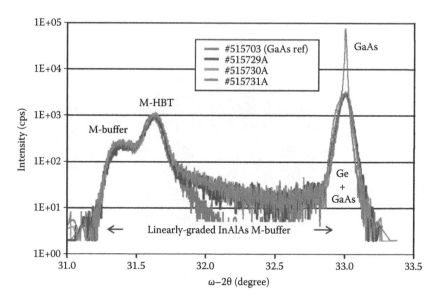

FIGURE 6.9
X-ray spectra for InP HBT grown in windows on SOLES for multiple wafers.

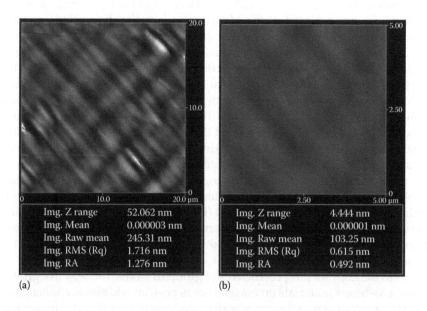

FIGURE 6.10
20 μm × 20 μm (a) and 5 μm × 5 μm (b) AFM image of InP HBT grown in windows on SOLES. Note the smooth surfaces (1.7 nm RMS roughness for 20 μm × 20 μm scan) and the cross-hatch pattern characteristic of high-quality metamorphic growth.

windows, while maintaining good quality GaAs on Ge nucleation and an InP device growth on the Ge surface inside these windows. A benefit of epitaxial growth in windows is that the limited area within the Ge windows reduces the likelihood of misfit dislocations in the active layers by limiting the misfit dislocation length and by reducing dislocation interaction and multiplication (Fitzgerald 1989). Potentially, small-area growth within the windows may also enhance dislocation filtering within the metamorphic buffer, and hence helps reduce the threading dislocation density (Fitzgerald 1989).

Therefore, optimization of the window etch and the epitaxial growth processes in windows are key to achieving high-quality device layers. In order to obtain device material quality on patterned substrates, the MBE growth conditions must satisfy several additional requirements beyond that of standard growth on unpatterned substrates. While smooth epi inside the windows is a prerequisite, problems can arise in the subsequent processing of the wafers if the polycrystalline growth on the BPSG mask area that protects the CMOS transistors is too rough. In addition, Ga droplet formation on exposed Si along the sidewalls of the windows can also promote polycrystalline, wirelike growth that can protrude into the window openings, thus shrinking the area for high-quality III–V growth (see Figure 6.11, left). Note the surface roughness, the poor edge definition, and the formation of nanowire material due to nucleation and growth of III–V material on impurities at the windows' edge.

To minimize the roughness of the polycrystalline material and nucleation along the window sidewalls, growth conditions must be optimized for patterned substrates. For example, the GaAs nucleation layer is grown at a reduced temperature to minimize Ga surface migration during epi

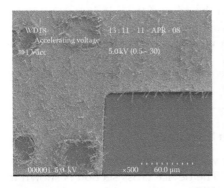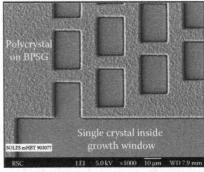

FIGURE 6.11

SEM image of InP HBT device epitaxial material grown in windows on SOLES for an unoptimized process (left—*Note*: nanowire growth) and an optimized process (right—*Note*: well-defined windows down to a 15 μm × 15 μm windows dimensions). (From Kazior, T.E. et al., A high performance differential amplifier through the direct monolithic integration of InP HBTs and Si CMOS on silicon substrates, in: *Microwave Symposium Digest, 2009. MTT'09. IEEE MTT-S International Microwave Symposium*, 7–12 June 2009, Boston, MA, 1113–1116, 2009a. With permission.)

FIGURE 6.12
Micrograph of InP HBT epitaxial growth in windows on a 100 mm diameter SOLES. *Note:* Uniform growth across entire wafer. The large area at the center of the wafer is the reflection high energy electron diffraction (RHEED) window for use during MBE growth.

growth and to inhibit droplet formation and wire nucleation. With optimized windows etch and growth conditions, well-defined window edges, flat, smooth epi with a high filling factor inside the growth windows, as well as low roughness polycrystalline material on the masked area are achieved (see Figure 6.11, right). The optimized process is highly repeatable (wafer to wafer) and uniform across a 100 mm diameter wafer in windows as small as 15 μm × 15 μm (Figure 6.12). With optimized growth conditions, a filling factor in the Ge windows is achieved that allows the placement of the active device junctions to within <5 μm from the edge of the window opening for intimate heterogeneous integration. A detailed report on the growth of high-quality InP HBT epitaxial material in windows on SOLES has been previously published (Liu et al. 2009).

6.5 InP HBT Fabrication and Performance

InP HBTs with various emitter finger widths were fabricated using a standard nonself-aligned emitter process. The device cross section is shown in Figure 6.13. An SEM image of an InP HBT in a 15 μm × 15 μm window in a SOLES wafer is shown in Figure 6.14. The electrical performance of InP HBTs fabricated on SOLES is comparable to HBTs grown directly on native InP substrates (Ha et al. 2008). Figure 6.15 shows the Gummel characteristics and the small signal parameters of a 0.5×5 μm^2 emitter HBT grown in a 15×15 μm^2 window on a SOLES substrate. Gain (beta), f_t and f_{max} of 40, >200 GHz and >200 GHz, respectively are achieved.

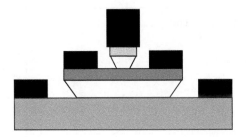

FIGURE 6.13
HBT device cross section.

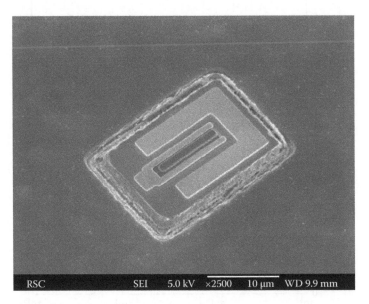

FIGURE 6.14
SEM image of an InP HBT in a 15 μm × 15 μm window in a SOLES wafer.

6.6 Circuit Demonstration

Using the InP HBT described above and standard CMOS, a differential amplifier test vehicle was designed and fabricated (Kazior et al. 2009a,b). Figure 6.16 shows an optical image of a completed differential amplifier circuit. In addition to the core differential amplifier, the circuit contains a bias circuit and an all HBT output buffer. The role of the output buffer is to attenuate the output of the core differential amplifier to facilitate characterization of the differential amplifier.

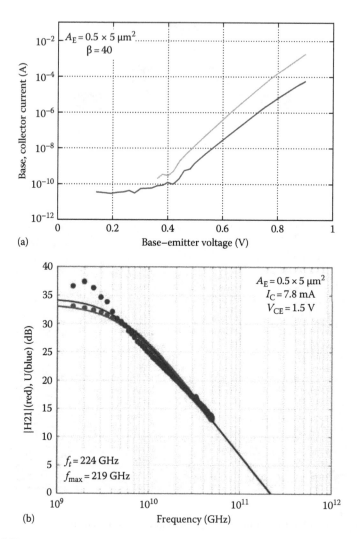

FIGURE 6.15

(a) Measured Gummel characteristics and gain of a $0.5 \times 5\ \mu m^2$ InP-HBT on the SOLES substrate. (b) Measured small-signal RF characteristics of a $0.5 \times 5\ \mu m^2$ InP-HBT on a SOLES substrate. (From Kazior, T.E. et al., A high performance differential amplifier through the direct monolithic integration of InP HBTs and Si CMOS on silicon substrates, in: *Microwave Symposium Digest, 2009. MTT'09. IEEE MTT-S International Microwave Symposium*, 7–12 June 2009, Boston, MA, 1113–1116, 2009a. With permission.)

The monolithically integrated, planar approach allows the inclusion of multiple design variants within a reticle on a wafer, effectively creating a circuit optimization design of experiments (DOE) within the reticle. Each design variant is step and repeated across the 100 mm SOLES wafer. The planar approach also facilities automated on-wafer probing for circuit

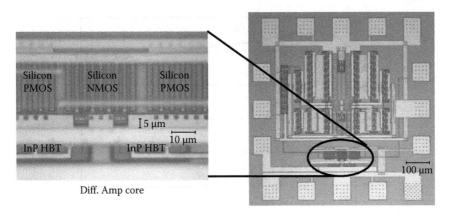

FIGURE 6.16
Optical image of a core differential amplifier with an output buffer and a bias circuit. (From Kazior, T.E. et al., A high performance differential amplifier through the direct monolithic integration of InP HBTs and Si CMOS on silicon substrates, in: *Microwave Symposium Digest, 2009. MTT'09. IEEE MTT-S International Microwave Symposium,* 7–12 June 2009, Boston, MA, 1113–1116, 2009a. With permission.)

characterization, and the collection of circuit performance and uniformity data for the different design variants.

The following test results are for one of these design variants that utilizes HBTs for each half of the differential pair and pMOS devices for the amplifier loads. For all the measurements that are shown, the differential amplifier core was biased at a $V_{ss} = 6$ V and $I_{ss} = 14$ mA ($P_{diss} = 84$ mW). Separate DC supply inputs are provided for the amplifier core and output buffer circuits to ensure an accurate measurement of the dissipated power of the core. Four-port S-parameter measurements were made to determine the low-frequency amplifier gain and unity-gain bandwidth. Measurements were made from 1 MHz to 20 GHz using on-wafer differential ground-signal-ground-signal-ground (GSGSG) probes. A probe tip calibration was performed using a GGB Industries calibration substrate.

Measurements from 1 to 50 MHz were used to extract the low-frequency gain of the differential amplifier. The low-frequency voltage gain of the differential amplifier core was determined by measuring the gain of the differential amplifier – the output buffer chain and by correcting (subtracting) the attenuation of the amplifier, such that Av,diff amp = S_{21}, chain – S_{21}, buffer. The output buffer amplifier has a low-frequency attenuation of ~25 dB—a value that agreed well with simulations.

Figure 6.17 shows the corrected low-frequency gain of the core differential amplifier. A peak low-frequency gain of 584 V/V was measured at 6 MHz. At lower frequencies, the gain is observed to decrease slightly due to device self-heating (increased output conductance of HBT).

The high-frequency gain measurements are extracted using a similar scalar approach for determining the core amplifier characteristics. Figure 6.18

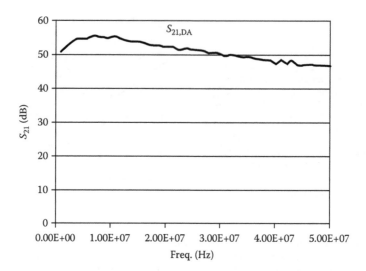

FIGURE 6.17
Corrected S_{21} of amplifier core at low frequencies. Peak low-frequency gain of 55.3 dB (584 V/V). $I_{ss} = 14$ mA, $V_{ss} = -6$ V. (From Kazior, T.E. et al., A high performance differential amplifier through the direct monolithic integration of InP HBTs and Si CMOS on silicon substrates, in: *Microwave Symposium Digest, 2009. MTT'09. IEEE MTT-S International Microwave Symposium*, 7–12 June 2009, Boston, MA, 1113–1116, 2009a. With permission.)

shows the corrected high-frequency characteristics for the same amplifier as shown in Figure 6.17. A deviation in the slope of the roll-off was observed at the higher end of the frequency band. The cause of this discrepancy has not been determined. To determine the unity-gain cutoff frequency of the amplifier, this portion of the frequency response was not utilized. Instead, the unity-gain frequency was extrapolated from the intercept of data taken from 1 to 15 GHz. A unity-gain frequency of 22.3 GHz was extracted from the measurement shown in Figure 6.18. From the DC-gain measurement, the DC-gain × unity-gain bandwidth product is measured to be 1.3×10^4 V/V GHz.

Slew-rate measurements were made on the same amplifier shown in Figures 6.17 and 6.18. For the slew-rate measurement, a 400 MHz input signal was provided from a signal generator. A differential input signal was generated using a 180° balun. Both outputs from the amplifier were provided to a high-speed Agilent sampling oscilloscope that was used to determine the differential amplifier output. Figure 6.19 shows the measured output waveform of the amplifier when driven to saturation. A peak output swing of 412 mV was measured from the output buffer stage. Correcting the measured attenuation of the output buffer (25.5 dB from S-parameters), the corresponding voltage swing of the amplifier core is 7.76 V.

The average rise time and fall time (10%–90%) of the amplifier, as measured on the Agilent sampling scope (Figure 6.19), was determined to be 489 ps. Based on the signal swing of the amplifier core, this corresponds to a measured slew rate of 1.27×10^4 V μs.

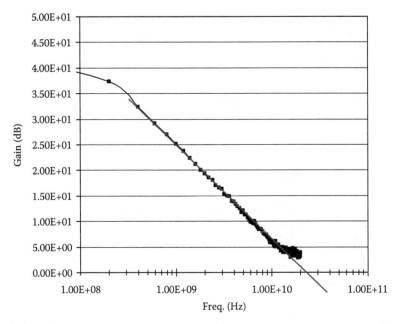

FIGURE 6.18
High-frequency gain response of the amplifier core. Extrapolated unity-gain cutoff frequency is 22.3 GHz. $I_{ss} = 14$ mA, $V_{ss} = -6$ V. (From Kazior, T.E. et al., A high performance differential amplifier through the direct monolithic integration of InP HBTs and Si CMOS on silicon substrates, in: *Microwave Symposium Digest, 2009. MTT'09. IEEE MTT-S International Microwave Symposium*, 7–12 June 2009, Boston, MA, 1113–1116, 2009a. With permission.)

The differential amplifier design variant, whose performance is shown in Figures 6.17 through 6.19, was step and repeated across a 100 mm diameter SOLES wafer (each design variant appears once per reticle). To demonstrate the manufacturability of this planar integrated approach, a wafer map was generated of the DC gain and the unity-gain bandwidth for this design variant (Figure 6.20). Highlighted cells are for differential amplifiers that have DC gain × unity-gain bandwidth products > 1×10^4 V/V GHz. Similar results were achieved for other differential amplifier design variants and for different wafers highlighting the manufacturability of this approach.

6.7 Other III–V Devices Integrated with Silicon

Palicios et al. have demonstrated the integration of GaN-based devices and Si electronics using a process similar to the COSMOS or III–V BiCMOS process described earlier in this chapter (Chung et al. 2009a,b). The process starts by fabricating a virtual Si (001)/GaN/Si (001) substrate by wafer bonding with a SiO_2 bonding interlayer (Figure 6.21) (Chung et al. 2009a). Due to the high

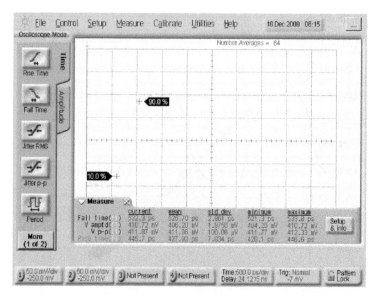

FIGURE 6.19

Measured output waveform for slew-rate measurements of the amplifier (same amplifier as that measured in Figures 6.17 and 6.18). 400 MHz differential input signal is provided to saturate the amplifier. A peak output voltage swing of 412 mV was measured from the output buffer corresponding to an internal input swing of 7.76 V. An average rise/fall time of 489 ps is measured. (From Kazior, T.E. et al., A high performance differential amplifier through the direct monolithic integration of InP HBTs and Si CMOS on silicon substrates, in: *Microwave Symposium Digest, 2009. MTT'09. IEEE MTT-S International Microwave Symposium*, 7–12 June 2009, Boston, MA, 1113–1116, 2009a. With permission.)

thermal stability of GaN, Si CMOS electronics can then be processed in these new substrates without affecting the nitride layers underneath the surface. After the Si devices are fabricated, windows are etched in the silicon layer to expose the buried GaN epilayer. Then, the nitride devices (transistors, light emitting diodes (LEDs), lasers, or sensors) are processed, and finally, an interconnection layer forms the final hybrid circuits. Figure 6.22 shows an optical micrograph of a GaN HEMT adjacent to a silicon pMOS transistor. The current–voltage characteristics of the GaN power and silicon pMOS transistors are shown in Figure 6.23. As observed from the *I–V* measurements, the integration with Si (001) does not degrade the device performance. The Si transistor did not show any degradation either.

6.8 Potential Applications and Insertion Opportunities

III–V BiCMOS technology, when fully developed and matured, can be expected to have immense impact on how future systems are architected.

	469 24.9 1.17E+04	411 27.2 1.12E+04	NF		
	461 22.6 1.05E+04	NF	445 27 1.20E+04	435 23.7 1.03E+04	
467 23 1.07E+03	439 23.8 1.04E+04	RHEED window		441 23 1.02E+04	
454 23.2 1.05E+04	325 22.8 7.42E+03			403 20.5 8.27E+03	486 20 9.72E+03
596 23.4 1.40E+04	596 23.4 1.40E+04	NF	271 20 5.43E+03	NF	
	NF	277 24.1 6.67E+03	429 21 9.03E+03		

DC Gain (V/V)   Major flat down
UGBW (GHz)   Gray cells not available for test
FOM (v/v—GHz)   NF—nonfunctional part

FIGURE 6.20

Wafer map of the DC-gain and the unity-gain bandwidth measurements for a differential amplifier design (P_{diss} = 72 mW). First number in each cell represents the DC gain (V/V), second number represents the unity-gain bandwidth (GHz), and the third number represents the figure-of-merit product (V/V GHz). (From Kazior, T.E. et al., A high performance differential amplifier through the direct monolithic integration of InP HBTs and Si CMOS on silicon substrates, in: *Microwave Symposium Digest, 2009. MTT'09. IEEE MTT-S International Microwave Symposium*, 7–12 June 2009, Boston, MA, 1113–1116, 2009a. With permission.)

In the short term, there are several circuit applications that can successfully leverage the performance/cost advantages of this technology. In addition to the converter circuits (DAC and ADC) that are the demonstration vehicles of the DARPA COSMOS program (InP HBT and CMOS), circuits include (but are not limited to): High-performance direct digital synthesizers (DDS) for the direct conversion of digital to RF (microwave) signals (InP HBT and CMOS); transceiver circuits (pHEMT and CMOS, GaN and CMOS), the building block of active electronically scanned arrays (AESAs) or phased arrays for radar, and communication systems; and power regulators or conditioning circuits (GaN and CMOS). In the long term, III–V BiCMOS technology will enable new circuit and system architectures.

6.8.1 Direct Digital Synthesizer

A DDS generates arbitrary waveforms from a clock and a small number of digital control inputs. III–V BiCMOS technology will enable DDS

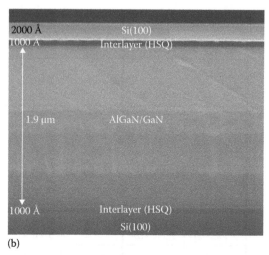

(a)   (b)

FIGURE 6.21
Schematic (a) and scanning electron micrograph (b) of the cross section of a Si/GaN/Si hybrid wafer. (From Chung, J.W. et al., On-wafer integration of nitrides and Si devices: Bringing the power of polarization to Si, in: *2009 Microwave Symposium Digest (MTT'09. IEEE MTT-S International Microwave Symposium)*, 7–12 June 2009, Boston, MA, 1117–1120, 2009a. With permission.)

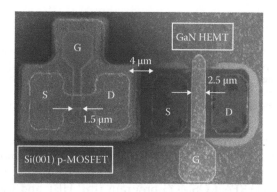

FIGURE 6.22
Optical image of a GaN HEMT and a Si CMOS on a Si/GaN/Si wafer. (From Chung, J.W. et al., On-wafer integration of nitrides and Si devices: Bringing the power of polarization to Si, in: *2009 Microwave Symposium Digest (MTT'09. IEEE MTT-S International Microwave Symposium)*, 7–12 June 2009, Boston, MA, 1117–1120, 2009a. With permission.)

circuits with increased capabilities, such as high-frequency outputs from ultra high frequency (UHF) through the Ku-band and beyond, and compact receiver/exciter (REX) architectures that eliminate mixers plus other hardware, and results in lower cost, size, weight, and power requirements. In some DDS applications, the high-frequency capability of III–V BiCMOS technology will enable high ratio over sampling to reduce the quantization

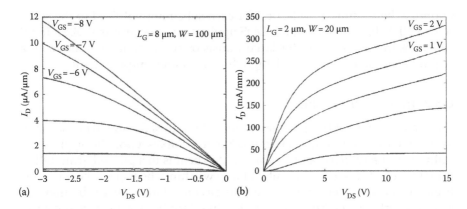

FIGURE 6.23
I–V characteristics of pMOS transistors (a) and GaN HEMT fabricated on same wafer (b). (From Chung, J.W. et al., On-wafer integration of nitrides and Si devices: Bringing the power of polarization to Si, in: *2009 Microwave Symposium Digest (MTT'09. IEEE MTT-S International Microwave Symposium)*, 7–12 June 2009, Boston, MA, 1117–1120, 2009a. With permission.)

noise floor. DDS designs with high transistor counts and featuring dual (i.e., frequency and phase) accumulators will enable wideband chirps and other complex waveforms with high spurious free dynamic range (SFDR), which are high-frequency outputs. The DDS is digitally programmable, thereby enabling dynamically reconfigurable arrays, and features an output that is frequency agile.

The direct integration of high-density, high-speed silicon CMOS for the digital circuits with high-speed, high-breakdown-voltage, high-transconductance (for higher drive capability) InP HBT transistors for the output DAC is key for high SFDR DDSs. Compared to SiGe BiCMOS, III–V BiCMOS offers a number of advantages. First, InP HBTs are significantly faster than currently available or future SiGe HBTs. Second, for a given cutoff frequency, InP HBTs offer a higher breakdown voltage than silicon or SiGe devices; a high breakdown voltage results in a higher full-scale voltage output from the DAC, and thus enables a higher dynamic range output. Third, high-speed InP divider circuits have demonstrated a lower dissipated power to frequency ratio than similar high-speed SiGe circuits.

A DDS developed entirely of InP HBTs will achieve the highest DDS clock speeds possible. However, by utilizing a silicon CMOS for the digital waveform generation portion of a DDS, a III–V BiCMOS-based DDS would achieve a much denser digital core than an InP-only DDS. Additionally, the 200 mm III–V BiCMOS wafer size would help minimize costs compared to a 100 mm InP wafer solution.

Thus, for a high-speed, high-SFDR DDS, III–V BiCMOS has specific advantages over competing semiconductor technologies. While there will be applications for which other technologies will offer a better solution for a specific DDS set of requirements, the degrees of freedom offered by III–V BiCMOS to optimize the technology selection for each subcircuit will lead to the selection of III–V BiCMOS for many DDS applications, because the DDS circuit combines the requirement for large transistor count digital circuits with a requirement for a mixed signal, high dynamic range output circuit.

6.8.2 Transceiver Circuits

The main purpose of transceiver circuits is to control the signal phase for an individual transmit/receive (T/R) module that makes it possible to electronically steer the received and/or transmitted beams in a phased array. Other functions of a transceiver circuit include switching the module between the transmit and the receive modes; amplitude control for array calibration and receive aperture weighting; I/O digital interface between external control signals and the RF path components of the module (both internal: phase shifters, attenuators, switches; and external: limiter; to the T/R circuit).

For a semiconductor technology to be considered a viable choice for a monolithically integrated transceiver circuit, it has to satisfy the following requirements: RF switching capability with low power consumption; availability of high-Q passive components (capacitors, inductors, transmission lines); acceptable gain/bandwidth, linearity, noise figure transistor characteristics; power handling; digital/logic functionality; cost effectiveness.

Traditionally, transceiver circuits, as well as all the other components in the RF path of the T/R modules have been manufactured using GaAs-based technologies, taking advantage of high-frequency capabilities of III–V transistors and semi-insulating properties of GaAs substrates. The digital part of the T/R circuit has historically been of relatively low complexity in part due to a somewhat limited infrastructure of logic components in the existing production-qualified GaAs technologies.

The recent advancement of silicon-based technologies into RF, microwave, and millimeter-wave domains has sparked interest in implementing silicon-based transceiver circuits. The premise behind this interest is to significantly lower the cost, which would be one of the steps required to make AESAs more affordable than they are today. This, in turn, would create new opportunities both in DoD and commercial markets. However, this cost advantage comes at the expense of degradation in the performance of RF components (phase shifters, attenuators, switches, amplifiers) in the transceiver circuit due to the use of silicon instead of high-performance III–V in the RF path. A main advantage of silicon-based transceiver circuits is the ability to integrate much more sophisticated digital circuitry (including integrated FPGAs and memory) than their III–V counterparts, allowing for greater architectural flexibility.

If one is to conceive an optimum technology for transceiver circuits, it would contain III–V devices in the RF path, taking advantage of their superior high-frequency properties; and silicon devices in its digital part making use of its high-speed capabilities and extensive existing infrastructure.

Therefore, the III–V BiCMOS process can be viewed as the optimum technology choice for transceiver circuit implementation. In addition to the performance advantages, III–V BiCMOS technology, when matured and fully transitioned into a high-volume state-of-the-art silicon foundry, would also provide a cost-effective solution and enable the proliferation of AESAs.

6.8.3 Regulator Circuitry

Power regulation is an important challenge in modern microprocessors due to the trade-off between power dissipation, operating voltage, and input bias current. To keep a constant power dissipation in current and future microprocessors, the operating voltage has to be decreased, which increases the input current to levels well above 100 A per microprocessor. This very high input current increases conductive power losses and reduces the number of I/O pins available in traditional microprocessor packaging. One of the most promising solutions to this problem is to introduce the power into the microprocessor and distribute it at high voltages (and low currents), and then down convert it to the required voltage, locally, in highly integrated on-wafer power regulator circuits. The fabrication of an all-Si solution is very challenging due to the low breakdown voltage and frequency performance of Si power electronics. By integrating GaN HEMTs and Si CMOS one can realize highly compact DC–DC power converters for advanced power distribution in Si microprocessors. In a hybrid GaN HEMT—Si CMOS regulator or a DC–DC converter, GaN switches are used in the high voltage stress/low-current part of the circuit and Si power MOSFETs are used in the low-voltage/high-current regions of the power circuit. A very high-switching frequency (>300 MHz) allows for on-chip integration of the entire converter. Simulations show that the use of GaN switches in the high-voltage part of the circuit instead of Si devices reduces the circuit losses and significantly increases conversion efficiency—highlighting the great potential of the hybrid GaN/Si power electronic circuit to enable local power conversion in high-performance Si electronics.

6.9 Summary

In this chapter, the direct monolithic integration of InP HBTs with Si CMOS on a silicon substrate was discussed. The direct growth approach yields InP HBTs with a similar RF performance to HBTs fabricated on InP substrates. Similar results were achieved for GaN HEMTs integrated with silicon CMOS. The planar III–V BiCMOS approach allows a tight device placement (III–V—Si CMOS transistor separation as small as 2.5 μm) and the use of standard wafer-level multilayer interconnects. The approach is equally applicable to a wide range of III–V electronic (FETs, HEMTs, HBTs) and opto-electronic (photodiodes, VSCELS) devices and opens the door to a new class of highly integrated, high-performance, mixed signal circuits.

Acknowledgments

This work was supported in part by the DARPA COSMOS Program (Contract Number N00014-07-C-0629). The authors would like to thank Mark Rosker (DARPA), Harry Dietrich (ONR), and Karl Hobart (NRL).

References

Chilukuri, K. et al. (2007) Monolithic CMOS-compatible AlGaInP visible LED arrays on silicon on lattice-engineered substrates (SOLES). *Semiconductor Science and Technology* 22: 29–34.

Chung, J.W. et al. (2009a) On-wafer integration of nitrides and Si devices: Bringing the power of polarization to Si. In: *2009 Microwave Symposium Digest (MTT'09. IEEE MTT-S International Microwave Symposium)*, 7–12 June 2009, Boston, MA, pp. 1117–1120.

Chung, J.W. et al. (2009b) On-wafer seamless integration of GaN and Si (100) electronics. In: *2009 Compound Semiconductor Integrated Circuit Symposium (CISC 2009)*, 11–14 October 2009, Greensboro, NC, pp. 1–4.

Dohrman, C.L. et al. (2006) Fabrication of silicon on lattice-engineered substrate (SOLES) as a platform for monolithic integration of CMOS and optoelectronic devices. *Materials Science and Engineering B*, 135: 235–237.

Fang, S.F. et al. (1990) Gallium arsenide and other compound semiconductors on silicon. *Journal of Applied Physics* 68: R31–R58.

Fitzgerald, E.A. (1989) The effect of substrate growth area on misfit and threading dislocation densities in mismatched heterostructures. *Journal of Vacuum Science and Technology* B, 7: 782.

Ha, W. et al. (2008) Small-area InP DHBTs grown on patterned lattice-engineered silicon substrates. In: *66th Device Research Conference Digest*, Santa Barbara, CA, IV.B-9.

Herrick, K.J. et al. (2008a) Engineering substrates for 3D integration of III–V and CMOS. *Electrochemical Society Transactions* 16(8): 227–234.

Herrick, K.J. et al. (2008b) Direct growth of III–V devices on silicon. *Materials Research Society Proceedings* 1068: 203–216.

Herrick, K.J. et al. (2008c) Direct growth of compound semiconductors on silicon. In: *33rd Annual GOMACTech Conference*, Las Vegas, NV.

Kazior, T.E. et al. (2009a) A high performance differential amplifier through the direct monolithic integration of InP HBTs and Si CMOS on silicon substrates. In: *Microwave Symposium Digest, 2009. MTT'09. IEEE MTT-S International Microwave Symposium*, 7–12 June 2009, Boston, MA, pp. 1113–1116.

Kazior, T.E. et al. (2009b) Progress and challenges in the direct monolithic integration of III–V devices and Si CMOS on silicon substrates. In: *IEEE International Conference on Indium Phosphide and Related Materials (IPRM'09), Digest*, 10–14 May 2009, Newport Beach, CA, pp. 100–104.

Li, J.C. et al. (2008) 100 GHz+ gain-bandwidth differential amplifiers in a wafer scale heterogeneously integrated technology using 250 nm InP DHBTs and 130nm CMOS. In: *2008 Compound Semiconductor Integrated Circuits Symposium (CSICS'08)*, 12–15 October 2008, Monterey, CA, 10.1109/CSICS.2008.53.

Liu, W.K. et al. (2009) Monolithic integration of InP-based transistors on Si substrates using MBE. *Journal of Crystal Growth* 311(7): 1979–1983.

Lubyshev, D. et al. (2004) Comparison of As- and P-based metamorphic buffers for high performance InP heterojunction bipolar transistor and high electron mobility transistor applications. *Journal of Vacuum Science and Technology B* 22: 1565.

Maleville, C. et al. (2004) Smart Cut™ technology: From 300 mm ultrathin SOI production to advanced engineered substrates. *Solid State Electronics* 48: 1055–1063.

Rosker, M.J. et al. (2008) The DARPA compound semiconductor materials on silicon (COSMOS) program. In: *2008 Compound Semiconductor Integrated Circuits Symposium (CSICS'08)*, 12–15 October 2008, Monterey, CA, 10.1109/CSICS.2008.6.

Sieg, R.M. et al. (1998) Anti-phase domain-free growth of GaAs on off-cut 001 Ge substrates by molecular beam epitaxy with suppressed Ge out-diffusion. *Journal of Electronic Materials* 27: 900.

Ting, S.M. and Fitzgerald, E.A. (2000) Metal-organic chemical vapor deposition of single domain GaAs on $Ge/Ge_xSi_{1-x}/Si$ and Ge substrates. *Journal of Applied Physics* 87: 2618.

7

Optoelectronic Devices Integrated on Si

Di Liang and John E. Bowers

CONTENTS

7.1 Introduction

Silicon (Si), the second most abundant element on earth, is one of the key components of contemporary advanced microelectronic circuits. On the other hand, modern complementary metal oxide semiconductor (CMOS) technology has also enabled it to be an ideal platform for some optoelectronic applications, primarily passive components. Incomparably better purity and defect density plus much more mature fabrication over its compound semiconductor (CS) counterparts allow Si waveguides to demonstrate low-loss characteristics even with nanometer device dimensions. Additional advantages over most compound semiconductors include better thermal and mechanical properties. As elaborated in previous chapters, the biggest hurdle for Si in competing with conventional GaAs- and InP-based CS in active optoelectronics is its direct bandgap, which results in extremely poor electron–hole direct recombination (i.e., photon generation). Fundamental material/process challenges and recent progress in monolithic integration of direct bandgap materials on the Si substrate are discussed elsewhere in this book. In this chapter, we discuss recently developed new hybrid integration approaches to provide electrically pumped optical gain medium on Si substrates (including Si-on-insulator [SOI] substrates). A number of competitive hybrid components, such as lasers, modulators, and photodetectors, have been demonstrated and are studied here.

The traditional approach to hybrid integration is to take prefabricated III–V lasers and amplifiers and die bond (i.e., flip chip bond) these elements onto a passive planar lightwave circuit (PLC). Since the waveguides on the host substrate, and the laser die are already defined, the alignment accuracy during placement needs to be a fraction of the mode width. This is typically within a few hundred nanometers for Si waveguides, making alignment a challenge for high volume manufacturing and leads to substantial variation in coupling power and back reflections between the two waveguides. Efforts have been made to reduce the sensitivity of this coupling by increasing the mode size through spot size converters [Yan 2009]. Precision cleaving the III–V active die and creating a perfectly matched trench in the PLC host substrate have been explored in order to create assembly methods that allow for self-alignment of the optical modes [Friedrich 1992]. Figure 7.1a illustrates a PLC utilizing hybrid integration for switching [Sasaki 2001]. In this work, self-alignment with ±1 μm precision is achieved by placing solder wettable pads on both the host PLC substrate and the III–V chips. During bonding, the surface tension of the Au–Sn solder bumps pulls the two chips into alignment. This demonstration yielded coupling losses of 4–5 dB when used with spot size converters. *Luxtera* has become a pioneer to deliver Si photonic integrated chips by mastering this conventional III–V-on-Si hybrid integration approach (Figure 7.1b). However, the production efficiency and chip yield result in expensive integration cost, preventing it from being adopted widely by the industry so far.

7.2 Hybrid III–V-on-Si Platform

7.2.1 Overview of New Hybrid Platforms

The recently developed novel hybrid platforms involve transfer of as-grown thin (<2 μm thick) crystalline III–V thin films, rather than finished III–V lasers with bulky substrate (>100 μm thick), to a SOI host substrate. The Si

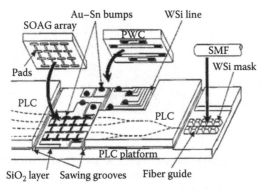

(a)

(b)

FIGURE 7.1
(a) Schematic of proposed hybrid integration scheme for switching. (From Sasaki, J. et al., *IEEE Trans. Adv. Pack.*, 24, 569, 2001. With permission). (b) Flip-chip-mounted lasers enables Luxtera's optical active cable [Laser Focus World: http://www.laserfocusworld.com/articles/305700]. (Courtesy of Luxtera.)

is typically patterned prior to the transfer, and the III–V films are processed after transfer allowing for the use of standard lithography-based patterning techniques to fabricate III–V lasers. A fraction of light generated in III–V is then evanescently coupled to the Si waveguide underneath.

Figure 7.2 highlights the critical steps to form a hybrid Si platform developed by a joint effort between University of California, Santa Barbara (UCSB), and *Intel* [Park 2005]. The hybrid structure is comprised of III–V epitaxial layers transferred to an Si waveguide through a low-temperature, O_2 plasma-assisted wafer bonding process (Figure 7.2a). The bonding mechanism is discussed in detail separately in Section 7.2.2. Upon removing the thick InP substrate selectively, the mesa structure to enable a carrier injection scheme similar to vertical-cavity surface-emitting laser (VCSEL) is then formed on the III–V region by standard photolithography and etching (Figure 7.2b). Fabry–Perot (FP), distributed-feedback Bragg (DFB), distributed

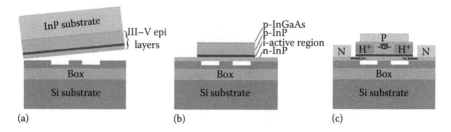

FIGURE 7.2
A simplified fabrication process to form the hybrid Si device platform. (a) Bonding of the III–V wafer to the patterned SOI wafer. (b) InP substrate removal and mesa etching. (c) Current confinement and metal contact formation.

Bragg reflector (DBR), and ring resonator structures can be realized easily to provide feedback for lasing. Typically, the III–V mesa width is larger than the Si waveguide (1–2 μm) so that the mode confinement is determined by the SOI waveguide and not the III–V mesa. This eliminates any issues with alignment of the III–V etch to the SOI etch, but it limits how small the bend radius can be. Amplifiers and lasers have a wide III–V mesa (12–14 μm) for better heat conduction and mechanical strength while a narrow III–V mesa (2–4 μm) is chosen for detectors and modulators for high-speed operation with a reduced capacitance. H$^+$ proton implantation is used if carrier confinement or electrical isolation between integrated devices is necessary. In the case of a wide III–V mesa, proton implantation is employed to confine the current flow for better overlap with the optical mode (Figure 7.2c). Detailed fabrication steps can be found in Ref. [Fang 2006]. The general structure of III–V layers consists of a p-type InGaAs contact layer, a p-type InP cladding, an optional p-type separated confinement heterostructure (SCH) layer, an undoped multiple quantum well layer, n-type contact layer, and n-type superlattice bonding layer. Examples of epitaxial structures for electrically pumped lasers can be found in Refs. [Fang 2007, Fang 2006].

Due to the similar refractive index (RI) of Si and III–V materials, the optical mode in this hybrid waveguide lies both in the Si waveguide and III–V mesa. A unique feature of this platform is the flexibility to widely adjust optical confinement factors in Si and III–V layers. High confinement factor is useful for low threshold in lasers, high gain, low noise amplification in optical amplifiers, and high quantum efficiency in photodetectors. Low confinement factor is useful for high-power lasers, high-saturation-current amplifiers, and high-power photodetectors. In many cases, the best performance can be obtained by changing the confinement factor along the length of the device. For example, changing the confinement factor from high to low results in high-gain, low-noise, high-saturation-power optical amplifiers. Changing the confinement factor from low to high is useful for high-saturation-current, high-quantum-efficiency photodiodes. This can be implemented in the hybrid silicon platform by changing the width of the waveguide (Figure 7.3 calculated from

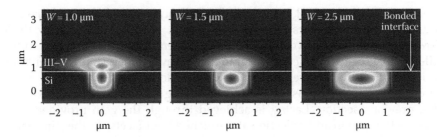

FIGURE 7.3
Mode profiles with different waveguide widths. The height of the Si waveguide is fixed at 0.7 μm. (From Park, H.A. et al., *Adv. Opt. Technol. 2008*, Article ID 682978, 2008. With permission.)

beam propagation method [BPM] simulations). The Si confinement factor is an important parameter in determining coupling efficiency when the device is integrated with Si passive devices. Mode profiles from different waveguide widths ($W = 1$, 1.5, and 2.5 μm) in Figure 7.3 show that narrowing the Si waveguide pushes the mode up to the III–V, resulting in a large III–V confinement factor, while a wide Si waveguide accommodates a larger portion of the optical mode. A taller Si waveguide also leads to larger Si confinement factor. This unique characteristic therefore allows different confinement factors in different regions of the hybrid waveguide for the same III–V epitaxial structure, catering to the requirements of different components on the same chip.

Another hybrid III–V-on-Si platform recently developed by Ghent University possesses similar device structure, though the III–V material and the Si waveguide perform relatively independent functions [Roelkens 2006]. A three-dimensional (3D) schematic is shown in Figure 7.4 to depict the photon generation, optical feedback, and coupling to the Si waveguide below [Roelkens 2005]. The III–V epitaxial layer transfer is achieved by thermosetting polymer divinylsiloxane-benzocyclobutene (DVS-BCB) adhesive

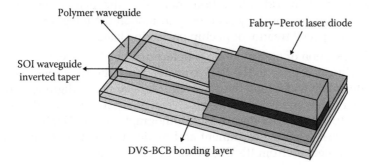

FIGURE 7.4
Schematic of the layout of the optical coupling scheme for efficient and fabrication tolerant coupling between a bonded Fabry–Perot laser diode and an underlying SOI waveguide circuit using an inverted adiabatic taper approach. (From Roelkens, G. et al., *Opt. Express*, 13, 10102, 2005. With permission.)

bonding [Roelkens 2006]. A typical DVS-BCB layer is on the order of several hundreds of nanometers with an RI ~1.5 at $\lambda = 1.55\,\mu m$. A relatively thick low-index medium between III–V and Si prevents photons generated in the III–V active region from coupling into the Si waveguide instantly. Therefore, it is akin to a III–V waveguide-to-Si waveguide asymmetric vertical directional coupler, in contrast to the hybrid waveguide structure discussed in the previous section. Lasing is achieved through the gain provided by a III–V active region and reflection at the etched laser facets. As the stimulated emission leaves the edge of the laser diode, an additional coupling structure is required for efficient coupling to the SOI waveguide. An optimal adiabatic inverted taper structure is employed to achieve good coupling efficiency and fabrication tolerance. The concept of the inverted adiabatic structure is to butt-couple the bonded laser diode to a polymer waveguide, after which the optical mode is gradually transformed into that of the SOI waveguide by increasing the cross-sectional area of the Si waveguide. The polymer waveguide is self-aligned to the laser ridge, eliminating a possible source of coupling efficiency reduction arising from the misalignment between the waveguides. The Si inverted taper structure is buried underneath the polymer waveguide. The inverted taper tip width has to be sufficiently small in order for the fundamental optical waveguide mode at the tip to resemble the waveguide mode of the polymer waveguide closely [Roelkens 2005]. The formation of III–V mesa and electrodes is similar to the first platform.

It is noted that the entire back-end process from patterning Si waveguide to III–V epitaxial transfer to final metallization for both platforms can be done using CMOS tool sets under low temperature (<350°C), making them completely CMOS compatible. These new methods are superior to the older approach of attaching III–V dies onto a Si wafer in multiple aspects, which can be summarized as follows:

1. Since only ~μm thick III–V epitaxial layers are transferred to Si wafer, this hybrid integration maintains the planar topography, an advantage normally seen only in monolithic integration. It also avoids special treatment in chip packaging later.

2. The position of optical mode is determined by Si waveguide, which is fabricated prior to the III–V transfer. Since the III–V mesa is typically much wider than the Si waveguide, a large misalignment in the order of few μm is tolerable, which can be regarded as a self-aligned structure under modern CMOS lithography tools. Negligible coupling loss from misalignment and mode mismatch and eliminated facet reflection from III–V to Si waveguide result.

3. Benefitting from the advanced CMOS integration technique, a huge number of active hybrid devices can be fabricated in a single process flow, dramatically enhancing the production efficiency and yields, subsequently lowering cost.

It is noted that in addition to the UCSB-Intel research team and the European research team led by Ghent University, several groups worldwide have made important contributions to this hybrid integration approach to active Si photonics [Maruyama 2006, Yariv 2007, Yuya 2008, L. Chen 2008]. This is important to developing multifunctional Si photonic integrated chips for a variety of immediate and emerging applications.

7.2.2 Bonding Techniques

Obviously, high-quality III–V epitaxial transfer through wafer bonding is the key to success for these two similar hybrid platforms, and also the fundamental advantage compared to present III–V-on-Si heteroepitaxy technology. "Wafer bonding" refers to the phenomenon that mirror-polished, flat, and clean wafers of almost any material, when brought into contact at room temperature, are locally attracted to each other by van der Waals forces and adhere or "bond" to each other [Gosele 1999]. It has been an increasingly popular approach for a variety of heterogeneous material system integration and microelectromechanical system (MEMS) applications [Gosele 1998, Gosele 1997]. Since the strong bond does not rely on one-to-one atomic connections at the interface, the number of threading dislocations (TDs) and misfits at the bonding interface are much smaller than heteroepitaxial growth [Fehly 2001, Pasquariello 2001]. High-quality *wafer bonded* SOI substrates up to 300 mm in diameter have been commercially available since 2000 [SOITEC: http://soitec.com/en/about/], showing wafer bonding is a highly manufacturable and high-yield fabrication process. Typically, a high-temperature anneal (>600°C) is desirable to enhance the bonding strength, and is routinely conducted to make commercial SOI wafers. However, this technique cannot be used in InP-to-Si bonding for several reasons: (1) thermal expansion coefficient (TEC) mismatch of Si ($\alpha_{Si} = 2.6 \times 10^{-6}$ K^{-1}) and InP ($\alpha_{InP} = 4.8 \times 10^{-6}$ K^{-1}); (2) thermal damage to CS; (3) dopant diffusion; and (4) CMOS incompatible back-end process. Equation 7.1 describes the thermal mismatch stress of the bonded wafers [Pasquariello 2002, Pasquariello 2001]:

$$\sigma = \frac{E}{1-\upsilon^2} \left(\frac{\dfrac{E_{InP}}{1-\upsilon_{InP}^2}\alpha_{InP}h_{InP} + \dfrac{E_{Si}}{1-\upsilon_{Si}^2}\alpha_{Si}h_{Si}}{\dfrac{E_{InP}}{1-\upsilon_{InP}^2}h_{InP} + \dfrac{E_{Si}}{1-\upsilon_{Si}^2}h_{Si}} - \alpha \right) \Delta T \tag{7.1}$$

where
  α is the thermal expansion coefficient
  h is the thickness of the substrate
  E is Young's modulus
  υ is Poisson's ratio
  ΔT is the difference the bonding temperature and room temperature

The critical stress required to generate dislocations in InP is empirically formulated by Pasquariello et al. using the theory of stress-induced dislocation generation [Pasquariello 2001] and formula as follows:

$$\tau_{critical} = 898e^{(5934.17/T)} \tag{7.2}$$

Figure 7.5 shows a plot of the critical stress and the sheer stress as a function of temperature. In order to prevent the generation of dislocations in the InP, the bonding temperature must be kept below 300°C. Figure 7.6 shows a Nomarski-mode microscopic image of the top surface of a ~2 μm thick InP epitaxial layers transferred to a SOI substrate at 600°C. Crosshatching can be seen for this high temperature direct wafer bonding, which can lead to degradation of material quality and scalability issues due to the accumulation of stress over larger sample sizes. Lowering the anneal temperature to 300°C results in successful transfer of InP epitaxial layers with smooth,

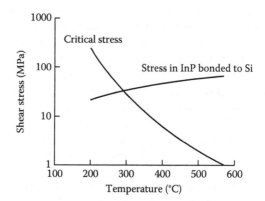

FIGURE 7.5
Critical dislocation generation stress in InP and sheer stress between InP bonded to Si versus temperature. (From Pasquariello, D. and Hjort, K., *IEEE J. Sel. Top. Quant. Electron.*, 8, 118, 2002. With permission.)

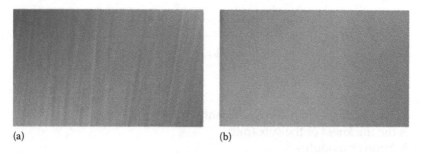

(a)   (b)

FIGURE 7.6
Nomarski microscope images of the transferred III–V surface at bonding temperatures of (a) 600°C and (b) 300°C. (From Fang, A.W. et al., *Mater. Today*, 10, 28, 2007. With permission.)

device-quality, surface morphology. Here, the three demonstrated low-temperature bonding techniques are discussed first before exhibiting the corresponding device results.

7.2.2.1 O_2 Plasma-Assisted Direct Bonding

Lacking the aid of enough thermal energy, strong bonding at low temperature is attainable with necessary proper surface treatment (e.g., O_2 plasma surface processing). An O_2 plasma surface treatment prior to the contact of SOI and InP enhances the bonding strength in both physical and chemical ways. By careful control of the discharge conditions (RF power, chamber pressure, gas flow rate, etc.), O_2 energetic ion bombardment can remove hydrocarbon and water-related species on the sample surface very efficiently. Figure 7.7 is the schematic process flow of the low-temperature

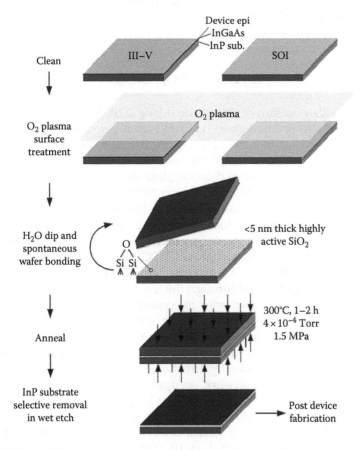

FIGURE 7.7
Oxygen plasma-assisted low temperature wafer bonding process flow. (From Fang, A.W. et al., *Mater. Today*, 10, 28, 2007. With permission.)

(300°C) oxygen plasma-assisted (LTOPA) wafer bonding. After thorough sample cleaning, the native oxides on SOI and InP are removed in standard buffered HF solution and NH_4OH, respectively, resulting in clean, hydrophobic surfaces. An ultrathin layer of oxide (<5 nm) grown by plasma oxidation turns the hydrophobic sample surfaces into very smooth (rms <0.5 nm) and extremely hydrophilic surfaces, which are less sensitive to the microroughness as compared to the hydrophobic bonding [Pasquariello 2002]. The Si–O–Si bonds of the oxide (SOI side) are also found more strained and active than conventional oxides formed in standard RCA-1 cleaning process or other hydrophilic wet-chemical treatment, indicating a higher readiness to break and form new bonds [Tong 1998]. The following DI water dip further terminates the oxide surface with polar hydroxyl groups OH^-, forming a bridge between mating surfaces to result in spontaneous bonding at room temperature. The 300°C annealing process enhances the formation of covalent bonds to achieve higher bond energy [Pasquariello 2002], following the polymerization reactions below:

$$Si-OH+M-OH \rightarrow Si-O-M+H_2O\,(g) \qquad (7.3)$$

$$Si+2H_2O \rightarrow SiO_2 + 2H_2\,(g) \qquad (7.4)$$

where M is either indium or phosphide [Liang 2008, Gosele 1998].

Bonding-related residual gases (H_2O, H_2) are inherent to low-temperature bonding mechanisms in general. Since gas diffusion energy at 300°C is too small for gas molecules to escape the bonding interface, they tend to reside in any microcavities or surface defects at interface, forming interfacial voids with a surface density >50,000 cm^{-2} [Liang 2008] as shown in the Nomarski-mode microscopic image (outside the dash-line box) of Figure 7.8b. These local delaminations largely reduce the usable device area and cause potential contamination after voids burst during fabrication. Among several proposed solutions, vertical outgassing channel (VOC) is one of the most efficient methods [Liang 2008]. VOCs are essentially an array of through-holes a few μm in dimension and etched through the top Si device layer to the underlying buried oxide (BOX) layer prior to contact with the III–V material. The generated gas byproduct molecules along with a small amount of trapped air molecules, including any gaseous impurities, can migrate to the closest VOC and promptly be trapped inside the BOX as the schematic in Figure 7.8a shows. Thereafter, they gradually diffuse out and are even absorbed by the BOX layer, due to its open network, occupying only 43% of the lattice space, and its large diffusion cross section for 0.3–3 μm thick BOX layer in general. The outgassing effectiveness depends primarily on the spacing of VOCs [Liang 2008]. Figure 7.8b demonstrates a clear contrast of area with and without VOCs on Si when only 2 μm thick III–V material is bonded. An optimal VOC size of 8 × 8 μm with a spacing of 50–100 μm is used, resulting in void-free epitaxial transfer. No mechanical instability issue (i.e., deformation) is

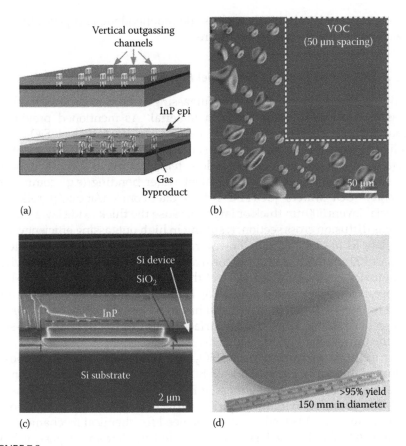

FIGURE 7.8

(a) Schematic of VOCs on Si and its functionality during InP-to-Si bonding. (b) Nomarski-mode microscopic image of thin III–V epitaxial layer bonded on a Si sample where only the dash-line highlighted area has VOCs. (c) SEM cross-sectional image of a VOC. (d) 150 mm in diameter InP-based III–V device epitaxial layers transferred to a SOI wafer with the same diameter. (From Liang, D. and Bowers, J.E., *J. Vac. Sci. Technol. B*, 26, 1560, 2008; Liang, D. et al., *J. Electron. Mater.*, 37, 1552, 2008; Liang, D. et al., *Electrochem. Solid-State Lett.*, 12, H101, 2009. With permission.)

observed as shown in a cross-sectional scanning electron microscopy (SEM) image in Figure 7.8c. Waveguide circuits can be placed in the rest of blank area freely. Due to small dimensions of Si waveguides and strong optical confinement, VOCs do not pose any negative impact even very close to the waveguide. If the same design scheme is applied to the entire wafer, the outgassing issue is completely eliminated and wafer-scale epitaxial transfer can be accomplished without prolonged processing. A 150 mm in diameter multiple quantum well (MQW) laser epitaxial layer has been bonded to a same-size, patterned SOI wafer successfully in Figure 7.8d, showing >95% yield, void-free, low-strained epitaxial transfer [Liang 2009b]. It indicates the bonding process is essentially wafer scale-independent, and paves the way for wafer-level production in the future.

7.2.2.2 SiO₂-Based Covalent (Molecule) Bonding

Another method to circumvent the outgassing issue is to sandwich a layer of gas absorption/diffusion medium material. As mentioned previously, porous SiO_2 is a CMOS-compatible natural choice. Covalent SiO_2-based bonding is hydrophilic, sharing the same polymerization reactions at the interface and outgassing issues at low temperature as LTOPA bonding above. Conventional SiO_2-based covalent wafer bonding (e.g., commercial Unibond™ SOI wafers [SOITEC http://soitec.com/en/about/]) relies on interfacial layers 500 nm thick or larger because the thick oxide layer offers a large gas diffusion cross section, resulting in high outgassing efficiency. Due to the low RI ($n \sim 1.45$) of SiO_2, however, the interfacial SiO_2 layer has to be less than 100 nm thick in order to form a hybrid waveguide structure where the same device design philosophy of the first hybrid platform (UCSB-Intel developed) can be applied [Liang 2008]. For the second platform, SiO_2 is analogical to DVS-BCB polymer, and still has to be less than 300 nm for decent evanescent coupling. Thus, special surface treatment is still required to suppress the formation of interfacial voids.

As an example, a 60 nm interfacial SiO_2 layer is targeted, which is composed of 30 nm of thermally grown oxide on SOI surface, and 30 nm PECVD on the III–V sample. The surface roughness of PECVD SiO_2 on as-grown InP needs to be below 1 nm, a maximum empirical surface roughness value allowed for good bonding [Gui 1999] eliminating the need for chemical mechanical planarization (CMP). The special surface treatment starts from a degassing bake of PECVD SiO_2-covered III–V sample. Due to the inhomogeneous nature of the PECVD process [Tan 2003, Tong 2006], a degassing bake at 250°C for 1–2 h is done after deposition to drive undesirable gas molecules out of the thin film. Next, an O_2 plasma surface treatment on SOI and III–V samples is conducted as it is found to suppress the interfacial void formation [Liang 2008], although is reported not to be contributive to bonding strength enhancement in SiO_2–SiO_2 covalent bonding of Si wafers [Tong 2006]. Immediately after the O_2 plasma surface treatment, the second surface activation step is to dip SOI and III–V samples in very dilute HF solution (0.025%) for 1 min to form a more porous fluorinated oxide network [Tong 2006, Liang 2008]. The last activation step involves further converting Si–OH to Si–NH_2 bonds in NH_4OH solution or vapor [Chao 2005, Liang 2008]. Upon room-temperature contact, a low-temperature (e.g., 300°C) anneal is conducted to form O–Si–O bonds at the two SiO_2 surfaces, resulting in strong integration.

As shown by the Nomarski-mode microscopic image in Figure 7.9a, an interfacial void-free bond is achieved after removing the InP substrate. Around 2.63 J/m² surface energies of bonded pairs are measured by the conventional crack-opening method [Maszara 1988]. This is over four

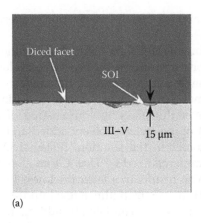 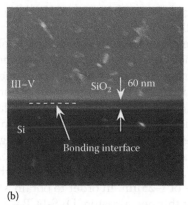

(a)   (b)

FIGURE 7.9

(a) Nomarski-mode microscopic (200×) top-view of a diced bar showing void-free bonding and ≤15 μm III–V chip due to dicing. (b) SEM cross-sectional view of a polished bar, showing III–V epitaxial layers tightly bonded on the SOI substrate with a 60 nm interfacial SiO$_2$ layer.

times larger than the bulk InP fracture energy of 0.63 J/m^2 [Tong 2004]. III–V-SOI bonded samples also experience a harsh dicing test, a standard process for fabricating the Fabry–Perot cavity devices. The bonded samples with only InGaAs and InP epitaxial layers on SOI substrate are cut by a 100 μm thick blade with over 10,000 rpm. Though the III–V side faces the blade and there is no surface protection during dicing, the chipping of III–V epitaxial layer is no more than 15 μm and follows the SOI fringe consistently in Figure 7.9a. Figure 7.9b represents SEM cross-sectional view of a further polished bar, showing III–V epitaxial layers tightly bonded on the SOI substrate with a 60 nm interfacial SiO$_2$ layer. The survival of harsh dicing and polishing also indicate strong bonding between III–V material and SOI substrate.

7.2.2.3 Polymer-Based Adhesive Bonding

Adhesive bonding is a material integration process in which a liquid or semiliquid substance is applied to adjoining work pieces to provide a long-lasting bond. Thermoplastic, elastomeric, and thermosetting materials have typically been used for adhesive bonding. The former two adhesives are not temperature stable enough for back-end III–V device processing after epitaxial transfer; therefore, thermosetting material is chosen for this hybrid integration purpose [Roelkens 2006]. Among typical choices of thermosetting materials, such as DVS-BCB, epoxies, spin-on-glasses, and polyimides, DVS-BCB is the optimal material based on the particular criteria. They include high bonding strength [Niklaus 2001], low optical loss at telecommunication wavelengths, low shrinkage upon cure (as this can be the origin of void formation at the bonding interface), high glass transition temperature (allowing a large post-bonding thermal budget), and excellent planarization properties [Roelkens 2009].

The divinylsiloxane-bis-benzocyclobutene (BCB) monomer is a symmetrical molecule consisting of a Si backbone terminated by two benzocyclobutene rings [Roelkens 2006]. The monomer can be B-staged; this means that it is partially cured to form an oligomer. An oligomer solution is made by adding mesitylene. The achievable layer thickness by spin coating of the solution is determined by the amount of mesitylene solvent added and the degree of polymerization of the oligomer in the solution. B-staged DVS-BCB solutions are formulated and commercialized by *Dow Chemical* as the CYCLOTENE 3022 product series, which results in a layer thickness in the range of 1–25 μm. In order to obtain a thin DVS-BCB film, a few hundreds of nm or thinner, a custom DVS-BCB solution was formulated by adding mesitylene to the CYCLOTENE 3022-35. The effect of this dilution on the layer thickness is shown in Figure 7.10a, in which the resulting layer thickness for a spin speed of 5000 rpm is plotted as a function of the amount of added mesitylene. By adjusting both spin rate and diluting agent, sub-100 nm thick film has been achieved [Roelkens 2009]. Upon curing, the benzocyclobutene ring thermally opens to form *o*-quinodimethane. This very reactive intermediate readily undergoes a so-called Diels–Alder reaction with an available vinylsiloxane group, to form a 3D network structure [Niklaus 2006]. No by-products are created during the polymerization.

A standard bonding process at Ghent University starts from thorough surface cleaning of both SOI and III–V samples, though the tolerance to surface particles is slightly better than with the previously discussed two bonding techniques. Then the DVS-BCB is deposited on the SOI sample surface by spin coating. Although most of the mesitylene solvent evaporates during the spin coating process, some solvent remains in the spin coated film, which is

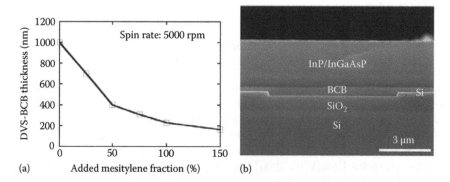

(a) Added mesitylene fraction (%)   (b)

FIGURE 7.10

(a) Influence of the DVS-BCB dilution on the achievable bonding layer thickness. (b) SEM cross-sectional image of a DVS-BCB bonding to integrate InP thin epitaxial film on the SOI substrate. (From Roelkens, G. et al., *Mater. Today*, 10, 36, 2007; Roelkens, G. et al., *J. Electrochem. Soc.*, 153, G1015, 2006. With permission.)

evaporated by a thermal treatment at 150°C for 1 min. This thermal treatment also causes a reflow of the DVS-BCB, which improves its planarizing properties. Next, initial mating of III–V and Si samples are conducted at 150°C as DVS-BCB has the lowest viscosity at this temperature [Niklaus 2006] and keeping the DVS-BCB at this temperature does not significantly increase the degree of polymerization for at least an hour. After attachment of the III–V die, the DVS-BCB film has to be cured. The curing is performed at 250°C to obtain a fully polymerized DVS-BCB layer. This step has to be done in an atmosphere containing less than 100 ppm oxygen to prevent the oxidation of the DVS-BCB. To achieve this, nitrogen is purged through the curing chamber [Roelkens 2009]. Figure 7.10b shows a SEM cross-sectional image of a thin III–V epitaxial layer intimately bonded on a patterned SOI sample. It is noted that DVS-BCB fills the etched trenches next to waveguides, preventing liquids from flowing in and attacking the semiconductor. It is one processing-related advantage over direct bonding, although this can be solved in the direct bonding approach by depositing silicon oxynitride and CMP to planarize prior to direct bonding.

The three bonding techniques above all share the advantages of strong, void-free, robust integration of dissimilar materials at low temperature (<400°C). Each one has intrinsic advantages and disadvantages and should be adopted based on particular applications. All the hybrid devices in this chapter are fabricated by either LTOPA bonding or DVS-BCB bonding.

7.3 Integrated Optoelectronic Devices

Though a number of hybrid Si optoelectronic devices have been demonstrated recently, only light emitters, modulators, and photodetectors are selected to discuss in this chapter as they are the basic active components required to realize a hybrid Si transceiver.

7.3.1 Hybrid Si Light Emitters

7.3.1.1 Hybrid Si Fabry–Perot Resonator Lasers

As the simplest type of laser, Fabry–Perot (FP) resonator devices were demonstrated first. Since a (001) Si wafer breaks along the <110> direction, cleaving does not form an FP cavity as most III–V wafers do. Dicing and facet polishing have to be employed, and devices suffer from facet roughness and misalignment.

Figure 7.11a demonstrates the cw *LI* characteristic of InAlGaAs-MQW, LTOPA-bonded, hybrid Si lasers with emission wavelength around 1550 nm [Park 2008]. The waveguide height, width, rib etch depth, and cavity length were 0.7, 2, 0.5, and 850 μm, respectively. The calculated confinement factors

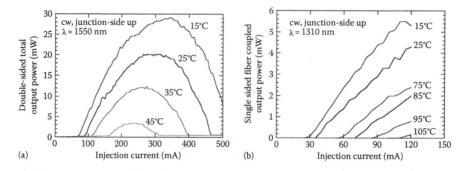

(a)   (b)

FIGURE 7.11
(a) *LI* curves for a 1550 nm FP hybrid Si laser. (From Park, H.A. et al., *Adv. Opt. Technol. 2008*, Article ID 682978, 2008. With permission.) (b) *LI* curves for 1310 nm FP hybrid Si lasers. The secondary *y*-axis represents the estimated total laser output power. (From Chang, H.-H. et al., *Opt. Express*, 15, 11466, 2007. With permission.)

in the Si and the quantum well (QW) region were 63% and 4%, respectively. The cw single-sided output power for this device is collected on one side with an integrating sphere. Total power is calculated by doubling measured power, assuming equal outputs at both sides. It can be seen that the maximum laser output power, threshold, and differential efficiency at 15°C are 32.8 mW, 70 mA, and 26%, respectively. Figure 7.11b shows the device performance of counterpart devices in the 1310 nm emission wavelength regime. The threshold current at 15°C is 30 mA [Chang 2007]. Single-side output power is measured by coupling the light to a single-mode lensed fiber. Total power can be estimated by taking ~5 dB coupling loss and then doubling the result. Lasing is observed up to a stage temperature of 105°C. The higher temperature performance is due to a higher confinement factor, a larger conduction band offset than the 1550 nm epitaxial design in Figure 7.11a and the intrinsic advantage of 1310 nm lasers in reduced intravalence band absorption and Auger scattering [Chang 2007].

7.3.1.2 Hybrid Si Mode-Locked Lasers

Hybrid Si mode-locked lasers (MLLs) are also fabricated on the same platform by forming separate gain and saturable absorption (SA) sections [Koch 2007]. MLLs are capable of generating stable short pulses that have a corresponding wide optical spectrum of phase-correlated modes. Their short pulses with high extinction ratios (ERs), low jitter, and low chirp make them an excellent choice for potential applications in optical pulse generation, optical time division multiplexing (OTDM), wavelength division multiplexing (WDM), and regenerative all-optical clock recovery [Koch 2008]. In passive mode locking, the locking occurs automatically from the dynamic interaction between the gain and absorber sections. In hybrid mode locking, a passively MLL has an RF signal injected into the saturable absorber section (in addition to the reverse DC bias) with a repetition rate close to the passive mode locking repetition rate. This allows the MLL to be synchronized to an external source. The repetition rate and the directly related mode spacing of an MLL are given approximately by Equation 7.5:

$$f \approx \frac{cm}{2n_g L} \tag{7.5}$$

where
   c is the speed of light
   n_g is the group index of refraction
   L is the length of the cavity
   m is the number of pulses in the cavity

If the laser is a self-colliding pulse design, in which the saturable absorber is located next to a mirror, then only one pulse exists in the cavity ($m = 1$) and it collides with itself after reflecting from the mirror. If the laser is a colliding pulse laser, which is the case for ring designs and for linear cavities with the saturable absorber placed in the middle of the cavity, then there are two pulses propagating in the cavity in opposite directions that collide in the absorber. In this case, $m = 2$.

An FP hybrid Si MLL is shown in Figure 7.12a and its test setup is shown in Figure 7.12b. The 39.4 GHz FP ML-SEL had a total cavity length of 1060 μm and a SA length of 70 μm. Separate gain and SA sections were electrically isolated using proton implantation. For these devices, the Si waveguide had a width of 2.5 μm, height of 0.69 μm, and rib etch depth of 0.39 μm, resulting in a Si confinement factor of 67.4% and QW confinement factor of 4.3%. These devices were temperature stabilized on a stage held at 13°C. Passive mode locking was achieved for a range of gain currents between 195 and 245 mA with similar output characteristics. For bias conditions Gain = 216 mA and SA = 0.4 V, this device emitted 3.7 ps hyperbolic secant square-shaped pulses (5.7 ps autocorrelation FWHM) shown in Figure 7.13a and b. The extinction ratio was over 18 dB and the time bandwidth product was 0.4. This is fairly close to the transform limited value of 0.32 for this pulse shape. These pulses are short enough and have high enough extinction to be time division multiplexed to 80 GHz. The jitter (1 kHz–100 MHz) was 1.0 ps and the locking range was 5 MHz for an RF power of 17 dBm at 20 GHz (the first subharmonic of 40 GHz). Locking to an RF signal at 40 GHz with the same RF power would provide lower jitter and a larger locking range.

The self-colliding pulse design is realized on a racetrack hybrid Si MML with 30.4 GHz repetition rate in Figure 7.14a [Fang 2008c]. A racetrack MLL has the advantage that its cavity is defined by lithography. Since the repetition rate is determined by the cavity length, this allows for the repetition rate to be precisely determined and repeated across different devices.

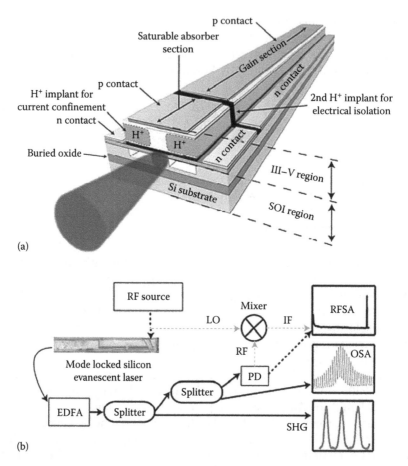

FIGURE 7.12
(a) SEM top-view of one end of an FP ML laser. (b) Schematic of the general experimental setup.
(From Koch, B.R. et al., *Opt. Express*, 15, 11225, 2007. With permission.)

Racetrack lasers can also be integrated monolithically with other components. This laser has a cavity length of 2.6 mm, an absorber length of 50 μm,
and two separate gain sections. The gain sections were biased together at
410 mA and the absorber was biased with −0.66 V. This resulted in Gaussian-
shaped output pulses with 7.1 ps FWHM pulse width, 0.5 nm FWHM spectral width, and approximately 10 dB ER. The time-bandwidth product was
0.43. The peak power was 6.8 dBm on chip, determined from on-chip photocurrent measurements [Fang 2008].

Figure 7.14b shows the jitter under hybrid mode locking versus the upper
integration limit. For the frequency range of 1 kHz–100 MHz offset from the
repetition frequency, the laser has an absolute jitter of 364 fs, which is well
below the extrapolated ITU specification of 3.3 ps for digital transmission at
a 30 GHz repetition rate (a typical ITU spec is 0.1 times the bit period). The

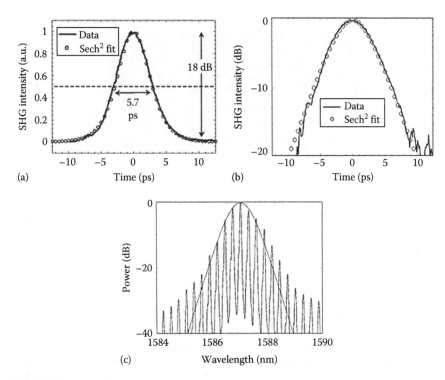

FIGURE 7.13
(a) Linear, (b) logarithmic autocorrelation traces, and (c) logarithmic optical spectra for the 40 GHz hybrid Si MLL. (From Koch, B.R. et al., *Opt. Express*, 15, 11225, 2007. With permission.)

locking range is the range of frequencies over which the ML-SEL can synchronize to an input RF signal. The locking range can be increased by injecting more RF power. Similarly, the timing jitter decreases with higher injected RF power. Figure 7.14c shows the locking range and the jitter as a function of injected RF power under hybrid mode locking.

7.3.1.3 Hybrid Si DFB/DBR Lasers

In conventional III–V photonic integrated circuits (PICs) for telecommunications, grating-based lasers, such as distributed feedback (DFB) and DBR devices, are very popular on-chip light sources. Both have been realized on the direct bonded hybrid platform for optically pumped carrier injection at Tokyo Institute of Technology (TIT) [Maruyama 2006] and electrically pumped versions have been realized at UCSB [Fang 2008a,b]. As shown in Figure 7.15a, electrically pumped hybrid DFB lasers with a ~25 nm surface corrugated grating pattern with a 238 nm pitch and 71% duty cycle were formed on Si for a grating stop-band designed at around 1600 nm [Fang 2008]. The 50 dB side-mode suppression ratio (SMSR), over 100 nm single-mode

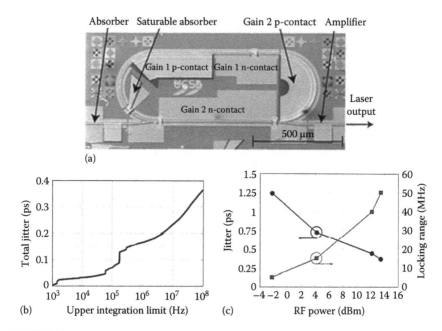

FIGURE 7.14
(a) SEM top-view of the racetrack mode-locked hybrid Si laser, (b) jitter versus upper integration range in the offset frequency for the laser output, and (c) jitter and locking range
as a function of RF input power. (From Fang, A.W. et al., *Opt. Express*, 16, 4413, 2008. With
permission.)

operation span, and 3.5 MHz linewidth are comparable to III–V DFBs.
Figure 7.15b shows the *LI* curves of the device showing lasing up to 50°C. The
output power is measured by integrated photodetectors with a responsivity
~1 A/W [Park 2007]. More than 4 mW output power was obtained.

The hybrid DBR counterpart (Figure 7.16d) includes two passive (i.e., Si)
Bragg reflector mirrors (Figure 7.16a) placed 600 μm apart to form an optical
cavity. Two 80 μm long tapers sandwich the 440 μm long hybrid III–V-on-Si
gain region (Figure 7.16b) [Fang 2008a]. The back and front mirror lengths
are 300 and 100 μm, respectively. The power reflectivities of back and front
mirrors are calculated to be 97% and 44%, respectively. The Si waveguide
has a width, height, and rib etch depth of 2, 0.7, and 0.5 μm, respectively. This
results in Si and quantum well confinement factors of 66% and 4.4% in the
hybrid region. The surface-corrugated gratings have an etch depth and duty
cycle of 25 nm and 75%, respectively, with an upper cladding of SU-8 leading
to a grating strength, κ, of 80 cm^{-1}.

The laser output power is measured with an integrating sphere at the
front mirror of the laser. The front mirror *L–I* characteristic is shown in
Figure 7.17a. The device has a lasing threshold of 65 mA and a maximum
front mirror output power of 11 mW, leading to a differential efficiency

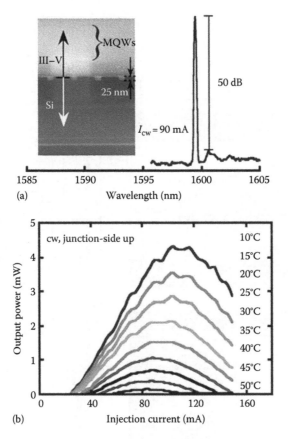

FIGURE 7.15

(a) Experimental spectrum of a Si DFB laser at 90 mA cw injection current. Inset: image of longitudinal cross section. (b) *L–I* curve for stage temperatures of 10°C–50°C. (From Fang, A.W. et al., *Opt. Express*, 16, 4413, 2008. With permission.)

of 15%. The laser operates up to a stage temperature of 45°C. The kinks in the *LI* are present because of the longitudinal mode hop caused by the laser wavelength redshifting due to self-heating. The lasing spectrum is shown in Figure 7.17a inset with a lasing peak at 1597.5 nm and 50 dB side-mode suppression ratio (SMSR) when driven at 200 mA. Figure 7.17b shows the photodetected electro-optic (EO) response of the laser under small signal modulation of –10 dBm. A 2 pF device capacitance was extracted from the S_{11} measurement, resulting in an resistance-capacitance (RC) limited bandwidth of 7 GHz. Figure 7.17b inset shows the resonance frequency versus the square root of DC drive current above threshold, which has a roughly linear dependence as expected. Under higher modulation powers, the resonance peak becomes significantly dampened. The 3 dB electrical bandwidth at 105 mA is ~2.5 GHz.

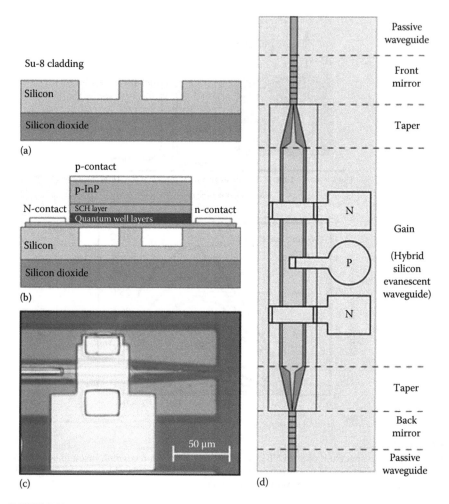

FIGURE 7.16
(a) Passive Si rib, (b) hybrid Si waveguide cross section, (c) microscope image of a hybrid to passive taper, and (d) DBR-SEL top-view topographical structure. (From Fang, A.W. et al., *IEEE Photon. Tech. Lett.*, 20, 1667, 2008. With permission.)

7.3.1.4 Hybrid Si Micro-Disk/Ring Lasers

Another type of on-chip laser with the capability to form even shorter resonator cavities are micro-disk or ring lasers. They are ideal candidates for applications requiring low power consumption and high-speed direct modulation. The inset in Figure 7.18b exhibits the schematic of a compact hybrid microring laser structure integrated with on-chip photodetectors [Liang 2009d]. Figure 7.18a shows the cw, temperature-dependent *LI* characteristic of a 50 μm diameter device with a 150 nm coupling gap between the resonator and bus waveguide. Finite-difference time-domain (FDTD) simulations

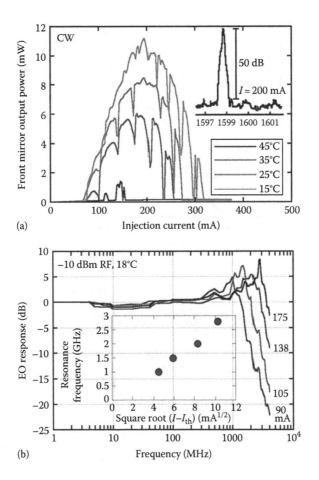

FIGURE 7.17

(a) DBR laser *L–I* curve for various temperatures measured out of the front mirror. Inset: The lasing spectrum at 200 mA injection current, showing a single-mode operation with 50 dB SMSR. (b) Photodetected frequency response of the DFB-SEL for three different bias currents with a stage temperature of 18°C and (inset) plot of resonance frequency versus the square root of current above threshold. (From Fang, A.W. et al., *IEEE Photon. Tech. Lett.*, 20, 1667, 2008. With permission.)

indicate 1%–2% outcoupling in this geometry. The output power is the sum of the photocurrent measured at both photodetectors for various stage temperatures. The minimum threshold current is 8.37 mA at 10°C, corresponding to a threshold current density of 2.24 kA/cm². A threshold current as low as 4 mA and a lasing temperature as high as 55°C have been observed with further reduction of the outcoupling by increasing the coupling gap to 250 nm, although with lower output power as shown in Figure 7.19a. The lasing spectrum at $2 \times I_{th}$ is shown in the inset of Figure 7.19b. Over 40 dB extinction ratio (ER) and <0.04 nm FWHM (limited by the measurement system) are obtained.

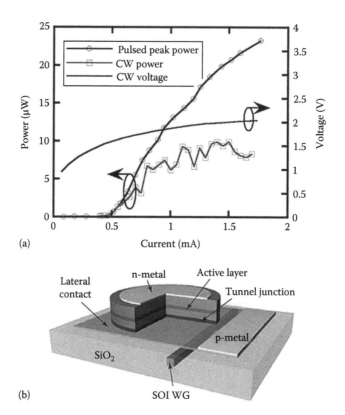

(a)

(b)

FIGURE 7.18
(a) Pulsed and cw light–current–voltage (*LIV*) characteristics at 20°C for a 7.5 μm disk laser. (From Van Campenhout, J. et al., *Opt. Express*, 15, 6744, 2007. With permission.) (b) Schematic of the disk laser with a vertically coupled SOI waveguide underneath.

Sub-mA threshold operation is recently demonstrated in a DVS-BCB bonded disk laser with 7.5 μm disk diameter by Van Campenhout et al., as shown schematically in Figure 7.18b [Van Campenhout 2007]. A threshold of 0.5 mA is achieved under both pulsed and cw operation at 20°C in Figure 7.18a, leading to a threshold current density of 1.13 kA/cm² for uniform injection. A recent optimized design leads to a further threshold reduction to 0.35 mA [Spuesens 2009]. The output power was collected at one end of the output SOI waveguide, using a fiber grating coupler. The maximum cw and pulsed output power are 10 and 100 μW, respectively. The early thermal rollover in the cw regime is caused by a high thermal resistance, which was measured to be 10,000 K/W [Van Campenhout 2007].

A multi-wavelength laser (MWL) source with four cascaded microdisk lasers was also demonstrated [Van Campenhout 2008]. Shown by a top-view microscopic image in Figure 7.20 (top), the microdisk diameters were 7.632, 7.588, 7.544, and 7.5 μm for lasers L1, L2, L3, and L4, respectively, such that

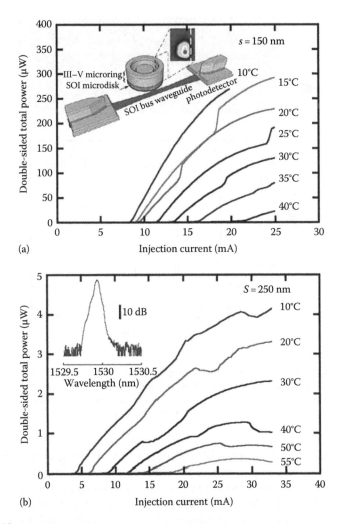

FIGURE 7.19
(a) Schematic of the Si microring laser with integrated photodetectors, (b) cw *LI* of the device as a function of stage temperature. (From Liang, D. et al., *The 6th IEEE International Conference on Group IV Photonics*, San Francisco, CA, June 9–11, 2009. With permission.)

the laser peaks are uniformly spread within the free-spectral range of a single microdisk. The uniform MWL output spectrum is achieved by tuning the individual microdisk lasers with different bias currents as shown in Figure 7.20 (bottom). Although the measured thermal impedance was about 6400 K/W, the wavelength shift of one laser from changing the bias does not lead to the drifting of the others, indicating negligible thermal crosstalk [Van Campenhout 2008].

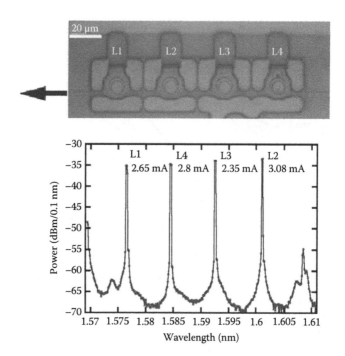

FIGURE 7.20

Top: Microscope image of the multi-wavelength laser source consisting of four microdisk lasers coupled to a common, underlying SOI waveguide. Bottom: Balanced output spectrum obtained by adjusting the individual microdisk drive currents. (From Van Campenhout, J. et al., *IEEE Photon Tech. Lett.*, 20, 1345, 2008. With permission.)

To summarize, a Si laser with the possibility of integration with other CMOS-compatible components is the most important component in the Si photonics toolbox. It may become a crucial component for the microprocessor industry to deploy low-cost, high-speed optical interconnects in microelectronic chips for larger communication bandwidth and lower power consumption. The highest performance to date has been demonstrated with hybrid Si approaches, compared with other monolithic approaches, and shows great possibilities for commercial usage.

7.3.2 Hybrid Si Modulators

While direct modulation is conceptually the simplest means to produce a modulated light source, it has drawbacks of (1) limited bandwidth, (2) excess chirp, and (3) lower efficiency at larger (10 Gb/s) bandwidths. A lower power approach is to bias the laser to close to the loss budget for the link being considered and use an external modulator to encode data. Si-based modulators are of interest because of their potential to be manufactured in high volume and at a low cost using standard CMOS processing technology. These modulators typically utilize the electroabsorption (EA) and EO effects inside the Si waveguide to deliver the necessary phase shift (or extinction ratio) needed for modulation. Exciting results recently reported on modulating light in Si include electroabsorption modulators (EAM) based on the Franz–Keldysh effect in strained Ge/Si [Roth 2007], and Mach–Zehnder (MZ) or ring modulators based on carrier injection plasma effects [Xu 2005]. However, Ge/Si EAMs need to overcome additional loss caused by the indirect bandgap absorption of Ge under zero bias, and modulation bandwidth for carrier injection designs is fundamentally limited by carrier life time (~ns) due to the relatively slow recombination process. Mach–Zehnder modulators (MZM) with RI shift introduced by the carrier depletion effect can increase the electrical bandwidth up to 40 Gb/s and have reasonable optical bandwidths at the expense of a higher voltage–length product: 40 V mm [Liu 2007]. The multiple advantages of high bandwidth, large modulation efficiency, high power, and wide optical bandwidth can be achieved by hybrid III–V-on-Si platforms [Chen 2008a,b, Kuo 2008, Liu 2008]. Here, we focus on the hybrid Si MZM, which is reverse biased for depletion-mode operation, and can modulate different wavelengths channels by changing the MZI bias, where the EO effect is introduced through the III–V material.

In contrast to the weak plasma effect in Si, once the carriers deplete out of the MQW in III–V, several physical effects, such as the band-filling, plasma, Pockels, and Kerr effects, all contribute to the index change. Among these, the Pockels effect is the only phenomenon, which is sensitive to crystal orientation. In other words, this effect can be additive to other effects if the optical signal propagates along the right direction, or it can reduce the overall index change otherwise. On the hybrid Si platform, the direction of patterned Si waveguide needs to be aligned to the [011] direction of the III–V material so that the phase shift is maximized, or at the correct angle to make it polarization independent. Figure 7.21 shows the simulation results with consideration of all these effects [Chen 2009b]. As can be seen, the index change is proportional to the magnitude of the reverse bias. Moreover, the introduced index shift with orientation match is approximately 1.5 times larger than in the mismatch case.

Under the same reverse-bias operation, the 16-period MQW active region is designed with a PL peak at 1.36 μm in order to achieve high optical confinement in MQWs and low absorption at 1.55 μm [Chen 2008a]. Both the top SCH layer and MQW are n-type doped in order to introduce free carriers. The thickness and doping of the top SCH layer are carefully designed to result in the complete depletion of this layer at zero bias. Thus, all applied bias voltage will be used to deplete carriers in the MQW region only. As shown in Figure 7.22a, a capacitively loaded (CL) traveling wave electrode (TWE) based on a slotline architecture is adopted to prevent the electrical field from overlapping with the doped semiconductor underneath the metal electrode. The device capacitance is reduced by half by using a push-pull

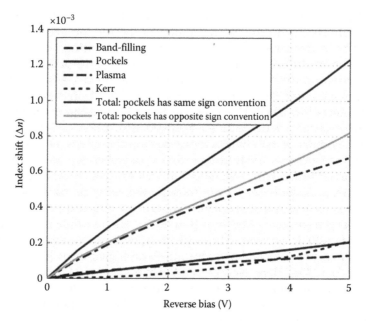

FIGURE 7.21
Estimated index shift (TE polarization) of a carrier depletion phase modulator in InAlGaAs
MQW. (From Chen, H.-W. et al., *Chin. Opt. Lett.*, 7, 280, 2009. With permission.)

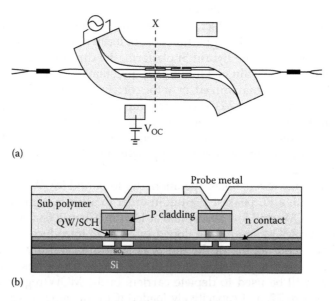

FIGURE 7.22
(a) Top view of a device with a CL-slotline electrode. (b) Cross section (along *x*) of the hybrid
waveguide. (From Chen, H.-W. et al., *The 6th IEEE International Conference on Group IV Photonics*,
San Francisco, CA, June 9–11, 2009. With permission.)

scheme because the two diodes on both arms are in series [Chen 2009a]. The cross section of the loaded region is depicted in Figure 7.22b. The signal and ground of the slotline are on top of each arm. The two arms have a common ground formed by connecting the n-contact layers together. The cladding mesa is 4 μm wide, and the active region is intentionally under-cut to 2 μm to reduce the device capacitance. The Si waveguides have a height of 0.67 μm, a rib etch depth of 0.2 μm, and a width of 1 μm.

The normalized transmission as a function of reverse bias for two devices is shown in Figure 7.23a. As can be seen, the $V\pi$ of a 250 and a 500 μm long modulators is 6.3 and 4.8 V, respectively. This results in voltage length products of 2.4 and 1.6 V mm, respectively. The ER of a 250 and a 500 μm long modulator are 12.2 and 18.4 dB, respectively. The frequency response of the 250 and 500 μm long modulators with −3 V bias across the diode is shown in Figure 7.23b. The transmission curve indicates 3 dB cutoff frequencies around 20 and 12 GHz, respectively. Good eye diagrams are attainable up to 25 Gb/s with 11 dB ER.

On the other hand, the transferred InAlGaAs-based MQWs also make strong quantum confined stark effect (QCSE) available for electroabsorption modulation due to the large conduction band offset for high ER. A 100 μm long hybrid Si EAM with 16 GHz 3 dB bandwidth, 10 dB ER at 4 V bias was demonstrated as well. More details can be found in Ref. [Kuo 2008].

7.3.3 Si Photodetectors

7.3.3.1 Hybrid Si Waveguide Photodetectors

Conventional Si photodetectors have been widely used as high-efficiency ($\eta_i > 60\%$) devices in the wavelength span of 400–850 nm. Nevertheless, devices compatible with telecommunication wavelengths require integration with materials with smaller bandgaps (e.g., III–V, Ge). Hybrid Si waveguide photodetectors [Park 2007] are therefore demonstrated simultaneously with hybrid Si lasers [Fang 2006], sharing the same III–V epitaxial structure and similar device geometries. The devices consisted of a ~100 μm passive Si waveguide coupled to a 400 μm long hybrid photodetector region. At the junction of the hybrid waveguide and the passive Si waveguide, the III–V region is tilted by 7° to reduce the reflection at the waveguide transition (Figure 7.24a). The waveguide height, width, and rib etch depth are 0.69, 0.19, and 0.5 μm, respectively. The passive to active junction is shown by a SEM close-up image in Figure 7.24b.

The photodetector responsivity was measured by launching light into the passive waveguide through a lensed fiber. The transverse electrical (TE) responsivity versus reverse bias is shown in Figure 7.25a. With a 5.5 dB measured coupling loss from the fiber, the device responsivity is ~1.13 A/W. The quantum efficiency is ~90% as shown on the right axis of Figure 7.25a. It can be seen from Figure 7.25b that the responsivity is relatively flat from 1500 to

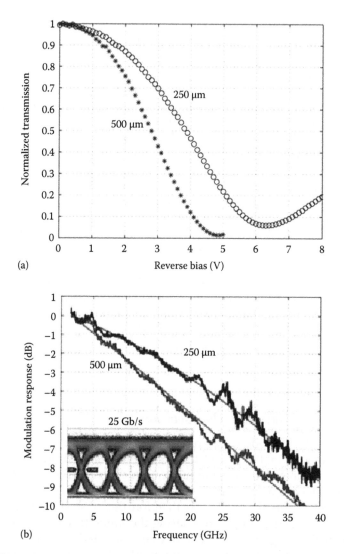

FIGURE 7.23
(a) Modulation efficiency of 250 and 500 μm devices at 1550 nm. (b) EO modulation response of 250 and 500 μm long devices. Inset: Optical eye diagram of the device at 25 Gb/s data transmission rate. (From Chen, H.-W. et al., *The 6th IEEE International Conference on Group IV Photonics*, San Francisco, CA, June 9–11, 2009. With permission.)

1600 nm under 3 V reverse bias. The advantage of bonding III–V layers over SiGe photodetectors is the uniform responsivity across the S, C, and L bands. From measurements of the output power from the Si output waveguide, the TE material absorption is estimated to be 1594 cm^{-1}. TM responsivity was measured to be 0.23 A/W which is, as expected, substantially lower than the TE responsivity due to the compressively strained quantum wells.

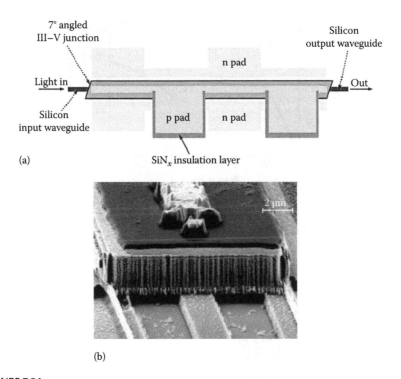

(a)

(b)

FIGURE 7.24

(a) Schematic top view of a hybrid Si photodetector. (b) SEM close-up view of a junction between the input Si waveguide and the hybrid photodetector. (From Park, H. et al., *Opt. Express*, 15, 6044, 2007. With permission.)

For the hybrid platform in which the Si waveguide is separated from the III–V active material by a layer of 200–300 nm low-index polymer (e.g., DVS-BCB bonding) or dielectric (e.g., SiO_2 covalent bonding), a synchronous coupler design is employed to allow the optical signal in the Si waveguide to transit up to the III–V absorbing region gradually as shown schematically in Figure 7.26a [P. Binetti 2009, P. R. A. Binetti 2009]. The synchronous coupler consists of a transparent InP membrane input waveguide on top of a Si waveguide buried in polymer or dielectric (insets in Figure 7.26). Coupling efficiency is calculated by overlapping the fundamental mode guided by the InP membrane input waveguide and the distributed optical field excited by the Si waveguide mode over the coupler length [P. Binetti 2009]. By properly choosing the InP waveguide dimensions, mode matching is achieved with the Si waveguide. A coupling efficiency of more than 80% is achieved with a fabrication tolerance of 30 nm for the bonding layer thickness, a deviation of 10 nm for the InP waveguide layer thickness, and 70 nm for the InP waveguide width from the optimum simulated geometry, as shown in Figure 7.26b.

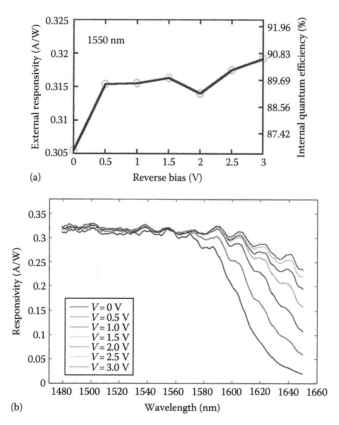

FIGURE 7.25
(a) TE responsivity with different biases at 1550 nm, (b) spectral response for TE polarization. (From Park, H. et al., *Opt. Express*, 15, 6044, 2007. With permission.)

The photodetector structure is built as an not intentionally doped (n.i.d.) 700 nm InGaAs absorption layer sandwiched between a highly p-doped 50 nm InGaAs contact layer and a highly n-doped 250 nm InP layer, which is also used for realizing the membrane waveguide, and has a footprint of $5 \times 10 \, \mu m^2$. A total detector thickness of 1 μm is chosen as a trade-off between device efficiency and speed. The structure shown in Figure 7.26 also allows the fabrication of laterally tapered membrane waveguides, which provide an increase in alignment tolerance between the waveguides without additional processing steps [P. R. A. Binetti 2009].

The device DC characterization was performed by using a tunable laser source to couple λ = 1550 nm light into the Si waveguide and measure photocurrent generated in reverse-biased photodetectors. A low 1.6 nA dark current is obtained at −4 V bias [P. R. A. Binetti 2009]. Taking the losses of polarization controller, fiber connection, and fiber grating coupler into account, photocurrent as a function of input powers of 0, 25, 50 mW is presented in

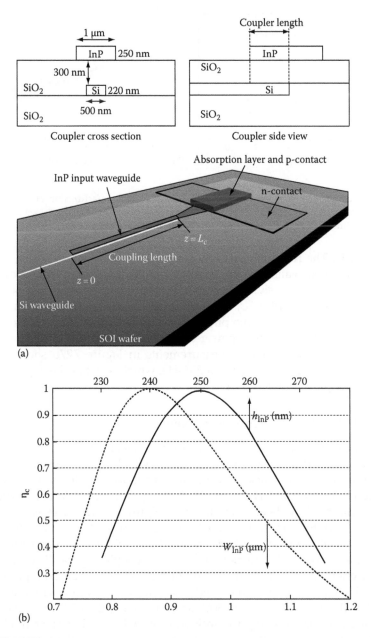

FIGURE 7.26
(a) Hybrid photodetector schematic structure. Insets: Front and side cross sections of the coupler. (From Binetti, P.R.A. et al., *The 6th IEEE International Conference on Group IV Photonics*, San Francisco, CA, June 9–11, 2009. With permission.). (b) Coupling efficiency as a function of the InP waveguide width (dashed line) and thickness (solid line). (From Binetti, P., *IEEE Photon. Tech. Lett.*, 21, 337, 2009. With permission.)

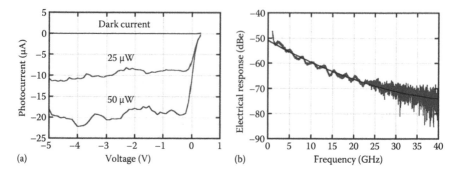

FIGURE 7.27

(a) Measured photocurrent for various optical input powers as a function of the detector applied bias voltage. (b) Detector RF frequency response: normalized transmission parameter S21. (From Binetti, P., *IEEE Photon. Tech. Lett.*, 21, 337, 2009. With permission.)

Figure 7.27a. The responsivity is thus calculated to be $R = 0.45\,\text{A/W}$, which is a conservative value, as the grating coupler maximum efficiency was assumed. Such responsivity corresponds to a quantum efficiency $\eta = 35\%$, which includes the efficiency of the InP membrane coupler and the internal quantum efficiency of the pin-detector itself. Consistent response from -0.3 to $-4\,\text{V}$ and input optical power-dependent linearity are noticed as well. Optical-to-electrical dynamic measurements in Figure 7.27b show a 3 dB cutoff frequency response of about 33 GHz, which matches the expectations quite well [P. R. A. Binetti 2009].

7.3.3.2 Hybrid Si Normal Incidence Photodetectors

A normal incidence type of hybrid Si photodetector is also fabricated on DVS-BCB bonded platform [Roelkens 2005], which is also particularly useful in the situation of thick bonding interfacial layer up to ~µm. A surface grating is used to convert the in-plane waveguide signal to one vertically incident on the III–V detector as shown in Figure 7.28, though some portion of light is reflected back toward the Si substrate, most of which is lost.

The exact location of the SOI grating inside the cavity determines its directionality (being the fraction of the power coupled upward) and its coupling strength (determining the length of the grating). As an example, the Si waveguide layer is 220 nm thick and a second-order grating with a grating period of 610 nm, a duty cycle of 50%, and an etch depth of 50 nm is assumed. Simulation results showing the influence of BCB bonding layer thickness and buried oxide layer thickness on the fraction of absorbed power for a 50 µm long detector at 1.55 µm are shown in Figure 7.29a. Optimal device operation is achieved for a BCB bonding layer thickness of 3 µm and a 1.4 µm thick buried oxide layer. The influence of device length on absorbed power in the photodetector for these optimal parameters is shown in Figure 7.29b. Close to 65% of the input optical power can be absorbed [Roelkens 2005].

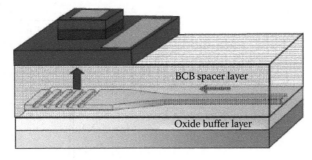

FIGURE 7.28
Coupling scheme for III–V photodetectors bonded to SOI waveguide circuitry. (From Roelkens, G. et al., *Opt. Express*, 13, 10102, 2005. With permission.)

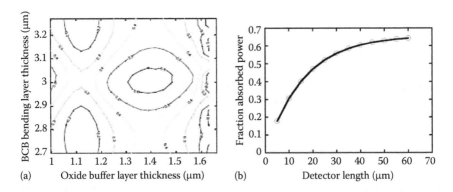

FIGURE 7.29
(a) Influence of BCB bonding thickness and SiO_2 buffer layer thickness on detector efficiency, and (b) the influence of device length in the optimum case. (From Roelkens, G. et al., *Opt. Express*, 13, 10102, 2005. With permission.)

The wavelength dependence of the absorbed power fraction for the optimal device parameters depends both on the wavelength-dependent properties of the grating coupler used to diffract the light and on the wavelength dependence of the absorption coefficient of the transferred InGaAs absorption layer. The absorbed power fraction, the spurious reflection back into the SOI waveguide, and the wavelength dependence of the absorption coefficient are shown in Figure 7.30. A strong reflection peak around 1650 nm is visible. This is due to the second-order Bragg reflection of the grating. Efficient detection up to a wavelength of 1600 nm is achieved [Roelkens 2005].

In addition to the hybrid integrated III–V-on-Si photodetectors, Ge, as a material with <4% lattice mismatch to Si, has been monolithically grown on Si. The high quality has recently enabled high-performance Ge/Si PIN photodetectors and avalanche photodetectors (APDs) to be realized. Many device figures of merit have outperformed their III–V counterparts, and new underlying device physics has been studied for the first time. Thus, we

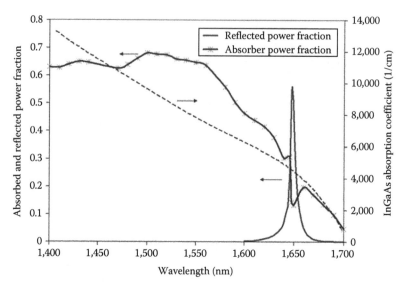

FIGURE 7.30
Wavelength dependence of the absorbed power fraction in the detector and reflected power back into the SOI waveguide. A 50 mm long device and optimum SiO_2 buffer layer thickness and BCB bonding layer thickness are assumed. The wavelength dependence of the InGaAs absorption coefficient is also shown. (From Roelkens, G. et al., *Opt. Express*, 13, 10102, 2005. With permission.)

believe that it is necessary to include them here although this book primarily focuses on III–V CS integration on Si.

7.3.3.2.1 Monolithic Ge/Si Avalanche Photodetectors

APDs are an attractive choice to detect weak optical signals because their high sensitivities achieved by using internal gain can be used to reduce the noise from a first-stage electrical amplifier. APDs integrated on Si could be crucial components, in particular for Si optical interconnect applications (inter-, intra-chip, and rack-to-rack), where the coded optical signal in the Si waveguide can be very weak. It helps to loosen the requirement for losses introduced by the waveguide, modulators, (de)multiplexers, etc., and requires low laser output power, which is also desirable for low power-consumption operation. It might be possible to avoid on-chip amplifiers as well. For conventional telecommunications, Si APDs can be very competitive due to their CMOS compatibility if comparable performance to III–V counterparts is attainable.

The APDs are typically characterized through two functions that can be optimized separately. The first function is the absorption of light and its conversion to an electrical signal. The second function is the amplification of this electrical signal through the avalanche process. Small bandgap material (e.g., Ge) integrated on Si extends its functionality to telecommunication wavelength regime. The avalanche multiplication process that generates the internal gain determines the frequency performance of the APD due to the avalanche buildup time, which is related to the effective ratio of the electron and hole ionization coefficients k_{eff} [Emmons 1967]. The inherent advantage in the large asymmetry of electron and hole ionization coefficients in Si ($k_{eff} < 0.1$) makes it an ideal material for APDs, superior to conventional InP ($k_{eff} = 0.4–0.5$) and InAlAs ($k_{eff} = 0.2$) [McIntyre 1972]. As a result, APDs with an InGaAs [Hawkins 1997] or Ge [Kang 2008b] absorption layer and a Si multiplication layer can achieve very good performance with high quantum efficiency and low noise.

Figure 7.31a illustrates the cross section of a normal-incident illuminated Ge/Si APD with a separate-absorption-charge-multiplication (SACM) structure. Thin epitaxially grown intrinsic Ge and Si are separated by a p-doped Si charge layer, which controls the electric field distribution in the device [Kang 2008b]. Details on device fabrication can be found in Refs. [Kang 2008b, Kang 2008a]. Figure 7.31b is the SEM cross-sectional image for a 30 μm diameter SACM APD. It is noted that a slight modification in Si and Ge dimensions allows implementation of the same idea as an in-plane waveguide type of Ge/Si APD [Shi 2004, Kang 2009], which is favorable for inter-chip optical interconnects.

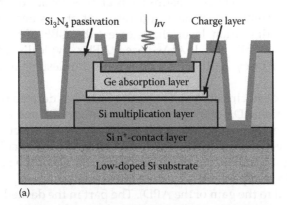

(a)

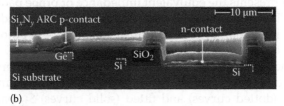

(b)

FIGURE 7.31

(a) Schematic configuration (From Dai, D. et al., *IEEE J. Sel. Top. Quant. Electron.*, 2009. With permission.) (b) cross-sectional SEM image of the Ge/Si SACM APD device. (From Kang, Y. et al., *Nat. Photon.*, 3, 59, 2008. With permission.)

These SACM APDs demonstrated a punch-through voltage of about –22 V, with a responsivity of 5.88 A/W at 1310 nm [Kang 2008b]. The breakdown voltage V_{bd} occurs at –25 V, corresponding to a dark current of 10 μA. The breakdown voltage thermal coefficient (defined as $\delta = (\Delta V_{bd}/V_{bd})/\Delta T$, where ΔV_{bd} and ΔT are the increment of V_{bd} and temperature, respectively) is 0.05%°C^{-1} over a temperature range from 200 to 380 K [Kang 2008b]. This value is about 70%, and 30%–50% that of the InAlAs- and InP-based APDs with similar layer structures [Ma 1995, Rouvie 2008], respectively. This decreased thermal sensitivity is another major advantage of Si-based APDs, as it will facilitate temperature stabilization of Ge/Si APDs [Kang 2008b].

The product of multiplication gain and bandwidth represents a key figure of merit to determine the performance of APDs. Extremely high gain-bandwidth-product (GBP) at high bias voltages has been observed on Ge/Si APDs experimentally by multiple groups [Kang 2008b, Zaoui 2009, Shi 2007]. Figure 7.32a shows the GBP up to 845 GHz (gain = 65 and a 3 dB bandwidth = 13 GHz) of the APD under different input powers [Zaoui 2009]. The GBP is proportional to the gain increase at low gain values due to the constant parasitic limited bandwidth. As the gain increases, the bandwidth drops due to the multiplication time, and the GBP saturates. Beyond the gain peak, the GBP starts to increase again dramatically because of bandwidth enhancement [Dai 2009a] and the relatively slow decrease in gain. The highest GBP, 845 GHz, is measured in high bias voltage operation where space charge current is dominant and resonant effects become apparent. For similar multiplication layer thicknesses, the GBP of Si-based APDs surpasses that of InP- and InAlAs-based APDs (typically 100–300 GHz) [Yasuoka 2003, Kinsey 2001], which represents a milestone in the advance of Si photonics.

A valid equivalent circuit model in Figure 7.32b was derived recently from transport equations [Dai 2009a] to explain the high GBP and address interesting experimental phenomena, such as the resonance effect when SACM Ge/Si APDs are biased close to breakdown voltage. In the circuit model, the RC input circuit results from the transit time roll-off of the absorber. The avalanche gain is represented by the source current $I_{in} = gI_t$ (where g is a constant related to the gain of the APD). The part in the dashed rectangle in Figure 7.32b represents the equivalent impedance of probe pads and cables. The Ge/Si APD is represented by an RLC-circuit (R_A, L_A, and C_A), similar to the impact ionization avalanche transit-time (IMPATT) diode structure [Dai 2009b, Shi 2007]. R_l results from the leakage of the diode.

By fitting the experimentally measured S_{22} value, parameters for the elements included in the equivalent circuit can be extracted. Figure 7.33b shows the measured (dotted curves) and fitted (solid curves) S_{22} parameters for an 80 μm diameter device at $V_b = -26$ V (corresponding $I = 5.36$ mA) with an optical power of –14 dBm. Excellent agreement between experimental data and model is achieved. It is noted that nearly constant products of $L_A \times I$ are obtained for various bias voltages. It indicates that the inductance L_A

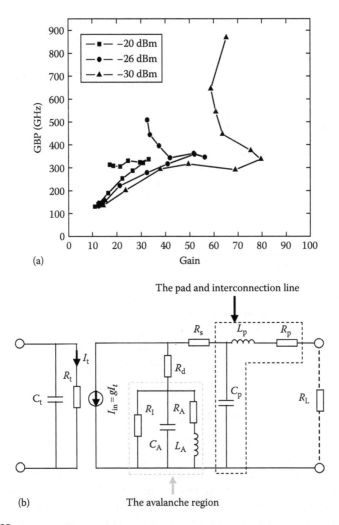

(a)

(b)

FIGURE 7.32
(a) Calculated GBP versus gain under input power of −20, −26, and −30 dBm at 1310 nm. (From Zaoui, W.S. et al., *Opt. Express*, 17, 12641, 2009. With permission.) (b) The equivalent circuit of Ge/Si SACM APD. (From Dai, D. et al., *IEEE J. Sel. Top. Quant. Electron.*, 2009. With permission.)

is almost inversely proportional to the current density, which is similar to the characteristic of an IMPATT diode in the circuit model, but has not been seen in III–V APDs previously. Having the equivalent circuit extracted, the device frequency response can be modeled as well. Figure 7.33a shows the comparison of experiment and modeling for the same parameters. The parameters for the part of transit-time circuit are $g = 150$, $R_t C_t = 26.0\,\Omega \cdot \text{pF}$, and the primary responsivity is about 0.55 A/W. A clear peak enhancement occurs at the frequency where the inductance resonates the capacitance and

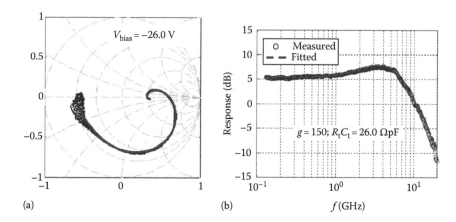

FIGURE 7.33

(a) The measured and fitted (a) reflection coefficients and (b) frequency response of Ge/Si APD under –14 dBm optical illumination and –26 V bias.

the RF current is delivered to the load [Bowers 2009]. Such a peak enhancement is beneficial to increase the bandwidth, contributing to the extremely high GBP, and a rise of 3 dB or less does not cause too much eye closure in a digital system [Dai 2009b]. This resonant behavior has been seen in waveguide SiGe APDs [Dai 2009b] and in shorter-wavelength APDs with 20% Ge [Shi 2007].

7.3.3.3 Monolithic Ge/Si PIN Photodetectors

The same Ge-on-Si (including SOI) monolithic integration technique has also been applied to fabricate PIN photodetectors. In order to compensate the relative small absorption coefficient of Ge at 1550 nm, Intel designed waveguide-type PIN photodetectors as shown in Figure 7.34. By increasing the detector length, high responsivity of Ge waveguide detectors can occur out to 1550 nm. The detector speed can also be simultaneously optimized by using a thin Ge layer to reduce transit time limitations. The overall reduction in area and capacitance in going from a normal incidence to a waveguide-based photodetector helps to improve receiver sensitivity and enable greater than 20 GHz operation [Yin 2007].

Figure 7.35a shows the Ge/Si PIN photodetector responsivity for ~126 μW input power as a function of wavelength. The Ge and Si thicknesses and Si waveguide rib etch are 1.3, 1.5, and 0.6 μm, respectively [Yin 2007]. Two device areas (length × width) are selected to be A: 7.4 × 50 μm² and B: 4.4 × 100 μm². Their responsivity values are measured to be 0.89 and 1.16 A/W at –2 V, respectively, corresponding to quantum efficiencies of 71% and 93% at 1550 nm. A flat responsivity was visible for wavelengths out to 1560–1570 nm, beyond the absorption edge of bulk Ge due to the tensile-strain-induced bandgap narrowing [Jifeng 2005]. Their responsivities at 1610 nm

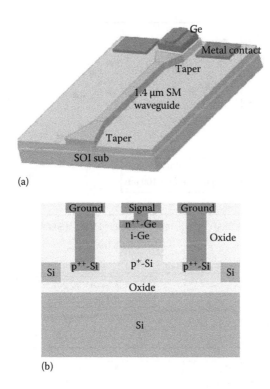

FIGURE 7.34

(a) Schematic layout for the Ge detector integrated with a passive waveguide. (b) Cross-sectional schematic of the Ge/Si n-i-p waveguide photodetector. (From Yin, T. et al., *Opt. Express*, 15, 13965, 2007. With permission.)

are 0.34 and 0.41 A/W at −2 V, respectively. The dark currents of A and B at the same bias are 169 and 267 nA, respectively [Yin 2007].

The device frequency response in Figure 7.35b is measured by launching a modulated optical signal at 1550 nm with an average power of 278 µW into the waveguide detector. The optical bandwidths of detectors A and B are 31.3 and 29.4 GHz, respectively, at −2 V. The corresponding electrical (power) bandwidths are 26 and 24.1 GHz. As shown in Figure 7.35b inset, detectors A and B still achieve optical bandwidths of 15.7 and 10 GHz at 0 V bias, respectively. It can be seen that the bandwidths decrease with increasing device area due to increased RC delay [Yin 2007].

Furthermore, power handling and temperature stability are also intrinsically better than in III–V-only detectors. High current operation in a photodiode is usually limited by (1) space charge effects and (2) thermal effects [Williams 1999]. The former can be minimized by proper device design and higher voltages, but the latter is largely related to material properties. Si has a much larger thermal conductivity (1.5 W/cm·K), which is multiple times better than conventionally used III–V materials, such as InP (0.68 W/cm·K),

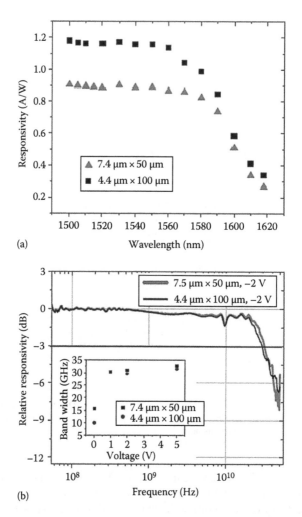

FIGURE 7.35
(a) Responsivity versus wavelength and (b) frequency response for detector 7.4×50 μm, and detector 4.4×100 μm, respectively, at −2 V. Inset in (b) shows optical bandwidth versus supply voltage for both detectors. (From Yin, T. et al., *Opt. Express*, 15, 13965, 2007. With permission.)

InGaAs (0.05 W/cm · K), and InGaAsP (0.05 W/cm · K) [Adachi 2007]. For an InP-based uni-traveling-carrier photodetector, failure can occur for a dissipated electrical power of 300 mW. Ge/Si photodetectors should be capable of handling higher input power fundamentally [Ramaswamy 2009].

Figure 7.36a shows the saturation characteristics of device A (7.4 × 50 μm²) at 1550 nm for different reverse bias values. It can be seen that at a lower bias, the output current saturates faster because of carrier screening effects [Williams 1999]. A maximum photocurrent of 125 mA under 8 V of reverse bias was observed. This corresponds to 1 W of electrical power dissipation in

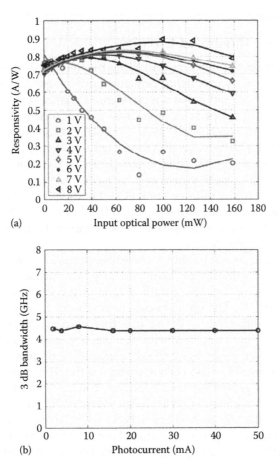

(a)

(b)

FIGURE 7.36

(a) Photocurrent as a function of optical power at different reverse bias voltages and (b) 3 dB bandwidth as a function of photocurrent at a fixed reverse bias (5 V) for a $7.4 \times 500\,\mu m$ device. (From Ramaswamy, A. et al., *The 6th IEEE International Conference on Group IV Photonics*, San Francisco, CA, June 9–11, 2009. With permission.)

the device [Yin 2007]. The 3 dB bandwidth at high input optical power levels were also studied [Ramaswamy 2009]. Although the bandwidth decreases at high input optical power levels as expected [Williams 1992], fairly constant 3 dB bandwidth of ~4.38 GHz up to 50 mA photocurrent is noticed at a bias of −5 V, indicating strong heat dissipating efficiency. Further improvement depends on circumvention of the thermal barrier of the buried oxide layer in SOI [Sysak 2007].

The selected hybrid and monolithic integrated Si photodetectors here all show impressive performance in certain aspects, though they are in their first generation. With further improvement in design and fabrication, they should find commercial applications.

7.4 Si Photonic Integrated Circuits

The driving force of Si photonics (including the hybrid III–V-on-Si photonics discussed here) are its promising prospects for integration with Si microelectronics in parallel and on an unprecedentedly large scale [Liang 2009a].

While almost all key optical components in the Si toolbox have been demonstrated using the hybrid III–V-on-Si approach, one of this techniques' key drawbacks is that devices requiring multiple III–V bandgaps cannot be realized without complex epitaxial base structures or multiple bonding steps. One solution to this problem involves using a quantum well intermixing (QWI) technology to shift the as-grown QW bandgap across the III–V wafer before bonding. Using the bandgap shifted III–V material, components such as EAM and low loss passive regions can be integrated together with hybrid Si DBR, DFB, or tunable lasers.

The QWI technique used combines selective removal of an InP buffer with implant-enhanced intermixing [Skogen 2003]. An outline of the as-grown III–V QWI base structure along with details of the intermixing process is shown in Figure 7.37. The QWI process begins with a blanket phosphorus implant into the patterned III–V base structure as shown in Figure 7.37a. The implant is used to generate vacancies that act as a catalyst in the QWI process. A 400 nm thick plasma-enhanced chemical vapor deposition (PECVD) SiN_x dielectric is used to mask certain regions of the III–V during the implant to preserve the as-grown material bandgap. The implant energy (200 kV) is chosen to generate vacancies that are localized in the InP buffer layer without penetrating into the QW active region. Following the implant, the vacancies generated during the implant are diffused through the QWs and barriers via a rapid thermal anneal (RTA) (Figure 7.37b). The vacancy diffusion causes

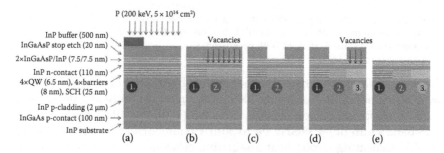

FIGURE 7.37
Overview of the hybrid laser QWI process. The three bandgaps realized are numbered 1, 2, and 3: (a) Implantation of P into InP buffer with SiN_x mask to preserve the as-grown bandgap. (b) Diffusion of vacancies through QWs and barriers for bandgap 2. (c) Removal of InP buffer layer to halt intermixing. (d) Diffusion of vacancies via RTA for bandgap 3. (e) Removal of InP buffer layer and InGaAsP stop-etch layer before bonding. (From Sysak, M.N. et al., *Opt. Express*, 16, 12478, 2008. With permission.)

atomic interdiffusion between the wells and barriers, modifying their potential profile, and hence bandgap. Once a desired bandgap has been reached, the InP buffer layer can be selectively etched to remove the vacancies from the intermixing process (Figure 7.37c). Additional RTA steps are then used to continue shifting the bandgap in regions where the InP buffer remains (Figure 7.37d and e).

Using an RTA temperature of 725°C, a 90 nm PL shift from the as-grown material bandgap can be achieved after a 330 s anneal. Halting the anneal after 45 s and removing the InP buffer allows an intermediate bandedge to be defined for integrated EAM (Figure 7.38a). For the standalone sampled-grating (SG) DBR, the two bandgaps employed have PL peaks at 1520 nm (in the gain regions) and at 1440 nm (mirror regions). For the integrated SGDBR-EAM schematically shown in Figure 7.38c, the gain region bandgap is at 1520 nm, the mirror bandgap is at 1440 nm, and the EAM bandgap is at 1480 nm. Figure 7.38b shows PL spectra from the final three bandgaps in the SGDBR-EAM [Sysak 2008]. Good uniformity of the PL FWHM for all three bandgaps, indicates consistent material quality [Nie 2005].

Fabrication of the QWI sampled grating DBR lasers is divided into pre-bonding, bonding, and post-bonding steps. Prebonding includes QWI, patterning and etching first-order gratings into the bandgap-shifted III–V

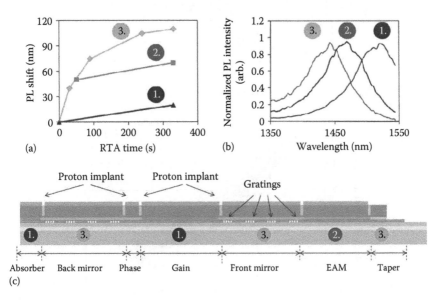

FIGURE 7.38
(a) PL shift as a function of RTA time for III–V regions that have been protected from the implant (region 1), partially intermixed (region 2), and fully intermixed (region 3) at 725°C. (b) Normalized PL spectra from the three bandgaps utilized in the SGDBR-EAM devices. (c) Schematic cross-sectional of a hybrid Si SGDBR-EAM shown with four front mirror and back mirror grating bursts. (From Sysak, M.N. et al., *Opt. Express*, 16, 12478, 2008. With permission.)

passive regions, and patterning shallow-etched waveguides into the SOI. Following prebonding, the SOI and III–V wafers are bonded together. Then, standard III–V fabrication is done including patterning and etching III–V mesas, depositing n and p metal contacts, proton implantation, and probe metallization. Modulators include an additional lithography to etch away the n-InP contact layer to minimize parasitic capacitance. Proton implantation is used to electrically isolate the laser sections and provide current confinement. After backside processing, the laser waveguides are diced at a 7° angle, polished, and antireflection (AR) coated. Additional descriptions of the fabrication process can be found in Ref. [Sysak 2008].

SGDBR-EAM cw *LI* characteristics are shown in Figure 7.39a. cw operation up to 45°C is achieved with output power up to 0.5 mW at 10°C.

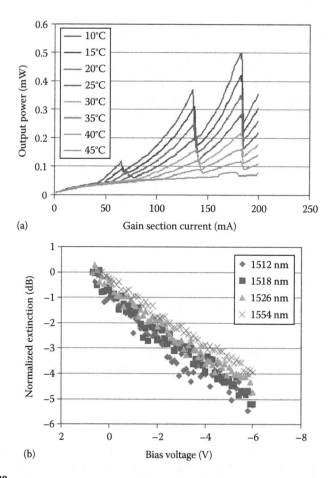

(a)

(b)

FIGURE 7.39

(a) cw *LI* characteristic of SGDBR lasers, (b) DC extinction characteristics of integrated EAM in the SGDBR-EAM with a 2.5 μm wide Si waveguide. (From Sysak, M.N. et al., *Opt. Express*, 16, 12478, 2008. With permission.)

Discontinuities in the output power characteristics are a result of temperature-induced cavity mode hops [Amann 1989]. DC extinction characteristics from the integrated EAM are shown in Figure 7.39b. Greater than 5 dB extinction is achieved at −6 V reverse bias depending on the wavelength of operation. Shorter wavelengths show more efficient operation than longer wavelengths due to the proximity between the modulator bandedge and the operating wavelength. The bandwidth of the integrated modulators is 2 GHz with a −6 V bias [Sysak 2008].

In the situation where different components require different III–V epitaxial structure (e.g., different QW design plus QW-free region), nonplanar bonding [Geske 2001] or bonding individual III–V dies with different designated structure [Fedeli 2008] is inevitable. It is also the solution to push individual component to their optimal performance on top of fulfilling the requirement of low power consumption, small footprint, high speed, and low fabrication cost. Figure 7.40 shows multiple InP epitaxial dies transferred to a 200 mm in diameter SOI wafer after the InP substrate is removed.

Without a doubt, III–V-on-Si hybrid integration technology remedies the fundamentally poor light emission efficiency in Si. It stands out in its full CMOS compatibility and robustness for a diversity of devices. Marching along with the successful and successive demonstration of individual hybrid and monolithic integrated components, Si microelectronic giants, such as Intel, IBM, HP, etc., have established their own vision to deploy the Si photonic platform [INTEL: http://techresearch.intel.com/articles/Tera-Scale/ 1419.htm, IBM: http://domino.research.ibm.com/comm/research_projects.nsf/pages/photonics.index.html, HP: http://www.hpl.hp.com/news/2008/oct-dec/photonics.html]. As shown in Figure 7.41a, one approach to achieve 1 Tb/s Si transmitter can be realized if signals from 25 hybrid DFB Si lasers are modulated by 25 40 Gb/s Si modulators and combined by a passive multiplexer [Rong 2009]. One step further is to realize a Si transceiver controlled by Si microelectronics from Intel's version in Figure 7.41b [http://techresearch.intel.com/articles/Tera-Scale/1419.htm]. Owing to the outstanding performance

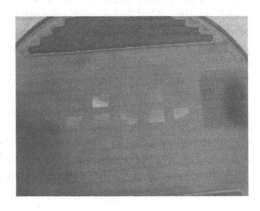

FIGURE 7.40
A 200 mm in diameter SOI wafer with InP-bonded dies after InP substrate removal. (From Fedeli, J.M. et al., *Adv. Opt. Technol. 2008*, Article ID 412518, 2008. With permission.)

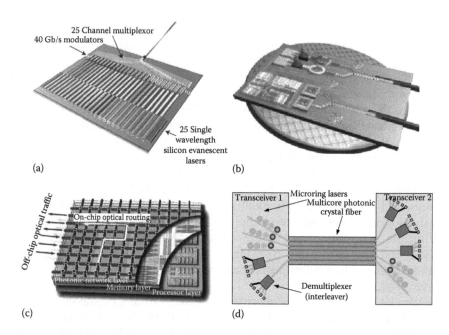

FIGURE 7.41
(a) A Si terabit transmitter with 25 single wavelength hybrid Si lasers externally modulated at 40 Gb/s. (b) Intel's version of CMOS-compatible Si photonics. (c) IBM's concept of 3D Si processor chip with optical IO layer featuring on-chip nanophotonic network. (d) HP's active optical cable based on hybrid integration technology. [IBM: http://domino.research.ibm.com/comm/ research_projects.nsf/pages/photonics.index.html, INTEL: http://techresearch.intel.com/ articles/Tera-Scale/1419.htm, HP: http://www.hpl.hp.com/news/2008/oct-dec/photonics. html]. (From Liang, D. et al., *Opt. Express*, 17, 20355, 2009. With permission.)

of monolithically integrated Ge/Si photodetectors (PIN and APD schemes), a compound integration that employs both III–V-on-Si hybrid and Ge-on-Si monolithic integration for optimal functionality is likely. Very recently, Intel has unveiled a new type of high-speed optical cable named "Light Peak." It is designed to connect electronic devices to each other and delivers high bandwidth starting at 10 Gb/s with the potential ability to scale to 100 Gb/s over the next decade [http://techresearch.intel.com/articles/None/1813.htm]. IBM has proposed a 3D-integrated chip consisting of several layers connected with each other with very dense and small pitch interlayer vias [http://domino. research.ibm.com/comm/research_projects.nsf/pages/photonics.index. html]. The lower layer is a processor itself with many hundreds of individual cores. A memory layer (or layers) will be bonded on top to provide fast access to local caches. At the top of the stack is the photonic layer with many thousands of individual optical devices (modulators, detectors, switches) as well as analog electrical circuits (amplifiers, drivers, latches, etc.). The key role of a photonic layer is not only to provide point-to-point broad bandwidth optical links between different cores and/or the off-chip traffic, but also to route this traffic with an array of nanophotonic switches [Vlasov 2008]. HP sees the potential in high-capacity Si active optical cables. A schematic of a possible cable configuration is shown in Figure 7.41d [Liang 2009c]. The optical engines consist of bundles of microring lasers that are coupled on a single waveguide. The hybrid Si lasers are designed to emit light at different wavelengths (with a spacing of ~5 nm) and can be directly modulated to create a 10 Gb/s on-off keying (OOK) signal. The modulated light from the lasers is coupled into one of the cores of a multicore photonic crystal fiber (PCF) [Taylor 2006, Taylor 2008]. PCFs together with vertical grating couplers [Taillaert 2002] allow high-density connections while reducing the cost of pigtailing the photonic circuit. At the other end of the cable, light is coupled onto a second PIC and sent to an interleaver-based demultiplexer [Song 2008] that separates the various wave-length components and sends them to hybrid detectors that recover the data. Each fiber core can therefore carry multiple (typically 4–5 for current active medium gain spectrum) 10 Gb/s channels in one direction. The fiber consists of up to 19 cores [Taylor 2008] that can be configured (at the hardware level) for upload or download depending on the traffic needs.

Regardless of which architecture will make the debut of practical hybrid Si PICs on a microelectronic chip, they all aim to eliminate bandwidth and distance limitations, allowing for radically new flexible architectures capable of processing data more efficiently. The hybrid and monolithically integrated optoelectronic components may even have applications beyond digital com-munications, including optical debugging of high-speed data, expanding wireless networks by transporting analog RF signals, and enabling lower-cost lasers for certain biomedical applications.

Acknowledgments

The authors thank Alexandra Fang, Brian Koch, Hui-Wen Chen, Hsu-Hao Chang, Hyundai Park, Jag Shah, Mario Paniccia, Matthew Sysak, Mike Haney, Ray Zanoni, Richard Jones, and Ying-Hao Kuo for valuable discus-sions, and Molly Piels for careful proofreading. We thank DARPA MTO, Intel, Hewlett Packard, and Rockwell Collins for supporting this research.

References

[Adachi 2007] Adachi, S. *Journal of Applied Physics* 102, 063502, 2007.

[Amann 1989] Amann, M.-C., S. Illek, C. Schanen et al. *Applied Physics Letters* 54, 2532–2533, 1989.

[P. Binetti 2009] Binetti, P., R. Orobtchouk, X. Leijtens et al. *IEEE Photonics Technology Letters* 21, 337–339, 2009.

[P. R. A. Binetti 2009] Binetti, P. R. A., X. J. M. Leijtens, T. de Vries et al. *The 6th IEEE International Conference on Group IV Photonics*, San Francisco, CA, June 9–11, 2009.

[Bowers 2009] Bowers, J. E. *The 36th International Symposium on Compound Semiconductors*, Santa Barbara, CA, 2009.

[Chang 2007] Chang, H.-H., A. W. Fang, M. N. Sysak et al. *Optics Express* 15, 11466–11471, 2007.

[Chao 2005] Chao, Y. L., Q. Y. Tong, T. H. Lee et al. *Electrochemical and Solid-State Letters* 8, G74–G77, 2005.

[H.-W. Chen 2008a] Chen, H.-W., Y.-H. Kuo, and J. E. Bowers. *IEEE Photonics Technology Letters* 20, 1920–1922, 2008.

[H.-W. Chen 2008b] Chen, H.-W., Y.-H Kuo, and J. E. Bowers. *Optics Express* 16, 20571–20576, 2008.

[Chen 2009a] Chen, H.-W., Y.-H. Kuo, and J. E. Bowers. *The 6th IEEE International Conference on Group IV Photonics*, San Francisco, CA, June 9–11, 2009.

[Chen 2009b] Chen, H.-W., Y. Kuo, and J. E. Bowers. *Chinese Optics Letters* 7, 280–285, 2009.

[L. Chen 2008] Chen, L., P. Dong, and M. Lipson. *Optics Express* 16, 11513–11518, 2008.

[Dai 2009a] Dai, D., H.-W. Chen, J. E. Bowers et al. *Optics Express* 17, 16549–16557, 2009.

[Dai 2009b] Dai, D., M. Rodwell, J. E. Bowers et al. *IEEE Journal of Selected Topics in Quantum Electronics*, 2009.

[Emmons 1967] Emmons, R. B. *Journal of Applied Physics* 38, 3705–3714, 1967.

[Fang 2008a] Fang, A. W., B. R. Koch, R. Jones et al. *IEEE Photonics Technology Letters* 20, 1667–1669, 2008.

[Fang 2008b] Fang, A. W., E. Lively, Y.-H. Kuo et al. *Optics Express* 16, 4413–4419, 2008.

[Fang 2008c] Fang, A. W., B. R. Koch, K.-G. Gan et al. *Optics Express* 16, 1393–1398, 2008.

[Fang 2006] Fang, A. W., H. Park, O. Cohen et al. *Optics Express* 14, 9203–9210, 2006.

[Fang 2007] Fang, A. W., H. Park, Y.-H. Kuo et al. *Materials Today* 10, 28–35, 2007.

[Fedeli 2008] Fedeli, J. M., L. Di Cioccio, D. Marris-Morini et al. *Advances in Optical Technologies 2008*, Article ID 412518, 2008.

[Fehly 2001] Fehly, D., A. Schlachetzki, A. S. Bakin et al. *IEEE Journal of Quantum Electronics* 37, 1246–1252, 2001.

[Friedrich 1992] Friedrich, E. E. L., M. G. Oberg, B. Broberg et al. *Journal of Lightwave Technology* 10, 336–340, 1992.

[Geske 2001] Geske, J., Y. L. Okuno, J. E. Bowers et al. *Applied Physics Letters* 79, 1760–1762, 2001.

[Gosele 1997] Gosele, U., M. Alexe, P. Kopperschmidt et al. *International Semiconductor Conference, CAS '97 Proceedings*, Sinaia, Romania, 1997.

[Gosele 1999] Gosele, U., Y. Bluhm, G. Kastner et al. *Journal of Vacuum Science & Technology A: Vacuum, Surfaces, and Films* 17, 1145–1152, 1999.

[Gosele 1998] Gosele, U. and Q. Y. Tong. *Annual Review of Materials Science* 28, 215–241, 1998.

[Gui 1999] Gui, C., M. Elwenspoek, N. Tas et al. *Journal of Applied Physics* 85, 7448–7454, 1999.

[Hawkins 1997] Hawkins, A. R., W. Wu, P. Abraham et al. *Applied Physics Letters* 70, 303–305, 1997.

IBM: [http://domino.research.ibm.com/comm/research_projects.nsf/pages/photonics.index.html] SOITEC: [http://soitec.com/en/about/]

INTEL: [http://techresearch.intel.com/articles/None/1813.htm]

INTEL: [http://techresearch.intel.com/articles/Tera-Scale/1419.htm]

HP: [http://www.hpl.hp.com/news/2008/oct-dec/photonics.html]

Laser Focus World: [http://www.laserfocusworld.com/articles/305700]

[Jifeng 2005] Jifeng, L., D. C. Douglas, W. Kazumi et al. *Applied Physics Letters* 87, 011110, 2005.

[Kang 2009] Kang, Y., M. Morse, and M. J. Paniccia. *The 6th IEEE International Conference on Group IV Photonics,* San Francisco, CA, June 9–11, 2009.

[Kang 2008a] Kang, Y., M. Zadka, S. Litski et al. *Optics Express* 16, 9365–9371, 2008.

[Kang 2008b] Kang, Y., H.-D. Liu, M. Morse et al. *Nature Photonics* 3, 59–63, 2008.

[Kinsey 2001] Kinsey, G. S., J. C. Campbell, and A. G. Dentai. *Photonics Technology Letters, IEEE* 13, 842–844, 2001.

[Koch 2008] Koch, B. R., A. W. Fang, H. N. Poulsen et al. *Optical Fiber Communication/ National Fiber Optic Engineers Conference 2008* (OFC/NFOEC 2008), San Diego, CA, 2008.

[Koch 2007] Koch, B. R., A. W. Fang, O. Cohen et al. *Optics Express* 15, 11225–11233, 2007.

[Kuo 2008] Kuo, Y.-H., H.-W. Chen, and J. E. Bowers. *Optics Express* 16, 9936–9941, 2008.

[Liang 2009a] Liang, D. and J. E. Bowers. *Electronics Letters* 45, 578–581, 2009.

[Liang 2008a] Liang, D. and J. E. Bowers. *Journal of Vacuum Science and Technology B* 26, 1560–1568, 2008.

[Liang 2009b] Liang, D., J. E. Bowers, D. C. Oakley et al. *Electrochemical Solid-State Letters* 12, H101–H104, 2009.

[Liang 2008b] Liang, D., A. W. Fang, H. Park et al. *Journal of Electronic Materials* 37, 1552–1559, 2008.

[Liang 2009c] Liang, D., M. Fiorentino, T. Okumura et al. *Optics Express* 17, 20355–20364, 2009.

[Liang 2009d] Liang, D., M. Fiorentino, D. T. Spencer et al. *The 6th IEEE International Conference on Group IV Photonics,* San Francisco, CA, June 9–11, 2009.

[Liu 2007] Liu, A., L. Liao, D. Rubin et al. *Optics Express* 15, 660–668, 2007.

[Liu 2008] Liu, L., J. Van Campenhout, G. Roelkens et al. *Optics Letters* 33, 2518–2520, 2008.

[Ma 1995] Ma, C. L. F., M. J. Deen, L. E. Tarof et al. *IEEE Transactions on Electron Devices* 42, 810–818, 1995.

[Maruyama 2006] Maruyama, T., T. Okumura, S. Sakamoto et al. *Optics Express* 14, 8184–8188, 2006.

[Maszara 1988] Maszara, W. P., G. Goetz, A. Caviglia et al. *Journal of Applied Physics* 64, 4943–4950, 1988.

[McIntyre 1972] McIntyre, R. J. *IEEE Transactions on Electron Devices* 19, 703–713, 1972.

[Nie 2005] Nie, D., T. Mei, H. S. Djie et al. *Journal of Vacuum Science and Technology B* 23, 1050–1053, 2005.

[Niklaus 2001] Niklaus, F., H. Andersson, P. Enoksson et al. *Sensors and Actuators A: Physical* 92, 235–241, 2001.

[Niklaus 2006] Niklaus, F., R. J. Kumar, J. J. McMahon et al. *Journal of the Electrochemical Society* 153, G291–G295, 2006.

[Park 2005] Park, H., A. W. Fang, S. Kodama et al. *Optics Express* 13, 9460–9464, 2005.

[Park 2007] Park, H., A. W. Fang, R. Jones et al. *Optics Express* 15, 6044–6052, 2007.

[Park 2008] Park, H., A. W. Fang, D. Liang et al. *Advances in Optical Technologies* 2008, Article ID 682978, 2008.

[Pasquariello 2002] Pasquariello, D. and K. Hjort. *IEEE Journal of Selected Topics in Quantum Electronics* 8, 118–131, 2002.

[Pasquariello 2001] Pasquariello, D., M. Camacho, F. Ericsson et al. *Japanese Journal of Applied Physics* 40, 4837–4844, 2001.

[Ramaswamy 2009] Ramaswamy, A., N. Nunoya, L. A. Johansson et al. *The 6th IEEE International Conference on Group IV Photonics*, San Francisco, CA, June 9–11, 2009.

[Roelkens 2006] Roelkens, G., J. Brouckaert, D. Van Thourhout et al. *Journal of the Electrochemical Society* 153, G1015–G1019, 2006.

[Roelkens 2007] Roelkens, G., J. Van Campenhout, J. Brouckaert et al. *Materials Today* 10, 36–43, 2007.

[Roelkens 2005] Roelkens, G., J. Brouckaert, D. Taillaert et al. *Optics Express* 13, 10102–10108, 2005.

[Roelkens 2009] Roelkens, G., L. Liu, D. Liang et al. *Laser & Photonics Reviews*, 1–29, 2010.

[Rong 2009] Rong, H. and M. Paniccia. CLEO/EQEC, Munich, Germany, 2009.

[Roth 2007] Roth, J. E., O. Fidaner, R. K. Schaevitz et al. *Optics Express* 15, 5851–5859, 2007.

[Rouvie 2008] Rouvie, A., D. Carpentier, N. Lagay et al. *IEEE Photonics Technology Letters* 20, 455–457, 2008.

[Sasaki 2001] Sasaki, J., M. Itoh, T. Tamanuki et al. *IEEE Transactions on Advanced Packaging* 24, 569–575, 2001.

[Shi 2007] Shi, J. W., Y. S. Wu, Z. R. Li et al. *IEEE Photonics Technology Letters* 19, 474–476, 2007.

[Shi 2004] Shi, J.-W., Y.-H. Liu, and C.-W. Liu. *Journal of Lightwave Technology* 22, 1583, 2004.

[Skogen 2003] Skogen, E. *Quantum Well Intermixing for Wavelength Agile Photonic Integrated Circuits*, University of California, Santa Barbara, CA, 2003.

[Song 2008] Song, J., Q. Fang, S. H. Tao et al. *Optics Express* 16, 8359–8365, 2008.

[Spuesens 2009] Spuesens, T., L. Liu, T. de Vries et al. *The 6th IEEE International Conference on Group IV Photonics*, San Francisco, CA, June 9–11, 2009.

[Sysak 2008] Sysak, M. N., J. O. Anthes, J. E. Bowers et al. *Optics Express* 16, 12478–12486, 2008.

[Sysak 2007] Sysak, M. N., H. Park, A. W. Fang et al. *Optics Express* 15, 15041–15046, 2007.

[Taillaert 2002] Taillaert, D., W. Bogaerts, P. Bienstman et al. *IEEE Journal of Quantum Electronics* 38, 949–955, 2002.

[Tan 2003] Tan, C. S., A. Fan, K. N. Chen et al. *Applied Physics Letters* 82, 2649–2651, 2003.

[Taylor 2006] Taylor, D. M., C. R. Bennett, T. J. Shepherd et al. *Electronics Letters* 42, 331–332, 2006.

[Taylor 2008] Taylor, D.M. and T.J. Shepherd, (personal communication, 2008).

[Tong 2006] Tong, Q.-Y., G. Fountain, and P. Enquist. *Applied Physics Letters* 89, 042110–042112, 2006.

[Tong 1998] Tong, Q.-Y. and U. Gösele. *Semiconductor Wafer Bonding: Science and Technology*, John Wiley & Sons Inc., New York, 1998.

[Tong 2004] Tong, Q. Y., Q. Gan, G. Hudson et al. *Applied Physics Letters* 84, 732–734, 2004.

[Van Campenhout 2008] Van Campenhout, J., L. Liu, P. R. Romeo et al. *IEEE Photonics Technology Letters* 20, 1345–1347, 2008.

[Van Campenhout 2007] Van Campenhout, J., P. Rojo Romeo, P. Regreny et al. *Optics Express* 15, 6744–6749, 2007.

[Vlasov 2008] Vlasov, Y., W. M. J. Green, and F. Xia. *Nature Photonics* 2, 242–246, 2008.

[Williams 1999] Williams, K. J. and R. D. Esman. *Journal of Lightwave Technology* 17, 1443–1454, 1999.

[Williams 1992] Williams, K. J. and R. D. Esman. *Electronics Letters* 28, 731–732, 1992.

[Xu 2005] Xu, Q., B. Schmidt, S. Pradhan et al. *Nature* 435, 325–327, 2005.

[Yan 2009] Yan, L., L. Yan, F. Zhongchao et al. *Journal of Optics A: Pure and Applied Optics* 085002, 2009.

[Yariv 2007] Yariv, A. and X. Sun. *Optics Express* 15, 9147–9151, 2007.

[Yasuoka 2003] Yasuoka, N., H. Kuwatsuka, and M. Makiuchi. *Proceedings of the 16th IEEE Annual Meeting of LEOS*, Tucson, AZ, 2003.

[Yin 2007] Yin, T., R. Cohen, M. M. Morse et al. *Optics Express* 15, 13965–13971, 2007.

[Yuya 2008] Yuya, S., M. Tetsuya, Y. Hideki et al. *Applied Physics Letters* 92, 071117, 2008.

[Zaoui 2009] Zaoui, W. S., H.-W. Chen, J. E. Bowers et al. *Optics Express* 17, 12641–12649, 2009.

8

Reliability of III–V Electronic Devices

Anthony A. Immorlica Jr.

CONTENTS

8.1 Introduction

Integration of compound semiconductor (CS) devices with silicon (Si) integrated circuits (ICs) offers an exciting opportunity to meld the advantages of superior electronic and optical properties of III–Vs with the high integration level, low cost, and pervasive infrastructure of Si technologies. Some recent developments, namely, epitaxial growth of GaN on Si substrates, and heterogeneous integration of GaAs, InP, or InSb devices on Si ICs, will soon be coming to fruition. The integration of established technologies, however, carries with them the threat of upsetting finely honed processes, which can compromise reliability. Large bodies of knowledge of reliability of Si and GaAs technologies exist in the literature and, as new generations and processes evolve, investigation of reliability and failure mechanisms continues to be ongoing. It is not the intent of this chapter to review the state of the art in Si or GaAs reliability. Instead, we will put into perspective what is important in determining device reliability and discuss in an illustrative manner the status of GaN reliability, and potential issues in integration of dissimilar technologies on a common platform.

8.2 Reliability Defined

Discussions of semiconductor reliability generally focus on measurement of lifetimes and wear-out mechanisms. Such a focus, usually manifested by a figure of merit defined as the median time to failure (MTTF), can be quite misleading for reasons that will hopefully become apparent in this chapter. If we view the definitions of "reliability" and the word's roots in *Webster's New Collegiate Dictionary*, we find the following:

- Rely—*to have CONFIDENCE based on EXPERIENCE*
- Reliably—1. *Suitable or fit to be relied on—DEPENDABLE*
- Reliability—2. *Extent to which an experiment, test, or measuring procedure yields the same results on REPEATED trials.*

We can define the reliability of a semiconductor, or for that matter any other product, as the *PROBABILITY*—a bet—that a device

- Of a specific technology
- From a particular manufacturer
- Using a given fabrication and packaging process
- With an agreed set of operating parameters
- In a specified environment

will perform as desired and in the same manner over many instances of use for at least a specified time.

Since it is generally not possible to test the reliability of a given device, predictions are made based on tests of a similar family of devices and the result is attributed to the actual device used in the application. As we know from any *Wall Street* prospectus "Past performance is no guarantee of future performance," we must associate any measurement of reliability with a *confidence level*, which can be statistically determined from the sampling and measurement procedures. Thus, it is vitally important for the user to understand the details of the measurements used by the manufacturer for the determination of reliability assertions and to insure that the test conditions are reasonably close to or exceed in severity the contemplated conditions in the actual application of the device.

Reliability figure of merit is often expressed as a median time to failure (MTTF) or mean time between failures (MTBF). While these figures are often used interchangably they are distinctly different metrics and are not identical. MTTF is a *median time* estimated from laboratory trials and is meant to determine the wear-out time for the device under a given, controlled set of conditions, whereas MTBF is the observed mean, or *average time* between failures in the field under actual use conditions. MTTF is the most often quoted figure, and for mature semiconductor technologies is typically millions of hours. On the other hand, while MTTF is a prediction of the normal wear-out time, MTBF accounts for *all* possible failure modes during the device operational life.

8.3 Determination of MTTF

The rates of many chemical processes that are associated with semiconductor failure, such as diffusion, metal migration, etc. are exponential with temperature. Since wear-out times are on the order of 1×10^6 h or more (i.e., ~a *century*) at typical maximum operating temperatures, testing needs to be carried out under highly accelerated conditions in order to acquire failure statistics in a reasonable period of time.

For temperature-accelerated testing, an Arrhenius model is typically applied,

$$\text{MTTF} = A \exp\left(\frac{E_a}{kT}\right) \tag{8.1}$$

where

E_a is an activation energy associated with the failure mechanism
T is the device temperature
k is Boltzmann's constant
A is a constant

Many semiconductor failure mechanisms have been shown to follow the Arrhenius model with activation energies ranging from 0.5 to 2.0 eV. For example, failure modes such as gate sinking and ohmic contact degradation in GaAs field-effect transistors (FETs) ideally follow the Arrhenius model and activation energies for these and a host of other mechanisms have been published. Some failure modes are activated not only by temperature but also by other operational or environmental factors. For example, electromigration is strongly dependent on current density and there is evidence that gate sinking may also be accelerated by electric field. Failure modes, which are dependent on electric field, current density, and even radiation effects, are often ameliorated by specification on limits of operation by the manufacturer. These failure modes typically involve power law dependencies while degradation with temperature is exponential. Thus, device temperature has been the primary predictor of semiconductor wear-out.

A plot of Equation 8.1 for X-band GaN HEMTs [Mittereder et al. 2008] is shown in Figure 8.1, in which each datum is the MTTF of a number of devices tested under identical conditions. In order to determine the lifetime at a particular *use* temperature, one simply extrapolates the high-temperature data to the temperature of the application. This requires two underlying assumptions:

1. The failure mechanism is identical at the three test temperatures
2. The failure mechanism at the elevated test temperatures is the same at the *use* temperature

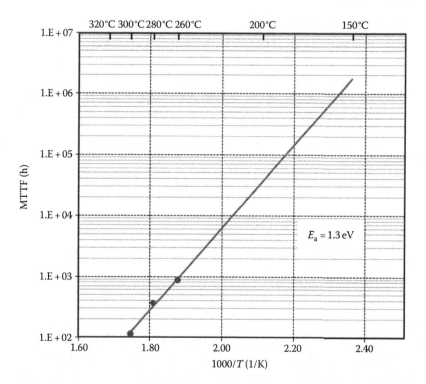

FIGURE 8.1
Arrhenius plot of median time to failures of GaN HEMTs. (After Mittereder, J.A. et al., RF Arrhenius life testing of X-band GaN HEMTs, *Proceedings of the Reliability of Compound Semiconductors (ROCS) Workshop*, Monterey, CA, October 2008. Reproduced with permission from JEDEC.) Note the large extrapolation to typical *use* temperatures, which is usual for such highly accelerated test conditions.

This brings another consideration to the forefront—namely, what constitutes a failure. New semiconductor technologies often exhibit catastrophic failures that occur relatively early in device operation. Such failures are frequently caused by defects in processing, incompatible metallurgies, poor design, etc. Once the technology is matured, failures tend to be associated with parametric changes, or as discussed above in "Reliability Defined," failure to perform as expected. Manufacturers define failure as a threshold percent change in an easily measurable DC device parameter or an RF operating parameter. Typical failure definitions are as follows.

For DC parameters:

- 10% change in saturation current of an FET—generally applied to power devices
- 10% change in transconductance of an FET—generally applied to small-signal devices
- 10% change in current gain of a heterojunction bipolar transistor (HBT)

For RF parameters:

- Change in output power of ~0.5–1 dB
- Change in gain of ~0.5–1 dB

MTTF is most often measured using DC bias. While this may be sufficient for small-signal devices, power transistors operate over the full range of the current–voltage plane and will experience additional stresses due to the applied electric field and higher current densities. Thus, it is prudent to be skeptical of MTTF results obtained with DC bias if the application involves large-signal operation. Customers with applications requiring high reliability often specify accelerated test conditions that are equivalent to actual *use* conditions. Additionally, failure criteria can also be specified by the customer and can vary with the specific application.

Figure 8.2 presents accelerated life test data [Menozzi, private communication] for a group of GaAs heterostructure field effect transistors (HFETs) that were tested at constant temperature and bias. In this case, failure is defined as a 20% change in the drain-source saturation current (I_{dss}) of the device. Note the fairly large spread of failure times, which stretch over a period of two orders of magnitude in time. The ramifications of such spreads, which are not atypical, will be discussed below.

The analysis of life test data involves fitting the failure times at a particular temperature and test condition to a mathematical model. Commonly used models include the exponential, normal, lognormal, and Weibull distributions. Generally, one chooses a distribution that best fits the data and the lognormal and Weibull distributions have been found to be applicable for

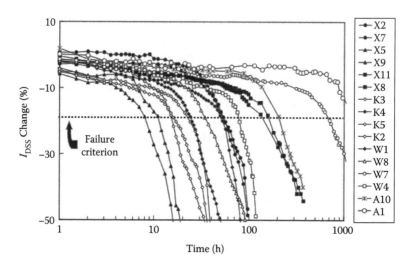

FIGURE 8.2

Change of drain-source current [I_{dss}] for a group of GaAs HFETs during accelerated life test-ing at elevated temperatures. (Courtesy of Professor Roberto Menozzi, Universita' di Parma, Parma). Times to failure, defined in this work as a 20% change in I_{dss}, are used to determine the median time to failure for this population at the test temperature.

wear-out data that is characterized by an increasing failure rate. (For a dis-cussion of different distributions, see [Crowe and Feinberg 2001]). Figure 8.3 shows a lognormal plot of the failure times that were used to generate the Arrhenius curve of Figure 8.1.

Much can be learned from the lognormal plot and it is unfortunate that such plots are rarely shown along with the Arrhenius plots. Salient features of the plots in Figure 8.3 are

1. The number of data points at each temperature—the confidence level increases with an increasing number of samples

2. The standard deviation [sigma] of each family of data—low sigma values [<1.0] imply a tight distribution and results in fewer "early" failures

3. The slope [sigma] at each temperature—similarity indicates that a single failure mechanism is operative

Because the MTTF is a statistically derived metric, it is important to be aware of the confidence level, or uncertainty, of the measurement due to a relatively small sampling of the true population. Figures 8.4 and 8.5 present results of an Arrhenius plot based on fictitious data generated from Monte Carlo trials of 16 devices at each of three temperatures, along with the 90% confidence limits for two different values of sigma. For sigma = 0.6, the confidence inter-val spans 1.7 orders of magnitude at 400 K, while for sigma = 1.0, the span

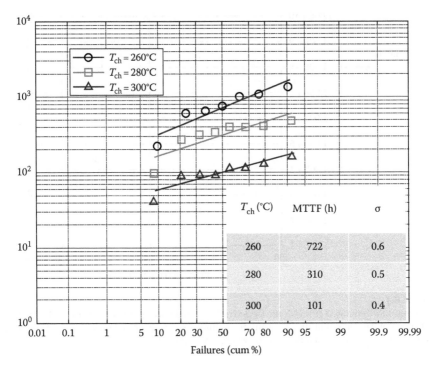

FIGURE 8.3

Plot of the failure times at each of three test temperatures for a population of devices. The median times for these data were used to construct the Arrhenius plot of Figure 8.1. (After Mittereder, J.A. et al., RF Arrhenius life testing of X-band GaN HEMTs, *Proceedings of the Reliability of Compound Semiconductors (ROCS) Workshop*, Monterey, CA, October 2008.)

increases to 2.7, attesting to the desirability of having a tight distribution. Similarly, the confidence limits shrink with an increasing number of samples, and expand with a smaller sample size. Note also that the confidence intervals increase dramatically the further the data is extrapolated towards the *use* temperature.

Since, as expressed in Equation 8.1, MTTF is exponential with device temperature, it is important to understand the spatial dependence of temperature and accurately determine its value. For FETs, the peak temperature occurs in a narrow region along the gate on the drain side of the gate electrode. This is the point where the electric field and current density are at a maximum in a standard design and this temperature is referred to as the "junction" temperature (a carryover from bipolar pn-junction transistors in which the hot spot occurs at the collector-base junction).

Techniques to empirically determine junction temperature are shown in Table 8.1; most of these have evolved over a period of time when semiconductor critical dimensions were measured in mils, rather than a fraction of a micron. With today's submicron devices, most of the techniques suffer

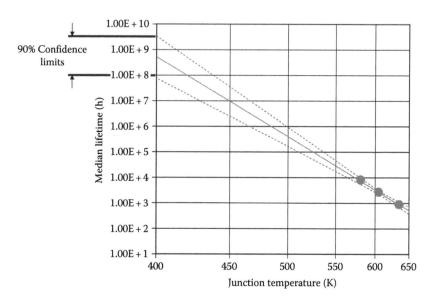

FIGURE 8.4
Fictitious data generated from Monte Carlo trials for 16 devices per temperature and a lognormal sigma of 0.6. Note that the 90% confidence limits extend over ~1.7 orders of magnitude at 400 K. (Courtesy of Roger Wallace, BAE Systems, Nashua, NH.)

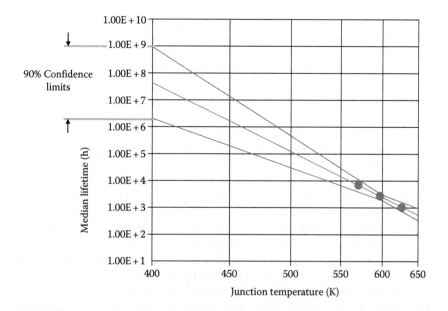

FIGURE 8.5
Fictitious data generated from Monte Carlo trials for 16 devices per temperature and a lognormal sigma of 1.0. Note that the 90% confidence limits extend over ~2.7 orders of magnitude at 400 K. (Courtesy of Roger Wallace, BAE Systems, Nashua, NH.)

TABLE 8.1

Empirical Methods to Determine Device Junction Temperature and Thermal Impedance

Method	Technique	Limitations	Uses/Advantages
Infrared imaging	Employ IR detector in optical microscope or IR camera to determine surface temperature; requires calibration of emissivity of surface	Resolution limited to ~5 μm	Fast screening tool for large hot spots, die-attach integrity; holistic view of entire circuit
Liquid crystal	Coat device with liquid crystal solution; power up slowly and observe transition temperature	Invasive—requires coating of device with liquid crystal solution; limited range of transition temperatures/crystals	R&D
Electrical	Apply constant current to a calibrated forward-biased diode. Temperature is proportional to forward voltage	Requires sense diode in vicinity of device; does not measure device directly	In situ monitor of circuit temperature
Raman spectroscopy	Observe shift in Raman spectrum after application of power	Slow; large integration time for best resolution	Combine with modeling to determine peak channel temperatures

shortcomings for accurate temperature measurement. For example, IR techniques generally average over a lateral 1–5 um spot size, which is equivalent to the source-drain spacing of today's microwave and millimeter-wave transistors. Raman spectroscopy has recently been developed to enable temperature measurements in lateral dimensions of well under a micron, but the beam can probe beyond epitaxial structure (normal to the surface); care must be taken to insure that the response is from the epitaxial material where the heat is generated, and not the substrate material that typically has a much higher thermal conductivity than ternary epitaxial layers employed in modern devices. Many researchers have resorted to finite element modeling to theoretically determine junction temperature. While models can generate precise results, the accuracy is only as good as the knowledge of the thermal properties of the semiconductor layers and there is considerable uncertainty in thermal conductivity of some materials, particularly the ternary semiconductor compounds. Nevertheless, by correlating thermal modeling with spatially precise techniques such as Raman spectroscopy, reasonably accurate estimates of the junction temperature are obtained.

In assessing reliability claims, it is important for the user to understand how these were established by asking

- How recent is the data? Processes continually evolve, particularly for new or nascent technologies such as GaN.
- Under what conditions were the data measured? Will user conditions be more stressful or more benign; did the test incorporate RF stress?
- How was the temperature determined, and what is the reference point—junction or baseplate?
- What was the failure criterion?
- What were the sample size, standard deviation, and confidence level?

One entrapment that can occur in accelerated testing involves the existence of a low activation energy failure mechanism that would not be detected at high test temperatures. This is illustrated in Figure 8.6 where the dash and dash-dot lines represent activation energies for two different, independent failure mechanisms. It is seen that extrapolation of the high-temperature data would result in an extremely optimistic (an incorrect) estimate of MTTF in presence of the lower activation energy mechanism. Such behavior was observed in 1989 with GaAs MMICs in hermetically sealed, metal ceramic packages [Camp et al. 1989]. It was determined that H_2 outgassing from the package walls was responsible for degradation of devices that incorporated

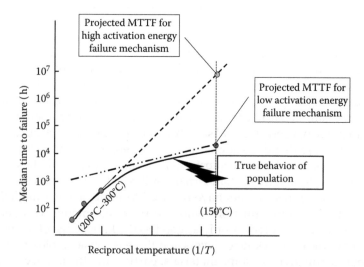

FIGURE 8.6
Illustration of MTTF behavior of two populations with different activation energies and failure mechanisms (broken lines). If both failure mechanisms are operative in the population of tested devices, the true behavior is shown by the solid line. Testing at temperatures exceeding the intersection of the broken lines would not reveal the lower activation failure and would result in a highly optimistic (and incorrect) value of MTTF at the *use* temperature.

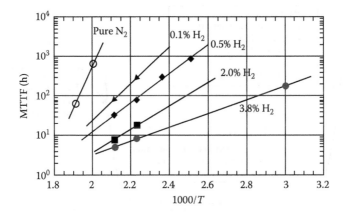

FIGURE 8.7
MTTF for GaAs PHEMTs as a function of ambient atmosphere. Note that the activation energy is a strong function of hydrogen partial pressure. (After Hu, W.W. et al., Reliability of GaAs PHEMTs under hydrogen containing atmosphere, *Proceedings of the GaAs IC Symposium*, Philadelphia, PA, pp. 247–250, 1994. With permission.)

Ti and Pt metallizations in the gate electrode [Ragle and Kayali 1997]. Life test data for pseudomorphic high electron mobility transistors (PHEMTs) in pure N_2 and as a function of H_2 partial pressure are shown in Figure 8.7 [Hu et al. 1994]. The low activation energy failure mechanism associated with H_2 results in parametric failures in several hundred, rather than several million hours. The H_2-poisoning effect was the subject of much research in the 1990s and a number of approaches, including use of a H_2 getter in the package, were devised to remediate this problem [Immorlica et al. 1999].

The "rule of thumb" for a "reliable" semiconductor is an MTTF $>1 \times 10^6$ at 150°C, and it is not uncommon to see reported MTTF values of 1×10^7–1×10^9 h. To put this into the "cosmic perspective," recall that the transistor was invented in 1947, which is "only" 5×10^5 h ago! So, is it really necessary to have MTTF hours which would stretch back to the end of the last ice age (9×10^7 h, or 10,000 years ago) or the end of the late Pleistocene Epoch (1×10^9 h, or 125,000 years ago)?

Actually, what matters is the application and the tolerance for failures. For a satellite in earth orbit or on a deep space mission, what counts is not the time when 50% of the devices have failed (the *median* time) but when the *first failure* occurs, because that is what jeopardizes the mission (notwithstanding on-board spares). Confidence for this application is gained through field experience—i.e., "heritage"—and risk is reduced by redundancy. For phased array antennas, which have thousands of devices, the failure rate is a more appropriate measure, since these arrays exhibit a "graceful degradation" and can tolerate as much as 10% failures depending on the array design and specifications. Confidence in this case is gained though statistical testing of a large number of representative samples. Finally, consider

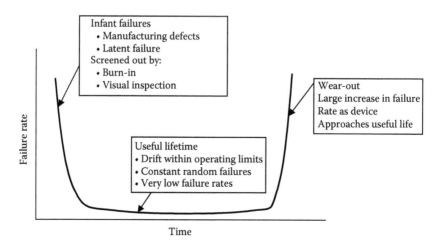

FIGURE 8.8

Typical failure rate behavior versus time for semiconductor devices. MTTF predicts wear out while MTBF is a more appropriate measure of failures during the flat or useful life portion of the behavior. Infant failures must be screened out during the manufacturing process.

our cell phones in which technology obsolescence occurs in as little as 18 months (1.3×10^4 h). We will leave it to the reader to decide what an appropriate level of reliability is.

Up to this point, we have been addressing reliability in terms of end-of-life predictions. A more comprehensive view of reliability should include consideration of the entire operating life of a device from the point of manufacture. Whole life models of failure rates can be described by a "bathtub" curve, Figure 8.8, with three distinct regions: (1) infant failures that occur early in life, (2) useful life, characterized by a constant, relatively small failure rate (i.e., 1/MTBF), and (3) wear-out, with a large increase in failure rate as the device approaches its useful life (i.e., 1/MTTF). The reciprocal of MTBF and MTTF is the failure rate and is often expressed as "failures in time" (FITs). An FIT is defined as one failure in 10^9 h. Thus, an MTTF of 1×10^7 h is equivalent to 100 FITs. The advantage of expressing reliability in FITs is that the reliability of a system can be expressed as the summation of the FITs of the individual components that comprise the system.

Early failures, or infant mortality, are generally related to manufacturing defects such as metal voids, step coverage issues, ionic contamination, cracks induced during the dicing process, die-attach or wire bonding defects. Such defects can generally be screened by visual inspection and burn-in at operating conditions and/or elevated temperatures. Screening can add significantly to the costs of a device and is generally dictated by the price point for the product and the intended application.

Failures during useful life can be due to a number of causes, including tails of the distribution from the infant mortality and wear-out regions, but can also be user induced. This includes overstress during testing or operation, improper die attach or heatsinking, and electrostatic discharge due to improper handling. Some manufacturers are very sensitive to the causes of mid-life failures and encourage customers to return failed product for failure analysis. Analysis of root-cause failures for 2941 field returns for 2005–2006 at a commercial foundry [Roesch and Brockett 2006] showed that only ~50% of failures were due to fabrication process defects or electrical overstress, while most of the balance failed due to assembly, packaging, test or design issues. (Interestingly, no faults were found in 19% of the cases.) Such data is useful for improving manufacturing technologies and gaining confidence in the product.

Up to this point, we have inferred that reliability is related to the quality of the intrinsic transistor. It is certainly true that in an IC, the transistor is most often the prime component associated with long-term failure. However, it would be remiss if manufacturers did not address reliability of each component and structure in an IC, including capacitors, resistors, metal lines, crossovers, vias, interconnects, and even packaging-related issues as we have mentioned above with regard to H_2 poisoning. Circuit components are generally evaluated separately from a circuit by using special test chips, and there is a body of knowledge with regard to reliable component design that crosses over many technology platforms. An excellent comprehensive review with guidelines for IC reliability can be found in [Kayali et al. 1996], and industry standards for reliability testing can be downloaded from www.jedec.org.* In this chapter, we have limited discussion to reliability of the intrinsic transistor.

8.4 Compound Semiconductor Reliability—Overview

Since the development of the first GaAs MESFET over 40 years ago, performance of CS devices has improved primarily through engineering of materials and material systems. This is in contrast to improvements in silicon technologies, which have been driven mainly by shrinking lithographic dimensions, although more recently, materials engineering such as strained silicon and SiGe heterojunctions have been pursued in conjunction with lithography advances. CS devices for electronic application are, for the most part, constructed from elements in columns III and V of the periodic chart, with GaAs and InP being the most prevalent binary compounds. Since the development of the high-electron-mobility transistor (HEMT) in the late

---

* JEDEC develops standards for the semiconductor industry, which are available for free download at www.jedec.org. Relevant standards include: JEP143B.01 *Solid State Reliability Assessment and Qualification Methodologies*, JEP122E *Failure Mechanisms and Models for Semiconductor Devices*, and Publication 118 *Guidelines for GaAs MMIC and FET Life Testing*.

1970s and early 1980s, however, leading-edge transistors increasing rely on ternary and even quaternary mixtures of column III and V elements, because the band structure of the epitaxial layers can be engineered to a great degree to achieve desirable electronic transport properties. There are very few advanced devices today that do not incorporate ternary layers in the active epitaxial stack.

From a reliability perspective, use of ternary compounds poses a few issues. First and foremost, the ternary epitaxial layers must be grown on a binary compound host substrate, and the mismatch of lattice constant between the two materials results in mechanical strain [Li et al. 2009] adversely effecting electrical properties. Additionally, while the electronic transport properties of a ternary are intermediate between the two binary compounds of which the ternary is composed, phonon scattering, which is a determinant of thermal conductivity, can increase substantially in the ternary mix, with the result that most ternary layers exhibit a considerable decrease in thermal conductivity compared to the host substrate.

With the evolution of each new material system, the process technology needs to be reengineered. Ohmic contacts, metallizations, and, particularly, passivation must be tailored to the material system, and each can have reliability ramifications and must be tested and optimized for both performance and reliability. It is not uncommon for a new system or device design to exhibit very low lifetimes—on the order of hours or days—which is manifested by very poor short-term stability. To resolve this, it is imperative that the failure mode be identified and in the course of the investigation, one also endeavors to determine an acceleration mechanism for the failure. Typical stresses include temperature, electric field, current density, mechanical stress, or a combination of these. Short-term stability testing of a powered device is usually the first line of investigation for a new device. Figure 8.9 shows results of short-term RF powered stability testing of two sets of GaN power HEMTs [Joh and del Alamo 2006]. While details are proprietary, note the significant improvement after a change in the design of the epitaxial material stack.

8.5 Reliability of Gallium Nitride

In contrast to GaAs and InP, which both have a zincblende lattice structure, the GaN lattice is a wurtzite structure and is highly piezoelectric. This can have significant ramifications on the electrical and mechanical properties and reliability of GaN devices. Additionally, while GaAs and InP single-crystal boules are commercially available, it is very difficult to synthesize a GaN single-crystal substrate, and GaN devices must be fabricated on a compatible host substrate. The most prevalent substrate in use today is the

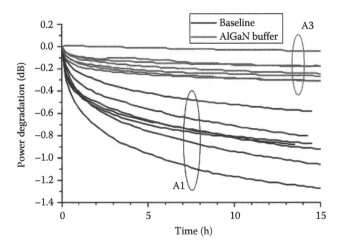

FIGURE 8.9
Results of room temperature RF short-term stability testing, also called power soak, of two sets of GaN HEMTs. Significant improvement is seen by changing the buffer layer. (From Joh, J. and del Alamo, J.A., *IEEE Electron Dev. Lett.*, 29, 4, 287, April 2008. With permission.)

4H or 6H polytype of single-crystal SiC, which can be grown in 3″ and 4″ diameter boules. Gallium nitride-based epitaxial layers are also being grown on silicon substrates by a few manufacturers, offering the promise of large-diameter substrates and processing infrastructure of silicon-based foundries.

Interest in GaN has been driven by fundamental material properties of high breakdown electric field and high thermal conductivity compared to any other III–V compound. A comparison of GaAs and GaN is given in Table 8.2 along with reliability implications that must be considered in transistor and IC design. In spite of the material advantages, GaN lagged development of GaAs by several decades due to the lack of a native substrate and epitaxial material growth issues. Significant research funded primarily by the Defense Advanced Research Projects Agency (DARPA) in the United States [Rosker et al. 2009] and a European consortium, KORRIGAN [Gauthier and Reptin 2006] have made inroads into GaN electronic devices that are now commercially available.

8.5.1 Short-Term Instabilities

As is prevalent in nearly any new technology, early devices are plagued by short-term instabilities. For GaN, these have included

- Collapse of DC current–voltage characteristics
- Short-term degradation and drift of RF output power
- High leakage current
- Parametric changes with applied field

TABLE 8.2

Comparison of Selected Fundamental Materials Parameters of GaAs and GaN

Parameter	Advantage	GaN	GaAs	GaN/ GaAs	Reliability Implication
Bandgap (eV)	High operating temperature	3.5 eV	1.4 eV	2.5	Many failure mechanisms are accelerated by temperature
Breakdown field (MV/cm)	High voltage	3.3 MV/cm	0.4 MV/cm	8.25	Passive components must sustain equivalent sustaining voltages
Sheet density (cm^{-2}) times saturated electron velocity (cm/s)	High current capability	2×10^{13}/cm^2 times 1.5×10^7 cm/s; electrons from spontaneous polarization	4×10^{12}/cm^2 times 1.0×10^7 cm/s; electrons from intentional doping	7.5	Interconnects, bias circuits must be designed for higher currents
Thermal conductivity (W/cm K)	Thermal management	1.7 W/cm K	0.45 W/cm K	3.8	Facilitates management of 10× higher power density capability of GaN over GaAs

Following is a brief summary of these short-term effects and current findings on causes and remedies. While research of these instabilities is ongoing, screening techniques at the wafer and device level have been established, which can be employed to improve quality and reliability of manufactured products.

8.5.1.1 DC Current Collapse

Collapse of the DC current–voltage characteristics is a phenomenon whereby the value of the drain current measured on a dynamic, or pulsed basis, is significantly less than the static or DC measured values. This is illustrated in Figure 8.10 and significantly impedes the ability of the device to operate over the full *I–V* plane, compromising power performance.

Collapse has been associated with electron traps at the surface and buffer interfaces [Faqir et al. 2008], and is a phenomenon that has been observed during the evolution of other device and materials technologies, including Si-MOSFETs [Ning et al. 1977], CdSe thin-film transistors [Wysocki 1982], and GaAs HEMTs [Kastalsky and Kiehl 1986]. Several different traps can be responsible for this behavior, all with characteristic energy levels and time constants. The impact of the relatively long time constants with respect to

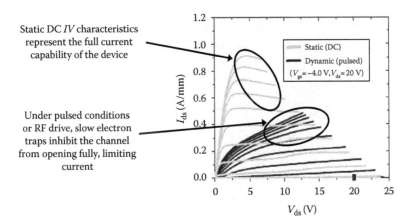

Static DC *IV* characteristics represent the full current capability of the device

Under pulsed conditions or RF drive, slow electron traps inhibit the channel from opening fully, limiting current

FIGURE 8.10
Illustration of current collapse in the current–voltage characteristics of a GaN HEMT. The higher-valued curves are measured with a DC bias, while the lower family of curves results upon pulsing the device from an "off" state ($V_{ds} = 20$ V, $V_{gs} = -4$ V) using short pulses. (After Lossy, R., AlGaN/GaN HEMTs on silicon carbide substrates for microwave power operation, *International Conference on Compound Semiconductor Manufacturing*, 2003. With permission.)

microwave frequencies is clearly seen to impact both the maximum power capability and power stability of the device. What might not be so obvious, from a reliability assessment standpoint, is that, depending on the energy level of the traps, they can be empty at elevated temperatures resulting in recovery of the *I–V* performance and leading to an erroneous conclusion regarding the soundness of the device. This can occur in accelerated temperature testing as well as CW power testing.

Figures of merit for pulse *I–V* performance have been established as a routine wafer level screening tool. One such test involves first measuring the DC saturation current and then DC biasing the device in the off-state at a high drain bias; this is followed by a fast, positive rising gate voltage pulse to drive the device to the on-state. The ratio of the on-state DC current (measured previously) to the on-state value during the fast pulse is the pulsed *I–V* figure of merit. Values of 0.9 or higher are desired.

8.5.1.2 RF Power Droop

Short-term power degradation, or droop, as illustrated in Figure 8.11, shows the result for a room temperature power stability test at a constant bias and RF input power. Such results have been observed by many researchers and have been found in various FET technologies, including GaAs MESFETs and PHEMTs as well as GaN HEMTs. Power droop has been associated with injection of hot electrons into the surface passivation and has been resolved by attention to passivation pretreatment and processing [Hwang 1999] in GaAs MESFETs and PHEMTs along with field plate designs and material

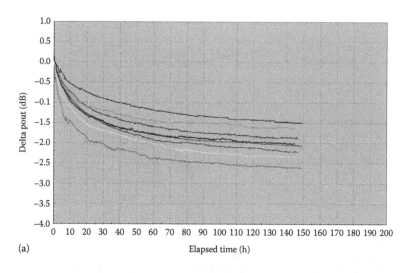

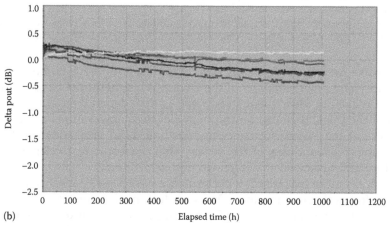

FIGURE 8.11
RF power soaks of GaN HEMTs at 10 GHz and 270°C. (a) First-generation devices exhibiting significant short-term power droop; (b) improved second-generation devices with relatively stable performance. (Note change in timescale.) This data is representative of a general trend in GaN device improvement over the past 5 years due to better material and field plate designs. (Courtesy of P.C. Chao, BAE Systems.)

improvements in GaN HEMTs. Short-term (e.g., 24 h) power stability tests are a good indicator of device stability and can be utilized as an effective wafer screening tool.

8.5.1.3 Leakage Current

Gate leakage current is the bane of many semiconductor devices and is usually associated with surface effects in the gate-drain region or the Schottky contact, buffer interface states, and/or substrate dislocations. Techniques for dealing with leakage in GaN at low applied bias can be borrowed from GaAs technology, and include passivation, passivation and gate metal pretreatments, and buffer layer design. Additionally, it has been shown that insertion of an interfacial Al layers under the Schottky contact to getter O_2 from the interface is effective in reducing leakage current and improving current collapse [Nanjo et al. 2006].

Leakage is very sensitive to the device fabrication processes and foundries have developed proprietary surface preparation and deposition techniques to minimize device leakage current along with leakage current screening for mass production [Yamaki et al. 2007]. In addition, leakage for GaN HEMTs in particular is impacted by inherent strain in the epitaxial layers and high dislocation densities in SiC substrates, which can propagate through the device layer. GaN HEMTs fabricated with GaN substrates have shown up to 1000× lower leakage [Chao et al. 2004], which has been attributed to the lower dislocation density of GaN substrates and an inherently better lattice match. As with GaAs, proprietary buffer, active, and cap layer designs have been implemented to minimize gate leakage current in GaN devices. However, at high drain bias, new defects can be generated in GaN HEMTs due to the piezoelectric nature of GaN; the resulting enhanced leakage that cannot be resolved by established techniques. This will be discussed further below.

8.5.1.4 Parametric Changes

As with RF power droop, DC device parameters have been observed to change with time upon application of applied bias. These changes, most notably a decrease of the drain current and increase in gate leakage, are often partially recoverable after a period of inactivity. This can be a false recovery, however, as after reapplication of bias, the device quickly returns close to its former degraded state. Such behavior is associated with several phenomena that are dependent on the magnitude of the applied drain bias. At low bias, hot electron injection into the passivation layer and resulting charge accumulation will pinch-off the channel, resulting in current degradation. As drain voltage is increased, a critical voltage is reached at which permanent changes occur due to mechanical stresses. Evidence of physical degradation has been revealed through transmission electron microscopy (TEM) studies of devices before and after a DC life test. The TEM analysis of the degraded devices showed a pit-like formation in the top AlGaN layer at the drain side of the gate, which is the position of the peak electric field [Jimenez et al. 2006, Park et al. 2009]. The resulting defects act as charge trapping centers. This phenomenon, termed the inverse piezoelectric effect [Joh and del Alamo 2008], is unique to GaN and is due to the strong polar nature of the crystal. The mechanism continues to be the subject of research and investigation.

8.5.2 Long-Term Life

A number of foundries have reported MTTF of GaN on SiC HEMTs using temperature-accelerated life tests with results ranging from 1×10^6 to 1×10^7 h [Li et al. 2009, Piner et al. 2006, Ohki et al. 2009] and one publication even reports MTTF exceeding 1×10^9 h [Conway et al. 2007]. While these results are commendable and useful for providing an initial technology assessment, they are by themselves inadequate to conclusively prove that these devices will last several centuries, or even several years, in practice. In addition to temperature acceleration, one needs to consider accelerated stresses imposed by electric field, high current density, and mechanical stress. This is particularly relevant for GaN, which is operated at high voltage and power levels, and in thermal environments, which can result in mechanical stress due to coefficient of thermal expansion (CTE) mismatches in both the device itself and the packaging technologies. Also, low activation failure mechanisms, such as susceptibility to H_2 poisoning and trapping-related power droop at low temperatures, need to be studied. Indeed, studies in these areas are underway and failure analysis and modeling are being pursued by many university, industrial, and government laboratories.

8.5.3 Infant Mortality

There have been few reports of catastrophic infant failures in GaN and most of the early problems are related to drift of parametric and performance parameters as discussed in Section 8.5.1. These effects can be screened out by the manufacturer and as the technology matures, we will continue to see a more defined set of infant failures and development of a screening methodology to deal with them.

8.5.4 Reliable Device Design

Short-term stability tests, along with accelerated life tests, serve as an excellent tool for assessing the integrity and maturity of a new technology, investigation of failure mechanisms, and development of solutions. A large body of knowledge exists with respect to reliable design of GaAs FET type devices and much has been adapted by GaN device designers. Indeed, GaAs and GaN share similar device structures and a common set of foundry fabrication tools. The significant differences between GaN and GaAs are

1. *Doping*—GaN doping is derived from spontaneous and piezoelectric polarization while GaAs doping is derived from intentional incorporation of specific impurities during epitaxial growth; thus, free carrier density in GaN can be impacted by varying stresses applied to the crystal.

2. *Crystal structure*—The wurtzite structure in GaN leads to a highly piezoelectric nature compared to the zincblende structure of

GaAs; this leads to failure mechanisms that are not observed in GaAs.

3. *Breakdown field and current density*—Higher current densities and applied voltages in GaN devices lead to higher stresses in operating devices.

4. *Dislocation density*—Dislocation density in GaN on SiC wafers is typically three to four orders of magnitude higher than GaAs and can impact device yield. GaN on GaN wafers exhibit much lower dislocation density but are not yet commercially available.

5. *Thermal conductivity*—The thermal conductivity of GaN and SiC (the substrate of choice for GaN epitaxy) is ~3.8 and ~7 times higher, respectively, than GaAs. While this is a significant advantage for thermal management, the much higher power levels that are achievable with GaN still results in a thermal management challenge.

Nevertheless, each material system is unique and while failure mechanisms may be similar for each system, foundries will develop proprietary approaches and techniques to improving performance and reliability. These include epitaxial and buffer layer designs, passivation pretreatment and formulation, ohmic and gate metallization, and gate structures.

Epitaxial and buffer layer designs are proprietary and unique to each foundry. Buffer layers consisting of undoped or Fe-doped GaN, and AlGaN have been reported and various ternary mixtures of Al, Ga, and N have been studied. GaN cap designs can also have a profound influence on reliability; in fact, a GaN cap of only a few nm thick can have dramatic impact on device stability [Ivo et al. 2009].

Likewise, passivation technology plays an important role in stability and lifetime, and can be effective in eliminating current collapse [Faqir et al. 2008]. Passivation results are very sensitive to application techniques, especially surface pretreatment [Edwards et al. 2005]. Vacuum-deposited Si_3N_4 is the most common passivation material, and while unpassivated devices have been reported, it is likely that a sound passivation will be required for the most reliable devices.

Ohmic contact and gate Schottky metallizations for GaN appear to be quite stable and there have been relatively few reports of metal-induced failures for GaN. Indeed, gate sinking, a predominant failure mechanism for GaAs- and InP-based devices, is not seen in GaN.

Perhaps the most prevalent feature of GaN devices is the inclusion of a field plate in the gate-drain region [Karmalkar et al. 2005]. This consists of an electrode isolated from the semiconductor surface by an intervening layer of dielectric, and can be integral with the gate electrode or connected to the source contact. The field plate reduces the electric field at the gate edge on the drain side of the device and spreads the field laterally over the gate-source region. This feature not only results in higher breakdown voltages, but reduces the surface charge, improving device stability and reliability. However, as is often the case, there is a trade-off between performance and reliability because a gate-connected field plate increases the gate-drain capacitance resulting in a compromise of high frequency gain. This can be mitigated by connecting the field plate to the source electrode [Wu et al. 2004]. A promising alternative to field plates involves "gate shaping" or tapering of the gate metal edge to reduce the electric field strength at the gate edge [Palacios et al.].

8.6 GaN on Alternative Substrates

While most GaN devices today are manufactured with epitaxial layers grown on SiC substrates, research is ongoing on growth and device fabrication of GaN layers on alternative substrates, the most prevalent of which are shown in Table 8.3.

GaN native substrates would clearly be the substrate of choice if they were commercially available in reasonably sized substrates. Epitaxial growth on native substrates is fundamentally straightforward, does not require special nucleation layers, and eliminates lattice mismatch and the resulting propagation of defects into the device layer. While GaN on GaN devices have exhibited orders of magnitude lower in leakage current and relatively good performance earlier in its development [Chao et al. 2004], research has slowed due to the lack of readily available commercial sources of GaN substrates. As an alternative to growth of single-crystal GaN boules in commercial quantities, some researchers are using GaN substrates, cut from a boule, as a template for vapor growth of a pure GaN layer, which is then separated from the original substrate and used for device fabrication. In this way, the precious boule-cut wafer can be reused. Reliability inferences of this technique have yet to be investigated.

Use of diamond as a substrate for GaN epitaxial layers is intriguing because diamond offers the highest thermal conductivity of any substrate candidate.

TABLE 8.3

Alternatives to SiC Substrates for Growth of GaN Epitaxial Device Layers

Substrate	Advantage	Issues
Gallium nitride	Lower relative defect density, lower leakage current, higher device yield	Difficult to synthesize single-crystal GaN; small-diameter substrates
Diamond	High thermal conductivity, highest operating power density and temperature capabilities	Large lattice mismatch causes severe strain and bowing
Silicon	Offers integration with Si IC technologies, large diameter substrates	Lower thermal conductivity

However, the large CTE mismatch between GaN and diamond results in unacceptable bowing for fabrication of submicron features and induced strain can significantly alter electrical transport properties in the device if not properly accounted for in design. Nevertheless, research on GaN on diamond is continuing [Francis et al. 2007] and holds promise for high-power devices once the materials issues are resolved. CTE mismatch remains the top technical issue for GaN on diamond substrates.

Although GaN and diamond substrates offer distinct technology advantages, GaN on Si has received the most attention due to both the significant cost advantages of fabricating devices on large-diameter substrates and the potential of direct integration of GaN devices with Si-CMOS circuits. Microwave power devices utilizing GaN on Si technology are now commercially available from several manufacturers and relatively good MTTF reliability statistics based on thermally accelerated life test have been published [Piner et al. 2006, Zanon et al. 2008]. It appears that from a device wear-out perspective, GaN-on-Si is as reliable as GaN-on-SiC for temperature-activated wear-out mechanisms. Reliability studies are continuing and investigation of other wear-out phenomena as discussed above will hopefully be forthcoming. As GaN-on-Si power devices find application in commercial infrastructure, particularly cell phone base stations, we will hopefully begin to see MTBF data on these devices.

8.7 Reliability Challenges for Heterogeneous Integration

The combination of dissimilar materials presents challenges in many areas. One must address compatibility of the fabrication processes employed in each technology, engineering, and sequencing of the unit process steps, interaction of the mating surfaces from both electrochemical corrosion and CTE considerations, and, of course, the impact of the combination on electrical transport properties.

This book presents integration of CS with silicon semiconductors from two perspectives: integration of separately fabricated CS devices onto silicon wafers and direct growth and processing of CS devices in situ with silicon processes. Integration of CS chips onto Si ICs is the more straightforward of the two in that device process compatibility issues are not as much of a factor. Challenges related to attach and electrical interconnects are the main issues along with thermal conduction and expansion considerations in design. Reliability assessment is also more straightforward as the device technologies can be individually assessed before being integrated, although reliability of the combined structure must also be assessed.

Direct growth and processing of CS with Si ICs is much more challenging. In addition to GaN, growth of GaAs and GaP on Si have been demonstrated and many other combinations are possible. To date, most of these studies have been motivated by the availability of low-cost, large-area Si substrates and the resulting lower costs of devices and circuits due to the economies of scale in wafer fabrication. From a reliability perspective, the main considerations have been the impact of the interface properties on the electrical, thermal, and mechanical properties of the CS device. However, true integration with *active* silicon circuits "compounds" the problem in that the materials, processes, and designs must contend with inherent incompatibility of two very dissimilar material systems, process techniques, chemistries, and thermal exposures during manufacturing, and device types.

It would be presumptive to offer an assessment of which of the many potential compound-silicon technology combinations would be most reliable, but based on decades of semiconductor research and development, it is likely that initial results would be discouraging and plagued with infant failures. "Virtual fabrication" through device and process modeling can be helpful in guidance of initial designs and processes. The key to developing a reliable technology lies in applying the same rigorous studies already employed in semiconductor reliability investigations to a heterogeneous integration technology in a systematic way. This includes

- Determination of potential stress conditions that may lead to failure
- Observation of failures under specific stress conditions
- Understanding failure mechanisms
- Modification of designs and processes to eliminate the cause of failures
- Most importantly, retesting to insure that design and process changes have not introduced any new failure mechanisms or compromising of performance

Thus, design and process improvements should continue hand in hand with reliability assessment to insure the technology is robust as manufactured. As technology is fielded, manufacturers should encourage return of failed devices for analysis of undetected or low activation failure mechanisms. It is only through continued reliability assessment and fielding that heritage and confidence in a new technology is attained.

8.8 Summary

This chapter has presented an overview of semiconductor device reliability from a testing and analysis perspective, along with a more detailed description of reliability studies and concerns in a nascent technology. While we have addressed reliability of GaN, the same considerations apply to all III–V compound integration with Si. One must first assess the reliability of the parent technologies, and then address the unique issues associated with integration of both. It is likely that changes in processing or design will be required to address compatibility. One rule of thumb is that one cannot ever make a single change in a semiconductor process—whatever is changed, no matter how benign it appears, will most likely have an effect on another parameter. It is the device and process engineers' challenge to find out what that other change is and to ameliorate any negative consequences.

References

[Camp et al. 1989] W. O. Camp Jr., R. Lasater, V. Genove, and R. Hume, Hydrogen effects on reliability of GaAs MMICs, *Proceedings of the GaAs Integrated Circuits Symposium*, San Diego, CA, pp. 203–206, 1989.

[Chao et al. 2004] P. C. Chao, K. M. Chu, and J. A. Windyka, Stable high power GaN-on-GaN HEMT, *Proceedings of the IEEE Lester Eastman Conference on High Performance Devices*, Rensselaer Polytechnic Institute, Troy, NY, pp. 114–120, 2004.

[Conway et al. 2007] A. M. Conway, M. Chen, P. Hashimoto, P. J. Willadsen, and M. Micovic, Failure mechanisms in GaN HFETs under accelerated RF stress, *CS ManTech Conference*, Austin, TX, pp. 99–102, 2007.

[Crowe and Feinberg 2001] D. Crowe and A. Feinberg (Eds.), *Design for Reliability*, CRC Press, Boca Raton, FL, 2001.

[Edwards et al. 2005] A. P. Edwards, J. A. Mittereder, S. C. Binari, D. S. Katzer, D. F. Storm, and J. A. Roussos, Improved reliability of AlGaN-GaN HEMTs using an NH_3 plasma treatment prior to SiN passivation, *IEEE Electron. Dev. Lett.*, 26(4), 255–227, April 2005.

[Faqir et al. 2008] M. Faqir, G. Verzellesi, A. Chini, F. Fantini, F. Danesin, G. Meneghesso, E. Zanoni, and C. Dua, Mechanisms of RF current collapse in AlGaN-GaN high electron mobility transistors, *IEEE Trans. Dev. Mater. Reliab.*, 8(2), 240–247, June 2008.

[Francis et al. 2007] D. Francis, J. Wasserbauer, F. Faili, D. Babic, F. Ejeckam, W. Hong, P. Specht, and E. Weber, GaN-HEMT epilayers on diamond substrates: Recent progress, *CS ManTech Conference*, Austin, TX, pp. 133–136, 2007.

[Gauthier and Reptin 2006] G. Gauthier and F. Reptin, KORRIGAN: Development of GaN HEMT technology in Europe, *Proceedings of the CS ManTech Conference*, Vancouver, Canada, pp. 49–51, 2006.

[Hu et al. 1994] W. W. Hu, E. P. Parks, T. H. Yu, P. C. Chao, and A. W. Swanson, Reliability of GaAs PHEMTs under hydrogen containing atmosphere, *Proceedings of the GaAs IC Symposium*, Philadelphia, PA, pp. 247–250, 1994.

[Hwang 1999] J. C. M. Hwang, Relationship between gate lag, power drift, and power slump of pseudomorphic high electron mobility transistors, *Solid State Electron.*, 43(8), 1325–1331, August 1999.

[Immorlica et al. 1999] A. A. Immorlica Jr., S. B. Adams, and A. R. Reisinger, Practical approaches to remediation of hydrogen poisoning in GaAs devices, *Proceedings of the Compound Semiconductor Manufacturing Technology Conference*, Vancouver, Canada, 1999.

[Ivo et al. 2009] P. Ivo, A. Glowacki, R. Pazirandeh, E. Bahat-Treidel, R. Lossy, J. Wurfl, C. Boit, and G. Trankle, Influence of GaN cap on robustness of AlGaN/GaN HEMTs, *Proceedings of the 47th International Reliability Physics Symposium*, Montreal, Canada, pp. 71–75, 2009.

[Jimenez et al. 2006] J. L. Jimenez, U. Chowdhury, M. Y. Kao, T. Balistreri, C. Lee, P. Saunier, P. C. Chao, W. W. Hu, K. Chu, A. Immorlica, J. A. del Alamo, J. Joh, and M. Shur, Failure analysis of X-band GaN FETs, *Presented at the Reliability Compound Semiconductors Workshop* (ROCS), San Antonio, TX, 2006.

[Joh and del Alamo 2006] J. Joh and J. A. del Alamo, Mechanisms for electrical degradation of GaN high electron mobility transistors, *International Electron Devices Meeting*, December 2006.

[Joh and del Alamo 2008] J. Joh and J. A. del Alamo, Critical voltage for electrical degradation of GaN high electron mobility transistors, *IEEE Electron. Dev. Lett.*, 29(4), 287–289, April 2008.

[Karmalkar et al. 2005] S. Karmalkar, M. S. Shur, G. Simin, and M. A. Khan, Field-plate engineering for HFETs, *IEEE Trans. Electron. Dev.*, 52(12), 2534–2540, December 2005.

[Kastalsky and Kiehl 1986] A. Kastalsky and R. A. Kiehl, On the low temperature degradation of AlGaAs/GaAs modulation-doped field-effect transistors, *IEEE Trans. Electron. Dev.*, ED-33, 414–423, 1986.

[Kayali et al. 1996] S. Kayali, G. Ponchak, and R. Shaw, editors, *GaAs MMIC Reliability Assurance Guideline for Space Applications*, JPL Publication, Pasadena, CA, pp. 96–25, 1996.

[Lee et al. 2008] S. Lee, R. Vetury, J. D. Brown, S. R. Gibb, W. Z. Cai, J. Sun, D. S. Green, and J. Shealy, Reliability assessment of AlGaN/GaN HEMT technology on SiC for 48V applications, *Proceedings of the International Reliability Physics Symposium*, Phoenix, AZ, pp. 446–449, 2008.

[Li et al. 2009] Y. Li, M. Krishnan, S. Salemi, G. Paradee, and A. Christou, Strain induced buffer layer defects in GaN HFETs and their evolution during reliability testing, *Proceedings of the 47th Annual International Reliability Physics Symposium*, Montreal, Canada, 2009.

[Lossy et al. 2003] R. Lossy, N. Chaturvedi, P. Heymann, K. Lohler, S. Muller, and J. Wurfl, AlGaN/GaN HEMTs on silicon carbide substrates for microwave power operation, *International Conference on Compound Semiconductor Manufacturing*, Scottsdale, AZ, 2003.

[Menozzi] R. Menozzi, Universita' di Parma, Italy, Private Communication.

[Mittereder et al. 2008] J. A. Mittereder, S. C. Binari, G. D. Via, J. A. Roussos, J. D. Caldwell, and J. P. Calame, RF Arrhenius life testing of X-band GaN HEMTs, *Proceedings of the Reliability of Compound Semiconductors (ROCS) Workshop*, Monterey, CA, October 2008.

[Nanjo et al. 2006] T. Nanjo, T. Oishi, M. Suita, Y. Abe, and Y. Tokuda, Effects of a thin Al layer insertion between AlGaN and Schottky gate on the AlGaN/GaN high electron mobility transistor characteristics, *Appl. Phys. Lett.*, 88(4), 043503, 2006.

[Ning et al. 1997] T. H. Ning, C. M. Osburn, and N. H. Yu, Effect of electron trapping on IGFET characteristics, *J. Electron. Mater.*, 6, 65–76, 1977.

[Ohki et al. 2009] T. Ohki, T. Kikkawa, Y. Inoue, M. Kanamura, N. Okamoto, K. Makiyama, K. Imanishi, H. Shigematsu, K. Joshin, and N. Hara, Reliability of GaN HEMTs: Current status and future technology, *Proceedings of the 47th Annual International Reliability Physics Symposium*, Montreal, Canada, pp. 61–70, 2009.

[Palacios et al.] T. A. Palacios, L. Shen, and U. K. Mishra, Fluorine treatment to shape the electric field in electron devices, passivate dislocations and point defects, and enhance the luminescence efficiency of optical devices, Patent pending, 2005.

[Park et al. 2009] S. Y. Park, C. Floresca, U. Chlowdhury, J. L. Jimenez, C. Lee, E. Beam, P. Saunier, T. Balistreri, and M. J. Kim, Physical degradation of GaN NEMT devices under high drain bias reliability testing, *Microelectron. Reliab.*, 49, 478–483, 2009.

[Piner et al. 2006] E. L. Piner, S. Singhal, P. Rajagopal, R. Therrien, J. C. Roberts, T. Li, A. W. Hanson, J. W. Johnson, I. C. Kizilyalli, and K. J. Linthicum, Device degradation phenomena in GaN HFET technology: Status, mechanisms and opportunities, *IEEE International Electron Devices Meeting*, San Francisco, CA, 2006.

[Ragle and Kayali 1997] D. Ragle and S. Kayali, Hydrogen effects on GaAs microwave semiconductors, *Reliability Workshop*, Anaheim, CA, pp. 66–71, 1997.

[Roesch and Brockett 2006] W. J. Roesch and S. Brockett, Natural failure mechanisms: Analysis of actual field returns, *Proceedings of the Reliability of Compound Semiconductors (ROCS) Workshop*, San Antonio, TX, 2006.

[Rosker et al. 2009] M. J. Rosker, C. Bozada, H. Dietrich, A. Hung, D. Via, S. Binari, E. Vivierios, E. Cohen, and J. Hodiak, The DARPA wide band gap semiconductor for RF applications (WBGS-RF) program: Phase II results, *2009 International Conference on Compound Semiconductor Manufacturing Technology*, Tampa, FL, May 2009.

[Wu et al. 2004] Y. F. Wu, M. Moore, T. Wisleder, P. M. Chavarkar, U. K. Mishra, and P. Parikh, High-gain microwave GaN HEMTs with source-terminated field-plates, *International Electron Devices Meeting*, San Francisco, CA, 2004.

[Wysocki 1982] J. J. Wysocki, Drain-current distortion in CdSe thin-film transistors, *IEEE Trans. Electron. Dev.*, ED-29, 1798–1805, 1982.

[Yamaki et al. 2007] F. Yamaki, K. Ishii, M. Nishi, H. Haematsu, Y. Tateno, and H. Kawata, Leakage current screening for AlGaN/GaN HEMT mass-production, *CS ManTech Conference*, Austin, TX, pp. 95–98, 2007.

[Zanon et al. 2008] F. Zanon, F. Danesin, A. Taxxoli, M. Meneghini, N. Ronchi, A. Chini, P. Bove, R. Langer, E. Zanoni, and G. Meneghesso, Reliability aspects of GaN-HEMTs on composite substrates, *Seventh International Conference on Advanced Semiconductor Devices and Microsystems*, Smolenice, Slovakia, pp. 23–30, 2008.

Part IV

Defect and Properties Evaluation and Characterization

9

In Situ Curvature Measurements, Strains, and Stresses in the Case of Large Wafer Bending and Multilayer Systems

Rainer Clos and Alois Krost

CONTENTS

General expressions for the curvature–incompatibility strain relationship in the case of large bending and arbitrary thickness ratios between film and substrate are developed, which can be solved largely analytically in the case of cylindrical symmetry. It is shown that the limits of the Stoney formula are easily reached in the case of, e.g., GaN on Si, large diameter wafers, thick layers, or vertical temperature gradients. As a consequence, the surface strain becomes anisotropic already at small bending with a possible impact on epitaxial growth. Moreover, analytical expressions for multilayer systems in linear approximation are developed and applied to real systems. The full nonlinear procedure as well as the multilayer evaluation can be easily implemented in a standard PC program to determine in situ the stress of, e.g., a growing epitaxial layer on a foreign substrate.

9.1 Introduction

Due to the lack of GaN wafers, so far, group-III nitrides are mostly grown on sapphire or SiC substrates. Silicon offers an attractive alternative because of its low cost, large wafer area, and physical benefits such as the possibility of chemical etching, lower hardness, good thermal conductivity, and electrical conduction or isolation for light-emitting devices or transistor structures, respectively. However, for a long time, a technological breakthrough of GaN-on-silicon has been thought to be impossible because of the cracking problem originating in the huge difference of the thermal expansion coefficients between GaN and silicon, which leads to tensile strain and cracking of the layers when cooling down. In the case of GaN-based compounds, stress is known to severely affect the device quality because of wafer bending, dislocation formation, or even cracking. The knowledge of the physical origins of stress is a key issue to overcome these problems. In recent years, several approaches to prevent cracking and wafer bowing have been successfully applied and nowadays, device-relevant, crack-free group-III-nitrides layers can be grown on silicon. To reach this goal, the most important issues were the identification of the physical origin of strains and its engineering by means of in situ monitoring during metal organic vapor phase epitaxy. Among the in situ techniques, most attention has been drawn to a quasi direct measurement of the stress by the wafer curvature method during deposition of thin solid films (Figure 9.1). This technique is based on the measurement of the substrate curvature κ (usually in the center of the wafer) induced by the mismatch because any mismatch strain ε_m (lattice, thermal, growth-induced mismatch, etc.) between film and substrate bends the whole system. Simultaneously, such measurements yield the actual thickness of the film via thickness interference fringes from the front- and backside of the growing film [Floro 2001, Floro 2002]. The in situ curvature method has already been successfully applied to follow the stress evolution during metal organic vapor phase epitaxy of GaN or AlGaN on sapphire [Hearne 1999, Terao 2001, Brunner 2007] or silicon [Krost 2003, Liu 2007] and for InGaAs/GaAs and InAs/InP [Lynch 2005, Fuster 2006]. Nowadays, in situ curvature monitoring has become a common tool in many epitaxy laboratories. The method always requires an equation, which correlates the curvature with the mismatch strain. Usually, the data are interpreted using the classical Stoney formula [Stoney 1909].

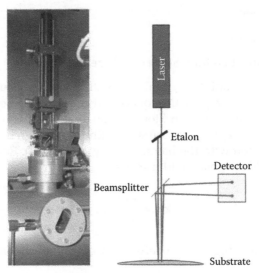

In situ curvature measurement

FIGURE 9.1
Schematic setup and photography for in situ curvature measurements during metal organic vapor phase epitaxy in horizontal reactor.

$$\kappa_{St} = \frac{6M_f \varepsilon_m h_f}{M_s h_s^2}.$$ (9.1)

Then, the film stress is given by $\sigma_f = M_f \varepsilon_m$. E_f, v_f, h_f and E_s, v_s, h_s are Young's modulus, the Poisson ratio, and the thickness of the film and the substrate, respectively. The related biaxial modules are $M_f = E_f/(1-v_f)$ and $M_s = E_s/(1-v_s)$. Equation 9.1 is based on several assumptions, e.g., isotropic (at least laterally) linearly elastic behavior of film and substrate, $h_f/h_s \ll 1$ and $\varepsilon_m =$ constant. Especially, it is supposed that the theory of small bending of thin plates is sufficient. Extensions for arbitrary thickness ratios were derived, e.g., by Freund [Freund 1999, Freund 1997] and Kroupa [Kroupa 1993] for small bending of thin plates. However, most publications are restricted to one layer systems and moreover to weak bending. So far, strong bending and multilayer systems have been hardly treated though practically all GaN-based devices consist of multilayer systems. In case of a strong bending, Stoney's formula is not further applicable. Instead, the theory of large bending including in-plane deformation is necessary for the interpretation of the curvature measurements.

Within the framework of the plate theory, we have developed general expressions for the curvature–strain relationship in the case of large bending and arbitrary thickness ratios between film and substrate, which can be solved nearly analytically. Moreover, for the case of multilayer systems, we have developed solutions, which will be given in the following too.

9.2 Single Layers

9.2.1 Small Wafer Bending: Stoney and Freund Equations

Inelastic (or "incompatibility") strains ε_{in} within layer and substrate (caused by lattice, thermal, or growth-induced mismatch, etc.) bent and induce stresses in the total system. In Stoney's equation given above, equi-biaxial inelastic strain $\varepsilon^m = -\varepsilon_{in} = \text{const.}$ is assumed (the minus sign has been used for a better comparison with the literature only, cf. Part III). Rearrangement of Equation 9.1 yields the biaxial film stress

$$\sigma_f = \frac{M_s h_s^2}{6h_f} \kappa,$$

assuming the following relation between stress and incompatibility strain:

$$\sigma_f = M_f \varepsilon^m. \tag{9.2}$$

It should be noted that Equation 9.2 is not generally valid but restricted to the case of very thin films (see below). Therefore, two questions should be clarified:

1. What is the general relation between incompatibility strain and the measurable curvature κ?
2. How can the elastic stresses in film and substrate be calculated knowing the incompatibility strain?

Applying the classical plate theory, both questions will be solved simultaneously. Stoney's equation can be considered as 0th approximation for the relation between incompatibility strain ε^m and wafer curvature. The limits of this equation have been investigated in detail, e.g., by Freund [Freund 1999]; particularly $h = h_f/h_s$ and ε^m must be sufficiently small.

A generalization of Stoney's equation for arbitrary thickness relations h_f/h_s (and $\varepsilon^m = \varepsilon^m(z)$, with z a coordinate perpendicular to the wafer plane) has been given by Freund et al. [Freund 1997], and Kroupa et al. [Kroupa 1993]. Freund's equation reads

$$\kappa_F = \frac{6mh(1+h)\varepsilon^m/h_s}{(1+mh(4+6h+4h^2)+m^2h^4)}$$

or

$$\kappa_F = \frac{\kappa_{st}(1+h)}{FD} \tag{9.3}$$

with $FD(m) = 1 + mh(4 + 6h + 4h^2) + m^2 h^4$ the "Freund-denominator," $h = h_f/h_s$ and $m = M_f/M_s$ and, e.g., $\varepsilon^m = (1 - a_s/a_f)$ in the case of pure lattice mismatch strain. The incompatibility strain has been assumed to be constant. It can be easily shown, that Equation 9.3 is fully symmetric with regard to the substrate and the layer, i.e., invariant against permutation of the layer and the substrate.

Isotropic, linear elastic behavior of film and substrate at least in transverse direction and the limit of small curvature were assumed when deriving Equation 9.3; otherwise analytical solutions are not possible. The limit of small curvature means that only the linear terms in the total strain tensor

$$\varepsilon_{ij} = \frac{1}{2}\left(\frac{\partial u_i}{\partial x_j} + \frac{\partial u_j}{\partial x_i}\right) + \frac{1}{2}\frac{\partial u_k}{\partial x_i}\frac{\partial u_k}{\partial x_j} \equiv \frac{1}{2}(u_{i,j} + u_{j,i}) + \frac{1}{2}u_{k,i}u_{k,j} \qquad (9.4)$$

are considered. The case of small bending requires $\zeta \ll (h_s + h_f)$ [Landau 1966]. In Equation 9.4, u_i are the components of the displacement vector, a comma means partial differentiation, double occurrence of indices requests for summation, and the description is in Cartesian coordinates; ζ holds for u_z, the strain component perpendicular to the wafer plane. Within plate theory, the real three-dimensional (3D) problem (which can be solved by numerical methods as finite element method only) is approximated by a two-dimensional (2D) one, that is why only the in-plane components of ε occur in the theory (cf. Part III). Consideration of nonlinear terms in Equation 9.4, i.e., the transition to strong bending, is necessary as has been pointed out already by Freund et al. [Freund 1999, Freund 1997]. Approximate solutions already show that significant deviations as compared to the linear treatment, Equation 9.3, may occur [Freund 1999, Harper 1990, Salamon 1995]. In principle, bending and in-plane displacement are coupled. For that reason, stress-free bending is impossible, even for a homogeneous plate. There are coupled nonlinear differential equations, which are not exactly analytically solvable ([Friedrichs 1939, 1942], Landau [Landau 1966]).

9.2.2 Strong Wafer Bending: Nonlinear Theory

In the following, we consider nonlinear wafer bending. For simplification, we assume cylindrical symmetry, i.e., a circular, elastically isotropic substrate and film as well as isotropic inelastic strains ε_{in}. The stress–(elastic) strain relations remain linear, i.e., all strains are $\ll 1$, but in the total strain tensor, nonlinear terms are taken into account. This means, that the theory is linear with respect to the material behavior but geometrically nonlinear. According to the classical plate theory, the total strain tensor is derived from a Kirchhoff–Love displacement field (cf. Part III). The relation between total strain ε and elastic strain ε_{el} reads

$$\varepsilon_{\alpha\beta} = \varepsilon_{el,\alpha\beta} + \varepsilon_{in,\alpha\beta},$$

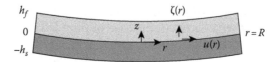

FIGURE 9.2
Cylindrical coordinate system for a film on a substrate.

α, $\beta = 1, 2$ denotes the in-plane components, $\varepsilon_{in,\alpha\beta}$ are the inelastic strains with $\varepsilon_{in,\alpha\beta} = \varepsilon_{in}\delta_{\alpha\beta}$ (cf. Part III, in the following we use $\varepsilon^{m'} = -\varepsilon_{in}$ for the same reason as above). Using cylindrical coordinates (r, φ, z; Figure 9.2) the nonvanishing components of the elastic strain tensor are

$$\varepsilon_{el,r}(r,z) = -z\zeta''(r) + u'(r) + \frac{\zeta'(r)^2}{2} + \varepsilon^m \quad \text{and} \quad \varepsilon_{el,\varphi}(r,z) = -\frac{z\zeta'(r)}{r} + \frac{u(r)}{r} + \varepsilon^m,$$

(9.5)

where
  ζ is the displacement perpendicular to the wafer plane (often denoted as deflection)
  u is the displacement in radial direction
  a streak means differentiation with respect to r

The incompatibility strain ε^m is assumed to be constant, being zero in the substrate and $\neq 0$ in the film (e.g., in the case of lattice mismatch $\varepsilon^m = -(a_f - a_s)/a_f$, in the case of thermal mismatch $\varepsilon^m = (\alpha_s - \alpha_f)\Delta T$, where $\alpha_{s,f}$ are the linear thermal expansion coefficients of the substrate and the film, respectively, and ΔT is the temperature difference, which is assumed to be independent on z in the moment). In the bending problem, the most important nonlinear term of the strain tensor is $\zeta'^2/2$, all other nonlinear terms can be neglected (as usual in the theory of strong bending of plates [Landau 1966]). The impact of nonlinear terms is twofold: on one hand, the local stresses are influenced; on the other hand, the overall deformation of the wafer is altered via the different differential equations and the boundary conditions. Basically, the choice of the plane $z = 0$ is arbitrary; however, in the case of a film–substrate system, it is favorable to choose the substrate–film interface (in the case of the naked wafer, the mid plane is more appropriate). The in-plane stresses can be written as

$$\sigma_r = E^*(\varepsilon_{el,r} + \nu\varepsilon_{el,\varphi}) \quad \text{and} \quad \sigma_\varphi = E^*(\varepsilon_{el,\varphi} + \nu\varepsilon_{el,r}),$$

(9.6)

with $E^* = E/(1 - \nu^2)$. The Euler–Lagrange equations for the unknown functions $\zeta(r)$ and $u(r)$ and the boundary conditions can be obtained as usual by varying the functional of the free elastic energy

$$F = \pi \int_{-h_s}^{h_f} \int_0^R E^* (\varepsilon_{el,r}^2 + \varepsilon_{el,\varphi}^2 + 2v\varepsilon_{el,r}\varepsilon_{el,\varphi})r\,dr\,dz.$$

As a result, the following differential equations for ζ and u are obtained:

$$\int_{-h_s}^{h_f} E^* \left[\begin{array}{l} z^2 r \Delta^2 \varsigma - z\dfrac{d}{dr}\left(\dfrac{d}{dr}(ru') + \dfrac{1}{2}\dfrac{d}{dr}(r\varsigma'^2) - r\varsigma'\varsigma'' - \dfrac{3}{2}v\varsigma'^2 - \dfrac{u}{r}\right) \\[4mm] -\dfrac{d}{dr}\left(\dfrac{1}{2}r\varsigma'^3 + ru'\varsigma' + vu\varsigma' + \varepsilon_m(1+v)r\varsigma'\right) \end{array} \right] dz = 0 \qquad (9.7)$$

and

$$\int_{-h_s}^{h_f} E^* \left(\dfrac{d}{dr}(ru') - \dfrac{u}{r} - z\left(\dfrac{d}{dr}(r\varsigma'') - \dfrac{\varsigma'}{r}\right) + \dfrac{1}{2}\dfrac{d}{dr}(r\varsigma'^2) - \dfrac{1}{2}v\varsigma'^2 \right) dz = 0, \qquad (9.8)$$

Δ is the Laplace operator, the accomplishment of the z-integration has been abandoned here. From the latter, several factors result containing h_s and h_f as well as the generally different elastic constants of substrate and film. The lengthy expressions for the boundary conditions are not given explicitly here. The differential equations are coupled and nonlinear; moreover, they contain the incompatibility strain and expressions depending explicitly on film and substrate thickness and the elastic constants.

We consider a substrate–film system being fixed at $r=0$ with $u(0)=\zeta(0)=\zeta'(0)=0$ and otherwise free then there remain two boundary conditions for $r=R$ and the order of the differential equation (9.7) for ζ is reduced by one. Irrespective of this, a direct solution of (9.7) and (9.8) does not make much sense because each material combination, actual thickness, and incompatibility strain would require a new calculation. Wittrick [Wittrick 1953] has given a procedure for the calculation of the stability of a bimetal disc, which can be adopted here. First, we introduce $\zeta'(r)/r$ and $\int \sigma(r,z)dz$ as new unknown functions instead of $\zeta(r)$ and $u(r)$. It is easy to see, that for $r=0$ $\zeta'(r)/r$ is the curvature in the center of the substrate–film system, t_r is the radial stress integrated over the thickness. After a suited normalization, the differential equations can be transformed in the following form [Clos 2004]:

$$4\frac{d^2}{d\rho^2}(\rho Y) = YT - \frac{1}{2}\gamma Y^2 \quad \text{and} \quad 4\frac{d^2}{d\rho^2}(\rho T) = -\left(1 - \frac{1}{2}\gamma^2\right)Y^2 - \gamma YT, \qquad (9.9)$$

$Y(\rho) = \alpha_1 \zeta/r$ and $T(\rho) = \alpha_2 t_r$ are dimensionless normalized functions of $\rho = r^2/R^2$, with $\gamma = (v_f - v_s)^* h^* f_0(e, h, v_f, v_s)$, $\alpha_1 = R(R/h_s)f_1(e, h, v_f, v_s)$, $\alpha_2 = (1/E_s h_s)(R/h_s)^2 f_2(e, h, v_f, v_s)$, $h = h_f/h_s$ and $e = E_f/E_s$. The remaining boundary conditions are

$$Y + 2\rho f_3(e,h,v_f,v_s)\frac{d}{d\rho}Y + 2\rho f_4(e,h,v_f,v_s)\frac{d}{d\rho}T - \varepsilon_m^* = 0 \quad \text{for } \rho = 1, \quad (9.10a)$$

$$T(1) = 0. \qquad (9.10b)$$

Equations 9.10a and b stand for the vanishing angular momentum and radial force at the wafer edge. $\varepsilon_m^* = \varepsilon^m (R/h_s)^2 f_5 (e,h,v_f,v_s)$ is the normalized incompatibility strain ε^m; f_0 up to f_5 are simple, but lengthy functions of these parameters. It should be noted, that the exact boundary condition $\sigma_r(R,z) = 0$, which holds in 3D, is replaced by $\int \sigma_r(R,z)dz = 0$ (Equation 9.10b), this is a consequence of the plate-approximation. For this reason the solution of the plate theory deviates from the exact solution near the point $r = R$. This concerns especially the stress distribution (we denote this as "edge-effect" in the following).

The differential equations (9.9) are coupled and nonlinear as well; however, their structure is much simpler than those of (9.7) and (9.8), and in particular the incompatibility strain does not enter the differential equations explicitly but only the boundary condition (Equation 9.10a).

The differential equations (9.9) can be solved by a power series expansion

$$Y(\rho) = A_0 + A_1\rho + A_2\rho^2 + \cdots + A_n\rho^n \quad \text{and} \quad T(\rho) = B_0 + B_1\rho + B_2\rho^2 + \cdots + B_n\rho^n.$$

$$(9.11)$$

The calculation of the coefficients A_k and B_k is done by choosing first A_0 and B_0; all other coefficients can then be determined recursively by the insertion of (Equation 9.11) in (9.9) with a proper choice of n, e.g., $> \approx 30$

$$4k(k+1)A_k = \sum_{i=0}^{k-1}(A_{k-1-i}B_i - 0.5\gamma A_{k-1-i}A_i)$$

$$(9.12)$$

$$4k(k+1)B_k = -\sum_{i=0}^{k-1}((1-0.5\gamma^2)A_{k-1-i}A_i + \gamma A_{k-1-i}B_i), \quad n \geq k \geq 1.$$

This procedure is repeated with different B_0 until the boundary condition (Equation 9.10b) is fulfilled yielding $B_0(A_0)$. Subsequently, the normalized incompatibility strain is calculated from the boundary condition (Equation 9.10a) resulting in ε_m^* as a function of A_0. In addition, all other expansion coefficients in Equation 9.11 are calculated as a function of A_0, which is the only arbitrary parameter. Evidently, the ansatz for Y corresponds to a series expansion of the deflection, because $\zeta'(r) \propto Yr$

$$\zeta(r) = \frac{1}{2}\kappa_0 r^2 + a_1 r^4 + a_2 r^6 + \cdots + a_n r^{2(n+1)} \tag{9.13}$$

(a constant term does not occur because of $\zeta(0)=0$).

As can be easily seen, κ_0 is the curvature at the position $r=0$, i.e., where the wafer curvature is optically determined. The expansion coefficient A_0 in Equation 9.11 is the normalized curvature κ_0. Therefore, after denormalization of (Equation 9.11), the solution drafted above directly yields the normalized incompatibility strain as a function of the measurable curvature and the involved geometrical and elastic parameters

$$\varepsilon_m = \varepsilon_m(\kappa_0, R, E_s, h_s, \ldots), \tag{9.14}$$

wherewith, question 1 from above is solved.

At this point it should be mentioned that the mean curvature κ of the deformed wafer at an arbitrary position r can be expressed as

$$\kappa(r) = \frac{1}{2}\left(\frac{1}{R_1} + \frac{1}{R_2}\right) = \frac{1}{2}\frac{\left(\dfrac{\zeta''}{1+\zeta'^2} + \dfrac{\zeta'}{r}\right)}{\left(1+\zeta'^2\right)^{1/2}} \tag{9.15}$$

with R_1, R_2 the cardinal curvature radii of plane.

Following the above scheme for the solutions of the differential equations (9.9) and taking into account the boundary conditions (Equation 9.10a,b), after a lengthy analysis, all interesting quantities such as the deflection $\zeta(r)$, the radial displacement $u(r)$, and the elastic stresses $\sigma_r(r)$ and $\sigma_\varphi(r)$ (in film and substrate) can be calculated as functions of the curvature κ_0 (as well as all other geometric and elastic parameters involved).

Because the differential equations (9.9) with boundary conditions (Equation 9.10) cannot be solved in a closed analytical form, no general equations similar to Equations 9.1 or 9.2 (Stoney, Freund) can be given. The same holds for the elastic stresses in the layer. However, the drafted calculation can be performed practically online with a modern PC after entering the material parameters, substrate thickness and radius; the solution is given then for measured time-dependent curvature $\kappa_0(t)$ and film thickness $h_f(t)$. The extension to multilayer systems can be done without any problems.

9.2.3 General Results

In the following, some general results of the nonlinear analysis are given; subsequently, for simplicity, we will concentrate on the case of equal Poisson numbers of film and substrate. As a matter of principle, for sufficiently small normalized incompatibility strains $\varepsilon^{m*} = A_0$ always holds, which is after denormalization exactly Freund's equation (Equation 9.3). In general, one obtains a representation

$$\varepsilon^{m*} = A_0 F_\varepsilon(A_0, \ldots) \tag{9.16}$$

and accordingly, the inverse function

$$A_0 = \varepsilon^{m*} F_\kappa(\varepsilon^{m*}, \ldots), \tag{9.17}$$

whereby $F_{\varepsilon,\kappa}$ are functions of the given variables (the dots mean geometric and elastic parameters, generally, the functions can only be obtained numerically). Both functions converge toward the value 1 for ε^{m*} or $A_0 \to 0$, respectively (independently of the other parameters). After denormalization (9.17) one obtains

$$\kappa_0 = \kappa_{st} F_\kappa,$$

where κ_{st} refers to the Freund equation (Equation 9.3), i.e., for sufficiently small normalized incompatibility strains the nonlinear calculation reduces itself to the linear theory, as expected. For arbitrary curvatures and misfit strains, the relations (Equation 9.16) and (9.17) considerably differ from the linear theory as has already been shown in our previous paper [Clos 2004]. Figure 9.3 shows the normalized curvature (Equation 9.15) as a function of

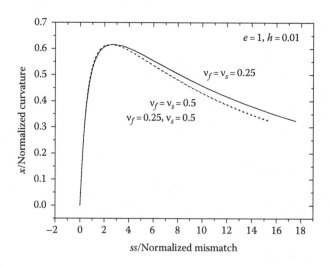

FIGURE 9.3
Normalized curvature vs. normalized mismatch for $r = 0$ showing the postbuckling behavior for large strains.

the normalized incompatibility strain for $r=0$, i.e., in the wafer center. The other parameters have been adopted from a finite-element calculation of this problem by Freund et al. [Freund 1997]. First of all, very good agreement is reached with Figure 9.3 in Ref. [Freund 1997] (not shown here). A most interesting behavior is observed, namely, some buckling behavior. The normalized curvature reaches a maximum and drops at higher strains. With increasing normalized incompatibility strain, the curvature reaches a maximum and then decreases in the wafer center (the wafer becomes more flat whereas the outer regions of the wafer are more strongly bent). It should be noted that this behavior is extremely interesting from a theoretical point of view (however, practical epitaxy occurs below the strain maximum, at least according to our observations). Moreover, approximate theoretical solutions [Freund 1999, Kroupa 1993, Harper 1990] and finite-element calculations for square- or rectangular-coated plates [Mezin 2006, Guyot 2004] show an instability of the rotational symmetry and a transition to roll-in of the wafer should occur (into a cylindrical shape). A priori, the exact location of this point cannot be determined from the present calculations and also the analytical estimations [Freund 1999, Harper 1990, Salamon 1997] are not exact enough. A comparison with finite-element calculations [Guyot 2004] shows, that the instability could occur between $x=0.5$ and $x=0.6$, i.e., close to the above mentioned curvature maximum.

Figures 9.4 through 9.6 show the normalized incompatibility strains as a function of the normalized curvature, i.e., Equation 9.16, (with $ss = \varepsilon^{m^*}/(4\sqrt{6})$ and $x = A_0/(4\sqrt{6})$, ss and x were introduced for a comparison with Ref. [Freund 1997]) for different parameters $h=h_f/h_s=0.001$, 0.5, and 2, an

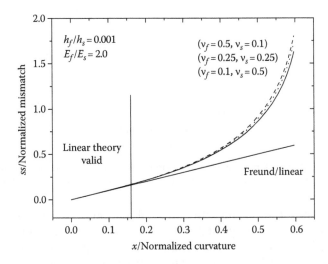

FIGURE 9.4
Normalized mismatch vs. normalized curvature in the nonlinear case for $h=0.001$, $e=2$, and different combinations of Poisson numbers.

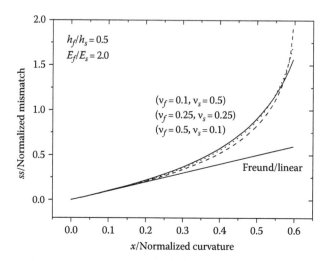

FIGURE 9.5
Normalized mismatch vs. normalized curvature in the nonlinear case for a very thick layer ($h=0.5$), $e=2$, and different combinations of Poisson numbers.

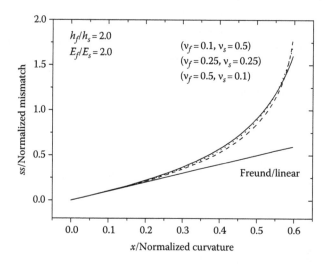

FIGURE 9.6
Normalized mismatch vs. normalized curvature in the nonlinear case for a thick layer on a thinner substrate $h=2$, $e=2$ and different combinations of Poisson numbers (see text).

e-module ratio $E_f/E_s=2.0$ and different combinations of the Poisson numbers (the wafer radius does only enter in the normalization of ss and x). For comparison the linear solution according to Freund, Equation 9.3, is also given in each case. It should be noted that other e-module ratios do not change the curves significantly. From these figures, it becomes clear that the linear approximation is valid up to normalized curvatures of about $x=0.15-0.2$. For higher curvatures, drastic deviations occur. In such cases, the linear theory generally yields smaller incompatibility strains than the nonlinear theory (at the same curvature), and therefore also lower stresses result.

For $v_f = v_s = v$, substantial simplifications result. In this case, $\gamma = 0$, $f_4 = 0$ and $f_3 = 1/(1+v)$, and the differential equations and boundary conditions are reduced to

$$4\frac{d^2}{d\rho^2}(\rho Y) = YT \quad \text{and} \quad 4\frac{d^2}{d\rho^2}(\rho T) = -Y^2, \tag{9.18}$$

and

$$Y + \frac{2\rho}{1+v}\frac{d}{d\rho}Y - \varepsilon_m^* = 0 \quad \text{for } \rho = 1, \tag{9.19}$$

$$T(1) = 0.$$

By this, a quasi universal solution is possible because only normalized quantities enter Equations 9.18 and 9.19 except for the Poisson number (the dependencies on the other elastic constants and the geometrical parameters are exclusively in the normalization functions).

Figure 9.7 shows the curvature–strain curves for different Poisson numbers as indicated in the figure. In most cases its value lies between 0.2 and 0.3, therefore, to a good approximation, the curves for $v = 0.25$ should be valid, e.g., for GaN. In Figure 9.8, a polynomial fit to the 0.25-curve was performed,

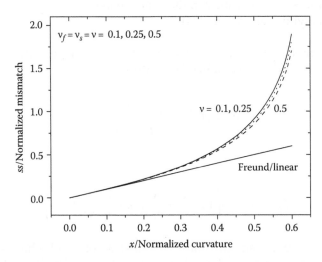

FIGURE 9.7
Normalized mismatch vs. normalized curvature for different Poisson numbers.

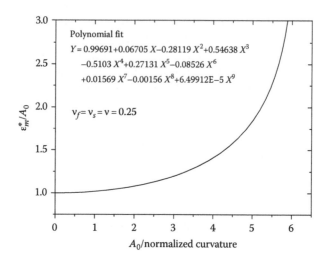

FIGURE 9.8
Normalized mismatch vs. normalized curvature for a fixed Poisson number 0.25 and a polynomial fit to the nonlinear theory curve.

the correction function F_ε in Equation 9.16 becomes a polynomial $p_\varepsilon(A_0)$ (no further parameters are involved). One obtains the denormalized incompatibility strain to be

$$\varepsilon^m = \kappa_0 h_s \frac{FN(m)}{6mh(1+h)} p_\varepsilon(A_0),\tag{9.20}$$

where
  κ_0 is the measured curvature in the wafer center
  p_ε is the polynomial given in Figure 9.8

After denormalization, Equation 9.16 has the same shape, only p_ε has to be replaced by F_ε, and the normalized curvature is

$$A_0 = \sqrt{6}\frac{R^2}{h_s}\kappa_0\left(\frac{FB}{FN(e^*)}\right)^{1/2},\tag{9.21}$$

with

$$FB = (1-v_s^2)\left(1+2\frac{1-v_f v_s}{1-v_s^2}e^* h + \frac{1-v_f^2}{1-v_s^2}e^{*2}h^2\right).$$

$FN(e^*)$ is the Freund-denominator with e^* instead of m, all other parameters are already given above in the context of Equation 9.3 (the Freund-equation (Equation 9.3) results from $p_\varepsilon = 1$). It should be noted that the normalization function was given here for the general case; for equal Poisson numbers of substrate and film simplifications occur. The wanted correlation between incompatibility strain ε^m and measured wafer curvature κ_0 is described with sufficient accuracy by Equation 9.20 for a large number of materials combinations. Somewhat more instructive than Figure 9.8 may be the inverse function

$$A_0 = \varepsilon^{m^*} p_\kappa(\varepsilon^{m^*}).$$

The polynomial p_κ is shown in Figure 9.9. In linear approximation (Equation 9.3) $p_\kappa = 1$. From the figure one immediately obtains the deviation of linear from nonlinear theory as a function of the normalized incompatibility strain ε^{m^*}. Analogous Figure 9.8 yields the deviation as a function of the normalized curvature. It should be noted, however, that in both cases, only normalized quantities are drawn; the denormalization is performed with nonlinear functions of elastic and geometrical parameters.

The elastic stresses in film and substrate are given by Equation 9.6 and can be calculated from our solution. In general, both the radial stress σ_r and the tangential stress (or circumference stress) σ_φ occur. Both are neither equal nor independent from the distance r from the wafer center (there is also a weak dependence on the coordinate z perpendicular to the wafer plane, which stems from the first terms of the strain tensor (9.5)). Strictly speaking, even in

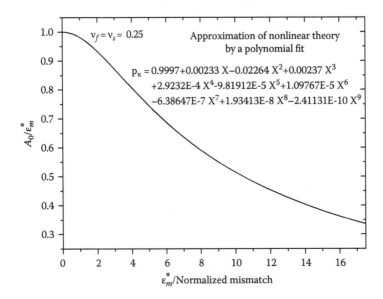

FIGURE 9.9
Normalized curvature vs. normalized mismatch for a fixed Poisson number 0.25 and a polynomial fit to the nonlinear theory curve (inverse function to Figure 9.7).

the case of ε^m = const. (equi-biaxial), not only one film stress is involved. Only for sufficiently thin films both stresses are nearly equal and can be calculated from the incompatibility strain ε^m in the framework of the simple equation (9.2), which is rewritten here

$$\sigma_f = M_f \varepsilon^m \tag{9.22}$$

(this condition seems to be fulfilled in numerous epitaxy applications). For the practical case of GaN on Si, the radial stress as a function of film thickness at a given incompatibility strain ε^m is shown in Figure 9.10. For simplicity, the calculation was performed with $v_f = v_s$. A small difference between the stresses in the wafer center ($r = 0$) and wafer edge ($r = R$), which increases with increasing layer thickness—in spite of this, the approximation $\sigma_f = M_f \varepsilon^m$ is still well applicable. Simultaneously, one recognizes that there are significant deviations already for very thin layers between the nonlinear theory and the approximations given by Stoney and Freund, Equations 9.1 and 9.3, respectively. The important message is that the linear theory underestimates the strain ε^m as compared to the nonlinear theory, and therefore, also the stress as calculated via Equation 9.22 is too low.

If the assumption of a thin layer is not fulfilled (this does not entail an abandonment of the nonlinear theory as shown in Figure 9.10), the elastic stresses in the film and substrate must be calculated within this framework from Equations 9.6. In general, σ_r^f will then be r-dependent, whereby the maximum value is always at $r = 0$. In such a case, with some effort, an approximate formula can be deduced, e.g., for $r = 0$.

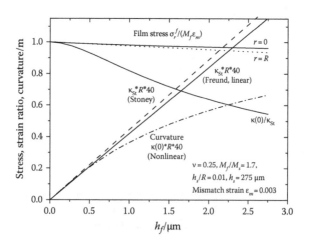

FIGURE 9.10
Stress at $r = 0$ and R, strain ratio, and curvature as a function of film thickness using the linear approximations by Stoney and Freund, and the nonlinear theory.

9.2.4 Discussion

9.2.4.1 Validity of the Linear Approach

As has been already mentioned above, one has to distinguish between the calculation of the inelastic strain from the measured curvature on one hand, and the determination of the stresses on the other hand. The inelastic strain–curvature relations have been given in Figures 9.3 through 9.7. For sufficiently low curvatures, one recognizes the transition to the linear Freund approximation. For a given tolerable error, the validity range of the linear approximation can be given quantitatively from the polynomial in Figure 9.8. For example, for $A_0=2$ (corresponding to $x=0.204$), the error for ε^m within the linear approach is about 8%, i.e., the strain calculated according to the Freund equation (Equation 9.3) is about 8% less than calculated within the nonlinear approach; the denormalization is done with Equation 9.21. For thin films (in the sense of Equation 9.23), this reduces to

$$h_s\kappa_0 \leq \frac{2}{\sqrt{6(1-v^2)}}\left(\frac{h_s}{R}\right)^2.$$

As long as the curvature is smaller than this value, the error of the linear approach is less than 8%. Because the numerical value in this equation is 0.84 (for $v=0.25$), roughly speaking, for thin layers, the error within the linear approximation is less than 10% as long as $\kappa_0\leq h_s/R^2$. This means that at a fixed value of h_s, the greater the radius of the wafer, the earlier the linear theory fails (this holds even for a fixed value of the ratio h_s/R).

9.2.4.2 Stresses

In linear approximation hold $\varsigma_{lin}(r)=1/2\ \kappa_0 r^2$, $\sigma_{r,lin}=\sigma_{\varphi,lin}$, $\varepsilon_{r,lin}=\varepsilon_{\varphi,lin}$ and $\sigma_{r,lin}=M\varepsilon_{r,lin}$ (with different $M=E/(1-v)$ and ε for film and substrate). Here, ε are the elastic strains and not equal to the equi-biaxial inelastic strains ε^m_{lin}. Moreover, they depend on the coordinate z perpendicular to the substrate surface. In the present case, one obtains

$$\sigma^f_{r,lin} = \sigma^f_{\varphi,lin} = \left(1-\left[6\frac{z}{h_s}(1+h)+4+3h+mh^3\right]\frac{mh}{FD(m,h)}\right)M_f\varepsilon^m_{lin}. \quad (9.23)$$

In the limit $h=h_f/h_s$ toward 0, Equation 9.23 reduces to Equation 9.2; in all other cases smaller stresses result (at a given value of $\kappa_0\ \varepsilon^m_{lin}$ means the inelastic strain calculated with Equations 9.1 or 9.3). If

$$4mh = \delta \ll 1, \quad (9.24)$$

the values resulting from Equation 9.23 are $\delta*100\%$ less than those from Equation 9.2, i.e., for a maximum error of 5%, h_f should be $\leq 0.0125\ h_s M_s / M_f$ (for $h_s = 275\ \mu m$ and $m = 1.7$ one obtains $h_f \leq 2.02\ \mu m$).

In this sense, thin films could be defined as layers that satisfy Equation 9.24 with an error of 5% or less and the stress calculation according to Equation 9.2 is applicable. Generally, the linear approach yields stresses lower than those from Equation 9.2. If h increases the thickness gradient must be taken into account, i.e., the values at the film/substrate interface are larger than those at the surface.

9.2.4.3 Impact of Nonlinear Strain on Surface Symmetry

As a consequence of the nonlinear approach, the elastic radial ε_r and azimuthal strain ε_φ components become different. Depending on the system and experimental conditions, tensile and compressive component may occur simultaneously.

In any case, the surface symmetry will be lowered. This is shown in Figure 9.11 for the system parameters of GaN/Si where the radial and azimuthal film stresses are drawn vs. the wafer radius as a function of film thickness. With increasing film thickness, a pronounced splitting and even a crossing between the radial and azimuthal strain and stress components is observed. The effect is shown in more detail for the strain components in Figure 9.12 for a 2 in. Si wafer with 0.5 and 2.75 μm GaN on top, respectively. For simplicity, equal Poisson numbers have been used in the calculation. Surface stress is well known to influence the surface reconstruction of, e.g., the (001) surface

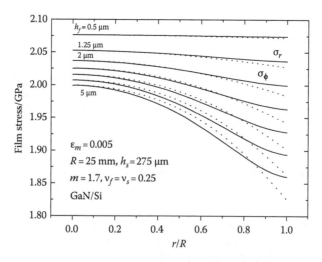

FIGURE 9.11
Radius-dependent radial (full lines) and azimuthal (dotted lines) film stresses as a function of film thickness for GaN/Si.

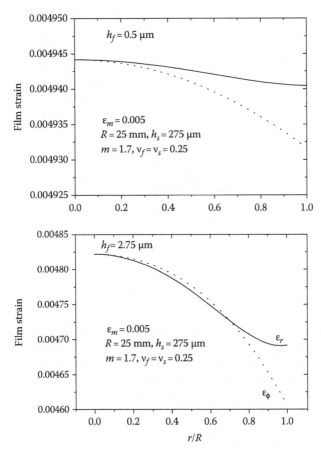

FIGURE 9.12
Splitting of radial (full curves) and azimuthal (dotted curves) strain components as a function of radial position for a 0.5 μm thick (top), and a 2.5 μm thick (bottom) GaN layer on a 2 in. Si wafer.

of Si [Swartzentruber 1990] or III–V compounds like InP [Fuster 2006]. The effect might be responsible for symmetry breaking and preferential orientations of GaN(0001) domains across a Si(001) wafer [Reiher 2009]. Due to the above mentioned edge-effect, neither the stresses nor the elastic strains are exact in a certain region near the wafer edge. The extension of this region should be in the order of some wafer thickness, i.e., small in comparison with the wafer radius.

9.2.4.4 Impact for Thick Layers

The linear approach completely fails in the case of very thick layers on large diameter substrates, e.g., when growing thick GaN layers by hydride vapor

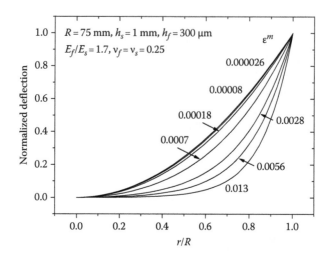

FIGURE 9.13
Normalized deflection vs. wafer radius for a 300 μm thick GaN layer on a 150 mm diameter Si(111) substrate for various mismatch strains. For simplicity, equal Poisson numbers have been chosen.

phase epitaxy on 150 mm Si or sapphire wafers. The deflection for 300 μm GaN layer on a 1 mm thick, 150 mm diameter Si as a function of the radius is shown in Figure 9.13 for different incompatibility strains. Remarkably, with increasing strain, the wafer flattens in the middle and the curvature increases in the outer regions. The overall shape changes from bowl like to pan like. Thus, curvature measurement across the wafer radius should be performed.

9.2.4.5 Temperature Gradients

With some effort, vertical and horizontal temperature gradients can be incorporated in the theory, if the inelastic strain can be written as $\varepsilon_{in} = f(z)\,g(r)$. If there is a radial dependence, additional terms occur in the differential equations, which are not given here. The problem of an axisymmetric temperature distribution for a thin circular disk has already been treated by Indenbom and Kaganer [Indenbom 1990] in linear approximation. In the case of a purely radial dependence of the inelastic strains, no bending occurs, and the remaining differential equation becomes linear. In such a case, our theory reproduces Equation 9.14 as given in Ref. [Indenbom 1990].

In conventional cold wall reactors for metal organic vapor phase epitaxy and also with molecular beam epitaxy, usually, the substrates are heated from beneath yielding a vertical temperature gradient from the bottom of the substrate to top. If we assume a linear temperature gradient, then the linear plate theory gives stress-free bending of the wafer. In contrast, in the nonlinear theory, bending is always coupled with in-plane stresses hence stress-free bending is impossible, even for a linear temperature gradient. This is a typical nonlinear effect.

Indeed, experimentally, a strong wafer bending is observed during the heating up of the wafers to epitaxy temperature, which, of course, depends on the wafer thickness and becomes stronger with increasing wafer diameters. As a consequence of the stresses caused by the bending, in the case of 150 mm c-plane sapphire substrates, the bending is that strong causing a wafer splitting. Thus, an upscale of GaN-based epitaxy toward large diameter sapphire substrates seems to be limited by this effect, at least in conventional reactors. In contrast, due to its lower thermal expansion coefficient (2.6×10^{-6} K^{-1} as compared to 8×10^{-6} K^{-1} for sapphire), the 150 mm silicon substrates withstand the heating-up process allowing epitaxy on such wafers.

In Figure 9.14, the radial and azimuthal stress components induced by a linear vertical temperature gradient are calculated for the case of both types of substrates. In both cases, a substrate thickness of 1.5 mm and diameter of 150 mm have been assumed. As can be seen in the figures, the stress values are mainly positive, i.e., tensile, with values roughly seven times higher in sapphire than in silicon.

9.3 Linear Theory for Multilayers

9.3.1 Basic Equations

We consider a multilayer system deposited on a circular substrate. There are inelastic strains in the layers causing residual stresses or so-called eigenstresses. Inelastic strains are the cause of residual stresses, eigenstresses (eigenstress is a generic name given to self-equilibrated internal stresses caused by one or several of these eigenstrains in bodies, which are free from other external forces or surface constraints [Mura 1987]). The inelastically strained layers are assumed to be tightly attached with respect to each other and the substrate and stress relaxation only being possible via bending. The z-coordinate is perpendicular to the substrate surface with its zero in the substrate plane (substrate: $-h_s \leq z \leq 0$, an arbitrary layer is located between h_{k-1} and h_k, with $h_k = h_{k-1} + d_k$ and $h_{k-1} < hh \leq h_k$ during growth with $h_0 = 0$ and $hh(t)$ the actual total layer thickness). In the substrate plane ($z = 0$) polar coordinates (r, φ) are used. The substrate and kth layer thicknesses are denoted h_s and d_k, respectively, and assumed to be independent of (r, φ); during growth, the thickness of the kth layer is ($hh(t) - h_{k-1}$). A schematic picture of a multilayer system on a substrate and the corresponding nomenclature are shown in Figure 9.15. All strains are restricted to be $\ll 1$. In this way, ε can be split additively in an elastic part ε_{el} and an inelastic one ε_{in} with the tensor components

$$\varepsilon_{ij} = \varepsilon_{el,ij} + \varepsilon_{in,ij},$$ (9.25)

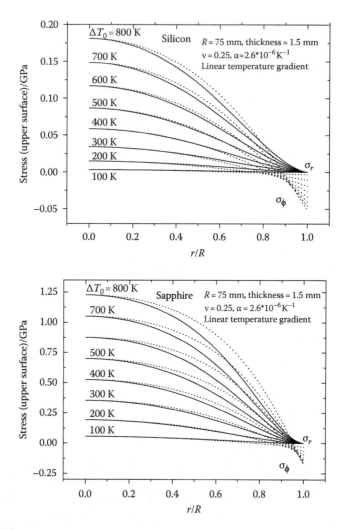

FIGURE 9.14
Radial (full lines) and azimuthal (dotted lines) stress induced by a linear vertical temperature gradient from the wafer bottom to the surface as caused by heating, e.g., in a conventional cold wall reactor for 150 mm diameter Si(111) (top), and sapphire wafers (bottom). Note the different scales. Experimentally, a breaking of the wafer is observed in the case of sapphire.

with $i,j = 1,2,3$. Using the classical theory of plate bending and cylindrical symmetry, the total strain results from a Kirchhoff–Love displacement field [Ciarlet 2004, Reddy 1999] as

$$u_r = -z\frac{d\xi}{dr} + u(r), \quad u_z = \xi(r), \quad u_\varphi = 0, \tag{9.26}$$

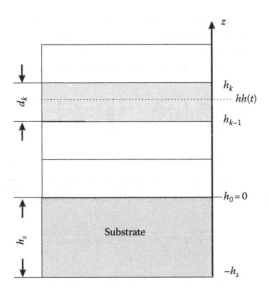

FIGURE 9.15
Schematic drawing of a multilayer system on a substrate and nomenclature used in the text.

because there is only bending allowed as a relaxation mechanism with u_r, u_z, and u_φ the r-, z-, and φ-components of the displacement vector in the r-, φ-, and z-direction (this form of the displacement field results from the basic assumptions of plate theory: $\varepsilon_{rz} = \varepsilon_{\varphi z} = \varepsilon_{zz} = 0$, with ε_{rz}, $\varepsilon_{\varphi z}$, ε_{zz} components of the strain tensor. By this, only tensor components $\varepsilon_{\alpha\beta}$ with $\alpha, \beta = 1, 2$ enter the theory, i.e., with coordinates in the plane $z = 0$; in addition, we assume a state of plane stress, $\sigma_{zz} = 0$). To begin with, when calculating the strains we restrict to linear theory

$$\varepsilon_{rr} = -z\frac{d^2\xi}{dr^2} + \frac{du}{dr}, \quad \varepsilon_{\varphi\varphi} = -\frac{z}{r}\frac{d\xi}{dr} + \frac{u}{r}, \quad \varepsilon_{r\varphi} = 0 \tag{9.27}$$

(in the following, the double indices will be mostly omitted). As a result, the elastic strain of the kth layer ($k = 1, 2, \ldots, n$; for the substrate $k = s$) is

$$\varepsilon^k_{el,\alpha\beta} = \varepsilon^k_{\alpha\beta} - \varepsilon^k_{in,\alpha\beta}. \tag{9.28}$$

The material properties are assumed to be linear elastic and isotropic; moreover, the inelastic strains can be written as

$$\varepsilon_{in,\alpha\beta} = \varepsilon_{in}(z)\delta_{\alpha\beta}, \tag{9.29}$$

i.e., at the moment, they do not depend on r but can be different in each layer and the substrate. The physical meaning of ε^k_{in} is that each free layer would be strained exactly by ε_{in}^k upon detaching from the other layers and the substrate and thereafter being stress-free; the value of the inelastic strain depends on the epitaxial process.

Under these assumptions, the in-plane stresses are

$$\begin{pmatrix} \sigma_r^k \\ \sigma_\varphi^k \end{pmatrix} = \begin{pmatrix} c_{11}^k & c_{12}^k \\ c_{21}^k & c_{22}^k \end{pmatrix} \begin{pmatrix} \varepsilon_r - \varepsilon_{in}^k \\ \varepsilon_\varphi - \varepsilon_{in}^k \end{pmatrix} \tag{9.30}$$

with $c_{11}^k = c_{22}^k = E_k/(1-v_k^2)$ and $c_{12}^k = c_{21}^k = v_k E k/(1-v_k^2)$, E_k and v_k are the elasticity modulus and the Poisson-number of the kth layer (for the substrate $k=s$). Without going into details, one can show the following relations:

$$\xi(r) = \frac{1}{2}\kappa r^2, \quad u(r) = \varepsilon_0 r, \tag{9.31}$$

κ and ε_0 being constants. κ denotes the curvature, which can be measured in situ experimentally. We assume the "plate" being free from external stresses, moments, or volume forces, and being fixed at $r=z=0$, or more precisely: $u(0)=\xi(0)=\xi'(0)=0$; in general, by variation of an energy functional, one obtains differential equations for $\xi(r)$ and $u(r)$, which are decoupled in the present case of linear approximation of the strain tensor. For the stresses, one obtains

$$\sigma_r^k = \sigma_\varphi^k = M_k(-z\kappa + \varepsilon_0 - \varepsilon_{in}^k(z)), \tag{9.32}$$

$M_k = E_k/(1-v_k)$. The constants κ and ε_0 are determined from the boundary conditions, i.e., that there are no integral stresses and moments in each slice plane $r=$const. [Finot 1996, Freund 2000].

$$\int_{-h_s}^{hh} \sigma\,dz = \int_{-h_s}^{hh} \sigma z\,dz = 0. \tag{9.33}$$

One obtains a linear equation system:

$$B_0\varepsilon_0 - C_0\kappa = P,$$

$$C_0\varepsilon_0 - D_0\kappa = Q,$$

B_0, C_0, and D_0 are stiffness matrices:

$$B_0 = \int M\,dz, \quad C_0 = \int Mz\,dz, \quad \text{and} \quad D_0 = \int Mz^2\,dz. \tag{9.34}$$

P and Q are the force in radial direction associated with the inelastic strains and the bending moment, respectively, integrated over the thickness

$$P = \int M\varepsilon_{in}\, dz, \quad Q = \int M\varepsilon_{in}z\, dz, \tag{9.35}$$

and the solution is

$$\kappa = \frac{C_0 P - B_0 Q}{B_0 D_0 - C_0^2}, \quad \varepsilon_0 = \frac{D_0 P - C_0 Q}{B_0 D_0 - C_0^2}. \tag{9.36}$$

In a one-layer system with thickness h_f and constant inelastic strains ε_{in}^s in the substrate and ε_{in}^f in the film and each independent from z it follows:

$$\kappa = -6 \frac{M_f(\varepsilon_{in}^f - \varepsilon_{in}^s)h_f(1 + h_f/h_s)}{M_s h_s^2 FN(m,h)}, \tag{9.37a}$$

$$\sigma^f(z) = -M_f\left(-6\frac{mh(1+h)}{h_s}z + 1 + 3mh^2 + 4mh^3\right)\frac{\varepsilon_{in}^f - \varepsilon_{in}^s}{FN(m,h)}, \tag{9.37b}$$

and

$$\sigma^s(z) = M_s\left(6\frac{mh(1+h)}{h_s}z + mh(4 + 3h + mh^3)\right)\frac{\varepsilon_{in}^f - \varepsilon_{in}^s}{FN(m,h)}. \tag{9.37c}$$

At the substrate to film interface there is a stress discontinuity

$$\Delta\sigma = \sigma^f(0) - \sigma^s(0) = -M_f\frac{1 + 4h + 3mh^2 + 4mh^3 + 3h^2 + mh^4}{FN(m,h)}(\varepsilon_{in}^f - \varepsilon_{in}^s),$$

$$\tag{9.37d}$$

the same effect occurs also in multilayers at all interfaces. In the equations above m is $m = M_f/M_s$, $h = h_f/h_s$ and $FN\,(m, h) = 1 + 4mh + 6mh^2 + 4mh^3 + m^2h^4$. The result of Equation 9.37a has also been obtained by, e.g., Freund et al. [Freund 2000], Röll [Röll 1976], or Kroupa et al. [Kroupa 1993]. Generally, in thin films with $h \ll 1$, the z-dependence of σ^f can be neglected because it depends on h^2; the same holds for all other similar or higher-order terms. FN can be approximated very well by $FN = 1 + 4\,mh$ (compare also to Equation 9.41).

When confining to the leading terms in h for κ and σ, one obtains the Stoney approximation [Stoney 1909]:

$$\kappa = -6\frac{M_f(\varepsilon_{in}^f - \varepsilon_{in}^s)h_f}{M_s h_s^2}, \quad \sigma^f = -M_f(\varepsilon_{in}^f - \varepsilon_{in}^s), \quad \sigma^s = 0, \tag{9.38}$$

and

$$\Delta\sigma = -M_f(\varepsilon_{in}^f - \varepsilon_{in}^s).$$

9.3.2 Inelastic Strains

For further considerations, it is necessary to specify the inelastic strains. In the layers as well as in the substrate, we split them in two parts:

$$\varepsilon_{in}^f = \alpha^k \Delta T + e^k, \tag{9.39}$$

α is the linear thermal expansion coefficient (we assume isotropy) and e the epitaxial strain, which in the case of heteroepitaxy is not only determined by the lattice mismatch but influenced by a variety of effects such as nucleation grain size, grain boundaries, number and type of dislocations, growth conditions, etc. In the substrate, there is only heat expansion and always $e^s = 0$. $\Delta T = T(t) - T_R$ is the difference between the actual growth temperature and a reference temperature, mostly room temperature. In the case of a temperature-dependent α, the thermal part must be replaced by

$$\int_{T_R}^{T} \alpha \, dT.$$

This is possible in all equations without difficulties. If there is a temperature gradient across the thickness, i.e., if there is a z-dependence in T, the exact equations (9.36) and (9.32) for κ and σ should be used. Then, the integrals in P and Q must be explicitly done. As a prerequisite, the functional dependence of the temperature gradient must be known. We have calculated that for the case of GaN on silicon a small temperature difference of $1°$–$2°$ between the bottom substrate side and the top layer side would yield a strong curvature that is actually not observed during in situ measurements. Thus, temperature gradients can be neglected in this system. However, in the case of GaN/sapphire, this effect seems to present due to the low heat conductivity of sapphire [Brunner 2008].

Obviously, in general, the epitaxial term e^k is not a constant but a function of thickness, e.g., when doping GaN with silicon during growth [Krost 2003]. For some applications, this can be considered as a higher-order effect and the calculations can be done piecewise with constant e^k whereby one layer k must be formally split in several "single layers." Therefore, it seems to be reasonable to provide approximate equations for such a case, also because this condition is fulfilled in most cases, at least by the thermal part of the inelastic strains—and with the decomposition as given in Equation 9.39, the additive splitting in a thermal and an epitaxial part is also valid for the curvatures and stresses

$$\kappa = \kappa_T + \kappa_e \quad \text{and} \quad \sigma = \sigma_T + \sigma_e. \tag{9.40}$$

For instance, when growth is started at a certain temperature $T = T_R$, then the sample is heated up to a higher temperature, and afterward cooled down again to T_R, only the κ_e- and σ_e-parts will be left over, irrespective of the concrete temperature characteristics. However, nearly no quantitative conclusions can be drawn from the final state alone. One needs the in situ monitoring of the growth process, which is considerably determined by the temperature.

9.3.3 Approximation for Piecewise Constant Inelastic Strains

The following approximations for κ and σ stem from the exact equations (9.36) and (9.32), if one takes into account only terms up to a certain order in $h = d/h_s$ whereby d is a length in the order of the layer system thickness. In the case of κ, only terms up to h^2 are considered because the leading term is already proportional to h; for σ, the leading term is of the order $h^0 = 1$ (Stoney approximation) and the approximation will be confined to the order h^1. One finds

$$\kappa = -\frac{6}{M_s h_s^2} \frac{\displaystyle\sum_{l=1}^{n} M_l d_l \left(1 + \frac{h_l + h_{l-1}}{h_s}\right)\left(\varepsilon_{in}^l - \varepsilon_{in}^s\right)}{1 + 4\displaystyle\sum_{l=1}^{n} \frac{M_l d_l}{M_s h_s}} \tag{9.41}$$

and

$$\sigma^k = -\frac{M_k}{1 + 4\displaystyle\sum_{l=1}^{n} \frac{M_l d_l}{M_s h_s}} \left[\varepsilon_{in}^k - \varepsilon_{in}^s + 4\sum_{l=1}^{n} \frac{M_l d_l(\varepsilon_{in}^k - \varepsilon_{in}^l)}{M_s h_s}\right] - M_k \kappa z, \tag{9.42}$$

$$\varepsilon_{in}^l - \varepsilon_{in}^s = (\alpha^l - \alpha^s)\Delta T + e^l, \quad \text{respectively,} \quad \varepsilon_{in}^l - \varepsilon_{in}^s = \int_{T_R}^{T}(\alpha^l - \alpha^s)dT + e^l, \quad e^s = 0$$

These equations are valid for the complete multilayer system with n layers of thickness $d_l = h_l - h_{l-1}$, the top side of the lth layer is located at $z = h_l$ ($h_0 = 0$ corresponds to the substrate surface). The last term of the layers in σ is of the order h^2 and can be neglected (but not in the substrate, $k = s$). During multilayer growth, the actual thickness $hh(t)$ runs from zero to h_n (end of growth), i.e., when simulating the growth, the last summation index is not

n but $n_1 + 1$, with n_1 the index of the last terminated layer and $d_{n1+1} = hh(t) - h_{n1}$ and $h_{n1+1} = hh(t)$. One recognizes the functional similarity for the curvature between Equation 9.41 and the solution by Freund for the one-layer system (Equation 9.37) if one factorizes it up to the same order. This is especially valid for the denominator; the nominator of Equation 9.41 is already the exact expression for the one-layer system but only an approximation for a multi-layer system because the relative contributions of the layers with respect to each other, which are $\propto (\varepsilon_{in}^l - \varepsilon_{in}^j)$ and at least of the order h^3, are missing. In the series expansion, the nominator of κ and σ leads besides a linear term in h also to higher-order terms, which should be, strictly speaking, omit-ted; however, the numerical results show that Equations 9.41 and 9.42 are a very good approximations. The numerical simulations of real AlGaN/GaN multilayers on silicon ($h_s = 275$ μm, hh up to 5 μm) show that one can further simplify Equation 9.41 without significant loss of exactness by neglecting the $(h_l + h_{l-1})/h_s$ terms (with an error of 1%–2%).

In the Stoney approximation, one obtains

$$\kappa = -\frac{6}{M_s h_s^2} \sum_{l=1}^{n} M_l d_l (\varepsilon_{in}^l - \varepsilon_{in}^s) \tag{9.43}$$

and

$$\sigma^k = -M_k (\varepsilon_{in}^k - \varepsilon_{in}^s). \tag{9.44}$$

During growth, with n_1 finished layers and $k = n_1 + 1$ the curvature is

$$\kappa = -\frac{6}{M_s h_s^2} \left(\sum_{l=1}^{n_1} M_l d_l (\varepsilon_{in}^l - \varepsilon_{in}^s) + M_k (hh(t) - h_{n_1}) (\varepsilon_{in}^k - \varepsilon_{in}^s) \right).$$

9.3.4 Determination of the Epitaxial Inelastic Strains

We assume that in the ideal case, the time-dependent parameters $T(t)$, $hh(t)$, and $\kappa(t)$ can be measured experimentally and the materials parameters like the temperature-dependent thermal expansion coefficient, the elastic mod-uli, and Poisson numbers are well known (this is actually not the case).

9.3.4.1 Piecewise Constant Inelastic Strains e^k

The evaluation of the epitaxial inelastic strains e^k of the kth layer can be performed by different approaches. The simplest way is for piecewise con-stant e^k; this is always the case if the experimental κ–$hh(t)$-curve consists of straight lines at least approximately (where the required one layer must be split into several layers). Then, the epitaxial elastic strains e^k can be evaluated successively via numerical simulation with the exact equations (Equation 9.36) for the curvature or the approximate equations (Equation 9.41), (9.43) starting with the first layer until the whole calculated and measured κ–$hh(t)$-curves of the multilayer system coincide. The numerical analysis shows that the exact equation (Equation 9.36) and the approximation equation (Equation 9.41) yield practically indistinguishable results. For AlGaN/GaN multilayers on Si(110) and (111) the inelastic strains amount to 1–$2*10^{-3}$, i.e., they are two orders of magnitude lower than the lattice mismatch, i.e., $(a_{Si}-a_{GaN})/a_{GaN}=17\%$. The experimental differences in the in situ curvature traces for the same GaN-based light-emitting diode (LED) structure grown on Si(110) and Si(111) can be calculated in this way. We measured in situ the curvature during the growth of InGaN/GaN LED structures on silicon (111) and (110) substrates by metal organic vapor phase epitaxy. The growth was performed at 1100°C. For more details on this topic, see Ref. [Reiher 2009a]. Due to the different surface symmetries, the nucleation of the first AlN nuclei leads to a slight difference in strain on the two substrates, which in turn results in completely different wafer curvatures at the end of the growth procedure and after cooling down. In Figure 9.16 the experimental results together with the calculated ones using the piecewise constant method are shown.

Most probably, the origin of the different behavior between the nitride growth on Si(111) and Si(110) is already located in the first seed layer the modeling of which seems to be the biggest problem.

Once the e^k are known the stresses can be calculated via Equations 9.42 or 9.32.

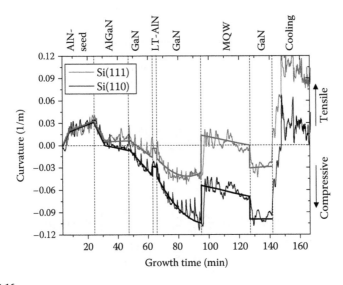

FIGURE 9.16

Experimental curvature traces for the growth of identical nitride multilayer structures on Si(111) (upper curves) and on Si(110) (lower curves) and simulations using the piecewise constant method as described in the text.

9.3.4.2 Variable Inelastic Strains e^k

When there is a curvature in the κ–hh-measurement of one or more layers, this is an indication of a variable z-dependent inelastic strain $e^k(z)$, which can be even geometrically nonlinear [Clos 2004].

We assume the layer k being the first one with a curved κ–hh-dependence and n_1 underlying layers with piecewise straight κ–hh-curves and known inelastic strains as determined by the method described above. Moreover, there should be no temperature gradient and during growth of the kth layer, the temperature remains constant. Because of the linearity of the basic equations in P and Q and Equation 9.39, κ can be separated in

$$\kappa(hh) = \kappa_{m_1}(hh) + \kappa_T^k(hh) + \kappa_e^k(hh), \qquad (9.45)$$

$$\kappa_e^k = \Delta\kappa = \kappa - \kappa_{m_1} - \kappa_T^k,$$

κ_e^k is the curvature part caused by the inelastic strain e^k of the kth layer (the variable thickness is $\zeta = hh(t) - h_{n_1}$, κ_T^k is the thermal part, κ_{n_1} the curvature part of the n_1 finished layers, and $\kappa(hh)$ is the experimentally measured curvature κ_{exp}). It should be pointed out that κ_{n_1} depends on ζ, thus $\kappa_{exp} - \kappa_{n_1}$ is not simply a zero point correction. However, the ζ-, respectively, the hh-dependence of κ_{n_1} stems only from the nominator in Equation 9.41. Let κ_a be the value of κ_{n_1} at the growth start of the kth layer (this is simply the experimental value κ_{exp} at this point), in the framework of Equation 9.41 it follows

$$\kappa_{m_1}(\zeta) = \frac{1 + s_{m_1}}{1 + s_{m_1} + 4\dfrac{M_k \zeta}{M_s h_s}} \kappa_a,$$

with

$$s_{m_1} = 4\sum_{l=1}^{n_1} \frac{M_l d_l}{M_s h_s},$$

and

$$\Delta\kappa = \kappa_{exp} - \frac{(1 + s_{n_1})\kappa_a - \dfrac{6M_k}{M_s h_s^2}\zeta\left(1 + \dfrac{\zeta + 2h_{m_1}}{h_s}\right)(\alpha^k - \alpha^s)\Delta T}{1 + s_{m_1} + 4\dfrac{M_k \zeta}{M_s h_s}}. \qquad (9.46)$$

The epitaxially caused curvature part κ_e^k can be expanded analogously to Equation 9.41, and in the same order of approximation, one obtains

$$\int_0^\zeta e^k(y)\left(1+2\frac{y+h_{m_1}}{h_s}\right)dy = -\frac{M_s h_s^2}{6M_k}\left(1+4\sum_{l=1}^{m_1}\frac{M_l d_l}{M_s h_s}+4\frac{M_k\zeta}{M_s h_s}\Delta\kappa(\zeta)\right) = f(\zeta),$$

$$(9.47)$$

$\zeta = hh(t) - h_{n_1}$, and for the inelastic strain $e^k(\zeta)$ of the kth layer

$$e^k(\zeta) = \frac{f'(\zeta)}{1+2\dfrac{\zeta+h_{m_1}}{h_s}},$$

$$(9.48)$$

f' is the first derivation of f. According to Equations 9.46 and 9.47, the function $f(\zeta)$ is calculated from the experimental curvature trace, which has been smoothed before and fitted, e.g., by a polynom. Thus, e^k can be completely calculated from the experimental values κ_{exp}, κ_a, the thicknesses and ΔT, as well the materials parameters M_s, M_l, α^s, and α^k. In particular, it is not necessary to know the inelastic strains of the underlying layers. Therefore, the layer k under consideration can be an arbitrary one; it must not be the first one with variable epitaxy strain e^k.

The stress in the layer that is calculated in the same approximation, which has been used for Equation 9.42 (the kth layer is the first with variable e^k, the n_1 underlying layers have piecewise constant e^l and must be known for the stress calculation) amounts to

$$o^k(x,\zeta) = -\frac{M_k}{1+s_{m_1}+4\dfrac{M_k\zeta}{M_s h_s}}$$

$$*\left[\varepsilon_{in}^k(x)-\varepsilon_{in}^s+\frac{4}{M_s h_s}\left(\sum_{l=1}^{m_1}M_l d_l(\varepsilon_{in}^k(x)-\varepsilon_{in}^l)+M_k(e^k(x)\zeta-\int_0^\zeta e^k dy)\right)\right].$$

$$(9.49)$$

The variable x denotes the position in the kth layer ($0\le x\le\zeta$), ζ is its actual thickness, in the substrate one has to add $-zM_s\kappa$. The quality of the approximation can be checked by additionally calculating the stress within the exact linear equations (Equations 9.32, 9.34 through 9.36). The results for a one-layer system, namely, a Si-doped GaN layer on Si(111) are shown in Figure 9.17. The thickness of the nucleation layer was omitted.

A considerable simplification results if one applies the Stoney approximation:

$\kappa_{n_1} = \kappa_a = $ const., i.e., independent of $hh(t)$

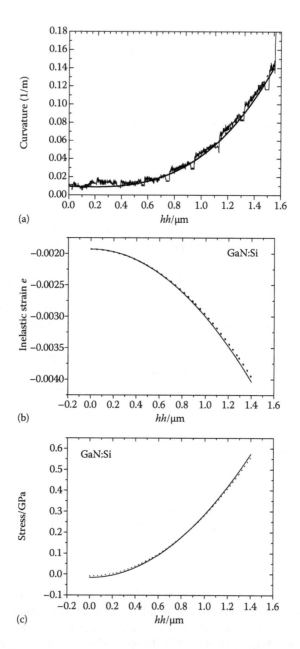

FIGURE 9.17

(a) Measured curvature (noisy) for a 1.55 μm Si-doped GaN layer on Si(111) and calculated ones using the exact (Equation 9.37a, dotted) and the approximation (Equation 9.41, full); (b) corresponding inelastic strain (full line) calculated with Equations 9.47 and 9.48, and within the Stoney approximation (dotted, Equations 9.50 and 9.51); (c) corresponding stress calculated on the basis of the inelastic strain formula (Equations 9.47 and 9.48) with the exact linear equations (full curve), and calculated on the basis of the Stoney approximation for the inelastic strain with the exact linear equations (dotted curve).

$$\Delta\kappa(\zeta) = \kappa_{exp}(\zeta) - \kappa_a + \frac{6M_k\zeta}{M_s h_s^2}(\alpha^k - \alpha^s)\Delta T \qquad (9.50)$$

and

$$e^k(\zeta) = -\frac{M_s h_s^2}{6M_k}\Delta\kappa'(\zeta). \qquad (9.51)$$

In the Stoney approximation the contributions κ^k (thermal and epitaxial ones) of the individual layers to the total curvature are independent with respect to each other. Therefore, the second term in Equation 9.50 causes only a zero point correction (it can be taken directly from the experimental curvature trace). According to Equation 9.51, in this approximation, a piecewise constant curvature exactly delivers piecewise constant e^k.

For one thin layer (1.4 µm, $h_s = 275$ µm) the Stoney approximation yields satisfactory results (Figure 9.17b, dotted curve).

9.4 Summary

In summary, expressions for the determination of inelastic strain and stresses from in situ curvature measurements are given for layers of arbitrary thickness and strong wafer bending. It is shown that the limits of the Stoney formula are easily reached in the case of, e.g., GaN on Si, large diameter wafers, thick layers, or vertical temperature gradients. In the case of GaN on silicon, strong bending and associated nonlinear geometrical effects are easily reached during epitaxial growth. The important message is that the linear theory underestimates the strain as compared to the nonlinear theory. Moreover, multilayer systems have been described in linear theory. Exact equations for an n-layer system and corresponding approximations have been developed.

Acknowledgment

The authors would like to thank the Deutsche Forschungsgemeinschaft for financial support in the framework of the Collaborative Research Center SFB787.

References

[Brunner 2007] F. Brunner, V. Hoffmann, A. Knauer, E. Steimetz, T. Schenk, J.-T. Zettler, and M. Weyers, Growth optimization during III-nitride multiwafer MOVPE using realtime curvature, reflectance and true temperature measurements, *J. Cryst. Growth* **298**, 202 (2007).

[Brunner 2008] F. Brunner, A. Knauer, T. Schenk, M. Weyers, and J.-T. Zettler, Quantitative analysis of in situ wafer bowing measurements for III-nitride growth on sapphire, *J. Cryst. Growth* **310**, 2432 (2008).

[Ciarlet 2004] P. G. Ciarlet, *Mathematical Elasticity* (Vol. 1: Three Dimensional Elasticity), North-Holland, Amsterdam, the Netherlands, 2004.

[Clos 2004] R. Clos, A. Dadgar, and A. Krost, Wafer curvature in the nonlinear deformation range, *Phys. Stat. Sol. (a)* **201**, R75 (2004).

[Finot 1996] M. Finot and S. Suresh, Small and large deformation of thick and thin-film multi-layers: Effect of layer geometry, plasticity and compositional gradients, *J. Mech. Phys. Solids* **44** (5), 683 (1996).

[Floro 2001] J. A. Floro and E. Chason, Curvature-based techniques for real-time stress measurements during thin-film growth, p. 191, in: *In Situ Real-Time Characterization of Thin Films*, edited by O. Auciello and A. R. Krauss, John Wiley & Sons, Inc., New York, 2001.

[Floro 2002] J. A. Floro, E. Chason, R. C. Cammarata, and D. J. Srolovitz, Physical origins of intrinsic stresses in Volmer–Weber thin films, *MRS Bull.* **27**, 19 (2002).

[Freund 1997] L. B. Freund, The mechanics of a free-standing strained film/compliant substrate system, *Mater. Res. Soc. Symp. Proc.* **436**, 393 (1997).

[Freund 1999] L. B. Freund, J. A. Floro, and E. Chason, Extensions of the Stoney formula for substrate curvature to configurations with thin substrates or large deformations, *Appl. Phys. Lett.* **74**, 1987 (1999).

[Freund 2000] L. B. Freund, Substrate curvature due to thin film mismatch strain in the nonlinear deformation range, *J. Mech. Phys. Solids* **48**, 1159 (2000).

[Friedrichs 1939] K. O. Friedrichs and J. J. Stoker, The non-linear boundary value problem of the buckled plate, *Proc. Natl. Acad. Sci.* **25**, 535 (1939).

[Friedrichs 1942] K. O. Friedrichs and J. J. Stoker, Buckling of a circular plate beyond the critical thrust, *J. Appl. Mech.* **9**, A7 (1942).

[Fuster 2006] D. Fuster, M. U. González, Y. González, and L. González, In situ measurements of InAs and InP (001) surface stress changes induced by surface reconstruction transitions, *Surf. Sci.* **600**, 23 (2006).

[Guyot 2004] N. Guyot, Y. Harmand, and A. Mezin, The role of the sample shape and size on the internal stress induced curvature of thin-film substrate systems, *Int. J. Solids Struct.* **41**, 5143 (2004).

[Harper 1990] B. D. Harper and C.-P. Wu, A geometrically nonlinear model for predicting the intrinsic film stress by the bending plate method, *Int. J. Solids Struct.* **26**, 511 (1990).

[Hearne 1999] S. Hearne, E. Chason, J. Han, J. A. Floro, J. Figiel, J. Hunter, H. Amano, and I. S. T. Tsong, Stress evolution during metalorganic chemical vapor deposition of GaN, *Appl. Phys. Lett.* **74**, 356 (1999).

[Indenbom 1990] V. L. Indenbom and V. M. Kaganer, X-ray analysis of internal stresses in crystals ii. Lattice distortions due to residual strains in crystals grown from melts, *Phys. Stat. Sol. (a)* **122**, 97 (1990).

[Krost 2003] A. Krost, A. Dadgar, G. Strassburger, and R. Clos, GaN-based epitaxy on silicon: Stress measurements, *Phys. Stat. Sol (a)* **200**, 26 (2003).

[Kroupa 1993] F. Kroupa, Z. Knesl, and J. Vlach, Residual stresses in graded thick coatings, *Acta Tech. CSAV* **38**, 29–74 (1993).

[Landau 1966] L. D. Landau and E. M. Lifschitz, *Theoretische Physik, Bd. VI*, Akademie-Verlag, Berlin, Germany, 1966.

[Liu 2007] W. Liu, J. J. Zhu, D. S. Jiang, and H. Yang, Influence of the AlN interlayer crystal quality on the strain evolution of GaN layer grown on Si (111), *Appl. Phys. Lett.* **90**, 011914 (2007).

[Lynch 2005] C. Lynch, E. Chason, R. Beresford, L. B. Freund, and K. Tetz, Limits of strain relaxation in InGaAs/GaAs probed in real time by in situ wafer curvature measurement, *J. Appl. Phys.* **98**, 073532 (2005).

[Mezin 2006] A. Mezin, Coating internal stress measurement through the curvature method: A geometry-based criterion delimiting the relevance of Stoney's formula, *Surf. Coat. Technol.* **200**, 5259 (2006).

[Mura 1987] T. Mura, *Micromechanics of Defects in Solids*, 2nd edn., Kluwer Academic Publishers, Dordrecht, the Netherlands, 1987.

[Reddy 1999] J. N. Reddy, *Theory and Analysis of Elastic Plates*, Taylor & Francis, Boca Raton, FL, 1999.

[Reiher 2009] F. Reiher, Wachstum von Galliumnitrid-Basierten Bauelementen auf silizium(001)-Substraten mittels metallorganischer Gasphasenepitaxie; Growth of Galliumnitrid-based devices on silicon(001) substrates by metal organic vapor phase epitaxy. Thesis, University of Magdeburg (in German), Magdeburg, Germany (2009).

[Reiher 2009a] F. Reiher, A. Dadgar, J. Bläsing, M. Wieneke, M. Müller, A. Franke, L. Reißmann, J. Christen, and A. Krost, InGaN/GaN light-emitting diodes on Si(110) substrates grown by metalorganic vapour phase epitaxy, *J. Phys. D: Appl. Phys.* **42**, 055107 (2009).

[Röll 1976] K. Röll, Analysis of stress and strain distribution in thin films and substrates, *J. Appl. Phys.* **47** (7), 3224 (1976).

[Salamon 1995] N. J. Salamon and C. B. Masters, Bifurcation in isotropic thin film/substrate plates, *Int. J. Solids Struct.* **32**, 473 (1995).

[Stoney 1909] G. Stoney, The tension of metallic films deposited by electrolysis, *Proc. R. Soc. London, Ser. A* **82**, 172 (1909).

[Swartzentruber 1990] B. S. Swartzentruber, Y.-W. Mo, M. B. Webb, and M. G. Lagally, Observations of strain effects on the Si(001) surface using scanning tunneling microscopy, *J. Vac. Sci. Technol.* **A8**, 210 (1990).

[Terao 2001] S. Terao, M. Iwaya, R. Nakamura, S. Kamiyama, H. Amano, and I. Akasaki, Fracture of $Al_xGa_{1-x}N$/GaN heterostructure—Compositional and impurity dependence, *Jpn. J. Appl. Phys.* **40**, L195 (2001).

[Wittrick 1953] W. H. Wittrick, D. M. Myers, and W. R. Blunden, Stability of a bimetallic plate, *Quart. J. Mech. Appl. Math.* **VI**, 15 (1953).

10

X-Ray Characterization of
Group III-Nitrides

Alois Krost and Jürgen Bläsing

CONTENTS

10.1 Introduction

So far, most GaN-based layers are grown heteroepitaxially on foreign substrates, such as sapphire, SiC, or silicon. In spite of a remarkable progress in the growth of high-quality group III-nitrides, the epitaxial films are far from being perfect as compared to lattice-matched structures in the conventional III–V heterosystems, such as (In,Ga,Al)As/InP or the nearly lattice-matched AlGaAs/GaAs. Large bulk single-crystal GaN substrates are not really available yet and, therefore, a large-scale homoepitaxial growth of GaN/GaN will not be possible in the near future. Therefore, apart from laser applications, heteroepitaxial growth will dominate the GaN growth in the near future. With all heterosubstrates, the large lattice mismatch leads to a high dislocation density of the order of 10^{10} cm^{-3} or higher near the interface. In addition, in the case of GaN on Si, the large difference in the linear thermal expansion coefficients leads to strong cracking problems. To a great extent, the real structure characterization of such defected layers relies on different x-ray diffraction methods. In addition, due to wafer bending during growth, there is an increasing demand for mapping methods to get an impression of the lateral homogeneity, especially in the case of large diameter wafers. Moreover, in recent years, the growth of nonpolar or semipolar nitrides has come into play with the *c*-axis of the wurtzite-type structure being perpendicular or inclined to the growth direction.

10.2 Crystal Structure and Mosaicity

10.2.1 Lattice Parameters and Thermal Expansion Coefficients

In contrast to the conventional III–V compounds that crystallize in the cubic zincblende-type structure, all III-nitrides (Al,Ga,In)N of technological importance crystallize in the hexagonal wurtzite-type structure. Figure 10.1 shows the hexagonal arrangement of the gallium and the nitrogen atoms as well as the primitive unit cell with the lattice parameters listed in Table 10.1. The volume of the primitive unit cell is 1/3 of the hexagonal one. It should be noted that the published lattice parameters differ from each other and strongly depend on the growth process as well as on the impurity concentration. In the wurtzite-type crystals, each metal atom is surrounded tetrahedrally by a nitrogen atom; however, there is an asymmetry of the four tetrahedrally bond lengths with the bond parallel to the *c*-axis somewhat different to the three others leading to spontaneous polarization. In a real

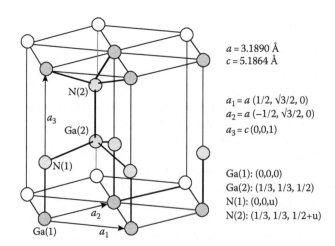

$a = 3.1890$ Å
$c = 5.1864$ Å

$a_1 = a\,(1/2, \sqrt{3}/2, 0)$
$a_2 = a\,(-1/2, \sqrt{3}/2, 0)$
$a_3 = c\,(0,0,1)$

Ga(1): $(0,0,0)$
Ga(2): $(1/3, 1/3, 1/2)$
N(1): $(0,0,u)$
N(2): $(1/3, 1/3, 1/2+u)$

FIGURE 10.1
Hexagonal arrangement of wurtzite-type GaN with unit cell indicated by shaded atoms.

TABLE 10.1

Lattice Parameters and Linear Thermal Expansion Coefficients

	Lattice Parameters (nm)	Linear Thermal Expansion Coefficient α_a (10^{-6} K^{-1})	Linear Thermal Expansion Coefficient α_c (10^{-6} K^{-1})
GaN on sapphire (hexagonal, wurtzite)	$a = 0.31892$ [Detchprohm 1992] $c = 0.51850$ [Detchprohm 1992]	5.59 (300–900 K) [Maruska 1969] 3.1 (300–350 K) [Leszczynski 1995]	3.17 (300–700 K) [Maruska 1969] 7.75 (700–900 K) [Maruska 1969]
GaN bulk crystals (hexagonal, wurtzite)	$a = 0.31879–0.31894$ [Leszczynski 1995, 1996] $c = 0.51856–0.51865$ [Leszczynski 1995, 1996]	6.2 (700–750 K) [Leszczynski 1995] 5.0 (250–600 K) [Kirchner 2000]	2.8 (300–350 K) [Leszczynski 1995] 6.1 (700–750 K) [Leszczynski 1995] 4.5 (250–600 K) [Kirchner 2000]
Sapphire (trigonal)	$a = 0.47577$ [Leszczynski 1995] $c = 1.29907$ [Leszczynski 1995] $\Delta a/a = (\sqrt{3} \cdot a_{GaN} - a_{sapph})/a_{sapph}$ $= 16.1\%$ [Kung 1994]	4.3 (300–350 K) [Leszczynski 1995] 9.2 (700–750 K) [Leszczynski 1995]	3.9 (300–350 K) [Leszczynski 1995] 9.3 (700–750 K) [Leszczynski 1995]
SiC (hexagonal)	$a = 0.30803$ [Leszczynski 1995] $c = 1.51071$ [Leszczynski 1995] $\Delta a/a = 3.54\%$	3.2 (300–350 K) [Leszczynski 1995] 4.2 (700–750 K) [Leszczynski 1995]	3.2 (300–350 K) [Leszczynski 1995] 4.0 (700–750 K) [Leszczynski 1995]
Si (cubic)	$a = 0.54301$ $a(111) = a/\sqrt{2} = 0.384$ $\Delta a/a = -16.9\%$	2.56 (300 K) [Landolt 1982] 4.42 (1400 K) [Landolt 1982]	

crystal, the *c*-axes of the individual crystallites constituting the macroscopic crystal are tilted with respect to each other due to grain boundaries and also rotated azimuthally with respect to each other. The corresponding macroscopic orientation distributions are called tilt and twist. Due to the lack of bulk GaN homosubstrates, so far most nitride films are grown heteroepitaxially on sapphire, SiC, or silicon substrates with all its drawbacks, especially the expitaxial and the thermal mismatch strain, the latter being the major difficulty in the case of the silicon substrate. In Table 10.1, the lattice parameters and linear thermal expansion coefficients of GaN, and the substrates are summarized. The temperature dependence of the linear thermal expansion coefficient was investigated by Roder et al. [Roder 2005] in the temperature range from 12 to 1025 K. Unfortunately, there are no available data on the linear thermal expansion coefficients at growth temperature (>1300 K) yet. Recently, the temperature dependence for AlN was published by Figge et al. [Figge 2009]. For a compilation of more data on group-III-nitrides, see Ref. [Edgar 1994]. For silicon, the recommended values in the temperature range from 0 to 1000 K are given by Swenson [Swenson 1983], and very recently by Mazur and Gazik [Mazur 2009], i.e., up to 1100°C. Most interestingly, they measured considerably different values for different crystal directions <$\bar{1}$10>, <–1–12>, and <111>. Such differences may contribute to the stress balance of a film and can be measured directly by in situ curvature measurements.

All layers show a more or less pronounced columnar hexagonal mosaic structure, which means that they consist of a large number of small nearly perfect hexagonal prisms with their basal (0001) plane parallel to the substrate surface. The diameter of the individual columns ranges from some nm to about some μm depending on the growth method and growth conditions. Thereby, the basal plane of the individual crystallites can be slightly twisted against each other and the individual columns' axes are slightly tilted with respect to each other. The degree of tilting and twisting and the diameter of the prisms determine the macroscopic crystal quality. Microscopically, the tilting is correlated to the mixed threading dislocation density and twisting to the edge threading dislocation density [Metzger 1998]. Typically, GaN mosaic blocks on sapphire or silicon have a tilting distribution over an angular range between 0.05° and 0.2° and a somewhat larger twisting distribution between 0.1° and 1°. For high-quality samples (for further discussion see Ref. [Ratnikov 2000]) these parameters can be evaluated by measuring the whole intensity spread (ω-scan), around two independent Bragg reflections, with an analyzer or a small slit in the front of the detector, e.g., the (0002) reflection parallel to the *c*-axis and the (10-10) or (11-20) reflection perpendicular to the *c*-axis. Alternatively, the information can be extracted from the broadening of corresponding reciprocal space maps (RSMs). Very often, only the full widths at a half maxima (FWHM) of the (0002) rocking curve is given, which contains no information on twisting.

Experimentally, the (0002) rocking curve is obtained by rotating the sample through the x-ray beam keeping the wide open detector fixed. Typically, the (0002) rocking curves of good quality GaN/sapphire show FWHMs between 300 and 500 arcsec. It should be noted, however, that the (0002) rocking curve of a thick, perfect GaN single crystal is only ~17 arcsec as calculated within the dynamical theory of x-ray diffraction. For the a-plane (11-20) and m-plane (10-10) reflections it is even less, around 10-11 arcsec. These values are strictly valid only for semi-infinite layer thicknesses, in practice around 3 μm (see Figure 10.2). The rocking curve of a thin layer is broadened due to the finite number of scattering planes, e.g., for the (0002) reflection of a 100 nm perfect GaN, one obtains a FWHM of 146 arcsec, and for 50 nm, the finite thickness leads to 292 arcsec (see Figure 10.3). In such a case, the intensity is modulated by interference oscillations that can be understood as the square of the Fourier transform of a single slit, and from which the layer thickness can be deduced. Until now, however, "single crystalline," thick GaN crystals grown, e.g., by hydride vapor-phase epitaxy, are far from being perfect. As a best value, we measured 41 arcsec for the (10-10)

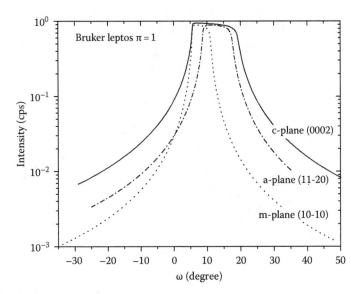

FIGURE 10.2
Theoretical rocking curves (π-polarized light) of perfect, semi-infinite GaN for different lattice planes as calculated within the dynamical theory of x-ray diffraction using the program Leptos from Bruker. A thick, perfect GaN crystal would have a rocking curve width of 10 (a-plane) to 17 (c-plane) arcsec.

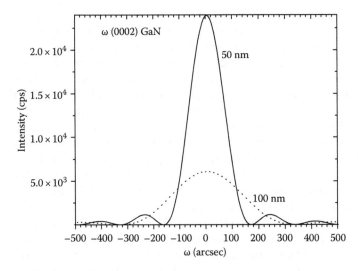

FIGURE 10.3
Theoretical rocking curve widths in the case of GaN thin films with thicknesses of 50 and 100 nm.

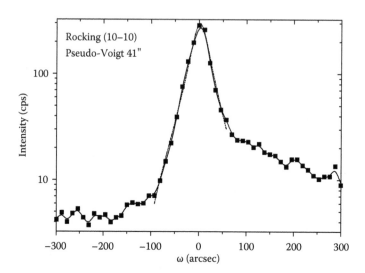

FIGURE 10.4
Experimental (10–10) rocking curve of a piece of 200 μm thick GaN thin film (squares). The measured curve was fitted by a pseudo-Voigt (dotted line).

reflection of a small (some mm²) piece of thick GaN grown by hydride vapor-phase epitaxy (see Figure 10.4).

In some cases, extremely low values, below 100 arcsec, have been reported for the (0001) reflection of the *c*-axis oriented GaN. However, one should be careful when looking for such data. First of all, one should verify whether a rocking curve was really measured and, secondly, one should look for the shape of the measured intensity distribution. Very often, there is in fact a very sharp correlation peak present that is superimposed on a broader background structure. As an example, the (0004) ω-scan of a ~100 nm AlInN layer on sapphire is shown in Figure 10.5. Clearly two components, namely, a sharp peak superimposed on a broader diffuse background is visible. The width of the sharp peak is limited by the resolution of the experimental setup and has nothing to do with the rocking curve width or the crystalline quality. Once such a feature is observed, care should be taken not to mix up the FWHM of the correlation peak as a direct measure for the structural quality in terms of tilt. Instead, both components must be separated in a broad background and a sharp peak. The true information on crystal quality is then given by the diffuse background. It should be mentioned that generally, such measured structures cannot be fitted with Gaussian, Lorentzian, or pseudo-Voigt profiles and that the material quality is overestimated when such correlation peaks come into play. The origin of the correlation peaks might be surface steps, stacking faults, edge dislocation, etc. For more information on this topic, the reader is referred to the paper of Bläsing et al. [Bläsing 2009] and references cited therein and reference [Moram 2009]. Care should also be taken

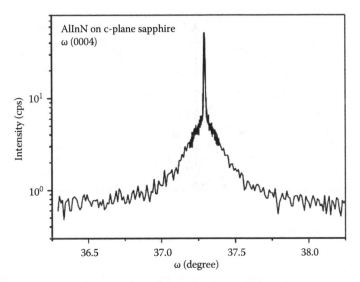

FIGURE 10.5
Experimental (0004) ω-scan of ~100 nm AlInN film grown on c-sapphire. The sharp peak superimposed on a broad background is called the correlation peak and stems from a correlation effect and is not an indication of good crystalline quality (see text).

not to mix up the term "rocking curve" with others. In a true rocking curve, the whole scattered intensity is gathered in a wide open detector at a fixed 2Θ position when rotating the sample through the incoming x-ray beam while in an ω-scan only a part of the totally diffracted intensity is selected through an analyzer crystal or a slit in front of the detector. The rocking curve may be also recorded in an $\omega/2\Theta$-scan without an analyzer crystal. However, it must be ensured that the scattered intensity is collected during the whole measurement in the coupled mode; otherwise, misleading structures due to the finite physical size of the detector opening may appear in the spectrum.

10.2.2 Ewald Sphere and Scans in Reciprocal Space

In Figure 10.6, the reciprocal space for a wurtzite-type GaN(0001) sample is drawn, whereby [0001] is the direction of the c-axis. The angle ω_i is the incoming angle of the primary x-ray beam $\mathbf{k}_i$ with respect to the sample surface, and 2Θ is the scattering angle included by $\mathbf{k}_i$ and the scattered beam $\mathbf{k}_f$. The scattering vector $\mathbf{q}$ ends up on the Ewald sphere. As can be seen from geometrical considerations, the shaded areas are not accessible in backscattering Bragg geometry but only in transmission Laue geometry. There is often confusion in the literature with the different scan modes that are shown schematically in Figure 10.7, as adopted from the book of Holý et al. [Holý 1999].

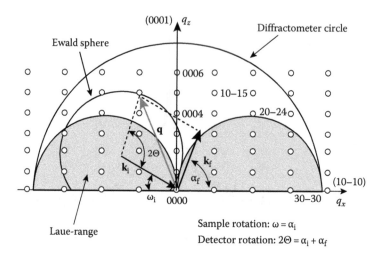

FIGURE 10.6
Schematic drawing of the reciprocal lattice and diffraction conditions for (0001)-oriented hexagonal material. The shaded areas are not accessible in Bragg geometry (see text).

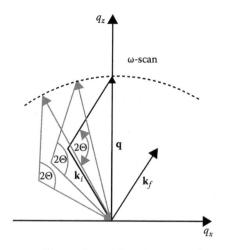

FIGURE 10.7
Schematic drawing of a symmetric ω-scan through a reciprocal lattice point. k_i, k_f, and q denote the incoming, outgoing, and transferred wave vector, respectively. The lattice planes are parallel to q_x (left); schematic drawing of different scan modes through a reciprocal lattice point (right). (After Holy, V. et al. *High Resolution X-ray Scattering from Thin Films and Multilayers* (Springer Tracts in Modern Physics 149), G. Höhler (managing editor), Springer, New York, 1999.)

In an ω-scan the sample is rotated at a fixed 2Θ position of the detector. Thereby, in coplanar geometry, i.e., when the primary beam k_i and the scattered beam k_f have a common scattering plane, the scattering vector $q = k_f - k_i$ moves azimuthally with a constant value around the origin of the reciprocal space. For a symmetric geometry, i.e., when the incoming angle

$\omega = \alpha_i$ of the x-rays with respect to the surface is identical to the outgoing angle α_f of the x-rays with respect to the surface, the scan runs nearly parallel to the surface.

10.2.3 Rocking Curve, ω-Scan, $\Theta/2\Theta$ Scan

In a pure ω-scan, a narrow slit or an analyzer crystal is in front of the detector, so that only a narrow band of the scattered intensity can reach the detector. The bandwidth is determined by the slit width or the acceptance angle of the analyzer crystal. Usually, the rocking curve is obtained in the ω-scan mode, when "rocking" the sample through the x-ray beam with the detector at a fixed position (2Θ) and wide open. This means that all intensity scattered by the sample is gathered during the whole measurement, i.e., integration is made over the wavelength band accepted and scattered by the sample. Wide open means that the physical size of the detector area must be greater than the acceptance angle needed to collect all the scattered intensity. For samples with low defect densities, this condition is usually fulfilled. However, care has to be taken when samples with a strong mosaicity are measured, since the angular spread of the scattered intensity may be very large.

In an $\omega/2\Theta$-scan, the scattering vector is moved radially, i.e., parallel to the reciprocal lattice vector of the Bragg reflection under consideration. This requires a coupled movement of the sample (ω) and the detector (2Θ) in the ratio $\omega/2\Theta = 1/2$. In the special case $\omega = \Theta$, the scan runs parallel to the surface normal. For a *c*-axis grown nitride layer, the *c*-lattice parameter and its distribution, i.e., the vertical strain, is measured by this. In a symmetric (0002) ω-scan, the *c*-axis tilt distribution, together with the in-plane finite size broadening is measured, which is usually interpreted as a measure of pure and mixed threading dislocations in the material. In order to separate rotational disorder and coherence lengths that are given by the lateral crystallite size, higher order reflections should be measured because the rotational disorder does not depend on the reflection order, whereas the coherence length does [Wieneke 2009]. In an (0002) ω-scan, more structural details of the sample are revealed as compared to a rocking curve because the diffuse scattering due to imperfections is not gathered during the measurement and the true broadening is hidden. The same holds for $\Theta/2\Theta$ scans that can be taken with and without an analyzing element in front of the detector. As an example, in Figure 10.8, the true (0002) rocking curve measured without an analyzer and a $\Theta/2\Theta$ scan with an analyzer of a 10-fold InGaN/GaN multiquantum well structure on a GaN buffer layer on a sapphire substrate is shown. Although the superlattice satellite peaks are well visible in both measurements, no finite thickness interference fringes in between. Two adjacent satellites are present in the measurement without an analyzer, indicating the convolution with the mosaicity (finite size and tilt) and a lot of residual diffuse scattering in this sample.

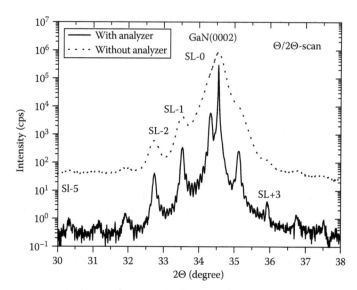

FIGURE 10.8

(0002) rocking curve and Θ/2Θ-scan of a 10-fold InGaN/GaN multiquantum well structure grown on GaN/c-sapphire. In the rocking curve, no finite thickness fringes are resolved.

10.3 Conventional High-Resolution X-Ray Diffraction

10.3.1 Reciprocal Space Mapping

A combination of ω/2Θ scans with different settings of ω in steps of Δω yields a RSM initially in ω/2Θ coordinates. The interpretation of the measurements, however, is much easier in the q-space. The transformation to the q-space is given by

$$q_x = k(\cos\omega - \cos(2\Theta - \omega)),$$
$$q_z = k(\sin\omega - \sin(2\Theta - \omega)),$$

q_x and q_z being the components of the scattering vector parallel and perpendicular to the surface respectively, and $k = 2\pi/\lambda$. For further reading, the reader is referred to [Krost 1996, Bowen 1998, Holý 1999]. Nowadays, a modern triple-axis diffractometer can be operated directly in the q-coordinates or a linear position sensitive detector can be used to get the whole 2Θ distribution in one ω-step, as shown in Figure 10.9. Depending on the reciprocal lattice point chosen, one obtains information on one or both lattice parameters. In literature, for c-axis grown nitrides, very often only a symmetric (0002) map is presented, from which only the c-lattice parameter and its distribution is obtained, and no information on the strain state is available. From its

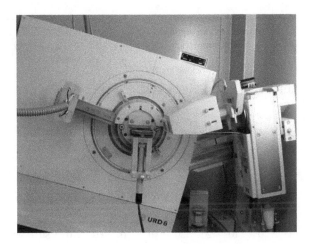

FIGURE 10.9
Diffractometer for fast reciprocal space mapping (GE/Seifert FPM), primary optics: horizontal x-ray source, channel cut compressor, detector: linear position sensitive detector (MBraun).

broadening, the tilt distribution of the *c*-axis is obtained. Tilt is generated by pure and mixed screw dislocations, whereas twist is caused by edge dislocations. In contrast, an RSM around an asymmetric reflection yields, in addition, the *a*-lattice parameter, and with it information on strain, e.g., (11-24) or (10-15). However, in general, the interpretation of the width of such a reflection is not straightforward because the individual contributions of twist and tilt cannot be separated easily. In order to measure the pure twist contribution, an in-plane reflection in grazing incidence diffraction (GID) geometry, e.g., (10-10) or (11-20) is required. However, due to lack of intensity in conventional laboratory x-ray equipment, such type of measurements are rarely performed. Alternatively, a set of measurements at the (0002) reflection and different asymmetric Bragg reflections can be performed in skew symmetric geometry and the results can then be fitted as a function of the tilt angle of the lattice planes with respect to the surface plane. As an example, in Figure 10.10, the FWHM of the rocking curves as well of the ω-scans of a 100 nm ZnO/sapphire measured in skew symmetric geometry are shown. It should be noted that ZnO is also wurtzitic like GaN, and that its growth behavior is very similar. Both sets of data points can be well fitted by a function

$$\text{twist} * \sin(\text{Chi}) + \text{tilt} * \cos(\text{Chi}),$$

whereby, in this example, the tilt values are extremely low and the twist values very large (uncorrelated case, see also Ref. [Sun 2002] and references cited therein). In the (0002) ω-scan, we measured a tilt value of 0.003° that is more than one order of magnitude lower than the corresponding value (0.041°) in the rocking curve. The twist value measured in the (10-10)-scan amounts to 0.694° while that of the rocking curve is 0.765°, as obtained from the fitting function.

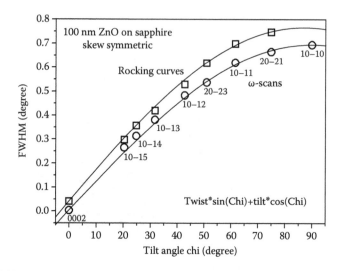

FIGURE 10.10
Full width at half maxima of a (0002) rocking curve and ω-scans of various Bragg reflections of a 100 nm ZnO/c-sapphire and according fits by a sum of sine and cosine functions.

10.3.2 Triple-Axis Diffractometry

In order to get more information on the structural quality, on the lattice parameters and its fluctuations, the strain state of the layer, and on the composition of ternary or quaternary compounds, high resolution x-ray diffraction measurements using a triple-axis diffractometer are indispensable. High-resolution x-ray diffraction and its possibilities are reviewed in detail in Refs. [Krost 1996, Bowen 1998, Holý 1999, Fewster 2000, Als-Nielsen 2001]. Therefore, only some important features are mentioned. A triple axis diffractometer allows the detailed inspection of the intensity distribution of a reciprocal lattice point and its neighborhood using a highly collimated x-ray beam with a narrow bandwidth (Figure 10.11). In most commercial instruments, collimation and monochromatization is reached by a Ge four-crystal monochromator as proposed by DuMond [DuMond 1937] and realized by Bartels [Bartels 1982] for high-resolution x-ray diffraction work. Hereby, the first two crystals serve as a collimator and allow the whole spectral distribution to pass through in a double crystal diffractometer. The third crystal acts as a monochromator with respect to the second one, and thus only a small wavelength range can pass. Finally, the fourth crystal reflects the x-rays in the original direction before entering the four-crystal monochromator. Therefore, the instrument can be introduced in the optical path with minimal readjustment of the optical setup. Using Ge(022) Bragg reflections, the angular divergence is 12″ and the bandwidth $\Delta\lambda/\lambda = 1.5 \times 10^{-4}$. The intensity can be enhanced by more than an order of magnitude by focusing the emitted x-ray beam by a parabolic graded multilayer x-ray mirror in front

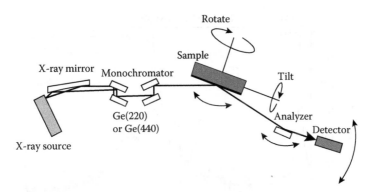

FIGURE 10.11
Schematic setup of a high-resolution triple-axis spectrometer.

of the x-ray tube [Schuster 1995]. In front of the detector, one or more analyzer crystals can be introduced acting as an extremely small slit and accepting scattered radiation only with a divergence <12". Alternatively, a simple Soller collimator can be inserted in the optical path. The benefits of such an instrument are revealed when comparing the rocking curve as described above and the $\omega/2\Theta$ scan obtained with the analyzer crystal in front of the detector (see Figure 10.8). From a symmetric (0002) $\omega/2\Theta$ scan, the distribution and the fluctuation of the c-axis lattice parameter can be evaluated. In a perfect single crystalline layer, both measurements would yield the same result. In Figure 10.12, an extended $\Theta/2\Theta$ scan in the vicinity of the (0002) Bragg reflections is displayed together with a simulation using the dynamical theory. The sample is an LED structure grown by metal organic vapor phase epitaxy (MOVPE) on 6" Si(111). It consists essentially of a thin ~100 nm (Al,Ga)N nucleation layer, followed by a ~4 μm thick GaN buffer layer separated by low temperature AlN interlayers and a 5-InGaN/GaN multiquantum well structure on top. The structure with the highest intensity at $2\Theta = 28.4°$ stems from the Si(111) substrate and the central peak at $2\Theta = 34.6°$ from the thick GaN buffer. The multiquantum well acts as a superlattice with well-pronounced superlattice peaks up to the order of ±5, as indicated. In between, the main superlattice peaks $(N-2) = 5$ finite thickness fringes stemming from the total superlattice thickness are visible indicating nearly perfect interfaces. The broadened structure around the SL + 2 peak is due to the (Al,Ga)N nucleation layer. From the simulation, it follows the well and barrier thicknesses as well as the indium concentration as indicated in the figure. With a triple axis diffractometer, different directions in the reciprocal space can be scanned and combined to a RSM. As an example for a RSM, in Figure 10.13, the intensity distribution around the asymmetric (11-24) Bragg reflection of a fivefold AlInN/GaN multiquantum well layer on GaN on Si(111) is shown. The central peak at $q_x = -6.249$ nm^{-1} and $q_z = 7.718$ nm^{-1} is due to GaN with the AlInN/GaN multiquantum well satellite peaks vertically

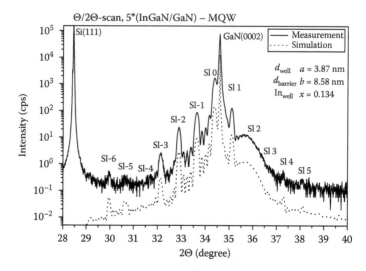

FIGURE 10.12
Θ/2Θ-scan of a fivefold InGaN/GaN multiquantum well structure grown on 150 mm GaN/Si(111) and simulation.

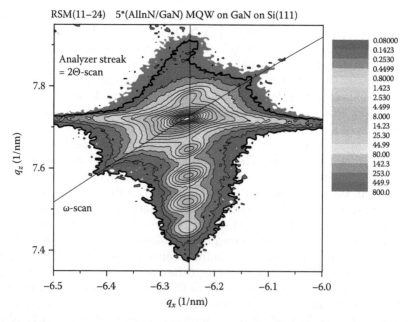

FIGURE 10.13
Reciprocal space map around the (11-24) Bragg reflection of a fivefold AlInN/GaN MQW on GaN/Si(111).

aligned to it, i.e., with the same in-plane lattice parameter as the underlying buffer layer. From such an asymmetric RSM, one immediately obtains information on the strain state of the multi quantum well (MQW) system; in this case, it was grown fully pseudomorphic on the buffer layer and a biaxial strain model should be valid. In addition, there is an inclined streak running from the lower left to the upper right that is a measure of the tilt orientation spread alone (ω-scan). In principle, in this RSM, the q_x-scan measures the in-plane finite size effects alone. Unfortunately, in the present (11-24) case, the 2Θ-scan or the analyzer streak [Holý 1999] is nearly parallel to q_x and is due to the finite angular acceptance of the diffracted beam optic used, and is not of interest here.

10.4 Grazing Incidence Diffractometry

In grazing incidence geometry, Bragg reflection from lattice planes perpendicular to the surface is performed, i.e., direct information on the in-plane lattice parameter is obtained (see Figure 10.14, left). The vectors $\mathbf{k}_i$ and $\mathbf{k}_s$ denote the incoming and scattered x-rays with the incoming angle α_i and the outgoing angle α_f, respectively. 2Θ is the scattering angle and $\mathbf{Q}$ the wave vector transform. For clarity, the directly reflected wave is not shown here.

Since the first report by Marra et al. [Marra 1979], grazing incidence in plane x-ray diffraction (GIIXRD, GID) has evolved a powerful tool for surface and near-surface analysis of all types of layers. For an overview on this topic, see, e.g., the book by Als-Nielsen and McMorrow [Als-Nielsen 2001]. Due to limited intensity with conventional laboratory equipment, most experiments at very thin layers in the nanometer range are performed at synchrotron

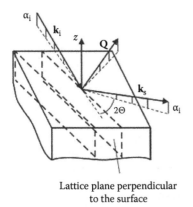

Lattice plane perpendicular
to the surface

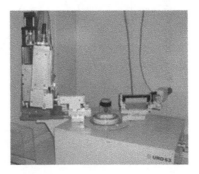

FIGURE 10.14
Schematics of grazing incidence diffraction (left) and photograph of the experimental setup for enhanced intensity (right).

facilities. A simple way to enhance the intensity in the laboratory with a conventional sealed x-ray tube is to make use of its line focus (12 mm × 1 mm) instead of its point focus (1 mm × 1mm). When the sample is put on a horizontal stage, this can be easily realized by installing the x-ray source upside down vertically (realized by GE/Seifert FPM).

Then, the full extension of the line focus can be used for the in-plane grazing incidence measurements as shown in the photograph of the apparatus. The result for such measurements at the (10-10) reflections is shown in Figure 10.15 for two AlInN layers grown on a GaN buffer on Si(111). The maxima near 32.4° are due to a superposition of scattering contributions of pseudomorphic AlInN on GaN. The shoulder in the dotted curve around 32.8° stems from a relaxed AlInN with less than 18% indium in the ternary alloy.

In order to vary the incoming angle α_i (see Figure 10.14) of the x-ray with respect to the sample surface, a relative linear motion of the x-ray source with respect to the sample is possible. We leave the sample stage fixed, and allow the complete x-ray tube housing a vertical movement. By this, the acceptance angle (resolution ~0.01°) of the x-rays with the sample, and thereby the penetration depth, is changed. In this way, direct access to the in-plane strain evolution is provided by depth sensitive GIIXRD $\Theta/2\Theta$ measurements at the (10-10) Bragg reflection; with increasing α_i, the penetration depth of the x-rays increases. It should be noted that each measurement at a certain α_i integrates over all the other layer parts with lower α_i, i.e., we do not measure at a certain depth but integrate over the whole layer thickness at a certain incoming angle. The integration effect could be corrected by taking into account the

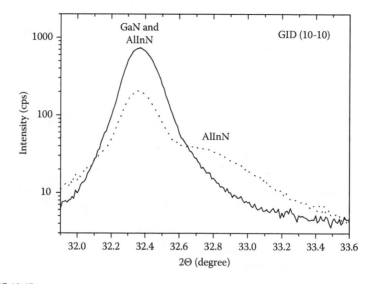

FIGURE 10.15

Grazing incidence (10-10) measurements of a pseudomorphic (full line) and a relaxed (dotted line) AlInN layer grown on GaN/Si(111).

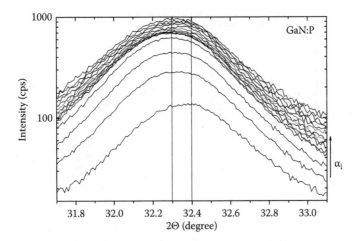

FIGURE 10.16
Depth-dependent grazing incidence (10-10) measurements of a phosphorous doped GaN layer grown on GaN/Si(111).

absorption coefficient of GaN, however, this is not done here. In Figure 10.16 depth-dependent measurements for a 1 μm phosphorous doped GaN sample on GaN/Si(111) are shown. Starting with an incoming angle of 0.2°, which is below the critical one of total external reflection (0.325°), the measurements were performed in steps of $\Delta\alpha = 0.1°$. For angles below the critical one, the amplitude of the evanescent wave decays at a short distance into the sample, typically at 5 nm. In this very-near surface region, the scattering angle shifts from 32.4° that corresponds to the angle of the nearly relaxed GaN toward lower values up to $2\Theta = 32.2°$. At the critical angle (third curve from below), the penetration depth is strongly increased by roughly an order of magnitude and the x-ray beam roughly penetrates the first 100 nm of the layer; at this position we observe no further shift of the maxima and the shape nearly remains the same. The final position of the maximum corresponds to a strain of about 3%. Clearly, these GID measurements indicate an increasing in-plane lattice parameter with depth due to phosphorous incorporation. The effect is opposite to that shown in our previous publication [Krost 2009], where we reported on an additional peak due to phosphorous incorporation at higher scattering angles. Obviously, there are different incorporation sites possible that lead to a shrinking or an increase of the in-plane GaN lattice parameter depending on the amount of phosphorous and growth conditions. However, in both cases, we do have a phosphorous gradient from the depth to the surface, i.e., a loss of phosphorous that leads to a self-healing of the lattice when approaching the surface.

Figure 10.17 shows a similar measurement on an UV-LED structure. In this example, the features right and left from the main AlGaN peak indicate the depth dependence of different relaxed parts. More examples on such kind of

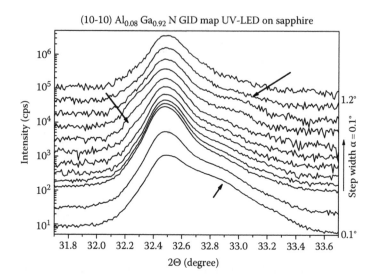

FIGURE 10.17
Depth-dependent grazing incidence (10-10) measurements on a c-axis UV-LED structure grown on sapphire. Arrows indicate relaxed parts.

measurements with our grazing incidence apparatus can be found in reference [Bläsing 2009].

Another useful application of GID is a quick evaluation of symmetry, e.g., when growing GaN on Si(100). On exactly oriented Si(100), the symmetry mismatch between cubic Si and hexagonal wurtzites leads to twofold nucleation sites for the nitride nuclei on the Si surface, which are rotated by 90° with respect to each other. As a consequence, for c-axis growth, there is a 12-fold symmetry instead of a 6-fold one. This can be checked very easily just by mounting the sample on the grazing incidence sample stage, choosing the (10-10) diffraction conditions and rotating the sample azimuthally around its surface normally during the measurement as shown in Figure 10.18. In the case of a single crystalline growth, only six peaks are observed during a full 360° rotation of the sample (cf. Fig. 4 in reference [Schulze 2007]).

The next step using an in-plane diffractometer is to define the angle of exit to the sample surface (α_e), in our case with the help of a fine horizontal slit, vertically movable together with the detector (see Figure 10.19).

Starting with ω and 2Θ at the position of the buffer of a (Al,Ga,In)N-based layer system, combined α_i/α_e scans prove the depth-dependent strain state with the help of well-defined x-ray standing wave (XSW) fields. In fact, this is a combination of x-ray reflectivity (XRR) and diffraction, benefit from the depth sensitivity of XRR and the in-plane lattice constant and twist sensitivity of GIID. Figure 10.20 shows a specular XSW scan on a 10-fold AlInN/GaNDBR, usable as a mirror for VCSEL structures. The scan shows clearly visible distributed Bragg reflector (DBR) thickness fringes (like in XRR or in

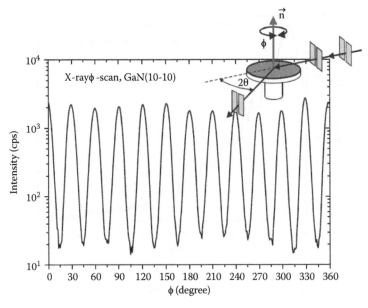

FIGURE 10.18
Grazing incidence (10-10) measurements on a *c*-axis GaN layer grown on Si(100) with two different nucleation sites rotated by 90°, leading to a 12-fold instead of a 6-fold symmetry.

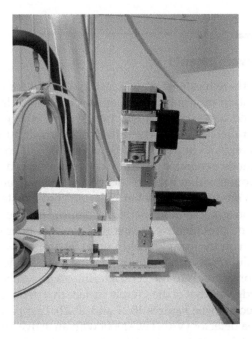

FIGURE 10.19
In-plane diffractometer (secondary optics): vertical movable detector with a slit in front.

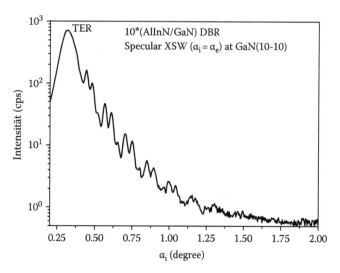

FIGURE 10.20
Specular XSW scan on a 10-fold AlInN/GaN DBR at the (10-10) Bragg reflection.

a symmetric $\Theta/2\Theta$-scan, not shown here), which indicate the fully pseudo-morphic growth and the sharp interfaces.

10.5 Latest Developments

In the last years, there has been an increasing demand for fast micrometer-scale characterization of large scale, nonperfect epitaxial layers. Therefore, we have developed a new x-ray diffraction system with convergent beam optics and fast parallel detection. The diffractometer was built by Bruker and is called the Bruker/AXS Super Speed Solution (SSS). It consists of a high-power rotating anode generator (up to 18 kW) and a curved Johannson-type Ge(111) monochromator that focuses the x-rays as a convergent beam onto the sample that is mounted on a Huber Eulerian cradle installed in a Bruker D8 high-resolution goniometer. Spatial resolution in line focus direction (Y-direction) can be provided by means of different Soller collimators that define the position on the sample within up to 50 µm.

The spatial resolution perpendicular to the line focus in the X-direction that simultaneously corresponds to the 2Θ scattering angle can be obtained by an adjustable knife edge in front of the sample surface that defines the illuminated area in the X-direction (see Figures 10.21 and 10.22). Fast parallel detection is enabled by an area detector (VANTEC-2000) with an active area of 14 cm².

The recording time of the new diffractometer is estimated to be more than 100 times faster than with a conventional system. Besides all other

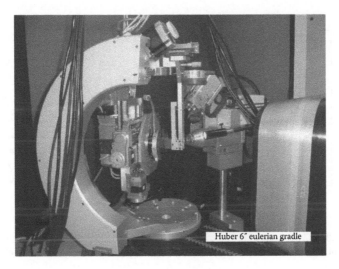

FIGURE 10.21
Photograph of Bruker/AXS Super Speed Solution (SSS).

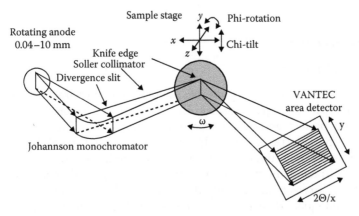

FIGURE 10.22
Schematic setup of the convergent beam diffractometer Bruker/AXS Super Speed Solution (SSS).

well-known x-ray techniques, such as $\omega/2\Theta$-scans, ω-scans, reciprocal space mapping, the system is especially well suited for a rapid evaluation of lateral structural homogeneity of large diameter wafers in grazing incidence as well as for conventional diffraction geometry. More details on the apparatus and the optical setup can be found in Ref. [Krost 2009]. In Figure 10.23, the structural quality in terms of twist across a 4 μm thick GaN layer grown on 100 mm Si(111) is presented. The measurements were taken in grazing incidence geometry at the (10-10) Bragg reflection along the line $Y = 0$, as indicated in the scheme. The information shown in the picture was gathered by

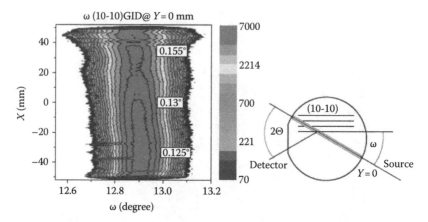

FIGURE 10.23
Twist distribution measured around $Y = 0$ across a GaN layer grown on 100 mm Si(111).

measuring subsequent ω-scans in steps of 0.01° and an integration time of 3 min each. It took us only 3 h to obtain the full information.

10.6 Conclusions

X-ray diffraction is an indispensable nondestructive characterization technique for epitaxially grown group-III nitrides. Such layers are always far from being perfect and suffer from strong mosaicity. The nondestructive structural characterization of such layers relies on high-resolution and conventional x-ray techniques. High-resolution triple-axis diffraction shows how the diffracted intensity is distributed in reciprocal space. When taking rocking curves from strongly defected samples, care has to be taken to choose an adequate detector acceptance angle. Besides rocking curve analysis, RSMs on symmetric and asymmetric Bragg reflection provide information on lattice relaxation, composition, and phase separation. Lattice tilts due to the mosaicity and strain effects can be easily separated. The finest resolution, i.e., an analyzer crystal in front of the detector is needed to resolve a fine structure in the diffraction scans, such as finite thickness fringes. From GID measurements, the pure twist component can be deduced. In a standard setup, laboratory diffraction in grazing geometry suffers from limited intensity. We have solved the problem by rotating the x-ray tube in order to benefit from the whole line focus. In addition, for depth-dependent measurements, the tube can be moved vertically thus changing the acceptance angle. Finally, a powerful diffractometer with convergent beam optics and fast parallel detection via an area detector has been presented. Besides all conventional applications, it allows for quick homogeneity measurements with a high lateral resolution for large-size GaN-on-silicon structures that come more and more into play for device applications.

Acknowledgment

We would like to thank the Deutsche Forschungsgemeinschaft for financial support in the framework of the Collaborative Research Center SFB787 and Research Group PolarCoN.

References

[Als-Nielsen 2001] J. Als-Nielsen and D. McMorrow, *Elements of Modern X-Ray Physics*, Wiley, Chichester, U.K., 2001.

[Bartels 1982] W.J. Bartels, Characterization of thin layers on perfect crystals with a multipurpose high resolution x-ray diffractometer, *J. Vac. Sci. Technol.* **B1**, 338 (1982).

[Bläsing 2009] J. Blaesing, A. Krost, J. Hertkorn, F. Scholz, L. Kirste, A. Chuvilin, and U. Kaiser, Oxygen induced strain field homogenization in AlN nucleation layers and its impact on GaN grown by metal organic vapor phase epitaxy on sapphire: An x-ray diffraction study, *J. Appl. Phys.* **105**, 033104 (2009).

[Bowen 1998] D.K. Bowen and B.K. Tanner, *High Resolution X-Ray Diffractometry and Topography*, Taylor & Francis, London, 1998.

[Detchprohm 1992] T. Detchprohm, K. Hiramatsu, K. Itoh, and I. Akasaki, Relaxation process of the thermal strain in the GaN/α-Al$_2$O$_3$ heterostructure and determination of the intrinsic lattice constants of GaN free from the strain, *Jpn. J. Appl. Phys.* **31**, L1454 (1992).

[DuMond 1937] J.W. DuMond, Theory of the use of more than two successive x-ray crystal reflections to obtain increased resolving power, *Phys. Rev.* **52**, 872 (1937).

[Edgar 1994] J.H. Edgar (ed.), *Properties of Group III Nitrides*, INSPEC, London, 1994.

[Fewster 2000] P.F. Fewster, *X-ray Scattering from Semiconductors*, Imperial College Press, London, 2000.

[Figge 2009] S. Figge, H. Kroencke, D. Hommel, and B.M. Epelbaum, Temperature dependence of the thermal expansion of AlN, *Appl. Phys. Lett.* **94**, 101915 (2009).

[Holý 1999] V. Holý, U. Pietsch, and T. Baumbach, *High-Resolution X-Ray Scattering from Thin Films and Multilayers* (Springer Tracts in Modern Physics 149), G. Höhler (managing editor), Springer, New York, 1999.

[Kirchner 2000] V. Kirchner, H. Heinke, D. Hommel, J.Z. Domagala, and M. Leszczynki, Thermal expansion of bulk and homoepitaxial GaN, *Appl. Phys. Lett.* **77**, 1434 (2000).

[Krost 1996] A. Krost, G. Bauer, and J. Woitok, High resolution X-ray diffraction, in: *Optical Characterization of Epitaxial Semiconductor Layers*, G. Bauer and W. Richter (eds.), Springer-Verlag, Berlin/Heidelberg, Germany, 1996, pp. 287–391.

[Krost 2009] A. Krost and J. Bläsing, Fast, micrometer scale characterization of group-III nitrides with laboratory X-ray diffraction, *Mat. Sci. Eng.* **A524**, 82 (2009).

[Kung 1994] P. Kung, C.J. Sun, A. Saxler, H. Ohsato, and M. Razeghi, Crystallography of epitaxial growth of wurtzite-type thin films on sapphire substrates, *J. Appl. Phys.* **75**, 4519 (1994).

[Landolt 1982] Landolt-Börnstein, Physics of group IV elements and III–V compounds, *New Series*, Vol. **17a**, O. Madelung (ed.), Springer, Berlin/Heidelberg, Germany, 1982, p. 62.

[Leszczynski 1995] M. Leszczynski, T. Suski, P. Perlin, H. Teisseyre, I. Grzegory, M. Bockowski, J. Jun, S. Porowski, and J. Major, Lattice constants, thermal expansion and compressibility of gallium nitride, *J. Phys. D. Appl. Phys.* **28**, A149 (1995).

[Leszczynski 1996] M. Leszczynski, H. Teisseyre, T. Suski, I. Grzegory, M. Bockowski, J. Jun, K. Pakula, J.M. Baranowski, C.T. Foxon, and T.S. Cheng, Lattice parameters of gallium nitride, *Appl. Phys. Lett.* **69**, 73 (1996).

[Marra 1979] W.C. Marra, P. Eisenberger, and A.Y. Cho, X-ray total-external-reflection-Bragg diffraction: A structural study of the GaAs-Al interface, *J. Appl. Phys.* **50**, 6927 (1979).

[Mazur 2009] A.V. Mazur and M.M. Gasik, Thermal expansion of silicon at temperatures up to 1100°C, *J. Mat. Process. Technol.* **209**, 723 (2009).

[Maruska 1969] H.P. Maruska and J.J. Tietjen, The preparation and properties of vapour-deposited single-crystal-line GaN, *Appl. Phys. Lett.* **15**, 327 (1969).

[Metzger 1998] T. Metzger, R. Höpler, E. Born, O. Ambacher, M. Stutzmann, R. Stömmer, M. Schuster, H. Göbel, S. Christiansen, M. Albrecht, and H.P. Strunk, Defect structure of epitaxial GaN films determined by transmission electron microscopy and triple-axis X-ray diffractometry, *Phil. Mag.* **A77**, 1013 (1998).

[Moram 2009] M.A. Moram, C.S. Ghedia, D.V.S. Rao, J.S. Barnard, Y. Zhang, M.J. Kappers, and C.J. Humpheys, On the origin of threading dislocations in GaN films, *J. Appl. Phys.* **106**, 073513 (2009).

[Ratnikov 2000] V. Ratnikov, R. Kyutt, T. Shubina, T. Paskova, E. Valcheva, and B. Monemar, Bragg and Laue x-ray diffraction study of dislocations in thick hydride vapor phase epitaxy GaN films, *J. Appl. Phys.* **88**, 6252 (2000).

[Roder 2005] C. Roder, S. Einfeldt, S. Figge, and D. Hommel, Temperature dependence of the thermal expansion of GaN, *Phys. Rev.* **B 72**, 085218 (2005).

[Schulze 2007] F. Schulze, O. Kiesel, A. Dadgar, A. Krtschil, J. Bläsing, M. Kunze, I. Daumiller, T. Hempel, A. Diez, R. Clos, J. Christen, and A. Krost, Crystallographic and electric properties of MOVPE-grown AlGaN/GaN-based FETs on Si(0 0 1) substrates, *J. Cryst. Growth* **299**, 399 (2007).

[Schuster 1995] M. Schuster and H. Göbel, Calculation of improvement to HRXRD system through-put using curved graded multilayers, *J. Phys. D. Appl. Phys.* **28**, A270 (1995).

[Sun 2002] Y.J. Sun, O. Brandt, T.Y. Liu, A. Trampert, K. Ploog, J. Bläsing, and A. Krost, Determination of the azimuthal orientation spread of GaN films by x-ray diffraction, *Appl. Phys. Lett.* **81**, 4928 (2002).

[Swenson 1983] C.A. Swenson, Recommended values for the thermal expansivity of silicon from 0 to 1000 K, *J. Phys. Chem. Ref. Data* **12**, 179 (1983).

[Wieneke 2009] M. Wieneke, J. Bläsing, A. Dadgar, P. Veit, S. Metzner, F. Bertram, J. Christen, and A. Krost, Micro-structural anisotropy of a-plane GaN analyzed by high-resolution X-ray diffraction, *Phys. Stat. Sol.* **c6**, 498 (2009).

11

Luminescence in GaN

Frank Bertram

CONTENTS

11.1 Introduction to Spectroscopy

Luminescence techniques belong to the most sensitive, nondestructive methods of semiconductor research. The most widely performed experiments in spectroscopy are transmission, reflection, and luminescence spectroscopy. Other techniques include, e.g., photoluminescence excitation spectroscopy, ellipsometry, modulation spectroscopy, measurements of the luminescence yield, or Raman spectroscopy. The standard setup for optical spectroscopy consists of a light source, the sample under investigation, usually placed in a cryostat, a monochromator to disperse the light, and a detection unit. Luminescence spectra usually give information on the deepest radiative states of a system including their optical phonon replica or acoustic phonon wings. At low temperatures, these are often defect states or localized states resulting from some disorder like spatial fluctuations of the width of a quantum well or of the composition of alloys. At higher temperatures also, extended states become accessible in luminescence. The absorption and reflection spectra give information of optically allowed transitions from the occupied ground state into excited states. Fabry–Perot modes, which appear often in quantum structures due to parallel surfaces or interfaces of substrate, buffer layers, etc., can be used to determine the optical thickness of the layers. Ellipsometry allows, in principle, to deduce the spectra of the real and imaginary parts of the dielectric function or of the complex index of refraction [Kli04a].

Spatially resolved luminescence techniques, like cathodoluminescence or micro-photoluminescence, provide a powerful tool for the optical nano-characterization of semiconductors, their heterostructures, as well as their interfaces due to their high spatial resolution.

11.2 Band Structure of GaN

The wurtzite-modification of GaN is a direct semiconductor. The band gap E_G is situated at the Γ-point [Din71a]. Figure 11.1 demonstrates the band structure of GaN schematically. At $T \to 0\,K$, the band gap of GaN is 3.504 eV after [Che96a]. Alternative values [Vur01a] vary up to $\pm 15\,meV$. The main origin for this variation is the different mechanical strain of heteroepitactic grown GaN [Gil95a], [Hof96a].

The conduction band of GaN is constituted of s-like states of gallium and exhibits Γ_7-symmetry. The valence band is built from p-like states of nitrogen [Rey00b]. The three-fold energetic degeneration of GaN is splitted in the wurtzite-structure. The splitting arises from the spin–orbit coupling as well as from the hexagonal wurtzite-structure of GaN and even more it is enhanced due to the difference of the GaN lattice parameters from the ideal wurtzite values $u = 3/8$ and $c/a = (8/3)^{1/2}$ [Wei96a]. Thus, in the center of the Brillouin zone, the valence band exhibits three energy levels A, B, and C with Γ_9- and Γ_7-symmetry. The highest Γ_9-valence band A forms the valence band edge [Ste99a]. The energetic distances of the valence bands are after [Wei96a], [Rod01a] $\Delta E_{CR} = 5.6\,meV$ (crystal field splitting) and $\Delta E_{SO} = 24.1\,meV$ (spin–orbit coupling).

The band gap E_G of the GaN crystal increases with higher compressive biaxial strain σ_{xx}. For the corresponding coefficient $dE_G/d\sigma_{xx}$, different references exist. In older works ([Ber99a], [Kas00a]), a coefficient of $dE_G/d\sigma_{xx} = -27\,meV/GPa$ for α-GaN [Kis96a] was used. A better estimate gives [Wag02a] with $-19\,meV/GPa$.

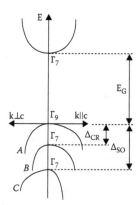

FIGURE 11.1
Schematic drawing of the band structure of wurtzite-GaN under consideration of the energetic splitting of the valence band due to the spin–orbit coupling and crystal field splitting. (After Rodina, A.V. et al., *Phys. Rev. B*, 64,115204, 2001.)

11.3 Optical Transitions in GaN without Impurities

Optical band–band transitions in GaN mostly occur around the Γ-point and therefore exhibit transition energies of around $h\nu \approx E_G = 3.504\,\text{eV}$ [Che96a].

For conduction band electrons and holes of the valence band exists an attractive coupling due to the Coulomb-interaction, which may lead to the formation of excitons. For the recombination of a free exciton X, one will find a transition energy of $E_X = E_G - E_X^B + E_X^{kin}$, with E_X^B as exciton binding energy and E_X^{kin} as kinetic energy of the center of mass of the free exciton.

The free exciton shows distinct differences in its discrete ground state energy depending on the different valance bands of the participating hole. The highest Γ_{9V}-band A takes part in the formation of the X_A-excitons. The lower Γ_{7V}-bands cause the formation of X_B-, respectively, X_C-excitons. The coupling $\Gamma_{7C} \times \Gamma_{9V}$ and $\Gamma_{7C} \times \Gamma_{7V}$, respectively, leads to a further splitting of the X_A-, X_B-, and X_C-levels with characteristic symmetries for the optical transitions [Pas01a].

The binding energy E_{XA}^B of the X_A exciton reveals an energetic value of around 25–26 meV (references in [Pas01a]), and the Bohr radius is around $a_{Bohr}^{Exciton} = 2.8\,\text{nm}$ [Ram98a], [Ram00a]. The binding energy of the free exciton depends on the participating valence band and after reference [Rod01a] exhibits $E_{XA}^B = 25.2\,\text{meV}$, $E_{XB}^B = 25.3\,\text{meV}$, and $E_{XC}^B = 27.3\,\text{meV}$.

The spectral distance of the free excitons X_A, X_B, and X_C distinctly varies with mechanical strain of the GaN lattice due to the different energetic shifts of the A-, B-, and C-valence bands [Gil95a], [Chi96a], [Vol96a], [Shi97a], [Ale98a], [Del01a]. In the best samples, the typical linewidth of the free exciton line is about 1 meV.

If we find a high concentration of free carriers, the attractive coupling of electrons and holes is screened. There is a transition from the excitonic luminescence into an electron–hole plasma luminescence above the мотт density (in GaN around 1.2–$3.8 \times 10^{18}\,\text{cm}^{-3}$ [Arn01a]).

The required free carriers can be produced by external excitation (generation of e–h pairs with high density $\rightarrow$ bipolar plasma [Hof98a]) or by high doping (high density of majority carriers $\rightarrow$ unipolar plasma [Arn99a]). The recombination of the (e,h)-plasma is characterized by a not so strict k-conservation of optical transitions due to scattering processes. The (e,h)-plasma-emission appears as a broad, unsymmetrical luminescence, in which the line shape mainly represents the population of the bands [Arn01a].

High carrier densities go along with nonlinear optical effects, like, e.g., the competing impact of band gap renormalization and Burstein–Moss shift [Lee99b], [Yos99a], [Skr99a] leading to a drastic change of transition energies.

11.4 Optical Transitions in GaN under the Impact of Impurities

Traps in GaN are generated by intrinsic defects as well as the incorporation of foreign atoms. Intrinsic defects can be vacancies (V_{Ga}, V_N), antisites (Ga_N, N_{Ga}), and interstitials (Ga_i, N_i) [Neu96a].

The dominating intrinsic defects in GaN are the nitrogen vacancy V_N in p-GaN and the gallium vacancy V_{Ga} in n-GaN, which act as acceptor and donor, respectively.

Nominally undoped GaN is mostly n-type [Hof96a]. Even without intentional doping, GaN owns a high concentration ($>10^{16}$ cm^{-3}) of intrinsic or extrinsic donors, whose exact origin is still unknown. The n-type conductivity cannot be explained due to the exclusive appearance of the dominating intrinsic donor V_N. Under n-type conditions, the energetically favored generation of V_{Ga} leads to a compensation and, furthermore, the generation energy for isolated V_N is too high to achieve the required concentrations [Neu94a].

The dominant extrinsic donors in nominally undoped GaN are Si_{Ga} and O_N, which is confirmed by secondary ion mass spectroscopy (SIMS)-measurements [Wet97a], [Fre01a], [Moo01a]. The values for the activation energies of the donors Si and O vary in the literature (around 30 meV [Kor99a], [Fre01a]) but Si exhibits the shallower donor [Res01a].

For p-type GaN, the acceptor is mostly generated by the incorporation of magnesium on a gallium site. In contrast to the Mg_{Ga}-configuration, Mg_N and Mg_i act as donors [Neu96a]. Depending on the experimental technique, the values for the activation energy vary in literature between 160 meV (determined from thermal activation) and 250 meV (spectral position of the characteristic luminescence) [Fis95a].

Only a small part of the Mg_{Ga}-acceptors is ionized at room temperature due to the relatively high activation energy. It is assumed, that Be_{Ga} has a smaller ionization energy with a comparable solubility. Theoretical works support this assumption [Wal01a], [Lee99a]. Nevertheless, the doping with Be has the problem, that Be_i acts as a donor and doping leads to compensation [Wal01a]. The ionization energy of Zn_{Ga} and Ca_{Ga} is higher than of Mg_{Ga} [Neu99a], in contrast, hydrogen acts under p-condition as donor [Wal97a].

The identification of defect-induced optical transitions occurs mainly on the basis of the spectral position, to some extent on the consideration of typical line shapes and of characteristic interactions (e.g., phonon-replica).

For pure band-impurity transitions, e.g., (D^0,h) or (e,A^0), the emission energy results from the band gap E_G reduced by the ionization energy E_D or E_A of the participating donor or acceptor, i.e., $h\nu = E_G - E_D$ or $h\nu = E_G - E_A$, respectively. For the assignment of a luminescence channel to a specific band-impurity transition or rather the determination of the ionization energy, an exact knowledge of the band gap E_G is necessary.

The coupling of four particles has to be considered for the binding of excitons on neutral donors or acceptors: i.e., the impurity D^+ or A^-, for the neutrality of the impurity-bound electron or hole as well as the electron and hole from the excitonic complex. Bound excitons associated with neutral donors and acceptors in GaN show prominent optical spectra, which are very useful signatures of the corresponding defects. The transition energy for the donor-bound exciton (analog for the acceptor-bound exciton) can be written as $E_X = E_G - E_X^B - E_{X-D0}^B$, with E_X^B as exciton-binding energy and E_{X-D0}^B as binding energy of the exciton X on the neutral donor D^0. Hence, the (D^0X)-luminescence is energetically sharply defined, since the kinetic term of the free excitons is missing for the bound exciton. The identification of the donor or acceptor from the emission energy requires the band gap E_G and the two parameters E_X^B and $E_{X-D0/A0}^B$.

The bound exciton spectra show narrow lines at low temperatures (<0.5 meV for residual doping $<10^{16}$ cm^{-3}) at an energy characteristic for each shallow donor.

The energetic distance of X_A to shallow D^0X_A is almost constant even in the presence of biaxial strain, i.e., the binding energy of the exciton to the shallow donor is almost independent of the strain [Rey00a]. For excitons bound to deep impurities, in particular, deep acceptors, a strong dependence of the binding energy $E_{X-D0/A0}^B$ with strain in GaN can be obtained [Mon01a].

In addition, there are optical transitions in GaN with impurities taking part in the initial and final state, the so-called donor–acceptor-pair transition (DAP) [Beb72]. The lattice sites of the participating donors and acceptors for a (D^0,A^0)-process can be close to (associated donor–acceptor pairs) or more distant than the exciton-Bohr radius $a_{Bohr}^{Exciton}$ (distant donor–acceptor pairs). For the second case, one can write $h\nu = E_G - E_D - E_A + (e^2/\varepsilon r)$. Here, E_G is the band gap, E_D and E_A are the activation energies of the participating donors D and acceptors A, and the term $E_C = (e^2/\varepsilon r)$ with the static dielectric constant ε and the donor–acceptor distant r describing the Coulomb-interaction of the ionized remaining impurities D^+ and A^-. For associated donor–acceptor pairs a similar expression can be found as long as the ionization energies of the donors and acceptors are clearly different. ε has to be replaced with the optical dielectric constant ε_{Opt}.

11.5 Interpretation of GaN Luminescence Spectra

Any kind of direct assignment of luminescence lines (in particular free excitons and bound excitons) in the spectral region of the near band gap (NBG) luminescence requires the exact value of the band gap E_G.

Here, we use $E_G = 3.504$ eV as reference for the band gap of relaxed GaN at $T \rightarrow 0$ K [Che96a] in agreement with [Mon74a]. However, the real value of E_G

is unknown or only approximately determinably because of the biaxial strain in heteroepitactic grown GaN samples. Hence, in general, the measurement of the emission energy of individual NBG transitions does not suffices for a reliable assignment to defined luminescence processes or the determination of binding and ionization energies.

A direct determination of E_G as well as of the biaxial strain from the line sequence of the free excitons X_A, X_B, and X_C [Vol96a], [Chi96a] is only possible with a sufficient dynamic range of the spectra. If this is not the case, like for instance spectra taken with some parallel detectors with low dynamic range [Ber99a], the assignment of E_G can only occur after a reliable interpretation of the spectra has happened or alternative techniques have to be taken.

For an assignment of luminescence lines, it is most practical to check for characteristic energy differences between the observed transitions.

In Table 11.1, a collection of experimentally determined low-temperature emission energies in the spectral region of the NBG for GaN from different groups can be found. The assignment of free and bound excitons in GaN as well as the energy difference of each emission to the emission of the free exciton X_A is shown. In particular, in the region of the bound excitons a number of lines can be observed indicating the participation of different donors and acceptors. Table 11.1 also exhibits a slight variance of the absolute energy positions of lines (i.e., for (X_A): 3.477–3.496 eV). On one hand, a clear tendency can be observed for thin heteroepitactic grown GaN samples on sapphire substrate to very thick, partly homoepitactic grown, GaN layers, strongly indicating the huge impact of strain. On the other hand, there is an energetic distance of X_A to the different donor-bound excitons (D^0X) with variations in the order of 1 meV.

The main conclusions of Table 11.1 are

1. The energetic shift between the different spectra is sometimes bigger than the distance of adjacent emission channels. Hence, the absolute value of measured luminescence energy is not sufficient for the identification of the recombination channel.

2. The luminescence line with a spectral distance of 2.2–3.7 meV with respect to (X_A) is identified as (D^0X) transition. This emission is observed only from some authors.

3. From all authors, a luminescence is observed with an energetic distance of 6–8 meV with respect to (X_A), which is interpreted as (D^0X) emission. At high spectral resolution, this (D^0X)-line can be separated in two spectral parts (1 meV energetic distance) with varying relative components. It is assumed that this effect is caused due to the participation of different donors [Skr99a], [Gra00a], [Res01a], whereas O_N is responsible for the high-energy component (at about 3.4714 eV) and Si_{Ga} for the low-energy component (at about 3.4723 eV) of the D^0X emission, respectively [Mon01a].

TABLE 11.1

Spectral Positions of Excitonic Luminescence Lines in GaN

(X_B)	(X_A)	(D_1^0X)	(D^0X)	(A^0X)	(A^0X)	(A^0X)	Reference
3.4968 eV (X_B) +8.2 meV	3.4886 eV (X_A) 0 meV		3.4823 eV (D^0X) −6.2 meV	3.4766 eV (Mg^0X) −12.0 meV			[Bea01a] MOVPE on two step MOVPE ELOG on sapphire
3.485 eV (X_B) +6 meV	3.479 eV (X_A) 0 meV		3.473 eV (D^0X) −6 meV		3.451 eV (A^0X) −28.1 eV		[Kas01a] 220 µm HVPE on ZnO/sapphire
3.4864 eV (X_B) +5.2 meV	3.4812 eV (X_A) 0 meV		3.4746 eV (D^0X) −6.6 meV	3.4696 eV (A^0X) −11.6 meV	3.4581 eV (A^0X) −23.1 eV	3.4433 eV (A^0X) −37.9 eV	[Kir01a,b] 4.9 µm HVPE on MOVPE on sapphire
3.4870 eV (X_B) +6.0 meV	3.4810 eV (X_A) 0 meV	3.4788 eV (D_1^0X) −2.2 meV	3.4745 eV (D_2^0X) −6.5 meV	3.4691 eV (A^0X) −11.9 meV			[Mar01a] 2 µm MOVPE on 300 µm freestanding HVPE-GaN
3.4909 eV (X_B) +6.6 meV	3.4843 eV (X_A) 0 meV		3.4777 eV (D_1^0X) −6.6 eV 3.4769 eV (D_2^0X) −7.4 meV	3.4724 eV (A^0X) −11.9 meV			[Pas01a] 80 µm HVPE on c-sapphire
	3.4779 eV (X_A) 0 meV		3.4714 eV (D_1^0X) −6.5 meV	3.4662 eV (A^0X) −11.2 meV	3.4546 eV (A^0X) −23.3 eV		[Pas01a] 130 µm HVPE on MOVPE on c-sapphire, substrate removed
3.484 eV (X_B) +5 meV	3.479 eV (X_A) 0 meV	3.4766 eV $(D_1^0X_B)$ −2.4 meV 3.4758 eV $(D_2^0X_B)$ −3.2 meV	3.4728 eV $(D_1^0X_A)$ −6.2 meV 3.4720 eV $(D_2^0X_A)$ −7.0 meV	3.4673 eV (A^0X_A) −11.7 meV			[Res01a] 1.5 µm MBE on 200 µm freestanding HVPE-GaN

(continued)

TABLE 11.1 (continued)

Spectral Positions of Excitonic Luminescence Lines in GaN

(X_B)	(X_A)	(D_1^0X)	(D^0X)	(A^0X)	(A^0X)	(A^0X)	Reference
	3.4814 eV[a] (X_A) 0 meV		3.4751 eV (D^0X) −6.3 meV		3.4589 eV (A_1^0X) −22.5 eV	3.4536 eV (A_2^0X) −27.8 eV	[Rey01a] 133 μm HVPE on c-sapphire, substrate removed
3.483 eV (X_B) +5 meV	3.478 eV (X_A) 0 meV		3.4720 eV (D_1^0X) −6.0 meV 3.4712 eV (D_2^0X) −6.8 meV	3.4664 (A^0X)? −11.6 meV			[Gra00a] MBE on GaN-bulk crystal
3.4832 eV (X_B) +4.3 meV	3.4789 eV (X_A) 0 meV	3.4752 eV (D^0X) −3.7 meV	3.4714 eV (D^0X) −7.5 meV	3.4654 eV (A^0X)? −13.5 meV 3.4664 (D^0X),(A^0X)? −12.5 meV	3.4556 eV (A^0X) −23.3 eV		[Kir00a] 1.5 μm MOVPE on GaN-bulk crystal

X_B	X_A	D^0_1X	D^0X / D^0_2X	A^0X / A^0_1X	A^0_2X	Reference
3.4817 eV (X_B) +4.6 meV	3.4771 eV (X_A) 0 meV		3.4709 eV (D^0X) −6.2 meV	3.4655 eV (A^0X) −11.6 meV		[Kor99a] 1.5 μm MOVPE on GaN-bulk crystal
3.492 eV (X_B) +6 meV	3.486 eV (X_A) 0 meV		3.4792 eV (D^0X) −6.8 meV; 3.4781 eV (D^0X) −7.9 meV	3.4732 eV (A^0_1X) −12.8 meV	3.4621 eV (A^0_2X) −23.9 eV	[Poz99a] 80 μm HVPE on a-sapphire
3.4832 eV (X_B) +4.7 meV	3.4785 eV (X_A) 0 meV		3.4718 eV (D^0_1X) −6.7 meV; 3.4709 eV (D^0_2X) −7.6 meV	3.4663 eV (A^0X) −12.2 meV		[May97a] MBE on GaN-bulk crystal
3.4860 eV (X_B) +3.9 meV	3.4799 eV (X_A) 0 meV	3.4762 eV (D^0_1X) −3.7 meV	3.4727 eV (D^0_2X) −7.2 meV			[Vol96a] 400 μm HVPE on c-sapphire
3.5050 eV (X_B) +8.8 meV	3.4962 eV (X_A) 0 meV	3.4935 eV (D^0_1X) −2.7 meV	3.4900 eV (D^0_2X) −6.2 meV			[Vol96a] 3 μm MOVPE on AlN on c-sapphire

[a] Binding energy of (X_A) from [Rey00b].

4. The incorporation of Mg is accepted to be the origin of the occasionally detected luminescence line (A_1^0X) at about 3.466 eV with a localization energy of 11–13 meV with respect to (X_A) [Bea01a]. But there is no consensus about the nature of the acceptor and its exclusive impact. The ratio of the localization energy $E_{X-D0/A0}^B$ of the exciton and the activation energy of the Mg_{Ga}-acceptor seems very small with a value of 0.05 for an assignment of (Mg^0X). The relatively strong coupling of LO-phonons intercedes for the acceptor nature of the participant impurity [Jay98a], because after [Mon01a] the intensity of LO-replicas of (A^0X)-transitions is typically one order of magnitude higher than of (D^0X)-recombination.

5. The lines with a localization energy around 22 and 27 meV are assigned to an additional (A^0X)-transitions. Also for this luminescence, a stronger LO-coupling can be detected [Poz99a]. Zn is assumed as origin for both lines [Mon01a].

The consideration of these typical localization energies of bound excitons (D^0X) and (A^0X) with respect to (X_A) allows in most cases a clear identification of luminescence spectra even with not knowing the exact value of the band gap E_G.

In Figure 11.2, low-temperature luminescence spectra of different GaN samples are plotted in normalized scale. The spectral resolution is around 1 meV. In Figure 11.2a exclusively the spectral region of the NBG is shown.

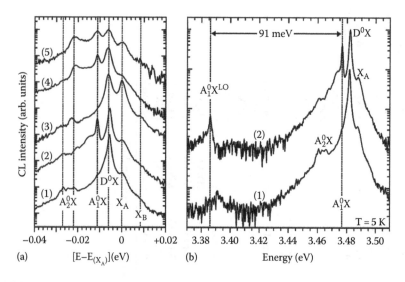

FIGURE 11.2

Different spectra of the excitonic luminescence in GaN: (a) plot with shifted energetic position each with respect to matching (X_A), (b) detection of phonon replica for impurity-bound excitons ($T < 6$ K).

The spectra were energetically shifted against each other to match in the spectral position of (X_A), partially reducing the impact of different strain on the band gap. In the spectra (1) and (2), the spectral position of (X_A) originally is at an absolute energy value of 3.487–3.488 eV. (X_A) is blueshifted here due to compressive biaxial strain. At spectrum (4) with $E(X_A) = 3.469$ eV, a clear redshift exists. The spectra (3) and (5) exhibit almost no difference in the spectral position.

In all spectra of Figure 11.2a, the emission of donor-bound excitons dominates. A distinct spectral division of the (D^0X)-line in two isolated components is not visible. Between the spectra, the position of the emission maximum varies of around 1 meV, whereas the position of (X_A) can only be determined with respect to the spectral resolution.

In agreement with the assignment of (A_1^0X) the spectrum (2) in Figure 11.2 is taken from a Mg-doped sample. Here (A_1^0X) is most intense. This assignment correlates with the distinct detection of a LO-replica for spectrum (2) in Figure 11.2b with respect to a replica of (D^0X) in spectrum (1). The (A_2^0X)-lines appear in the spectra of Figure 11.2a with different intensity ratios to each other as well as vary in their spectral positions with respect to (X_A) of about 4 meV.

Figure 11.3 shows a plot of the line positions from Figure 11.2 (open points) and the energies of Table 11.1 (filled points) with respect to the absolute position of the (X_A)-line. The lines indicate similar excitonic recombination channels. Clearly visible is the parallel characteristics of (X_A)-, (D^0X)-, and (A^0X)-positions (at least in the displayed spectral region).

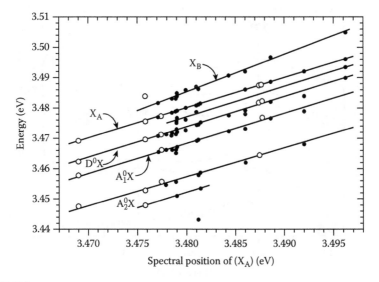

FIGURE 11.3
Energy of the (X_B)-, (D^0X)-, and (A^0X)-luminescence with respect to the spectral positions of the (X_A)-line in GaN. Filled dots: literature values; open dots: experimental data.

In the following we use the spectral position of the dominating (D^0X)-line for the determination of the biaxial strain of GaN. For this, we assume a negligible dependence of E^B_{X-D0} from the strain and a transition energy of (D^0X) at 3.472 eV (357.1 nm vacuum wavelength) as a reference for the relaxed GaN material. This energy value correlates emission energies from homoepitactic grown GaN (see Table 11.1) as well as measurements from the surface of thick GaN layers [Sie97a].

At the same time, it is consistent with a band gap of $E_G = 3.504$ eV [Che96a] under consideration of a binding energy for the ground state of the X_A exciton of $E^B_X = 26$ meV [Rod01a] and an averaged binding energy of the exciton to the dominating neutral donors of $E^B_{X-D0} = 6$–8 meV.

The disadvantage of this method is the general uncertainty with results from not knowing the value of the coefficient $dE_G/d\sigma_{xx}$.

In Figure 11.4 CL spectra of different samples in semilogarithmic scale in a broad spectral range are shown. The spectra are normalized with respect to the intensity of the NGB luminescence above 3.4 eV. Besides the NBG luminescence, there are in all spectra spectrally broad, defect-induced luminescence bands visible. The relative intensities and spectral positions clearly differ.

The transition (D^0, A^0) is labeled with DAP in Figure 11.4 (spectra (1)–(3)). At low doping with magnesium, a transition at 3.25–3.27 eV is obtainable after [Eck98a] and [Hof99a] often with a sequence of individual, well-separated

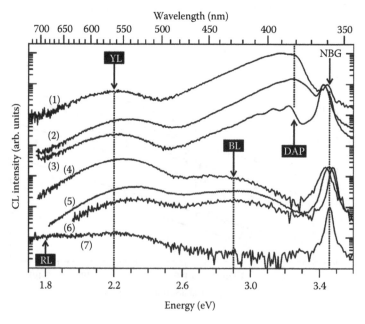

FIGURE 11.4
Defect luminescence in GaN (T < 6 K).

phonon replica on the low-energy side ($\Delta E_{LO} = 92\,meV$), which is assigned as a transition of a shallow donor and the magnesium acceptor.

Close to this spectral position the transition (e,Mg^0) of free electrons to the acceptor level of Mg is expected [Myo96a], since shallow donors D^0—which take part in the (D^0,Mg^0) transition—exhibit only a relatively small ionization energy of 30 meV [Res01a]. Carbon is attributed to be the origin of DAP transitions in nominal Mg-free GaN [Fis95a].

The blue luminescence (BL) around 2.9 eV (spectra (4)–(6) in Figure 11.4) also appears in nominally undoped GaN [Res00a] and n-doped GaN [Yan00b], but is characteristic for a high concentration of the Mg-acceptor [Res99a]. After [Ves00a], the BL occurs at least with the contribution of an energy level from a deep trap. According to [Res00a], the mechanism of the BL in undoped GaN consists of a transition of a shallow donor and a deep acceptor, possibly V_{Ga} or V_{Ga}–donor complexes [Tot99a], [Res01c].

A higher concentration of V_{Ga} was detected in n-doped GaN, as long as O instead of Si was the donor [Oil01a]. The higher stability of the V_{Ga}–O_N complex with respect to V_{Ga}–Si_{Ga} is supported in [Neu96b].

For Mg-doped GaN, the BL should be a transition of a deep donor (complex V_N–Mg_{Ga} after [Kau98a], which arises from self-compensation at higher doping concentrations) and a shallow acceptor Mg_{Ga}. After [Sha00a], the donor is created by the complex V_N–H. Alternative models contain (D^0,A^0)-transitions of shallow donors and acceptors in analogy to the DAP-band [Gla02a].

The yellow luminescence (YL) at 2.2 eV (spectra (1)–(7) in Figure 11.4) often appears as parasitic luminescence in nominally undoped GaN, but can also be obtained in samples with intentional doping.

As origin of the YL-band transitions of shallow donors and deep acceptors [Ogi80a], shallow donors and deep donors [Hof95a] as well as deep donors and shallow acceptors [Gla95a] are suggested.

The first model is supported from the energetic point of view with the participation of the isolated V_{Ga} as deep intrinsic acceptor and the formation of the complexes V_{Ga}–Si_{Ga}, V_{Ga}–O_N, or V_{Ga}–H with shallow extrinsic donors Si_{Ga} and O_N or rather hydrogen, respectively [Neu96b], [Mat97a], [Saa01a], [Wal97a]. Discussed are transitions inside the complexes as well as between spatially separated donor–acceptor pairs [Col99a]. After [Saa97a] correlations exist between the intensity of the YL and the concentration of V_{Ga}.

To some extent the YL can be separated in two components at 2.2 and 2.4 eV [Res01b]. There are hints for the accumulation of participating point defects near surfaces [Sha99a], interfaces [Sha00a], and extended defects [Pon96a].

The intensity of the YL exhibits a clear saturation at higher excitation densities with respect to the excitonic luminescence. Hence, the intensity ratio of the YL and the NBG luminescence strongly depends on the excitation conditions [Kuc01a], [Kir00b].

Red luminescence (RL) around 1.8 eV (spectrum (7) in Figure 11.4) appears due to the simultaneous doping of GaN with Mg and Si [Kau99a], but can also be obtained at very low doping concentration of Mg [Hof00a]. In general, transitions of deep donors and deep acceptors or excitons bound to neutral complexes [Gol01a] are assumed to be the origin of the RL.

The mechanisms of the DAP-band, of the BL, YL, and RL can be explained after [Kau99a] in a combined recombination model. Here, deep acceptors A_{deep} as well as deep donors D_{deep} act V_N–Mg_{Ga} and V_{Ga}–Si_{Ga} complexes, respectively.

11.6 Impact of Threading Dislocation in GaN

A direct correlation of transmission electron microscopy images with cathodoluminescence maps reveals that pure edge dislocations with $\mathbf{b} = 1/3 \langle 11\bar{2}0 \rangle$ [Che01a], mixed dislocations [Sug98a] and also pure screw dislocations with $\mathbf{b} = 1/3 \langle 11\bar{2}0 \rangle$ [Hac00a] act as a non-radiation recombination channel with a drastic reduction of quantum efficiency in their neighborhood.

The contrast in photoluminescence [Lee01a] or cathodoluminescence measurements [Ros97a], [Das01a], [Zha01a] can be used to determine the dislocation density.

11.7 Luminescence of GaN Layers Grown in Nonpolar Directions

Due to their wurtzite crystal structure and their polar bonding character, the nitrides are influenced by large spontaneous polarization fields. Moreover, huge piezoelectric fields arise in quantum wells if they are biaxially strained with respect to the c-plane. These strong internal electric fields cause the so-called quantum confined Stark effect (QCSE) and lead to both unwanted wavelength shifts and, even more importantly, a decrease of the oscillator strength due to a local separation of electrons and holes in c-plane quantum wells. To increase the efficiency, especially to realize efficient light emitters in the green spectral region, strong efforts are done to grow on semi- and nonpolar GaN.

The most common strategy to overcome the QCSE-problem is the growth of heterostructures in other directions than the c-axis. One approach to reduce the polarization fields is the growth in semi- or nonpolar directions. However, the epitaxial growth on such planes is by far less developed than the growth on the commonly used c-plane. Moreover, in ternary alloys and their heterostructures, nanoscale fluctuations of stoichiometry as well as interfaces have strong impact on the radiative recombination in light emitters.

Besides threading dislocations, the most prominent structural defects—in particular in a-plane GaN (nonpolar direction)—are stacking faults. Basal plane stacking faults (BSFs) can be treated as planar defects locally forming the ABC cubic structure within the usual ...ABABAB... wurtzite stacking sequence. Three different types of BSFs exist (for a review see [Zak05a]). Stacking faults are also possible on planes other than the *c*-plane, for example, formed on prismatic {1210} planes—the prismatic stacking faults (PSF)—terminating the BSFs [Sta98a]. The presence of these stacking faults results in intense luminescence peaks ranging from 3.27 to 3.42 eV. The origin of this emission has been attributed to the intrinsic crystalline structure as well as to the presence of impurities such as oxygen. The cubic stacking inclusion BSF can be regarded energetically as a quantum well in *c*-direction, providing a localization well for the excitons. As a matter of fact, the radiative recombination of this BSF-bound exciton is very efficient and usually dominates the (D^0,X) emission in nonpolar *a*-plane GaN by more than one order of magnitude [Dop99a], [Che04a], [Pas05a]. Each of the morphological defects listed above show their fingerprints in characteristic defect luminescence peaks. By correlating transmission electron microscopy (TEM) and cathodoluminescence microscopy, a clear-cut identification of the different stacking faults and dislocations giving rise to the various luminescence lines between 3.27 and 3.42 eV has been achieved [Liu05a].

Figure 11.5 displays a low-temperature photoluminescence spectrum of an *a*-plane, molecular organic vapor phase epitaxy (MOVPE)-grown GaN layer with a spectral range of 3.0–3.6 eV in semilogarithmic scale. The spectrum is divided in two parts:

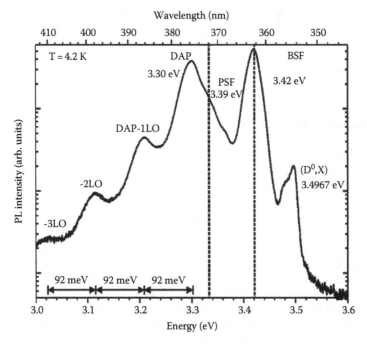

FIGURE 11.5
Defect luminescence in m-planar grown GaN (T < 6 K).

1. The spectrum shows a very pronounced near band edge luminescence at 3.49 eV, however, the line is formed asymmetrically indicating the involvement of more than one recombination channel. The
   near band edge luminescence is dominated by a (D^0,X) recombination at 3.497 eV and is 24 meV shifted to higher energies with respect
   to the emission energy of fully relaxed GaN. This value corresponds
   to a compressive strain of the crystal of 0.9 GPa. In addition, a (A^0,X)
   recombination is clearly observable at 3.478 eV.

2. The spectrum is dominated by the BSF luminescence at 3.42 eV with
   a full width half maximum (FWHM) of 27 meV [Liu05a].

The line at 3.30 eV with a FWHM of 45 meV is identified as the DAP band
which is 30 meV shifted to higher energies with respect to the literature values for (0001)-GaN. This shift is in the same order like the excitons lines
[Mon01a]. Up to three LO-phonon-replica of the DAP line are visible. A
detailed view on the intensity ratio of the LO-phonon-replica shows that the
lines do not follow the known ration of Huang–Rhys [Hua50a]

$$I_n = \frac{S^n}{n!} I_0$$

The literature value for the Huang–Rhys factor of the DAP recombination for
c-GaN is about $S \approx 0.7$ for GaN [Mon01a]. The DAP character of the recombination can be proofed with temperature depended and excitation density
depended measurements. A shoulder at the DAP-band can be observed at
3.339 eV, which is connected with luminescence from prismatic stacking
faults [Liu05a].

References

[Ale98a] A. Alemu, B. Gil, M. Julier, and S. Nakamura, *Phys. Rev. B* **57**, 3761 (1998).

[Arn99a] B. Arnaudov, T. Paskova, E. M. Goldys, R. Yakimova, S. Evtimova,
I. G. Ivanov, A. Henry, and B. Monemar, *J. Appl. Phys.* **85**, 7888 (1999).

[Arn01a] B. Arnaudov, T. Paskova, E. M. Goldys, S. Evtimova, and B. Monemar, *Phys.
Rev. B* **64**, 045213 (2001).

[Bea01a] B. Beaumont, P. Vennéguès, and P. Gibart, *Phys. Stat. Sol. (b)* **227**, 1 (2001).

[Beb72] H. B. Bebb and E. W. Williams, in: *Semiconductors and Semimetals* (Vol. 8:
Transport and Optical Phenomena), R. K. Willardson and A. C. Beer (eds.),
Academic Press, New York, 1972, Chapter 4.

[Ber99a] F. Bertram, T. Riemann, J. Christen, A. Kaschner, A. Hoffmann, C. Thomsen,
K. Hiramatsu, T. Shibata, and N. Sawaki, *Appl. Phys. Lett.* **74**, 359 (1999).

[Che96a] G. D. Chen, M. Smith, J. Y. Lin, H. X. Jiang, S.-H. Wei, M. Asif Khan, and
C. J. Sun, *Appl. Phys. Lett.* **68**, 2784 (1996).

[Che01a] D. Cherns, S. J. Henley, and F. A. Ponce, *Appl. Phys. Lett.* **78**, 2691 (2001).

[Che04a] L. Chen, B. J. Skromme, M. K. Mikhov, H. Yamane, M. Aoki, F. J. DiSalvo, B. Wagner, R. F. Davis, P. A. Grudowski, and R. D. Dupuis, *Mat. Res. Soc. Symp. Proc.* **798**, Y5.55.1 (2004).

[Chi96a] S. Chichibu, A. Shikanai, T. Azuhata, T. Sota, A. Kuramata, K. Horino, and S. Nakamura, *Appl. Phys. Lett.* **68**, 3766 (1996).

[Col99a] J. S. Colton, P. Y. Yu, K. L. Teo, E. R. Weber, P. Perlin, I. Grzegory, and K. Uchida, *Appl. Phys. Lett.* **75**, 3273 (1999).

[Das01a] S. Dassonneville, A. Amokrane, B. Sieber, J. L. Farvacque, B. Beaumont, and P. Gibart, *J. Appl. Phys.* **89**, 3736 (2001).

[Del01a] E. Deleporte, C. Guénaud, M. Voos, B. Beaumont, and P. Gibart, *J. Appl. Phys.* **89**, 1116 (2001).

[Din71a] R. Dingle, D. D. Sell, S. E. Stokowski, P. J. Dean, and R. B. Zetterstrom, *Phys. Rev. B* **3**, 497 (1971).

[Dop99a] D. Doppalapudi, E. Iliopoulos, S. N. Basu, and T. D. Moustakas, *J. Appl. Phys.* 85, 3582 (1999).

[Eck98a] L. Eckey, U. von Gfug, J. Holst, A. Hoffmann, A. Kaschner, H. Siegle, C. Thomsen, B. Schineller, K. Heime, M. Heuken, O. Schön, and R. Beccard, *J. Appl. Phys.* **84**, 5828 (1998).

[Fis95a] S. Fischer, C. Wetzel, E. E. Haller, and B. K. Meyer, *Appl. Phys. Lett.* **67**, 1298 (1995).

[Fre01a] J. A. Freitas Jr., G. C. B. Braga, W. J. Moore, S. K. Lee, K. Y. Lee, I. J. Song, R. J. Molnar, and P. Van Lierde, *Phys. Stat. Sol. (a)* **188**, 457 (2001).

[Gil95a] B. Gil, O. Briot, and R.-L. Aulombard, *Phys. Rev. B* **52**, 17028 (1995).

[Gla95a] E. R. Glaser, T. A. Kennedy, K. Doverspike, L. B. Rowland, D. K. Gaskill, J. A. Freitas Jr., M. Asif Khan, D. T. Olson, J. N. Kuznia, and D. Wickenden, *Phys. Rev. B* **51**, 13326 (1995).

[Gla02a] E. R. Glaser, W. E. Carlos, G. C. B. Braga, J. A. Freitas Jr., W. J. Moore, B. V. Shanabrook, R. L. Henry, A. E. Wickenden, D. D. Koleske, H. Obloh, P. Kzodoy, S. P. DenBaars, and U. K. Mishra, *Phys. Rev. B* **65**, 085312 (2002).

[Gol01a] E. M. Goldys, M. Godlewski, T. Paskova, G. Pozina, and B. Monemar, *MRS Internet J. Nitride Semicond. Res.* **6**, 1 (2001).

[Gra00a] N. Granjean, B. Damilano, J. Massies, G. Neu, M. Teissere, I. Grzegory, S. Porowski, M. Gallart, P. Lefebvre, B. Gil, and M. Albrecht, *J. Appl. Phys.* **88**, 183 (2000).

[Hac00a] P. Hacke, K. Domen, A. Kuramata, T. Tanahashi, and O. Ueda, *Appl. Phys. Lett.* **76**, 2547 (2000).

[Hof95a] D. M. Hofmann, D. Kovalev, G. Steude, B. K. Meyer, A. Hoffmann, L. Eckey, R. Heitz, T. Detchprohm, H. Amano, and I. Akasaki, *Phys. Rev. B* **52**, 16702 (1995).

[Hof96a] A. Hoffmann, Festkörperprobleme **36**, *Advances in Solid State Physics*, ed. by R. Helbig, 33 (1996).

[Hof98a] A. Hoffmann and L. Eckey, *Mat. Sci. Forum* **264–268**, 1259 (1998).

[Hof99a] D. M. Hofmann, B. K. Meyer, F. Leiter, W. von Förster, H. Alves, N. Romanov, H. Amano, and I. Akasaki, *Jpn. J. Appl. Phys.* Part 2 **38**, L1422 (1999).

[Hof00a] D. M. Hofmann, B. K. Meyer, H. Alves, F. Leiter, W. Burkhard, N. Romanov, Y. Kim, J. Krüger, and E. R. Weber, *Phys. Stat. Sol. (a)* **180**, 261 (2000).

[Hua50a] K. Huang and A. Rhys, *Proc. R. Soc. Lond. Ser.* **A 204**, 406 (1950).

[Jay98a] J. Jayapalan, B. J. Skromme, R. P. Vaudo, and V. M. Phanse, *Appl. Phys. Lett.* **73**, 1188 (1998).

[Kas00a] A. Kaschner, A. Hoffmann, C. Thomsen, F. Bertram, T. Riemann, J. Christen, K. Hiramatsu, H. Sone, and N. Sawaki, *Appl. Phys. Lett.* **76**, 3418 (2000).

[Kas01a] A. Kaschner, A. Hoffmann, and C. Thomsen, *Phys. Rev. B* 64, 165314 (2001).

[Kau98a] U. Kaufmann, M. Kunzer, M. Maier, H. Obloh, A. Ramakrishnan, B. Santic, and P. Schlotter, *Appl. Phys. Lett.* **72**, 1326 (1998).

[Kau99a] U. Kaufmann, M. Kunzer, H. Obloh, M. Maier, Ch. Manz, A. Ramakrishnan, and B. Santic, *Phys. Rev. B* **59**, 5561 (1999).

[Kir00a] V. Kirilyuk, A. R. A. Zauner, P. C. M. Christianen, J. L. Weyher, P. R. Hageman, and P. K. Larsen, *Appl. Phys. Lett.* **76**, 2355 (2000).

[Kir00b] V. Kirchner, H. Heinke, D. Hommel, J. Z. Domagala, and M. Leszczynski, *Appl. Phys. Lett.* **77**, 1434 (2000).

[Kir01a] V. Kirilyuk, P. R. Hageman, P. C. M. Christianen, W. H. M. Corbeek, M. Zielinski, L. Macht, J. L. Weyher, and P. K. Larsen, *Phys. Stat. Sol. (a)* **188**, 473 (2001).

[Kir01b] V. Kirilyuk, P. R. Hageman, P. C. M. Christianen, P. K. Larsen, and M. Zielinski, *Appl. Phys. Lett.* **79**, 4109 (2001).

[Kis96a] C. Kisielowski, J. Krüger, S. Ruvimov, T. Suski, J. W. Ager III, E. Jones, Z. Liliental-Weber, M. Rubin, E. R. Weber, M. D. Bremser, and R. F. Davis, *Phys. Rev. B* **54**, 17745 (1996).

[Kli04a] C. F. Klingshirn, *Semiconductor Optics*, Springer, Berlin, Germany. ISBN 3-540-21328-7.

[Kor99a] K. Kornitzer, T. Ebner, K. Thonke, R. Sauer, C. Kirchner, V. Schwegler, M. Kamp, M. Leszczynski, I. Grzegory, and S. Porowski, *Phys. Rev. B* **60**, 1471 (1999).

[Kuc01a] S. O. Kucheyev, M. Toth, M. R. Philips, J. S. Williams and C. Jagadish, *Appl. Phys. Lett.* **79**, 2154 (2001).

[Lee99a] S. R. Lee, A. F. Wright, M. H. Crawford, G. A. Petersen, J. Han, and R. M. Biefeld, *Appl. Phys. Lett.* **74**, 3344 (1999).

[Lee99b] I.-H. Lee, J. J. Lee, P. Kung, F. J. Sanchez, and M. Razeghi, *Appl. Phys. Lett.* **74**, 102 (1999).

[Lee01a] K. Lee and K. Auh, *MRS Internet J. Nitride Semicond. Res.* **6**, 9 (2001).

[Liu05a] R. Liu, A. Bell, F. A. Ponce, C. Q. Chen, J. W. Yang, and M. Asif Khan, *Appl. Phys. Lett.* 86, 021908 (2005).

[Mar01a] G. Martinez, C. R. Miskys, A. Cros, O. Ambacher, A. Cantarero, and M. Stutzmann, *J. Appl. Phys.* 90 5627 (2001).

[Mat97a] T. Mattila and R. M. Nieminen, *Phys. Rev. B* **55**, 9571 (1997).

[May97a] M. Mayer, A. Pelzmann, M. Kamp, K. J. Ebeling, H. Teisseyre, G. Nowak, M. Leszczynski, I. Grzegory, S. Porowsky, and G. Karczewski, *Jpn. J. Appl. Phys. Part 2* **36**, L1634 (1997).

[Mon74a] B. Monemar, *Phys. Rev. B* **10**, 676 (1974).

[Mon01a] B. Monemar, *J. Phys.: Condens. Matter* **13**, 7011 (2001).

[Moo01a] W. J. Moore, J. A. Freitas Jr., G. C. B. Braga, R. J. Molnar, S. K. Lee, K. Y. Lee, and I. J. Song, *Appl. Phys. Lett.* **79**, 2570 (2001).

[Myo96a] J. M. Myoung, K. H. Shim, C. Kim, O. Gluschenko, K. Kim, S. Kim, D. A. Turnbull, and S. G. Bishop, *Appl. Phys. Lett.* **69**, 2722 (1996).

[Neu94a] J. Neugebauer and C. G. Van de Walle, *Phys. Rev. B* **50**, 8067 (1994).

[Neu96a] J. Neugebauer and C. G. Van de Walle, *Mat. Res. Soc. Symp. Proc.* **395**, 645 (1996).

[Neu96b] J. Neugebauer and C. G. Van de Walle, *Appl. Phys. Lett.* **69**, 503 (1996).

[Neu99a] J. Neugebauer and C. G. Van de Walle, *J. Appl. Phys.* **85**, 3003 (1999).

[Ogi80a] T. Ogino and M. Aoki, *Jpn. J. Appl. Phys.* **19**, 2395 (1980).

[Oil01a] J. Oila, V. Ranki, J. Kivioja, K. Saarinen, P. Hautojärvi, J. Likonen, J. M. Baranowski, K. Pakula, T. Suski, M. Leszczynski, and I. Grzegory, *Phys. Rev. B* **63**, 045205 (2001).

[Pas01a] P. P. Paskov, T. Paskova, P. O. Holtz, and B. Monemar, *Phys. Rev. B* **64**, 115201 (2001).

[Pas05a] P. P. Paskov, R. Schifano, B. Monemar, T. Paskova, S. Figge, and D. Hommel, *J. Appl. Phys.* **98**, 093519 (2005).

[Pon96a] F. A. Ponce, D. P. Bour, W. Götz, and P. J. Wright, *Appl. Phys. Lett.* **68**, 57 (1996).

[Poz99a] G. Pozina, J. P. Bergman, T. Paskova, and B. Monemar, *Appl. Phys. Lett.* **75**, 4124 (1999).

[Ram98a] P. Ramvall, S. Tanaka, S. Nomura, P. Riblet, and Y. Aoyagi, *Appl. Phys. Lett.* **73**, 1104 (1998).

[Ram00a] P. Ramvall, P. Riblet, S. Nomura, Y. Aoyagi, and S. Tanaka, *J. Appl. Phys.* **87**, 3883 (2000).

[Res99a] M. A. Reshchikov, G.-C. Yi, and B. W. Wessels, *Phys. Rev. B* **59**, 13176 (1999).

[Res00a] M. A. Reshchikov, F. Shahedipour, R. Y. Korotkov, B. W. Wessels, and M. P. Ulmer, *J. Appl. Phys.* **87**, 3351 (2000).

[Res01a] M. A. Reshchikov, D. Huang, F. Yun, L. He, H. Morkoc, D. C. Reynolds, S. S. Park, and K. Y. Lee, *Appl. Phys. Lett.* **79**, 3779 (2001).

[Res01b] M. A. Reshchikov, H. Morkoc, S. S. Park, and K. Y. Lee, *Appl. Phys. Lett.* **78**, 3041 (2001).

[Res01c] M. A. Reshchikov and R. Y. Korotkov, *Phys. Rev. B* **64**, 115205 (2001).

[Rey00a] D. C. Reynolds, D. C. Look, B. Jogai, J. E. Hoelscher, R. E. Sherriff, and R. J. Molnar, *J. Appl. Phys.* **88**, 1640 (2000).

[Rey00b] D. C. Reynolds, D. C. Look, B. Jogai, A. W. Saxler, S. S. Park, and J. Y. Hahn, *Appl. Phys. Lett.* **77**, 2879 (2000).

[Rey01a] D. C. Reynolds, D. C. Look, B. Jogai, and R. J. Molnar, *J. Appl. Phys.* **89**, 6272 (2001).

[Rod01a] A. V. Rodina, M. Detrich, A. Göldner, L. Eckey, A. Hoffmann, Al. L. Efros, M. Rosen, and B. K. Meyer, *Phys. Rev. B* **64**, 115204 (2001).

[Ros97a] S. J. Rosner, E. C. Carr, M. J. Ludowise, G. Girolami, and H. I. Erikson, *Appl. Phys. Lett.* **70**, 420 (1997).

[Saa97a] K. Saarinen, T. Laine, S. Kuisma, J. Nissilä, P. Hautojärvi, L. Dobrzynski, J. M. Baranovski, K. Pakula, R. Stepniewski, M. Wojdak, A. Wysmolek, T. Suski, M. Leszczynski, I. Grzegory, and S. Porowski, *Phys. Rev. Lett.* **79**, 3030 (1997).

[Saa01a] K. Saarinen, T. Suski, I. Grzegory, and D. C. Look, *Phy. Rev. B* **64**, 233201 (2001).

[Sha99a] I. Shalish, L. Kronik, G. Segal, Y. Rosenwaks, Y. Shapira, U. Tisch, and J. Salzman, *Phys. Rev. B* **59**, 9748 (1999).

[Sha00a] F. Shahedipour and B. W. Wessels, *Appl. Phys. Lett.* **76**, 3011 (2000).

[Sha01a] F. Shahedipour and B. W. Wessels, *MRS Internet J. Nitride Semicond. Res.* **6**, 12 (2001).

[Shi97a] A. Shikanai, T. Azuhata, T. Sota, S. Chichibu, A. Kuramata, K. Horino, and S. Nakamura, *J. Appl. Phys.* **81**, 417 (1997).

[Sie97a] H. Siegle, A. Hoffmann, L. Eckey, C. Thomsen, J. Christen, F. Bertram, M. Schmidt, D. Rudloff, and K. Hiramatsu, *Appl. Phys. Lett.* **71**, 2490 (1997).

[Skr99a] B. J. Skromme, J. Jayapalan, R. P. Vaudo, and V. M. Phanse, *Appl. Phys. Lett.* **74**, 2358 (1999).

[Sta98a] C. Stampfl and C. G. Van de Walle, *Phys. Rev. B* 57, R15052 (1998).

[Ste99a] R. Stepniewski, M. Potemski, A. Wysmolek, K. Pakula, J. M. Baranowski, J. Lusakowski, I. Grzegory, S. Porowski, G. Martinez, and P. Wyder, *Phys. Rev. B* **60**, 4438 (1999).

[Sug98a] T. Sugahare, H. Sato, M. Hao, Y. Naoi, S. Kurai, S. Tottori, K. Yamashita, K. Nishino, L. T. Romano, and S. Sakai, *Jpn. J. Appl. Phys.* Part 2 37, L398 (1998).

[Tot99a] M. Toth, K. Fleischer, and M. R. Phillips, *Phys. Rev. B* **59**, 1575 (1999).

[Ves00a] S. Ves, U. D. Venkateswaran, I. Loa, K. Syassen, F. Shahedipour, and B. W. Wessels, *Appl. Phys. Lett.* **77**, 2536 (2000).

[Vol96a] D. Volm. K. Oettinger, T. Streibl, D. Kovalev, M. Ben-Chorin, J. Diener, B. K. Meyer, J. Majewski, L. Eckey, A. Hoffmann, H. Amano, I. Akasaki, K. Hiramatsu, and T. Detchprohm, *Phys. Rev. B* **53**, 16543 (1996).

[Vur01a] I. Vurgaftman, J. R. Meyer, and L. R. Ram-Mohan, *J. Appl. Phys.* **89**, 5815 (2001).

[Wag02a] J. M. Wagner and F. Bechstedt, *Phys. Rev. B* **66**, 115202 (2002).

[Wal97a] C. G. Van de Walle, *Phys. Rev. B* **56**, 10020 (1997).

[Wal01a] C. G. Van de Walle, S. Limpijumnong, and J. Neugebauer, *Phys. Rev. B* **63**, 245205 (2001).

[Wei96a] S.-H. Wei and A. Zunger, *Appl. Phys. Lett.* **69**, 2719 (1996).

[Wet97a] C. Wetzel, T. Suski, J. W. Ager III, E. R. Weber, E. E. Haller, S. Fischer, B. K. Meyer, R. J. Molnar, and P. Perlin, *Phys. Rev. Lett.* **78**, 3923 (1997).

[Yan00b] H. C. Yang, T. Y. Lin, and Y. F. Chen, *Phys. Rev. B* **62**, 12593 (2000).

[Yos99a] M. Yoshikawa, M. Kunzer, J. Wagner, H. Obloh, P. Schlotter, R. Schmidt, N. Herres, and U. Kaufmann, *J. Appl. Phys.* **86**, 4400 (1999).

[Zak05a] D. N. Zakharov, Z. Liliental-Weber, B. Wagner, Z. J. Reitmeier, E. A. Preble, and R. F. Davis, *Phys. Rev. B* 71, 235334 (2005).

[Zha01a] W. Zhang, S. Rösel, H. R. Alves, D. Meister, W. Kriegseis, D. M. Hofmann, B. K. Meyer, T. Riemann, P. Veit, J. Bläsing, A. Krost, and J. Christen, *Appl. Phys. Lett.* **78**, 772 (2001).

Part V

Device Structures and Properties

12

GaN-Based Optical Devices on Silicon

Armin Dadgar

CONTENTS

12.1 Introduction

Optoelectronics have been the driving force behind III–Vs on silicon and gallium-nitride in general. Despite the big success of GaN growth on sapphire and SiC in the early 1990s, GaN growth on silicon has been only a playground for researchers until the end of that decade. This is mostly because the growth on sapphire and SiC was simple and yielded a high output power for light-emitting devices. In contrast to that, device performance on Si was, if any, very poor. This changed a little in the early 2000s when thicker crack-free GaN on Si layers were demonstrated and companies as NITRONEX started with the commercialization of GaN on Si RF devices, leading to a change in general awareness of GaN on Si from academic research to a potential alternative to the growth on sapphire and SiC. Today, the general advancement in output power and substrate size as well as the drop in LED prices makes GaN on silicon more interesting than ever. In high-end applications high-brightness LEDs are usually (except, e.g., Nichia) based on a thin-film approach, where the grown layer is separated from the substrate and consequently independent on the optical properties of the substrate. Additionally, defect densities below 10^9 cm^{-2}, possible to achieve for GaN on silicon, are low enough to compete in internal quantum efficiency with devices grown on sapphire. Because of the possibility to grow on silicon substrate sizes of up to 300 mm, the technology is getting very interesting as a low cost route for general lighting applications based on GaN LEDs.

As one of the most important publications on GaN LEDs on Si possibly initiating many research groups to investigate GaN LEDs on Si were two publications on MBE grown LEDs by S. Guha and N.A. Bojarczuk in 1998 [Guha 1998a, Guha 1998b]. Their AlGaN buffer–based LEDs with GaN QWs and p-GaN top layer demonstrated that p-type doping is possible for GaN on silicon. Until then, a common assumption was that silicon will diffuse into the GaN layer during growth. However, the observation is that undoped layers are rather highly resistive than well conducting, and p-type doping is as simple as on other substrates. Although many researchers started investigating GaN on Si as an alternative to the growth on sapphire and SiC, it took until 2009 for the first report on a bright GaN on Si LED to be presented [Jiang 2009]. But big companies started investigating this technology: the second biggest LED manufacturer OSRAM, for example, bought the GaN on silicon process, originating on the development of the Otto-von-Guericke-University Magdeburg, from AZZURRO Semiconductors in 2009 [AZZURRO 2009]. At the current stage of development, where LEDs are discussed as energy saving replacement of incandescent light bulbs, the key to get LEDs into the general illumination consumer market is price. Price can be reduced by low-cost manufacturing using low-cost materials, large substrate diameters to achieve a higher number of devices per processing step, and high efficiency in terms of wall-plug efficiency or lumens per Watt of the LEDs. GaN on silicon is certainly an excellent candidate to achieve this as long as efficiency requirements are met by this material combination. In this chapter, growth techniques to achieve crack-free LEDs and methods to increase the efficiency are discussed.

12.2 Patterned and Planar Growth

12.2.1 Selective Epitaxy

Using selective epitaxy, one of the first LEDs on silicon by MOVPE was demonstrated [Yang 2000]. With this method, crack generation can be avoided

(Chapter 4). However, the first LED grown in that way did, although less than 500 nm thick, show cracks. A first crack-free LED on $100 \times 100 \,\mu m^2$ fields was presented in 2001 [Dadgar 2001a, Dadgar 2001b, Dadgar 2003]. This ~3.6 μm thick structure included an AlGaN/GaN multilayer in the buffer, which was helpful in reducing tensile strain. Honda et al. published a $200 \times 200 \,\mu m^2$ LED 1.5 μm thick in 2002 [Honda 2002]. A relatively bright crack-free LED was presented by Zhang et al. [Zhang 2007], with a total thickness slightly above 2 μm by using a strain engineering AlN interlayer. The device showed an output power of 0.7 mW at 20 mA and ~460 nm (Figures 12.1 and 12.2). For an identical structure on sapphire, the output power is found to be about a factor of 3–4 higher. The difference can be attributed to light absorption by the silicon substrate and is in agreement with the estimation for light extraction in Section 12.5.2.

In 2009, Sawaki et al. presented an LED based on GaN(10$\bar{1}$1) layers [Sawaki 2009]. Here, to avoid the formation of cracks in their structures, they applied patterning before growth to obtain $300 \times 300 \,\mu m^2$ large GaN LED structures.

Based on a similar structure as by Zhang et al. without AlN strain engineering interlayer, Latticepower announced an 18 mW blue LED in 2009 [Jiang 2009]. They claim up to 4 μm of n-GaN to be grown without further strain engineering after an AlGaN/AlGaN strain engineering buffer layer in fields up to $0.5 \times 0.5 \,mm^2$ with a crack-free chip yield >97%. The output power seems to be very high for an LED on Si. But to achieve this, a thin-film approach is used where the sample has been bonded p-side down with a reflecting mirror onto a carrier substrate and n-contact metallization and surface roughening is performed from the nitrogen-terminated side. This is the highest ever reported output power so far for GaN on Si-based LEDs. Presently, the LEDs on sale are much lower in intensity, with 8 mW

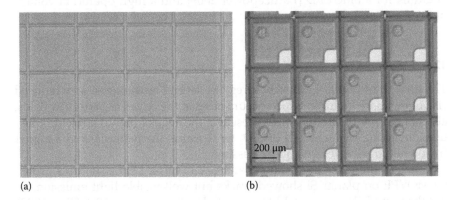

(a) (b)

FIGURE 12.1
Top view on $300 \times 300 \,\mu m^2$ as grown (a) and processed (b) patterned LEDs on silicon. (Reprinted from Zhang, B. et al., *J. Cryst. Growth*, 298, 725, 2007. With permission from Elsevier, © 2007 by Elsevier.)

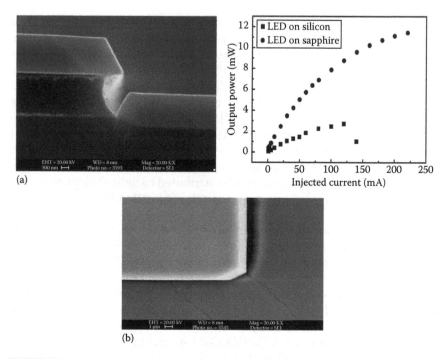

(a)

(b)

FIGURE 12.2

Cross-sectional (a) and top view (b) SEM of the LED structure. The etch patterned silicon substrate hinders the formation of a continuous layer with some stress related cracks being visible in (b) The output power of the device reached more than 2 mW at currents above ~70 mA, but a factor of 3 to 4 less than on sapphire (top right). (Reprinted from Zhang, B. et al., *J. Cryst. Growth*, 298, 725, 2007. With permission from Elsevier, © 2007 by Elsevier.)

at 20 mA for a blue LED [Latticepower 2009] and a high operation voltage of 3.6 V at 20 mA.

12.2.2 Planar LEDs

The growth of planar LED structures has several advantages, as a reduced risk of meltback etching, which occurs often at the edges of patterned fields, absence of growth enhancement present at the edge of patterned fields when using masking layers as SiN or SiO$_2$ for defining the patterns, and a higher freedom in device design and processing.

Although there are potential advantages, the first LED structures grown by MOVPE on planar Si showed cracks but well visible light emission. So did the first MOVPE grown LED presented by Tran et al. in 1999 [Tran 1999]. This LED structure had a thickness of over 4 μm but, as could be expected without any strain engineering applied, showed cracks. The XRD FWHM of the (0002) reflection showed a value of 600 arcsec and that of the asymmetric

($10\underline{1}2$) reflection only 590 arcsec indicating good material quality for this first report on an MOVPE-grown LED structure. Light output at ~460 nm was only reported to be well visible but no measured data was published. In PL and EL spectra, an interference pattern by reflection at the surface and the Si substrate, which is typically well pronounced for LEDs on Si and indicates a smooth surface, can also be seen.

In 2001, Feltin et al. presented a green (508 nm) LED with a thickness around 1.5 μm and an output power of 6 μW at 20 mA and 10.7 V. Series resistance was attributed to be high because of a thin n-GaN layer and poor p-type doping efficiency [Feltin 2001]. In 2002, the first crack-free LED significantly thicker than 1 μm was demonstrated [Dadgar 2002a, Dadgar 2002b]. The 2.8 μm thick structure was grown crack-free using two strain engineering LT-AlN interlayers and SiN masking layers for dislocation reduction. At 455 nm and 20 mA, drive current requiring a voltage ~4.5 V light output yielded 152 μW (Figure 12.3). Light output was low compared to today's standards but the blue LED light emission well visible in daylight. After this, a further jump in light output power was an LED that showed emission at 477 nm, with an output power of 0.9 mW at 20 mA and up to 3 mW for a current of 120 mA (Figure 12.4 left) [Dadgar 2004a], which marked the highest reported output power for several years. In comparison with an LED on sapphire identically grown for the functional parts of the LED, operation voltage was comparable for top contacted LEDs but light output power was about a factor of 4–5 lower. The benefit of the silicon substrate was much better thermal conductivity at high currents, visible in a later onset of a thermally induced redshift (Figure 12.4 right). Also, Egawa et al. observed an improved performance at high current levels for LEDs on Si in comparison to sapphire [Egawa 2005].

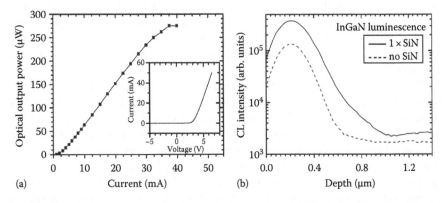

FIGURE 12.3
Output power and *I–V* curve of a crack-free planar LED on Si(111) (a). The usage of a SiN masking layer improved MQW luminescence in CL cross-sectional measurements (b). (Reprinted from Dadgar, A. et al., *Phys. Stat. Sol. (a)*, 192, 308, 2002. With permission from Wiley-VCH. © 2002 by Wiley-VCH.)

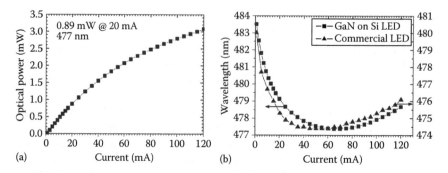

FIGURE 12.4
Output power of a top contacted LED on silicon (a). For higher currents first a blueshift by band-filling effects and then a redshift due to heating is observed (b). The onset of the heat-induced redshift is occurring earlier for the LED on sapphire than on silicon which can be attributed to a better thermal conductivity of the silicon substrate. (Reprinted from Dadgar, A. et al., *Advances in Solid State Physics* Vol. 44, B. Kramer editor, Springer, Heidelberg, Germany, 2004. With permission from Springer, © 2004 by Springer.)

12.3 Electrical Properties

In the early 2000s, GaN LEDs on conductive silicon were thought to be advantageous with regard to contacting, as is the case for GaN LEDs on SiC, and this is regarded as a major advantage. At this time, the most widely manufactured LEDs on sapphire had to be top contacted for p- and n-type contacts. This resulted in an additional bonding wire, loss of device area because of the area needed by the n-type bonding pad, reduced efficiency, and a reduction in the total amount of LEDs per wafer. But already in early 2000s data published on GaN on silicon LED operation showed a driving voltage of 8 V at 20 mA [Tran 1999] or even above 10 V for 20 mA [Yang 2000] indicating a severe contacting or conductivity problem, which can lead to an increased operation temperature.

The problem originates in the seeding and buffer layers of LED structures on silicon. They are usually required to be grown with a high bandgap AlN or at least Al-rich seeding layer, which is also very difficult to dope. N-type doping of AlN typically results only in low electron concentrations around 10^{16} cm^{-3} [Taniyasu 2008]. In addition to this, the band offset at the silicon interface is very high [Badylevich 2008]. The simplest approach would be the growth of a very thin AlN seeding layer, ideally a thickness of a few mono-layers, being thin enough for carrier tunneling. Here, the problem that occurs is that such a thin AlN layer would not be continuous, since typically the ini-tial nanometers of seeding layer growth are island-like growth. Thicker GaN layers on top are then likely to lead to meltback etching, at least in MOVPE processes. In 2003, Zhang et al. reported a low vertical resistance by applying a thin AlN/AlGaN seeding layer [Zhang 2003] and compared this to an LED on sapphire (Figure 12.6). While on sapphire the operation voltage at 20 mA yielded 3.5 V, it increased about 0.2 and 0.8 V for top and backside contacted LED on Si, respectively. For the top contacted LED on Si, it was assumed that the only 200 nm thick n-GaN layer lead to an increased series resistance in comparison to the top contacted LED on sapphire. Egawa et al. showed that reducing the thickness of the AlN seeding layer from 120 to 3 nm reduces the series resistance from 100 to 30 Ω [Egawa 2005]. However, 3 nm of AlN bear a significant risk of meltback etching for the growth of thicker layers. In a report by Egawa et al., the AlN seeding layer was covered with $Al_{0.27}Ga_{0.73}N$ followed by an AlN/GaN multilayer. All these Al-containing layers increase series resistance. Alternatively, an AlGaN seeding layer can be chosen, again with a significant meltback etching risk. A study on the series resistance of LED structures by Honda et al. [Honda 2007] showed that indeed lowering the Al-content or the thickness of the AlGaN seeding layer lowers the series resistance significantly. The limitation is given by the Al-content (Figure 12.5), thickness, and Si doping. From Figure 12.5, it can be concluded that using AlGaN seeding layers, even with low Al-content, will always lead to a higher operation voltage and additional heating. In addition, growing thick structures on such seeding layers bears a high risk of meltback etching.

Armitage et al. presented an MBE grown HfN buffer on Si(111), which enables subsequent GaN growth [Armitage 2002]. Since HfN, which crystallizes in the NaCl lattice, is metallic and well lattice matched to GaN, it potentially offers

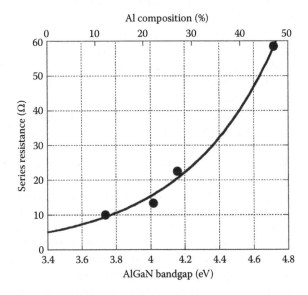

FIGURE 12.5
Series resistance of a vertical contact from the silicon substrate to an n-type top contact with varying Al-content of the seeding and buffer layer. (Reprinted from Honda, Y. et al., *Phys. Stat. Sol. (c)* 4, 2740, 2007. With permission from Wiley-VCH, © 2007 by Wiley-VCH.)

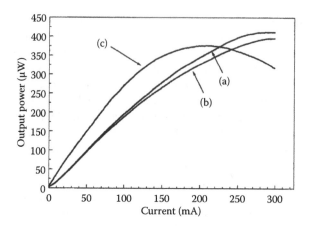

FIGURE 12.6

Output power vs. current for different LEDs on silicon and sapphire. Here a comparison between a diode on sapphire (c) and on silicon with vertical contacts (b) and top contacts (a) is shown. Also, here the maximum output power is higher on silicon than on sapphire because of a significant loss by the poor heat dissipation for the LED on the sapphire substrate. (Reprinted from Zhang, B. et al., *Jpn. J. Appl. Phys.*, 42, L 226, 2003. With permission from Japan Society of Applied Physics, © 2003 by Japan Society of Applied Physics.)

high material quality and low series resistance for vertical contacts. However, the lattice match to Si is not that good, leading to an average GaN material quality and deposition by MOVPE is difficult, with the risk of the formation of a highly resistive Hf_3N_4 phase [Wang 2006]. As will be discussed in Section 12.5.2, the reflectivity in the green–blue wavelength region is low and offers no improvement in comparison to pure Si. As an alternative to HfN, a metallic system which enables a graded transition layer from Si to GaN in-plane lattice parameters could in principle result in a high material quality.

A solution for the backside contacting problem is a bridged contact or a via contact from the back as shown in Figure 12.7. However, at the expense of additional processing steps and the disadvantage of an excellent current distribution in the first case (Figure 12.7a).

Modern concepts as thin-film LEDs avoid the above-mentioned electrical problems. With a p-type contact usually based on Ni/Ag [Chang 2009], also acting as a mirror, p-contacting layers are not an issue with regard to current spreading and resistivity anymore. Having the n-side up and the n-contact being alloyed after bonding the LED layer structure to a new carrier, the main problem in processing is the process temperature for n-type contacting, which should not exceed the alloying temperature of the p-type bond. This limitation in alloying temperature requires improved metallization and higher n-type doping levels compared to conventional standard processing to achieve low series resistance.

A typical property of most GaN LEDs on silicon is their higher operation voltage in addition to their lower output power when compared to those

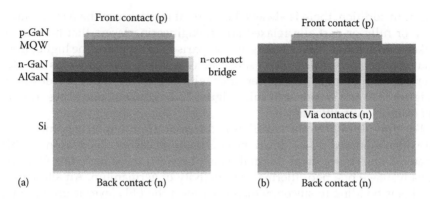

FIGURE 12.7
Scheme of backside contacted GaN on silicon LEDs avoiding current transport via a highly resistive AlN seeding layer. (a) Shows a simple contact bridge between the silicon substrate and the n-GaN layer, (b) a possible contacting route using via contacts.

on sapphire, for example, even when contacting is not performed through highly resistive Al(Ga)N layers. But many of these results stem from research activities in smaller labs which often do not have optimized contacting schemes, thus a comparison is difficult.

12.4 Thermal Properties

Silicon itself has a good thermal conductance, depending on the substrate type (FZ or CZ) and impurities; it ranges up to 1.5 W/cm K for standard (isotope impure) silicon and thus is comparable to GaN. Major disadvantage is the high thermal resistivity at interfaces [Kuzmik 2005]. Especially at the Si/AlN interface only about every second Al–N bond has a counterpart in the substrate leading to a poor thermal coupling. Nevertheless, when compared to sapphire, the situation at the interface is comparable. For thin-film LEDs mounted p-side down, the growth structure in the buffer is expected not to influence the thermal properties significantly. Interfaces in buffer layers may only limit the thermal performance for LEDs not transferred p-side down to a new carrier.

But LEDs on bare silicon substrates have a major disadvantage with regard to EQE, because of the omnipresent absorption of the substrate. This can be reduced, e.g., by suited SOI substrates [Tripathy 2007, 2008, 2009] or integrated DBR layers [Semond 2001], to a small part by reflective layers which should, however, be more reflective than HFN [Armitage 2002] in the blue spectral region. Of these options, at least the first two are cost intensive due to high substrate cost or an extended growth time. Especially in the case of SOI, an enhancement in thermal resistivity can be expected by the oxide layer. In addition, there is always the thermal resistance at the AlN/Si interface or fully or partially relaxed strain engineering layers that have to be taken into account. Nevertheless, in comparison to sapphire, the higher thermal conductance of the silicon substrate is still beneficial. It yields a better performance at high current densities with regard to output power droop (Figure 12.6) and wavelength shift (Figure 12.4, right) but, due to absorption, an overall lower efficiency.

For thin-film LEDs, the situation is different. If grown on sapphire, they are often mounted p-side down on a germanium carrier. The reason for this material choice is a good thermal match between these materials with the disadvantage of a poor thermal conductivity of germanium. An alternative to this is bonding to silicon, which is easier if the LED layer is grown on it due to a better thermal match when bonding. Compared to materials such as copper, the thermal properties are still inferior. But copper has a high thermal mismatch to GaN with approximately three times higher thermal expansion coefficient leading to thermal stresses during operation, which can significantly decrease the lifetime if the devices are often thermally stressed (switched on and off), especially in the case of large area devices. Metal carries as Cr or W are closely matched in thermal expansion coefficient but show inferior or only slightly higher thermal conductance, respectively, when compared to GaN or silicon (both ~1.5 W/cm K). An alternative to silicon and germanium, adding the benefits of both, are principal alloys of two materials such as Si and Al, which can be thermally matched to GaN. In the case of Si–Al this can be achieved with a Si content of 85%–90% with a high thermal conductivity comparable to pure silicon. In all cases it should be kept in mind that, e.g., via contact formation is an established process in silicon but not in other materials. Thus, if required, e.g., for top contactless LEDs, an etching process needs to be developed for all other carrier materials.

12.5 Optical Properties

For LEDs in the visible spectrum, the most relevant figure is lumens per watt or, in other wavelength regimes as UV, the external quantum efficiency (EQE) (and wall plug efficiency), and the total light flux that can be generated with a single device. But even if the best MQWs yield near 100% internal quantum efficiency (IQE), the light needs to be extracted to achieve a high EQE and high lumen flux.

12.5.1 Internal Quantum Efficiency

The internal quantum efficiency of MQW layers is known to be dependent on the defect density [Henley 2003, Zhu 2010], which should be targeted in the low 10^8 cm^{-2} range to be negligible. The often inferior quality of GaN on silicon in comparison to sapphire has long been thought to be inevitable and therefore LEDs on silicon were expected to be always significantly less efficient to those on sapphire. But material quality on silicon is improving with threading dislocation densities for best layers in the mid 10^8 cm^{-2} range, of which the screw type dislocation density is usually lower than the edge type dislocation density and around high 10^7 cm^{-2} to low 10^8 cm^{-2}. These values are comparable to average GaN layers on sapphire. Also, XRD proves that material with high-quality layers can be obtained on silicon. For GaN FWHMs of (0002) and ($10\bar{1}0$), reflections are found as low as 0.08–0.1 for both values.

For simultaneous growth on silicon and sapphire substrates in a multiwafer system, one typically observed difference is that often a higher In-content is found for MQWs on silicon in comparison to sapphire. Thus, the MQW layers need to be optimized for the growth on silicon to find the best fit of In-content and thickness for the target wavelength. With regard to piezoelectric fields, one can expect them to be different on silicon compared to those on sapphire since on silicon the layer typically tends to tensile strain while on sapphire it is under compressive strain at room temperature. However, this holds for the whole buffer layer and MQW stack, thus no principal change in the internal MQW polarization field changes can be expected.

12.5.2 External Quantum Efficiency—Extraction Bottleneck

As already mentioned, the main problem in LED efficiency on silicon is light extraction, which is inferior to LEDs on transparent substrates. The impact of it can be easily seen if an LED is cracked: Typically at cracks, a very strong light emission is observed while the surface of such an LED usually is, in comparison, not very bright. This indicates that a lot of light is generated but can't be extracted.

The difference for vertical extraction of a simple LED structure is shown in Figure 12.8. In addition to the upward emitted light, light emitted backward and light emitted laterally (Figure 12.9) in similar angle (~24/~38° for air/epoxy) needs to be taken into account. On silicon, only a small amount of the backward emitted light reflected at the substrate and of the laterally emitted light will be extracted with Si having a reflectivity <0.4 [Chelikowsky 1976]. For standard processed wafers, this leads to only a narrow edge region with high outcoupling efficiencies for laterally emitted light. But already about 50 μm from the edge, the laterally extracted light intensity reaching the facet is about a factor of 5 lower. In addition, all light emitted in the light cone in backward direction is lost to a large amount due to multiple reflections (<1%).

Of the totally emitted light within the structure, for simple standard LED processing assuming an ideally reflecting substrate, about 42% can be extracted by lateral emission. When including vertical emission, in total

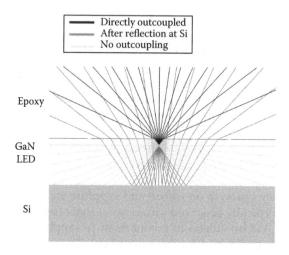

FIGURE 12.8

Cross-sectional view of light propagation in an LED structure on silicon (left): most of the directly extracted light (black) undergoes only reflection losses in the range of 1%–2%. Light being reflected at the silicon substrate within an angle below total reflection (medium gray) will also be extracted but undergoes a loss of over 60% at the GaN (AlN)/Si interface. Most light not within these light cones will be lost due to absorption losses by reflection at the Si interface except for a small portion <10% extracted at the edge of the device which originates in a narrow edge region of it (see Figure 12.9).

about 63% can be extracted in this ideal case. The other photons are trapped within the structure. For a real LED structure, all light will additionally undergo absorption by the LED structure itself or by non-ideally reflecting surfaces.

In contrast to LEDs on silicon, light propagation for LEDs on sapphire does also benefit from light propagation in the substrate and lateral light propagation in the GaN layer itself. Absorption losses, especially in the sapphire substrate, are low. Thus, aside from a difference in material quality, the differences can be mostly explained by losses for all light undergoing single, or multiple reflections at the silicon substrate. Therefore, the output power at identical driving current is always inferior for LEDs on silicon (see e.g., Figure 12.7). Assuming a 40% reflectivity at the AlN/Si interface, lateral light propagation is not extracting a significant amount of light (<10%). Only vertically emitted light in a light cone with an opening angle of ~48° (24° from the surface normal) or ~76° (38° from the surface normal) for the bare chip and when encapsulated in epoxy resin is extracted, respectively. In addition, <40% of the light emitted backward and reflected by the silicon substrate in a similar light cone are extracted. From the light surmounting all these extraction barriers, about 10%–20% is subsequently absorbed by the p-contact metallization. In sum, only a small fraction of light generated in the LED structure can be used. For sapphire, the situation is much better with about

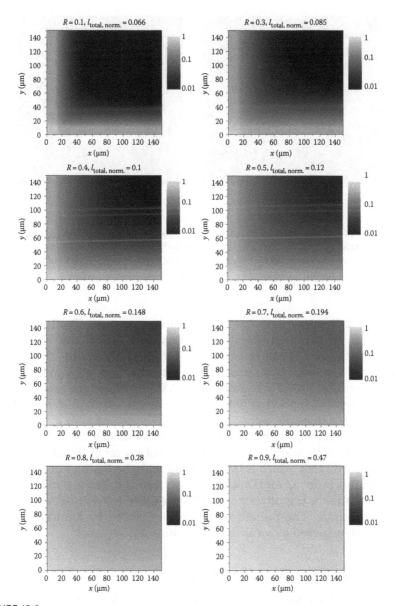

FIGURE 12.9
Amount and origin of laterally propagating light extracted at the side facets of a $300 \times 300\,\mu m^2$ LED assuming a GaN thickness of $4\,\mu m$, and a reflectivity R of the substrate ranging from 10% to 90%. Shown is only one quadrant of the LED with the outer facets at the left and bottom, taking into account light emitted into all lateral directions within a 38° extraction angle. In addition to the intensity shown here ~1/3 of the visible light of an LED stems from the vertical light cone (directly to the front facet and after reflection at the substrate). The difference between a Si substrate ($R \sim 0.3$–0.4) to a highly reflective substrate ($R > 0.9$) for the laterally extracted light is larger than a factor of five. Data is normalized to the case of a reflectivity of $R = 1$ neglecting any vertically emitted light.

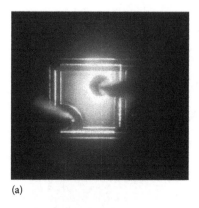
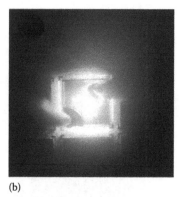

(a)   (b)

FIGURE 12.10

Photographs of top contacted LEDs on Si without (a) and (with) edge undercut etching (b) at identical driving current. The enhancement in light intensity at the edge is due to a significantly decreased absorption for backward propagating light rays, enhancing vertically and laterally emitted light intensities. (Processing and photographs by Tripathy, S. et al., IMRE, Singapore.)

a factor of 4–5 higher light output for simple structures. The higher value stems from light emitted laterally and backward, of which a much higher percentage than on Si can be extracted. But also on sapphire, the development of the last years has shown that light extraction is far from ideal on transparent substrates. For example, OSRAM has tripled the external quantum efficiency by improving light extraction from less than 25 for LEDs on SiC to 75% by a thin-film technique.

It has been already demonstrated that underetching an LED structure on silicon can significantly enhance the output power of the LEDs [Dadgar 2007b] (Figure 12.10), which is in agreement with the findings of Figure 12.9. Although this is not an ideal approach with regard to thermal management, it demonstrates the importance of modifying the reflectivity or better removing the substrate. The present output power record by Latticepower has indeed been achieved by a substrate free thin-film approach.

12.5.3 Technologies for Light Extraction

Since LEDs are always based on a high refractive index material most of the light generated is captured within the structure if it undergoes simple processing only. Even with textured surfaces, about 50% of the light (assuming the structure being above the MQW) will be extracted after reflection at the substrate with its known losses. If the light-emitting layer has a high IQE, the only remaining issue for improving LEDs is the improvement of light extraction. Light extraction can be achieved by several methods starting from techniques on wafer toward modified wafers to thin-film approaches where the LED layer structure is removed from the carrier substrate.

12.5.3.1 On-Wafer Approaches

LEDs on silicon are hampered by the low light-extraction efficiency. But they are interesting for integrated optoelectronics with Si electronics, e.g., for short distance data communication or as simple indicator lights. Then efficiency is not the highest priority but price, robustness, and in the case of data, communication also switching speed.

To enhance light emission when using the silicon growth substrate as LED carrier, several techniques can be applied:

1. Growth of thick layers; reduced lateral dimensions
2. Integration of a reflective layer in the silicon substrate
3. Integration of a Bragg mirror and formation of a resonant cavity LED
4. Formation of a photonic bandgap structure (PBG or photonic crystal, PC) to suppress lateral and enhance vertical emission

For topic (1), Figure 12.11 shows the impact of a thicker GaN layer that can increase lateral light extraction significantly. This is one reason why many devices on silicon suffer poor light-extraction efficiency, simply because only a thin GaN layer is grown. Then laterally propagating light rays undergo a higher number of reflections and thus a higher damping until they reach the side facet. Increasing the thickness from 1 to 4 μm increases the lateral light output for a standard $300 \times 300\,\mu m^2$ LED by a factor of four which is, however, only about 1/8th of the total light extracted. Alternatively, by decreasing the lateral dimensions multiple reflections are reduced and the light output from the side facets increased (Figure 12.11 bottom). For a $20 \times 20\,\mu m$ field size, the intensity of laterally extracted light is already >80% of what can be extracted with a simple structure. This can be easily understood by a view on Figure 12.9. There a ~10 μm wide region close to the facet is the origin of most of the light extracted laterally. Indeed even for a low substrate reflectivity this value is high because most light rays are directly extracted without being reflected at the substrate. Then laterally extracted light amounts to >50% of the total light extracted for an LED on silicon.

Nevertheless, the case of (1) is not ideal since even at 100% substrate reflectivity a significant amount of light is trapped within the structure. Therefore, also on sapphire LED efficiency can be increased. For the previous structures, the assumption was made that all light in an angle within 38 to the surface normal of the facet can be extracted, thus assuming the LED being embedded in a material with $n \sim 1.5$, e.g., an epoxy resin. Laterally propagating light can therefore account for about 2/3 of the light that can be extracted in simple device structures. This points out the importance of extracting laterally guided light to reach a high output power.

With regard to a reduced diameter, nanocolumns are an interesting approach. The main problem, as also for the small dimension LEDs, is the

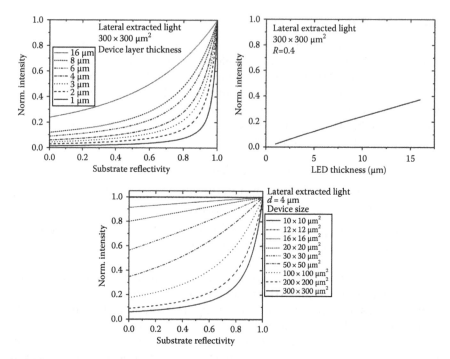

FIGURE 12.11

Calculated intensities of laterally extracted light (normalized to the maximum intensity of laterally emitted light) assuming a simple LED structure embedded in a transparent material with $n \sim 1.5$. By increasing the thickness of the grown layer (top) the number of multiple reflections at the substrate/seed layer interface is reduced for silicon with $R \sim 0.3–0.4$. Especially for thin LED structures the laterally extracted light intensity decreases to value of a few percent in comparison with an ideally reflecting substrate. Growing a 6–8 μm thick layer this intensity increases by a factor of 6–8 to above 10%, still low if compared to other substrates. For a significantly higher out-coupling efficiency of ~40% on silicon, a layer thickness above 15 μm is required, difficult and time consuming to grow. Reducing the lateral dimensions of the diode leads to an enhancement in laterally extracted light (bottom). Of all light generated in the diode, the portion of laterally extracted light amounts to max. 42% of the total light generated in the diode. Vertically emitted light, taking into account backward emitted light reflected at the substrate, amounts to max. 21%.

extraction of the light as only one nanocolumn or one small field is not sufficient to supply the required power for the application. Laterally extracted light will, to a significant amount, hit the facet of a neighboring LED field or nanocolumn if arrays of them are processed to obtain an overall high output power. Part of this light will then propagate in the neighboring field or nanocolumn, etc. Thus, to use the light, it is also required to be deflected vertically, which is a difficult task to solve. One possible approach is the growth of structures with inclined sidewalls, which will be beneficial in reflecting and/or refracting the photons in a vertical direction.

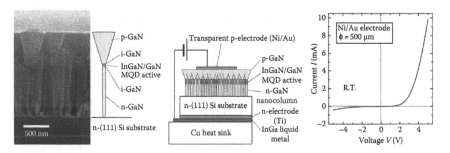

FIGURE 12.12
MBE grown nanorod LED structures which were coalesced by a Mg doped p-GaN layer on top. Such a coalesced layer is helpful in contacting the structures which is much more difficult for an uncoalesced structure. However, the I–V curve for vertically contacted LEDs shows a high series resistance. (Reprinted from Kikuchi, V. et al., *Jpn. J. Appl. Phys.*, 43, L 1524, 2004. With permission from Japan Society of Applied Physics, © 2003 by Japan Society of Applied Physics.)

A problem of nanocolumns is their contacting. An elegant solution might be a method presented by Kikuchi et al. [Kikuchi 2004]. They presented growth of nanocolumn structures and coalesced them to form a continuous p-GaN top layer for contacting (Figure 12.12). The method is based on the enhanced lateral growth rate when growing Mg-doped GaN. This is applied directly after the active region. However, light extraction is still an issue and series resistance for vertically contacted LEDs high.

In extension to a thicker layer and smaller device structures as in (1) the integration of a reflective layer (2) can be beneficial in lowering reflection losses. Such a layer can be achieved by several techniques. Starting with a reflective and conductive buffer layer on silicon as, e.g., HfN [Armitage 2002, Xu 2005] reflection losses can be in principle minimized. However, since the reflectivity is lowered from >80% in the red wavelength region to 30%–60% in the green–blue with a minimum of 20% around the GaN bandgap energy [Karlsson 1982, Perry 1988], this material, as also ScN [Moram 2006] or TiN is not ideally suited for GaN-based LEDs. In addition, deposition by MOVPE is often difficult, in the case of HfN with the risk of the formation of a highly resistive Hf_3N_4 phase [Wang 2006]. As can be seen in Figure 12.9, the reflectivity losses for laterally propagating light can be significantly reduced only for reflectivity values above ~90%. Other important factors, as already pointed out, are device thickness and device size.

Another method is the formation of a porous silicon layer or the usage of SOI as substrate with suited oxide and silicon overlayer thicknesses to achieve an enhanced reflection in vertical direction (Figure 12.13) [Tripathy 2007, Tripathy 2008, Tripathy 2009, Tripathy 2010]. Main disadvantage is the limited reflectivity of the reflective SiO_2 buffer layer and the silicon overlayer or, in the case of porous silicon, the porous silicon itself, which in all cases leads to absorption losses for laterally emitted light which usually undergoes several reflections. For example, for a photon energy of ~2.6 eV (~460 nm) the

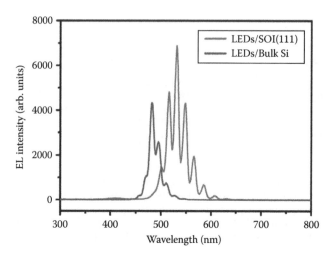

FIGURE 12.13

EL spectra of two LED structures grown identically on Si and SOI substrates. The LED on SOI yields a longer wavelength with higher intensity, the latter is assumed to be originating in the higher reflectivity of the SOI substrate. (Reprinted from Tripathy, S. et al., *Appl. Phys. Lett.*, 91, 231109, 2007. With permission from American Institute of Physics, © 2007 American Institute of Physics.)

absorption loss of each reflection will be 5% for a thickness of 10 nm [Sze 1981] assuming a 100% reflection at the Si/SiO_2 interface. For a $\lambda/2$ thick layer at 460 nm (~57 nm thick [Philipp 1960]), losses in the Si amount to about 25%. Thus, for LED applications, a proper design of thicknesses of the SiO_2 and Si layers, yielding high reflectivity and low absorption are required.

From the reflectivity viewpoint, the better option is the usage of dielectric mirrors with non-absorbing materials (3). This can be achieved on the basis of AlInGaN/(Al)GaN $\lambda/4$ layers, included, for example, in the buffer layers. Semond et al. demonstrated in 2001 a 10-fold MBE grown AlN/AlGaN Bragg mirror reaching a vertical reflectivity close to 80% [Semond 2001] (Figure 12.14). A full LED structure had been demonstrated by Ishikawa et al.

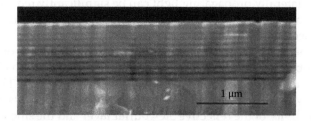

FIGURE 12.14

MBE grown AlN/AlGaN Bragg reflector on silicon. (Reprinted from Semond, F. et al., *Phys. Stat. Sol. (a)*, 183, 163, 2001. With permission from Wiley VCH, © 2001 by Wiley-VCH.)

[Ishikawa 2004a, Ishikawa 2004b]. Here, they were able to grow a three-fold AlN/AlGaN DBR crack-free and observed an enhancement in output power by a factor of two with an enhancement of the reflectivity of the structure from ~10% to 30% and even more for a (cracked) fivefold DBR, which showed a lower output power in electroluminescence, likely because of the cracks.

In principle, such dielectric layers can also serve as waveguides hindering laterally emitted light from being absorbed at the AlN/Si substrate interface when a sufficiently thick upper GaN waveguide below the MQW is grown and the propagation of optical modes into the lower part of the structure is inhibited. The low index contrast in the nitride systems makes this DBR approach a difficult task and to avoid cracking the lattice matched $Al_{82}In_{18}N$/GaN system [Feltin 2006] is a better choice than AlGaN/GaN.

Photonic bandgap or photonic crystal (PC) structures (4) are the two-dimensional (2D) or three-dimensional (3D) extension of Bragg mirror layers, being their one-dimensional (1D) representation. Two- and three-dimensional photonic bandgap structures are suited to suppress and/or enhance light modes in-plane or within a 3D structure. For LEDs on silicon, the simplest approach to enhance vertical light emission is the suppression of lateral emitted light by forming a forbidden optical bandgap within the range of quantum well emission, e.g., by structuring the surface or inter-face to silicon. To be efficient, it is best to have the photonic bandgap struc-ture as closely interfering with the electromagnetic wave of the photons, namely at the QW. Therefore, either a surface structuring method forming a refractive index contrast or etching hollows in between the MQW region is best suited to achieve an efficient suppression of laterally emitted light. Figure 12.15 shows some possible types of 2D photonic crystal structures for standard processing (a) and (b) or thin-film p-side down processing (c) and (d). In (a) and (c) only a surface index contrast by a 2D surface array, e.g., by etching or defining a low or high refractive index material by low-cost techniques as e.g., nanoimprint or holography lithography. In examples (b) and (d), the effect will be much more enhanced by etching deep into the structure. For the latter examples, care has to be taken to find suited sur-face passivation schemes within the p–n junction region not to degrade the device during operation. In (d), the design includes a PC structure at the MQW region and a roughened surface to optimize light extraction. Here, the overall effect of the PC will certainly not play a major role since such structures already can achieve an extraction efficiency of 75%. But it might further increase extraction thus being important for squeezing out the last 25% of efficiency and, depending on design, narrow the emitted spectrum. Tripathy et al. have demonstrated (Figure 12.16) that a structure compa-rable to Figure 12.15b can shift the dominant emission wavelength in PL measurements, thus can influence the emission wavelength significantly [Tripathy 2010].

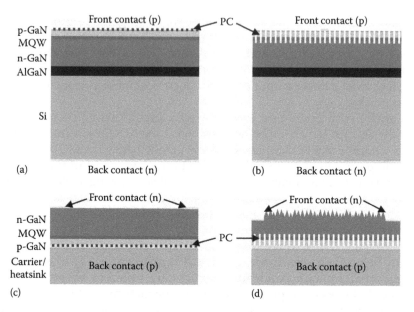

FIGURE 12.15

Different possible LEDs with integrated photonic crystal structures. In (a) only the metallization or an additional material with different refractive index to GaN is applied on the surface. In (b) holes are etched into the structure. For (b) a more efficient impact on lateral light emission can be expected to do the better overlap of laterally propagating light modes in contrast to (a). (c) and (d) are possible ways of realizing such structures on thin-film LEDs. (c) offers a possible way to realize a simple photonic crystal structure by modifying the p-contact mirror, e.g., nanoimprinting and subsequent filling with SiO_2 or another material with a refractive index different from GaN. As for the standard LEDs on Si it can be expected that an etched structure (d) is more efficient in suppressing lateral light modes.

12.5.3.2 Thin-Film Technology

Thin-film technology includes all technology approaches where the epitaxial III-nitride layer or such a layer including an additional thin substrate layer, e.g., from a SOI substrate, are removed from the growth substrate and transferred to a new carrier (Figure 12.17). The key idea is to remove the substrate and bond the samples p-side down onto an electrically and thermally well-conducting carrier, which ideally is highly reflecting or covered with a highly reflecting coating. Removing the substrate is commonly performed by laser-liftoff when grown on sapphire and by grinding and etching, e.g., with an $HF:HNO_3:CH_3OOH$-based etchant when grown on silicon. By the thin-film approach, first of all two efficiency limiting items, limited conductivity of p-GaN and absorption by a p-top-contact layer, can be avoided. In addition, when using sapphire substrates as growth substrate, thermal resistivity can be reduced with a suited carrier material. With regard to absorption losses a metallic p-contact mirror has no ideal reflectivity and is always

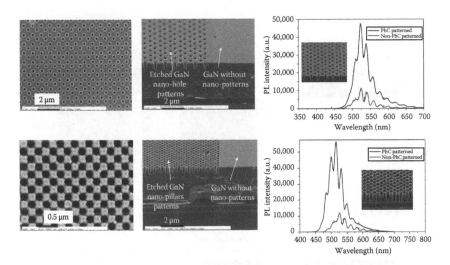

FIGURE 12.16
Two differently sized PC structures (left) on and their impact on PL spectra (right). The measured PL spectra with higher intensity originate from the structures regions. For the lower structure also a shift of the PL peak emission wavelength is observed, an indication for a photonic bandgap effect. (Reprinted from Tripathy, S. et al., *Phys. Status Solidi*, in print. With permission from Wiley-VCH, © 2010 by Wiley-VCH.)

well below 100% in reflectivity. However, in conventional LEDs a Ni/Au top contact layer leads to significant absorption losses and current spreading across the metal contact and the p-GaN layer is poor. For pure Ag, typically used as main p-contact metal and mirror layer in thin-film technology, the reflectivity yields ~96% at 460 nm, significantly higher than for Al with only ~87% at this wavelength. But good contacts require the addition of Ni at the p-GaN interface and this limits reflectivity, in the best case to 93% [Chang 2009]. In addition to the advantage of a p-side mirror for contacting and light extraction, an N-face up structure offers a simple route for surface texturing to improve light extraction significantly. By etching the N-face of the LED structure with an alkaline etchant as, e.g., KOH, GaN forms hexagonal pyramids. Size and to a smaller part also shape depends on the etching conditions as etchant concentration and temperature. An ideally roughened LED surface does then enable light extraction of all rays transmitted toward this surface with only a small fraction back reflected, and thus avoiding a high number of multiple reflections with its unavoidable losses.

At present thin-film LEDs still suffer from absorption at the n-type top contact metallization. OSRAM already presented a concept for their thin-film LEDs, which enables higher extraction efficiency by avoiding these contacts. Here, via etching through the carrier substrate with isolated vias enables n-type contacting from the backside. Absorption losses are further reduced by this concept. For such concepts Si carrier substrates onto which

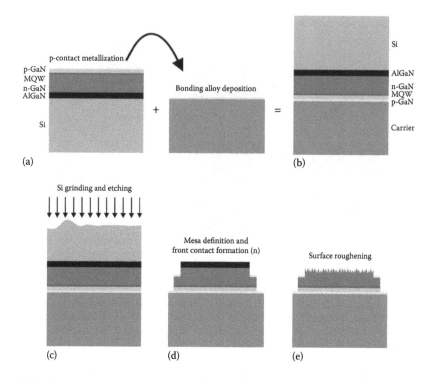

FIGURE 12.17

Process steps for fabricating a thin-film LED made from a GaN on Si structure. In principle, the steps are identical to the growth on sapphire with (a) p-contact and bonding metallization, (b) bonding, (c) substrate removal which is a laser lift-off process for GaN on sapphire and a mechanical/chemical process for GaN on Si, (d) mesa and n-contact etching and metallization, and finally a surface roughening step, (e) usually performed with an alkaline etchant as KOH.

the LED structure is transferred are to be preferred since VIA etching technology is a well-established process on Si.

Thin-film LEDs based on silicon were up to now only rarely reported. Zhang et al. reported an LED structure transferred onto a copper substrate with indium as bonding metal and Al/Au reflecting layer [Zhang 2005]. The devices show a significant enhancement in output power and *I–V* characteristics, but were not fully optimized with the top layer still not being roughened to enhance light extraction.

Latticepower which grows LED structures on patterned substrates is using a thin-film approach where they transfer the LEDs onto a new carrier substrate and perform backside roughening by an etching process [Jiang 2009]. Latticepower and Sanken Electric are at present the only LED companies known to sell LEDs grown on Si. But also big LED companies as OSRAM [AZZURRO 2009] and Lumileds are now exploring LED growth on silicon.

12.6 Growth on Large Diameter Substrates—Limitations of Today's Growth Systems

A major reason for the growth of GaN on silicon is the availability of large diameter substrates and with it the possible reduction of manufacturing cost. For sapphire, it has been found that indeed an increasing substrate diameter bears higher technological barriers than expected. The most difficult barrier which is present is homogeneity, especially of ternary and quaternary compound layers containing indium. The high vapor pressure of the binary InN compound leads to a strong dependence of the In-content upon growth temperature. Indeed the temperature is typically the dominant parameter controlling the In-content in InGaN layers. Even a substrate surface temperature change of a few degrees leads to a significant change in composition and with it the emission wavelength of LEDs.

It has been demonstrated that the growth on 150 mm silicon substrates is possible [Dadgar 2005], and LED structures [Li 2006] can yield a good homogeneity [Dadgar 2006a] in the latter case mostly influenced by the heater set up of the MOVPE system and by thickness interference shifting the observed PL peak emission wavelength. Curvature measurements revealed that the growth of the MQW can be performed at low curvature (Figure 12.18) and after cooling tensile stress is still low.

But there are limitations when further increasing the substrate diameter. Figure 12.19 shows the distance between the substrate and the substrate carrier for different substrate diameters vs. the radius of curvature. At realistic values, e.g., even for a low value of the radius as 100 m, the temperature gradient at the surface increases from about 1 K on a 2″ substrate to 5 K for 150 mm and to 8 K for 200 mm substrates (Figure 12.19). In reality, radii of curvature during the growth process are well below 100 m and temperature gradients can easily reach more than 20 K [Dadgar 2004b, Krost 2005].

The reason for such a behavior is that MOVPE systems are usually cold wall systems where a substrate is heated from underneath and on the opposite of it a liquid cooled metal or gas cooled quartz ceiling or gas inlet has a significantly lower temperature. This leads to a strong temperature gradient in the gas phase. This gradient has two effects on growth: a temperature gradient of the substrate resulting in a concave bowing. In addition to this, heteroepitaxy will lead to growth (lattice mismatch, island coalescence, etc.) and thermally induced stresses increasing concave bow [Dadgar 2004a, Dadgar 2007a]. Lowering the growth temperature to InGaN growth conditions will add tensile thermally induced stress increasing concave bow. Unless the bow during GaN epitaxy is exactly counterbalancing this thermally induced curvature, the wafer will be bowed. Currently, many LED manufacturers that are working on 100 mm substrate use, as already for 2″ wafers, thicker than SEMI thickness substrates and a curved substrate pocket which fits to

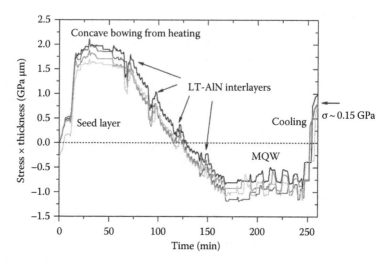

FIGURE 12.18

In situ curvature measured on four different points on the substrate circle 50 mm from the center of a 150 mm silicon substrate. A difference in bowing for the different measurement points is already observed at room temperature. Heating yields a concave curvature and when growing the ~4.5 μm thick structure, compressive stress induced by LT-AlN interlayers yields a convex curved wafer which has a low curvature during MQW growth and after cooling. (Adapted from Dadgar, A., *6th International Conference on Nitride Semiconductors (ICNS-6)*, Bremen, Germany, 2005.)

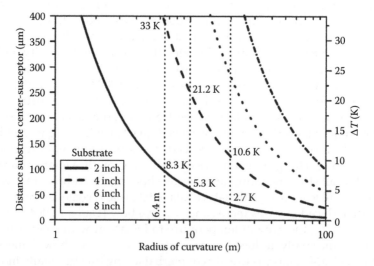

FIGURE 12.19

Distance of substrate to susceptor for different substrate diameters and bowing. The temperature difference was determined by a simultaneous measurement of silicon surface temperature and bow. (Reprinted from Dadgar, A. et al., *J. Cryst. Growth*, 272, 72, 2004. With permission from Elsevier, © 2004 by Elsevier.)

the curvature when growing the MQW portion of an LED structure. This method, however, is only suited for mass production and low high radii of curvature it can give problems in holding the substrate, especially for larger diameters this problem gets difficult to solve for all stages of epitaxial growth.

Substrate temperature gradients increase with increasing substrate thickness and decreasing ceiling temperature. In addition, highly anisotropic thermal stresses arise from inhomogeneous heating when the curved wafer has a highly non-uniform distance to the wafer carrier. For example, in the case of 150 mm diameter sapphire substrates with a 675 μm SEMI thickness this leads to cracking for *c*-axis oriented substrates upon heating [Dadgar 2006b] (see Figure 4.51). For silicon such a behavior has not been observed yet.

A first step toward a solution of this is already present in its early stages by temperature controlled reactor ceilings with the temperatures applied during growth increasing in the last years. But they are still too low to achieve a homogenous temperature profile for curved large diameter wafers. Here, a future development toward a hot or at least a warm wall concept with such temperatures being only a few hundred degrees below the substrate carrier temperature is best suited to increase homogeneity. This is also a topic for the growth of GaN electronics on large diameter substrates.

Another issue is throughput and conditioning of MOVPE reactors, which is also discussed in Chapter 4. Still many systems require manual loading and unloading between growth runs as well as conditioning of the growth chamber to remove deposits and maintain reproducibility. These time-consuming steps need to be minimized to lower cost, e.g., by a highly automated handling system or simplified handling and conditioning of large reactors.

12.7 Future Developments

A main aspect of GaN LEDs on silicon remains the improvement of material quality and the growth of thick, highly doped uninterrupted (by strain engineering layers) contacting and active layers, which is connected since high-quality layers allow thicker highly doped layers. For example, the LED process mentioned in Section 12.1 and which is tested by OSRAM for LED growth on silicon enables the growth of a 2 μm thick uninterrupted (interlayer free) and crack-free GaN:Si layer with a silicon doping concentration of 5×10^{18} cm^{-3}. To achieve this, the dislocation density must be well below 10^9 cm^{-2}. In addition to those parameters others, not specific to the growth on silicon, can be further improved. These are general LED topics as QW efficiency, electric contacting, and photon extraction, and also a further development of MOVPE systems for the requirements of LED mass production, for large substrate diameters and high throughput.

In summary, the main topics for achieving a high efficiency at low cost on the chip level, not taking into account luminescence conversion, are

1. Optimized internal quantum efficiency
   a. Low dislocation density
   b. Optimized quantum wells: QW and barrier thickness, composition, and reduced polarization fields within the MQW
   c. High doping levels for contacts and current spreading
2. Efficient light extraction
   a. Surface roughening
   b. Photonic bandgap structures
   c. Highly reflective mirror contacts
   d. Top contact free LED design
3. New MOVPE reactor concepts
   a. High throughput/automation
   b. "Heated-wall" concepts
   c. Combination with HVPE process

Acknowledgment

Part of this work has been and is still funded by BMBF and DFG within the framework of SPP "Group III-Nitrides and Their Heterostructures: Growth, Characterization and Applications," SFB787 and FOR957.

References

[Armitage 2002] R. Armitage, Q. Yang, H. Feick, J. Gebauer, E. R. Weber, S. Shinkai, and K. Sasaki, *Appl. Phys. Lett.* **81**, 1450 (2002).
[AZZURRO 2009] AZZURRO Semiconductors, Press release November 10, 2009.
[Badylevich 2008] M. Badylevich, S. Shamuilia, V. V. Afanas'ev, A. Stesmans, Y. G. Fedorenko, and C. Zhao, *J. Appl. Phys.* **104**, 093713 (2008).
[Chang 2009] L.-B. Chang, C.-C. Shiue, and M.-J. Jeng, *Appl. Surf. Sci.* **255**, 6155 (2009).
[Chelikowsky 1976] J. R. Chelikowsky and M. L. Cohen, *Phys. Rev. B* **14**, 556 (1976).
[Dadgar 2001a] A. Dadgar, M. Poschenrieder, J. Bläsing, K. Fehse, T. Riemann, A. Diez, J. Christen, and A. Krost, *MRS Fall Meeting*, Boston, MA, I4.7, 2001.
[Dadgar 2001b] A. Dadgar, A. Alam, T. Riemann, J. Bläsing, A. Diez, M. Poschenrieder, M. Straßburg, M. Heuken, J. Christen, and A. Krost, *Phys. Stat. Sol. (a)* **188**, 155 (2001).

[Dadgar 2002a] A. Dadgar, M. Poschenrieder, J. Bläsing, K. Fehse, A. Diez, and A. Krost, *Appl. Phys. Lett.* **80**, 3670 (2002).

[Dadgar 2002b] A. Dadgar, M. Poschenrieder, O. Contreras, J. Christen, K. Fehse, J. Bläsing, A. Diez, F. Schulze, T. Riemann, F. A. Ponce, and A. Krost, *Phys. Stat. Sol. (a)* **192**, 308 (2002).

[Dadgar 2003] A. Dadgar, A. Strittmatter, J. Bläsing, M. Poschenrieder, O. Contreras, P. Veit, T. Riemann, F. Bertram, A. Reiher, A. Krtschil, A. Diez, T. Hempel, T. Finger, A. Kasic, M. Schubert, D. Bimberg, F. A. Ponce, J. Christen, and A. Krost, *Phys. Stat. Sol. (c)* **0**, 1583 (2003).

[Dadgar 2004a] A. Dadgar, R. Clos, G. Strassburger, F. Schulze, P. Veit, T. Hempel, J. Bläsing, A. Krtschil, I. Daumiller, M. Kunze, A. Kaluza, A. Modlich, M. Kamp, A. Diez, J. Christen, and A. Krost, *Advances in Solid State Physics*, Vol. 44, B. Kramer, ed., Springer, Heidelberg, Germany, 2004.

[Dadgar 2004b] A. Dadgar, F. Schulze, T. Zettler, K. Haberland, R. Clos, G. Straburger, J. Bläsing, A. Diez, and A. Krost, *J. Cryst. Growth* **272**, 72 (2004).

[Dadgar 2005] A. Dadgar, *6th International Conference on Nitride Semiconductors (ICNS-6)*, Bremen, Germany, 2005.

[Dadgar 2006a] A. Dadgar, C. Hums, A. Diez, J. Bläsing, and A. Krost, *J. Cryst. Growth* **297**, 279 (2006).

[Dadgar 2006b] A. Dadgar, *Asia-Pacific Optical Communications Conference*, Gwanju, Korea, 2009.

[Dadgar 2007a] A. Dadgar, P. Veit, F. Schulze, J. Bläsing, A. Krtschil, H. Witte, A. Diez, T. Hempel, J. Christen, R. Clos, and A. Krost, *Thin Solid Films* **515**, 4356 (2007).

[Dadgar 2007b] A. Dadgar, *DPG-Spring Meeting 2007*, Regensburg, Germany, 2007.

[Egawa 2005] T. Egawa, B. Zhang, and H. Ishikawa, *IEEE Electron. Dev. Lett.* **26**, 169 (2005).

[Feltin 2001] E. Feltin, B. Beaumont, M. Laügt, P. de Mierry, P. Vennéguès, H. Lahrèche, M. Leroux, and P. Gibart, *Appl. Phys. Lett.* **79**, 3230 (2001).

[Feltin 2006] E. Feltin, J.-F. Carlin, J. Dorsaz, G. Christmann, R. Butté, M. Laügt, M. Ilegems, and N. Grandjean, *Appl. Phys. Lett.* **88**, 051108 (2006).

[Guha 1998a] S. Guha and N. A. Bojarczuk, *Appl. Phys. Lett.* **72**, 415 (1998).

[Guha 1998b] S. Guha and N. A. Bojarczuk, *Appl. Phys. Lett.* **73**, 1487 (1998).

[Henley 2003] S. J. Henley and D. Cherns, *J. Appl. Phys.* **93**, 3934 (2003).

[Honda 2002] Y. Honda, Y. Kuroiwa, M. Yamaguchi, and N. Sawaki, *Appl. Phys. Lett.* **80**, 222 (2002).

[Honda 2007] Y. Honda, S. Kato, M. Yamaguchi, and N. Sawaki, *Phys. Stat. Sol. (c)* **4**, 2740 (2007).

[Jiang 2009] F. Jiang, L. Wang, X. Wang, C. Mo, X. You, C. Zheng, W. Liu, Y. Zhou, C. Xiong, Y. Tang, W. Fang, and B. Lu, *H7, 8th International Conference on Nitride Semiconductors (ICNS-8)*, Jeju, Korea, 2009.

[Ishikawa 2004a] H. Ishikawa, K. Asano, B. Zhang, T. Egawa, and T. Jimbo, *Phys. Stat. Sol. (a)* **201**, 2653 (2004).

[Ishikawa 2004b] H. Ishikawa, B. Zhang, K. Asano, T. Egawa, and T. Jimbo, *J. Cryst. Growth* **272**, 322 (2004).

[Karlsson 1982] B. Karlson, R. P. Shimshock, B. O. Serapin, and J. C. Haygarth, *Phys. Scr.* **25**, 775 (1982).

[Kikuchi 2004] A. Kikuchi, M. Kawai, M. Tada, and K. Kishino, *Jpn. J. Appl. Phys.* **43**, L 1524 (2004).

[Krost 2005] A. Krost, F. Schulze, A. Dadgar, G. Strassburger, K. Haberland, and T. Zettler, *Phys. Stat. Sol. (b)* **242**, 2570 (2005).

[Kuzmik 2005] J. Kuzmík, S. Bychikhin, M. Neuburger, A. Dadgar, A. Krost, E. Kohn, and D. Pogany, *IEEE Trans. Electron. Dev.* **52**, 1698 (2005).

[Latticepower 2009] GaN-on-Si LED data http://www.latticepower.com/Upload/PicFiles/LP-SI08B07B-AL.pdf.

[Li 2006] J. Li, J. Y. Lin, and H. X. Jiang, *Appl. Phys. Lett.* **88**, 171909 (2006).

[Moram 2006] M. A. Moram, M. J. Kappers, T. B. Joyce, P. R. Chalker, Z. H. Barber, and C. J. Humphreys, *J. Cryst. Growth* **308**, 302 (2007).

[Perry 1988] A. J. Perry, M. Georgson, and W. D. Sproul, *Thin Solid Films* **157**, 255 (1988).

[Philipp 1960] H. R. Philipp and E. A. Taft, *Phys. Rev.* **120**, 37 (1960).

[Sawaki 2009] N. Sawaki, T. Hikosaka, N. Koide, S. Tanaka, Y. Honda, and M. Yamaguchi, *J. Cryst. Growth* **311**, 2867 (2009).

[Semond 2001] F. Semond, N. Antoine-Vincent, N. Schnell, G. Malpuech, M. Leroux, J. Massies, P. Disseix, J. Leymarie, and A. Vasson, *Phys. Stat. Sol. (a)* **183**, 163 (2001).

[Sze 1981] S. M. Sze, *Physics of Semiconductor Devices*, John Wiley & Sons, New York, (1981).

[Taniyasu 2008] Y. Taniyasu and M. Kasu, *Diamond Relat. Mater.* **17**, 1273 (2008).

[Tran 1999] C. A. Tran, A. Osinski, R. F. Karlicek, Jr., and I. Berishev, *Appl. Phys. Lett.* **75**, 1494 (1999).

[Tripathy 2007] S. Tripathy, V. K. X. Lin, S. L. Teo, A. Dadgar, A. Diez, J. Bläsing, and A. Krost, *Appl. Phys. Lett.* **91**, 231109 (2007).

[Tripathy 2008] S. Tripathy, T. E. Sale, A. Dadgar, V. K. X. Lin, K. Y. Zang, S. L. Teo, S. J. Chua, J. Bläsing, and A. Krost, *J. Appl. Phys.* **104**, 053106 (2008).

[Tripathy 2009] S. Tripathy, A. Dadgar, K. Y. Zang, V. K. X. Lin, Y. C. Liu, S. L. Teo, A. M. Yong, C. B. Soh, S. J. Chua, J. Bläsing, J. Christen, and A. Krost, *Phys. Stat. Sol. (c)* **6**, S822 (2009).

[Tripathy 2010] S. Tripathy, S. L. Teo, V. K. X. Lin, M. F. Chen, A. Dadgar, J. Christen, and A. Krost, *Phys. Stat. Sol., C* **7**, 88 (2010).

[Wang 2006] W. W. Wang, T. Nabatame, and Y. Shimogaki, *Thin Solid Films* **498**, 75–79 (2006).

[Xu 2005] X. Xu, R. Armitage, S. Shinkai, K. Sasaki, C. Kisielowski, and E. R. Weber, *Appl. Phys. Lett.* **86**, 182104 (2005).

[Yang 2000] J. W. Yang, A. Lunev, G. Simin, A. Chitnis, M. Shatalov, M. A. Kahn, J. E. Van Nostrand, and R. Gaska, *Appl. Phys. Lett.* **76**, 273 (2000).

[Zhang 2003] B. Zhang, T. Egawa, H. Ishikawa, Y. Liu, and T. Jimbo, *Jpn. J. Appl. Phys.* **42**, L 226 (2003).

[Zhang 2005] B. Zhang, T. Egawa, H. Ishikawa, Y. Liu, and T. Jimbo, *Appl. Phys. Lett.* **86**, 071113 (2005).

[Zhang 2007] B. Zhang, H. Liang, Y. Wang, Z. Feng, K. W. Ng, K. M. Lau, *J. Cryst. Growth* **298**, 725 (2007).

[Zhu 2010] D. Zhu, C. McAleese, M. Häberlen, C. Salcianu, T. Thrush, M. Kappers, A. Phillips, P. Lane, M. Kane, D. Wallis, T. Martin, M. Astles, and C. Humphreys, *Phys. Stat. Sol. C* **7**, 2168 (2010).

13

Conventional III–V Materials and Devices on Silicon

Edward Yi Chang

CONTENTS

13.1 Introduction

Materials with high channel mobility are essential as silicon CMOS technology approaches the 30 nm node and beyond. III–V compound semiconductors, with their high mobility and drift velocity under the influence of low electric field, are considered the most promising materials to realize next-generation nanoelectronic applications. Table 13.1 compares relevant channel material properties of Si, Ge, and main III–V semiconductors [Takagi 2008].

With the rapid growth of wireless communication industries, the frequency used in the wireless communication system is moving from the microwave toward millimeter-wave and submillimeter-wave range. Consequently, the requirements for the key components become more demanding. III–V based compound semiconductor devices, such as GaAs pseudomorphic high-electron-mobility transistors (PHEMTs), GaAs metamorphic HEMTs

TABLE 13.1

Bulk Carriers Mobility, Effective Mass, Bandgap, and the Permittivity of Si, Ge, and Main III–V Semiconductors

	Si	Ge	GaAs	InP	InAs	InSb
Electron mob. (cm²/Vs)	1,600	3,900	9,200	5,400	40,000	77,000
Electron effective mass (/m_0)	m_t: 0.19 m_l: 0.916	m_t: 0.082 m_l: 1.467	0.067	0.082	0.023	0.014
Hole mob. (cm²/Vs)	430	1,900	400	200	500	850
Hole effective mass (/m_0)	m_{HH}: 0.49 m_{LH}: 0.16	m_{HH}: 0.28 m_{LH}: 0.044	m_{HH}: 045 m_{LH}: 0.082	m_{HH}: 045 m_{LH}: 0.12	m_{HH}: 057 m_{LH}: 0.35	m_{HH}: 0.44 m_{LH}: 0.016
Bandgap (eV)	1.12	0.66	1.42	1.34	0.36	0.17
Permittivity	11.8	16	12	12.6	14.8	17

Source: Takagi, S. et al., *IEEE Trans. Electron. Dev.*, 55, 21, 2008. With permission.

(MHEMTs), conventional lattice-matched or pseudomorphic InAlAs/ InGaAs InP HEMTs with high indium mole fraction channel, which have demonstrated high performance at high frequencies in contrast to Si devices are the main devices used for millimeter-wave applications. Many efforts have been devoted to improve the high-frequency performance of the III–V HEMT devices by means of refined heterojunction structure, low resistance connection, T-shaped sub-nanometer gate, modified vertical scale of device, precise process control, etc.

On the other hand, size reduction for the CMOS has followed the Moore's law for over 30 years. The trend of transistor scaling is shown in Figure 13.1 [Chau 2005].

The gate length for the Si MOSFET is expected to reach approximately 10 nm by 2011, which is believed to be the ultimate limit for logic CMOS scaling. Candidates that are often mentioned to replace the Si CMOS include carbon nanotube (CNT) transistors, semiconductor nanowires, and spintronics. While the majority of the above-mentioned technologies are still in the primitive stage, recent developments of III–V FETs, especially the heterostructure HEMT, have shown considerable potential to be the next-generation high-speed logic technology due to their maturity in device fabrication and superb electronic performance.

Although the III–V HEMTs offer the highest device speed along with low power capability and low noise performance, these devices have only been built on high-cost GaAs and InP substrates. Recent researches have spotlighted on the seamless and robust integration of high-performance III–V devices on Si substrate, which will allow high-speed and low-voltage

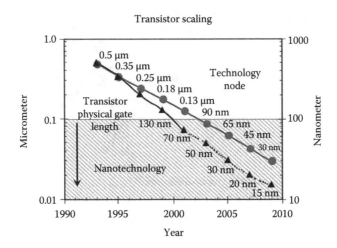

FIGURE 13.1
The trend of transistor scaling. (From Chau, R. et al., *IEEE Trans. Nanotechnol.*, 4, 153, 2005. With permission.)

III–V based circuits to couple with the main stream Si CMOS platform. In 2007, Intel [Hudait 2007] presented for the first time, the heterogeneous integration of $In_{0.7}Ga_{0.3}As$ HEMT device structure on Si substrate through the novel composite metamorphic buffer architecture with thin composite buffer thickness. These enhancement-mode $In_{0.7}Ga_{0.3}As$ HEMTs on Si substrate exhibit superior high-speed performance even at low supply voltage and their the results are shown in Figures 13.2 through 13.4. From the data shown in Figures 13.2 through 13.4, it can be seen that very high transconductance $In_{0.7}Ga_{0.3}As$ HEMT with cutoff frequency up to 300 GHz can be grown and fabricated on Si substrate. And much lower power consumption was observed for these HEMT devices grown on Si substrate compared to the Si MOSFET devices. They demonstrated much higher cutoff frequency at the same power level.

However, for conventional III–V HEMTs, the gate leakage current becomes too large as the gate length is reduced and the indium content in the channel layer is increased. Therefore, the insertion of high-quality oxide layer as the gate dielectric becomes a promising method to resolve these problems. In the past few years, more and more researchers have devoted to the III–V MOS structure study to effectively resolve the gate leakage issue.

In the following sections, the basic mechanism of the III–V device operation, the material growth techniques, current research results of III–V/Si integration for millimeter-wave and low-power logic applications, and emerging applications for high-performance III–V devices on Si will be introduced.

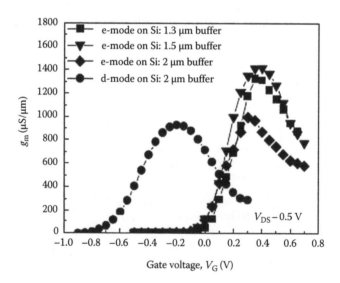

FIGURE 13.2

Transconductance characteristics of $L_g = 80$ nm e-mode and d-mode $In_{0.7}Ga_{0.3}As$ HEMTs on Si substrate with different composite buffer thickness at $V_{DS} = 0.5$ V. The e-mode devices can achieve g_m of 1400 mS/mm at low bias. (From Hudait, M.K. et al., *IEDM Tech. Dig.*, 625, 2007. With permission.)

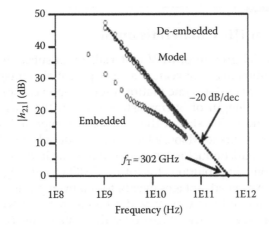

FIGURE 13.3

Current gain (h_{21}) versus frequency for the $L_g = 80$ nm e-mode In$_{0.7}$Ga$_{0.3}$As HEMTs on Si substrate with 1.3 μm composite buffer, showing the embedded and de-embedded data at $V_{DS} = 0.5$ V. After de-embed, the pad parasitic effect, the intrinsic current-gain cutoff frequency (f_T) can attain 302 GHz. (From Hudait, M.K. et al., *IEDM Tech. Dig.*, 625, 2007. With permission.)

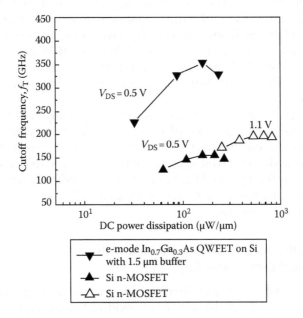

FIGURE 13.4

Cutoff frequency as a function of DC power dissipation for the $L_g = 80$ nm e-mode In$_{0.7}$Ga$_{0.3}$As HEMTs on Si substrate with 1.3 μm composite buffer at $V_{DS} = 0.5$ V, versus standard Si n-MOSFET with $L_g = 60$ nm at $V_{DS} = 0.5$ and 1.1 V. The In$_{0.7}$Ga$_{0.3}$As HEMTs can exhibit much lower power dissipation for the same speed. (From Hudait, M.K. et al., *IEDM Tech. Dig.*, 625, 2007. With permission.)

13.2 Growth of III–V Materials on Si

The heteroepitaxial growth of III–V materials on Si substrate has attracted substantial attention in recent years due to its potential of integrating Si- and III–V based devices on the same platform for low-power logic applications. The physical gate length of Si transistors used in the current generation node is about 50 nm. The size of the transistor is expected to reach 10 nm in 2011. In order to extend Moore's law into next decade, Si CMOS incorporated with III–V semiconductor compound materials in the device structure is one of the promising solutions for future CMOS technology [Chau 2005]. III–V materials exhibit advantages over Si for many applications due to their higher electron mobility as well as wider bandgap and direct bandgap. Si, however, reveals several advantages over III–V materials: (1) high thermal conductivity and mechanical hardness of Si substrate, (2) Si substrate would significantly reduce manufacturing costs compared to that of InP or GaAs substrates, and (3) Si substrate is capable of wafer-size expansion. However, III–V materials have excellent low-field and high-field electron transport properties. Therefore, ultrahigh-speed switching at very low supply voltages can be achieved. The integration of III–V devices on Si substrate will help to realize the ultimate vision of low-voltage high-speed III–V based logic circuit blocks coupled with the functional density advantages provided by the Si CMOS platform. Using proper buffer layers to accommodate the stress generated due to the large lattice mismatch between III–V and Si is a practical way to integrate III–V material on Si substrate. To date, several types of buffer layers include SiGe buffer, InAlAs buffer, etc., have been proposed and developed.

Two major problems need to be overcome for growing GaAs epitaxy: the relatively large lattice mismatch of 4% and the significant difference in the thermal expansion coefficients (TECs) (63%) between these two materials. To achieve high-quality epitaxial structure, it is necessary to reduce the dislocations density in the epitaxial GaAs layer. Several methods have been proposed for reducing the dislocation density in the III–V epitaxial layers. The following are the examples.

13.2.1 Selective Area Growth Combined with Thermal Cycle Annealing

Figure 13.5 describes the steps involved for selective area growth (SAG) of III–V material on Si. First, the substrate is oxidized using a typical oxidation furnace and patterned by photolithography. The region of SAG then undergoes ultra-high vacuum chemical vapor deposition (UHV-CVD) or metal-organic chemical vapor deposition (MOCVD) epitaxial growth. The thickness and width for the SAG can be controlled by different processing conditions.

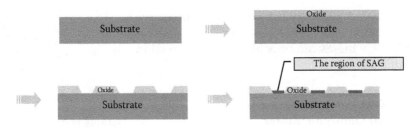

FIGURE 13.5
The evolution of selective area growth.

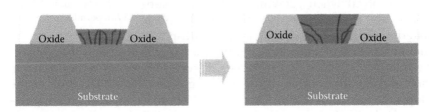

FIGURE 13.6
The mechanism of SAG structure blocking the penetration of threading dislocations.

During SAG, lateral threading dislocations (TDs) move; some TDs bend to the mesa sidewalls (dislocation sinks) without threading through the surface. As a result, the surface quality is improved (Figure 13.6).

Next, cyclic thermal annealing is introduced into the system. Due to the effect of annealing temperature on dislocation velocity, dislocations move quickly to the mesa sidewalls and their densities are then reduced dramatically, as shown in Figure 13.7a and b.

SAG combined with cyclic thermal annealing is more efficient in reducing threading dislocation densities (TDDs) than SAG combined with simple annealing. Cyclic thermal annealing reduces the thermal stress while

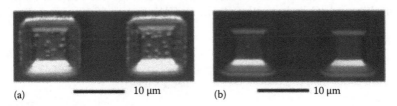

FIGURE 13.7
(a) Selective epitaxial growth using two-step UHV/CVD process followed by 10 min annealing at 900°C, the average EPD is $4.3 \pm 0.2 \times 10^7$ cm^{-2}. (b) Selective epitaxial growth using two-step UHV/CVD process followed by 10 annealing cycles between $T_H = 900$°C, and $T_L = 100$°C, and the average EPD is $2.3 \pm 0.2 \times 10^6$ cm^{-2}. (From Luan, H.-C. et al., *Appl. Phys. Lett.*, 75, 125187, 1999. With permission.)

dislocations moving to mesa sidewalls, which serve as dislocation sinks. Therefore, threading dislocation-free Ge mesas can be achieved as observed in Figure 13.7b.

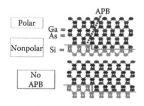

13.2.2 GaAs on Si Substrate Using Metamorphic Buffer

This method is developed by Intel with IQE Corporation. There are some problems that need to be resolved when growing GaAs on Si substrates. The first problem is that GaAs is a polar material while Si is nonpolar. Hence, anti-phase domains (APD) or anti-phase boundaries (APB) are likely to appear at the interface of these two materials. Figure 13.8 shows typical phenomenon of APD; for example, if the ordering is usually ABABABAB, an APB adopts the form of ABABBABA. Figure 13.9 is an example of growing III–V GaAs (polar material) on Ge (nonpolar material), from the transmission electron microscope (TEM) analysis, APD is clearly noticeable in the crystal. To date, methods for solving the APB problems include inserting thin AlAs buffer layers at the GaAs/Ge heterointerface, using low-temperature migration enhance epitaxy (MEE), or employing off-cut substrate to suppress APB information.

Intel Corporation grows III–V material on Si by using low-temperature InAlAs metamorphic buffer. Metamorphic buffer serves to (1) accommodate the large lattice mismatch between Si and GaAs, (2) form and trap misfit dislocation, and (3) prevent misfit dislocations to propagate into the active device layers. The $In_xAl_{1-x}As$ ($x = 0–0.52$) metamorphic composite buffer was then grown on low-temperature GaAs nucleation layer. The metamorphic buffer in the structure shows some advantages. It (1) minimizes APDs, (2) bridges lattice constant, (3) relaxes strain energy and glide dislocations, (4) eliminates parallel conduction, (5) provides large conduction band discontinuity, and (6) reduces the strain to QW. Figure 13.10 is the structure

FIGURE 13.9
TEM analysis of anti-phase domains shown in GaAs on Ge. (From Hudait, M.K. et al., *IEDM Tech. Dig.*, 625–628, 2007. With permission.)

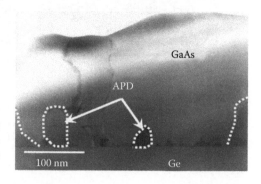

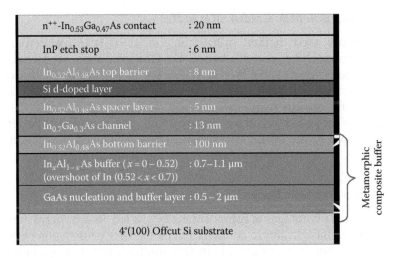

FIGURE 13.10
The structure of the integrated III–V material and Si by using metamorphic $In_xGa_{1-x}As$ on Si substrate. (From Hudait, M.K. et al., *IEDM Tech. Dig.*, 625–628, 2007. With permission.)

of an integrated III–V QWFET structure on Si using metamorphic buffer as demonstrated by Intel Corporation.

13.2.3 SiGe Buffer Layer between Si Substrate and GaAs

The SiGe buffer is widely used in III–V integration on Si. The first reason is GaAs and Ge crystals contain very low lattice mismatch, the second reason is the almost identical TECs between GaAs and Ge (Figure 13.11).

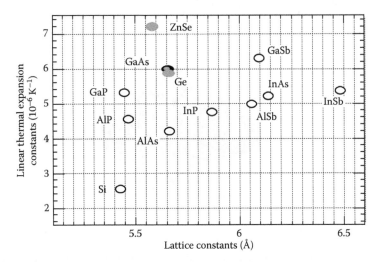

FIGURE 13.11
Linear thermal expansion versus lattice constant of various III–V semiconductor.

13.2.3.1 Compositionally Graded SiGe Buffer

This commonly used method utilizes $Si_{1-x}Ge_x$ ($x = 0-1$) to counterbalance the lattice mismatch between GaAs and Ge. However, there are two disadvantages. One of which is the $Si_{1-x}Ge_x$ buffer layer's thickness, the thickness of the SiGe buffer is thicker than 10 μm using this method. It is too costly to manufacture devices using this method. The other disadvantage is the occurrence of cross-hatch pattern (Figure 13.12). The cross-hatch pattern is possibly related to the lateral stress distribution developed over the entire layer of SiGe.

13.2.3.2 Two-Step SiGe Buffer

To avoid the disadvantages of compositionally graded SiGe buffer, a method using two-step SiGe buffer to counterbalance the lattice mismatch between GaAs and Ge has been proposed. In this method, two Si_xGe_{1-x} layers of $Si_{0.05}Ge_{0.95}$ and $Si_{0.1}Ge_{0.9}$ are used. Due to the lattice mismatch strains at the upper interfaces of $Si_{0.05}Ge_{0.95}/Si_{0.1}Ge_{0.9}$ and $Ge/Si_{0.05}Ge_{0.95}$, the upward-propagated dislocations can be bent sideward and terminated effectively. Many dislocations are generated while the interaction between them forms closed non-propagating loops and networks near the interface. This method controls the composition of the $Si_{1-x}Ge_x$ layers to produce a compressive stress field at each interface. The compressive stress at the interface could bend and terminate the upward-propagated dislocations very effectively. Additionally, the thermal annealing process, which was performed after growing each individual layer, could further reduce dislocation density in the epitaxial layers. The mechanism of TD reduction employed in this work is shown schematically in Figure 13.13.

The key issue of the method is controlling the composition of Si_xGe_{1-x} buffer layer to produce a compressive stress field at the top layer of each interface.

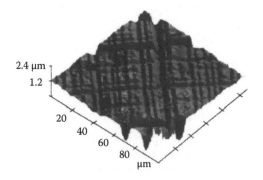

FIGURE 13.12
Cross-hatch pattern on compositionally graded $Si_{1-x}Ge_x$ buffer. (From Samavedam, S.B. and Fitzgerald, E.A., *J. Appl. Phys.*, 81, 3108, 1997. With permission.)

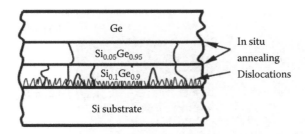

FIGURE 13.13
Mechanism of reducing threading dislocations. (From Luo, G. et al., *Jpn. J. Appl. Phys.*, 42, L517, 2003. With permission.)

The stress field at the interface must be high enough to block dislocation propagation. For example, if a $Si_{0.8}Ge_{0.2}$ film is grown on Si substrate (Figure 13.14). Because of the lattice mismatch between interfaces, a large number of dislocations are observed in $Si_{0.8}Ge_{0.2}$ layer.

If a $Si_{0.76}Ge_{0.24}$ layer is grown on the $Si_{0.8}Ge_{0.2}$ layer, some dislocations in the first $Si_{0.8}Ge_{0.2}$ are blocked at the interface of $Si_{0.76}Ge_{0.24}/Si_{0.8}Ge_{0.2}$, but others still penetrate through the interface to the top layer. It infers that the stress in the interface is insufficient to block all dislocations (Figure 13.15).

From the above observation, the composition of the top Si_xGe_{1-x} layer must be changed to increase the stress field in order to block the TDs; if the stress between the interface is increased by using $Si_{0.8}Ge_{0.2}/Si_{0.7}Ge_{0.3}$ layers, then the TDs that can penetrate through the interface are reduced. It suggests that the stress field at the interface is sufficient to block dislocations (Figure 13.16).

So, using the interface-blocking method to grow high-quality Ge epitaxial layers on Si is feasible. Variation on the Ge composition at the two strained

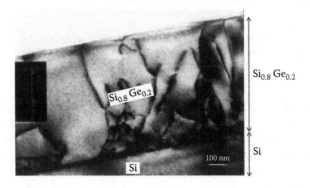

FIGURE 13.14
TEM picture of $Si_{0.8}Ge_{0.2}$ film grown on Si substrate. (From Yang, T.H. et al., *J. Vac. Sci. Technol. B*, L17, 2004. With permission.)

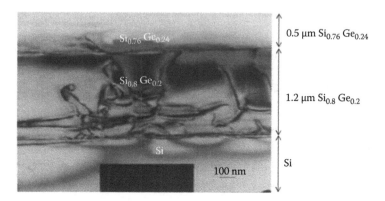

FIGURE 13.15
TEM micrograph of $Si_{0.76}Ge_{0.24}$ layer grown on top of $Si_{0.8}Ge_{0.2}$, it shows the stress in the interface is not enough to block all dislocations. (From Yang, T.H. et al., *J. Vac. Sci. Technol. B*, L17, 2004. With permission.)

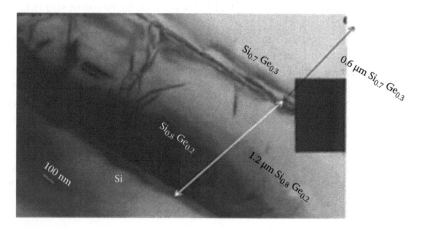

FIGURE 13.16
TEM micrograph of $Si_{0.8}Ge_{0.2}/Si_{0.7}Ge_{0.3}$ layers, the stress field at interface is large enough to block dislocation, so threading dislocation density is reduced. (From Yang, T.H. et al., *J. Vac. Sci. Technol. B*, L17, 2004. With permission.)

interfaces is set at 0.05. If the Ge composition variation is larger than 0.05, new dislocations would generate from the interfaces due to the relatively large lattice mismatch. If the Ge composition variation is less than 0.05, the mismatch strain formed at the interfaces is too small to terminate the dislocations effectively. From Figure 13.17, the interface of each layer terminates dislocation effectively.

From the double-crystal x-ray diffraction data at the [004] orientation of the Ge_xSi_{1-x} metamorphic buffer layer on Si substrate (Figure 13.18), the layer structure of the Si_xGe_{1-x} buffer can be seen clearly.

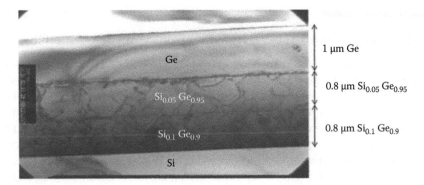

FIGURE 13.17
TEM micrograph of low dislocation density Ge film grown on Si using $Si_{0.05}Ge_{0.95}/Si_{0.1}Ge_{0.9}$ buffer. The Ge composition variation at the two strained interfaces is set at 0.05. (1) If the Ge composition variation is larger than 0.05: new dislocations will generate from the interfaces due to the relatively large lattice mismatch. (2) If the Ge composition variation is less than 0.05, the mismatch strain formed at the interfaces is too small to terminate the dislocations effectively. (From Chang, E.Y. et al., *J. Electron. Mater.*, 34, 23, 2005. With permission.)

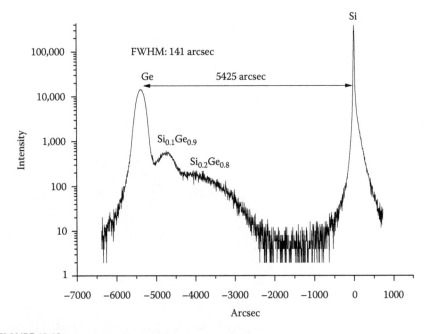

FIGURE 13.18
Double-crystal x-ray diffraction data indicating composition variations of the (004) orientation Ge_xSi_{1-x} metamorphic buffer layer on the Si substrate. (From Hsieh, Y.C. et al., *Appl. Phys. Lett.*, 90, 083507, 2007. With permission.)

13.3 Mechanism of III–V High-Electron-Mobility Transistor

High-electron-mobility transistor (HEMT), also known as modulation-doped field effect transistor (MODFET) or quantum-well field-effect transistor (QWFET) is one of the most mature III–V semiconductor transistors. The term "high electron mobility transistor" is applied to the device because the structure makes use of the superior transport properties (e.g., high mobility and velocity) of electrons in a potential well of the undoped semiconductor material. The first demonstration of the HEMT was made by Fujitsu Lab. in 1980 [Mimura 1980]. Figure 13.19 represents a cross-sectional view of a conventional HEMT structure. Three metal electrode contacts (source, gate, and drain) are made to the surface of the semiconductor structure. The source and drain electrode are ohmic while the gate is a Schottky barrier.

The epitaxial layers of the HEMT structure contain a doped high bandgap material in conjunction with an undoped low bandgap material to form a heterostructure. The structure is designed to form a two-dimension electron gas (2-DEG) channel layer in the low bandgap material. An undoped spacer about 30–50 Å high bandgap material is designed to prevent the electrons in the 2-DEG channel from the scattering by the ionized donors due to coulombs electric field. This scattering reduces electron mobility and, therefore, diminishes the effect exploited in this device. The band diagram of InAlAs/InGaAs metamorphic HEMTs is shown in Figure 13.20.

When small electron affinity and high-energy bandgap material (e.g., InAlAs) connects to large electron affinity and low-energy bandgap channel material (e.g., InGaAs), the Fermi-level, electron affinity, and the energy bandgap must follow the energy bang theory and reach balance; due to this effect, a discontinuity of conduction band at the junction will occur. This discontinuity will cause partial conduction band of channel material below

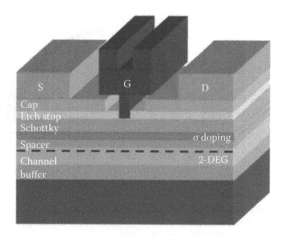

FIGURE 13.19
Conventional HEMT structure.

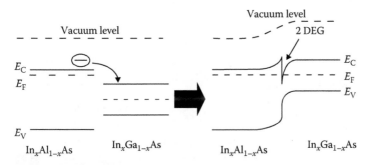

FIGURE 13.20
Band diagram of InAlAs/InGaAs MHEMTs.

the Fermi-level so the mass of free-electron concentration will accumulate in this 2-DEG quantum well and get quantified in terms of a sheet carrier density, n_s. Consequently, these electrons move faster in 2-DEG because they were accumulated in a channel of undoped high-intrinsic-electron-mobility material. As a result of these terrific bandgap engineering designs, HEMTs dominate in the category of superior carrier transport FETs and such transport properties in the potential well are favorable for fast response times as well as high-speed operation.

In general, the epitaxial layers of the high-speed $In_xGa_{1-x}As$ channel HEMTs are grown by molecular beam epitaxy (MBE) or metal organic chemical vapor deposition (MOCVD) on InP/GaAs substrates. A typical epitaxial structure of the $In_xGa_{1-x}As$ channel HEMTs grown on the InP substrate for high-frequency application is shown in Figure 13.21 [Kuo 2008a].

The InAlAs metamorphic buffer layer is used to release the lattice mismatch between InGaAs channel and GaAs/InP substrate. Therefore, a high indium content (50%–100%) $In_xGa_{1-x}As$ channel layer can be achieved in spite of the large lattice mismatch between the active channel layer and the GaAs/InP substrate. The addition of $In_xGa_{1-x}As$ sub-channels can enhance the electron confinement in the thin channel and low-energy bandgap main channel layer and further improve the electron transport properties. The InAlAs spacer layer is applied to form heterostructure interface with InGaAs-based channel so that the band diagram discontinuity occurs, and the 2-DEG forms. A δ-doped layer with Si doping concentration can provide carriers to the channel layer. And an InP layer can provide a good gate recess etching stop layer as well as a good surface passivation of $In_xAl_{1-x}As$ layer to avoid kink effect and reduce the hot-electron surface damage. Besides, with the use of the InP etching stop layer, the lateral recess length (L_r) can be easily controlled and RF performance can be improved. Finally, the highly Si-doped InGaAs cap layer is used to reduce the ohmic contact resistance of the source and drain contacts. Typically, the layer is etched away where the gate contact is to be formed.

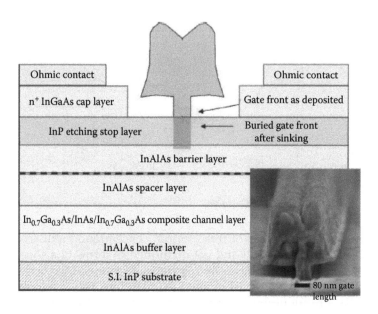

FIGURE 13.21
A typical high-speed InGaAs HEMT structure. (From Kuo, C.-I. et al., *IEEE Electron Dev. Lett.*, 29, 290, 2008. With permission.)

13.4 Basic Fabricating Steps of III–V High-Electron-Mobility Transistors

The standard fabrication process of HEMTs includes following steps:

1. Mesa isolation
2. Ohmic contact formation
3. Fabrication of T-shaped gate by E-beam lithography
4. Gate recess and gate formation
5. Airbridge formation

13.4.1 Mesa Isolation

Device isolation is the first step for the HEMT fabrication process, which is used to define the active region on the wafer. Unlike Si process, because the lack of good insulating native oxide, mesa trench etching is used for isolating individual devices in III–V wafer. In general, the mesa is etched to the buffer layer in order to provide non-interrupting environment for each device.

13.4.2 Ohmic Contact Formation

Traditional ohmic metals used for GaAs-based high-speed devices are multilayer Au/Ge/Ni/Au, from the bottom to the top. After the lift-off process, source and drain ohmic contacts are formed by annealing at appropriate temperature in forming gas atmosphere. Figure 13.22 exhibits an example of drain current as a function of annealing temperature at the fixed drain bias before gate recess. Apparently, there exists an optimized RTA condition, but for different epitaxial structures, the optimum condition may be different. During annealing, germanium atoms diffuse into the InGaAs cap layer and form a heavily doped layer so that reduced contact resistance can be achieved [Williams 1991]. The specific contact resistance can be analyzed by the transmission line model (TLM) in the specially designed process control monitor (PCM) pattern. In general, the measured contact resistance for high-speed HEMT must be less than 1×10^{-6} Ω cm^2.

13.4.3 Fabrication of T-Shaped Gate by E-Beam Lithography

Short gate length with low gate resistance is essential for high-speed HEMT devices. T-shaped gate structure is the most common approach for accomplishing low gate resistance and a small gate foot [Liu 1999]. T-shaped gate is achieved by employing a common multilayer photoresist (ZEP/PMGI/ZEP or PMMA/PMGI/PMMA) with electron beam lithography. Figure 13.23 illustrates the process flow for the fabrication of nanometer T-shaped gate.

The first E-beam exposure and development for the top two layers is used to define the head (Tee-top) of the T-shaped gate by modulating the exposure doses. Next, a single center exposure with high dosage is adopted to define

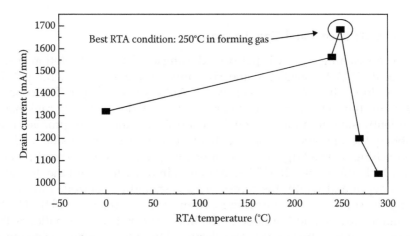

FIGURE 13.22
Optimization of annealing temperature for ohmic contact (depending on the epitaxial structure).

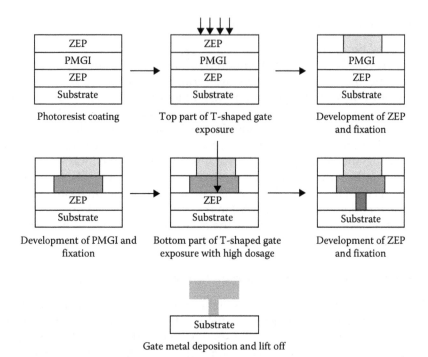

FIGURE 13.23
Electron beam lithography process flow of T-shaped gate.

the footprint of the bottom photoresistor layer. The nanometer T-shaped gate photoresist profile is illustrated in Figure 13.24.

13.4.4 Gate Recess and Gate Formation

The recess etching is usually performed using a PH-adjusted solution of succinic acid (SA)/citric acid (CA) and H_2O_2 mixture for selective etching of the heavily doped InGaAs cap layer over InAlAs Schottky layer. The concentration for the etchant should be adjusted to provide an etching rate that is sufficiently slow to allow reasonable control over the recess process, thus enabling the operation to approach the intended current value without over-etching it. The target current after the gate recess is a critical parameter, which affects the HEMT performance. In order to obtain the desirable recess depth and width, the recess process should be controlled by monitoring the source-to-drain ungated current. For the low-noise PHEMT, the saturation current and the slope of the linear region become smaller as the recess groove is etched deeper and deeper. After recess etching, gate metal is evaporated and lifted off to form T-shaped gate. The picture of T-shaped gate is shown in Figure 13.25 [Hsu 2007].

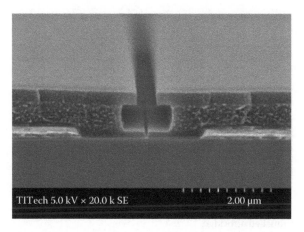

FIGURE 13.24
The tri-layer T-shaped gate photoresist profile.

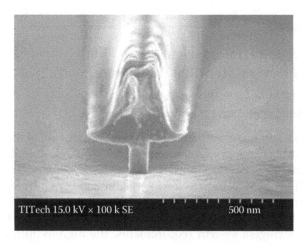

FIGURE 13.25
SEM images of 80 nm T-shaped gate. (From Hsu, H.T. et al., *Asia-Pacific Microwave Conference,* Bangkok, Thailand, pp. 1–4, TH A2-F4, 2007. With permission.)

13.4.5 Airbridge Formation

Airbridge is built using metal with air between the metal interconnection and the wafer surface underneath. Airbridges are used extensively in GaAs analog devices and monolithic microwave integrated circuits (MMICs) for interconnections. They may be used to interconnect sources of FETs, to cross over a lower level of metallization, or to connect the top plate of a MIM capacitor to adjacent metallization. The use of airbridge demonstrates several advantages including lowest dielectric constant of air, low parasitic capacitance, and the ability to carry substantial currents. A SEM image of the airbridge is shown in Figure 13.26.

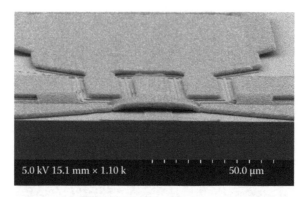

5.0 kV 15.1 mm × 1.10 k   50.0 μm

FIGURE 13.26
SEM image of the finished airbridge.

13.5 Electronic Performance of High-Electron-Mobility Transistors

13.5.1 III–V HEMTs for Millimeter-Wave and Submillimeter-Wave Applications

High-gain and low-noise amplifiers are critical components for many commercial and military applications such as wireless LANs, space-based radars, mobile millimeter-wave communications, and handheld imagers [Deal 2005] [Hacker 2004] [Jang 2002]. In addition to noise and gain requirements, low DC power consumption is also desired for large-scaled array applications since prime power is limited in such systems. Among all the possible technologies to meet such stringent system requirements, the high-indium-content InP-based InAs/In$_{1-x}$Ga$_x$As composite channel HEMTs are particularly promising because they provide high electron mobility, high saturation velocity, large Γ to L valley separation, and high sheet electron densities, which lead to superb speed-power device performance.

When it comes to HEMT for millimeter-wave process, gate length and gate-to-channel distance play rather critical role. Current-gain cutoff frequency can be expressed by using relevant circuit elements as follows:

$$f_T = \frac{g_m}{2\pi(C_{gs} + C_{gd})}$$

$$f_T = \frac{g_m}{2\pi C_g} = \frac{Z_G v_{sat}\varepsilon}{w} \cdot \frac{w}{\varepsilon Z_g L_g} \cdot \frac{1}{2\pi} = \frac{v_{sat}}{2\pi L_g}$$

From this relation, in order to achieve high f_T, enhancing g_m and decreasing total gate capacitance must be realized. Small total gate capacitance is accomplished by short gate length; so decreasing the gate length can increase the electronic field under the gate and then accelerate the channel electron transport property. Therefore, the shrinkage of gate length is an effective way to obtain high g_m and low C_g in order to attain high f_T. Additionally, through the recess modification to reduce the gate-to-channel distance can also effectively increase g_m and f_T because of the enhancement of average electron velocity underneath the gate electrode.

T. Suemitsu et al. [Suemitsu 2002] reported that by means of lateral recess control, the f_T performance of HEMTs can be optimized (Figure 13.27). Matsuzaki et al. [Matsuzaki 2005] have employed tiered-edge ohmic, modified recess structure, and applied low-k benzocyclobutene passivation to effectively minimize parasitic gate capacitance as well as to achieve relatively high g_m of 2 S/mm and f_T of 500 GHz (Figure 13.28). S.-J. Yeon et al.

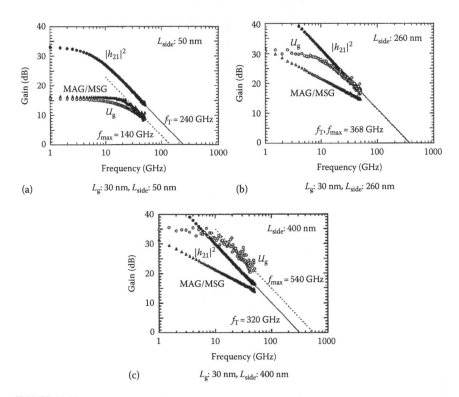

FIGURE 13.27

Typical current gain (h_{21}), maximum available/stable power gain (MAG/MSG), and unilateral power gain (U) of 30 nm two-step recess gate HEMTs with lateral widths of the gate recess of (a) 50 nm, (b) 260 nm, and (c) 400 nm at gate voltage: 0.05 V, drain voltage: 0.7 V. Obviously, lateral recess length (L_{side}) will affect the HEMTs RF performance. (From Suemitsu, T. et al., *IEEE Trans. Electron. Dev.*, 49, 1694, 2002. With permission.)

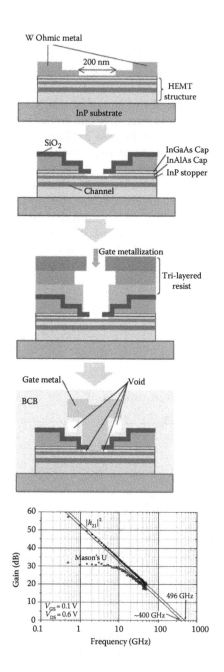

FIGURE 13.28

The fabricated 50 nm-gate HEMTs using the laterally scaled down ohmic structure along with a scaled-down gate length while employing an $In_{0.53}Ga_{0.47}As/InAs$ composite channel, which exhibited extrinsic f_T of 496 GHz with ultra-low ohmic resistance of 0.1 Ω mm and large drain current density. The ohmic structure ensuring low ohmic resistance features a markedly reduced distance of less than 100 nm between the gate and ohmic electrode. (From Matsuzaki, H. et al., *IEDM Tech. Dig.*, 775, 2005. With permission.)

[Yeon 2007] presented an ultrasmall 15 nm-gate-length HEMT with very high speed up to 600 GHz (Figure 13.29).

For the noise figure, NF_{min} can be approximated by the semiempirical equation given by Fukui [Fukui 1979] as shown below:

$$NF_{min} = 1 + k\left(\frac{f}{f_T}\right)[g_m(R_g + R_s)]^{1/2} = 1 + 2\pi k f(C_{gs} + C_{gd})[g_m(R_g + R_s)]^{1/2}$$

where k is a fitting parameter.

Reduction for the gate length (L_g) and parasitic resistances are essential to achieve extremely low noise figure of the devices. Although reduction

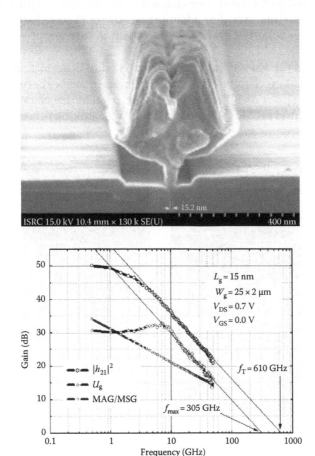

FIGURE 13.29
Frequency dependence of the extracted current gain (h_{21}), Mason's unilateral gain (U_g), and MAG/MSG of 15 nm $In_{0.75}GaAs$ HEMTs under a V_{DS} of 0.7 V and a V_{GS} of 0 V. (From Yeon, S.J. et al., *IEDM Tech. Dig.*, 613, 2007. With permission.)

of gate length appears to be a good approach, limitation of such method may render unnecessary degradation of performance caused by the short-channel effect. Thus, caution must be taken in obtaining the optimum physical parameters of devices for such applications. A Pt gate-sinking process is proposed to solve this problem. The Pt gate-sinking technology has been widely applied in the fabrication of HEMTs since it provides a promising solution that enables vertical scaling of gate-to-channel distance without increasing the access resistance. Meanwhile, the short-channel effect can also be minimized [Chen 1996]. By exact control of the annealing time, the Pt can react with As/Ga and move toward the channel. Figure 13.30 [Chu 2007] shows the TEM image of Pt/InGaP interface before and after the Pt sinking annealing. This function makes the gate electrode closer to the channel layer, and it enhances the device characteristics in high frequency range as well as logic circuits. Another advantage of using Pt-based structure is the relatively larger Schottky barrier height, which in turn suppresses gate leakage currents.

H.T. Hsu and C.I. Kuo report the 80 nm InAs HEMTs performance with and without gate sinking. Figures 13.31 and 13.32 present their experimental results of g_m and NF_{min} [Hsu 2008]. Table 13.2 lists the extracted capacitance, g_m, f_T, and f_{max} for 80 nm InAs HEMTs with and without gate-sinking process [Kuo 2008].

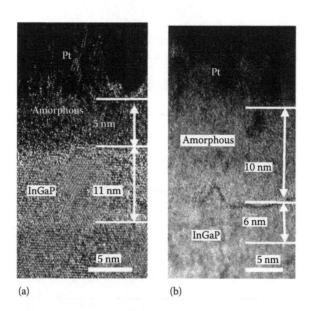

(a)   (b)

FIGURE 13.30
(a) Cross-sectional TEM image of the as-deposited Pt/InGaP interface. (b) Cross-sectional TEM image of Pt/InGaP interface after 325°C annealing for 1 min. InGaP here is the Schottky layer of HEMTs. After Pt sinking annealing, the Pt will react with Schottky materials, so the gate electrode is closer to the channel layer. (From Chu, L.H. et al., *IEEE Electron. Dev. Lett.*, 28, 82, 2007. With permission.)

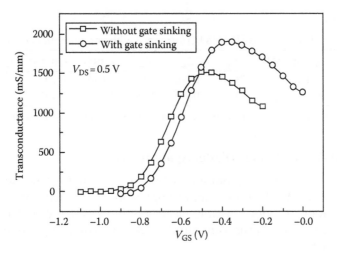

FIGURE 13.31
The transconductance g_m plotted as a function of V_{GS} for 80 nm InAs HEMTs with and without gate sinking; the peak g_m value increased from 1430 mS/mm for device without gate sinking to 1900 mS/mm for that with gate sinking, both measured at 0.5 V drain bias. (From Hsu, H.-T. et al., *Asia-Pacific Microwave Conference*, Hong Kong, pp. 16–19, 2008. With permission.)

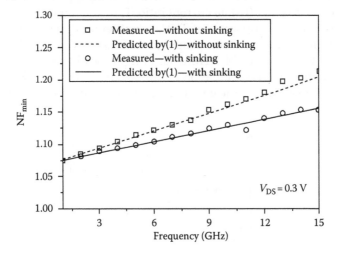

FIGURE 13.32
Measured minimum noise figure as a function of frequency for 80 nm InAs HEMTs with and without gate sinking at 0.3 V drain bias. Predicted NF_{min} were extracted by Fukui NF_{min} formula. (From Hsu, H.-T. et al., *Asia-Pacific Microwave Conference*, Hong Kong, pp. 16–19, 2008.)

Overall, the improvement in performance after gate sinking is mainly attributed to the increase of electron transport properties and the decrease in the corresponding capacitances of the device. Then, results demonstrate that superior HEMT performance for high-frequency, low-noise, and low-power applications can be achieved through a simple gate-sinking process.

TABLE 13.2

Summary of 80 nm InAs HEMTs Performance with and without Gate Sinking

InAs/In$_{0.7}$Ga$_{0.3}$AsHEMTs	C_{gs}	C_{gd}	C_{ds}	g_m (RF)	f_T (GHz)	f_{max} (GHz)
Without gale sinking	73.3 fF	16.3 fF	5.3 fF	201 mS	390	360
Gate sinking	60.5 fF	16.6 fF	3.7 fF	208 mS	494	390

Source: Kuo, C.-I. et al., *IEEE Electron. Dev. Lett.*, 29, 290, 2008. With permission.

13.5.2 III–V QWFETs for Digital Applications

Due to the low carrier mobilities of Si, the research of the alternative devices for Si MOSFET for digital applications has attracted more and more attention. Among possible electronic devices, III–V material-based QWFETs are the most promising candidates due to their superior electronic performances.

In digital applications, a transistor operates as a switch, which is different from microwave or millimeter-wave applications. Figure 13.33 exhibits the figures of merit of the transistor as a switch [Kim 2006] [Alamo 2007].

As shown from the figure, the relevant figures of merit for logic application are, for example, drain-induced barrier lowering (DIBL), subthreshold slope (S), on-state and off-state current ratio. They are important parameters for devices in digital applications. The evaluation of digital performance of such nonoptimized V_T device is proposed by Chau et al. [Chau 2005], which can be used to avoid possible erroneous and physically meaningless values of logic parameters. The evaluation methodology is shown in Figure 13.34. First, it is necessary to set the gate-to-source voltage at drain-source current

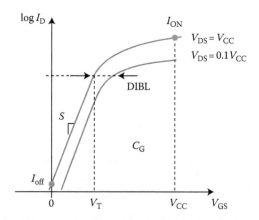

FIGURE 13.33

Electrical figures of merit of a transistor as a switch, including DIBL, S, I_{ON}, I_{OFF}, etc. (From del Alamo, J.A. and Kim, D.H., *Proceedings of the 19th International Conference on InP and Related Materials*, Matsue, Japan, 51, 2007. With permission.)

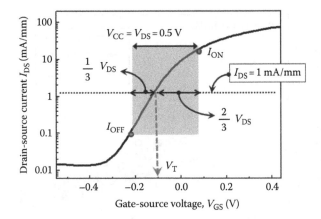

FIGURE 13.34
Evaluation methodologies for the logic performance of HEMTs. Define gate-to-source voltage at 1 mA/mm of drain-source current as the threshold voltage. I_{ON} as $2/3V_{CC}$ swing above V_T, and I_{OFF} as $1/3V_{CC}$ swing below V_T.

of 1 mA/mm as threshold voltage (V_T). Then, select I_{ON} as $2/3V_{CC}$ swing above V_T, and I_{OFF} as $1/3V_{CC}$ swing below V_T. Based on this definition, one can extract the device's logic parameters, such as subthreshold slope, DIBL, and I_{ON}/I_{OFF} ratio for the III–V HEMTs.

Another important figure of merit for high-speed operation of logic transistors is the intrinsic gate delay ($C_{total} V/I_{on}$), which has also been investigated. The C_{total} is the total gate capacitance including the gate-to-source capacitance (C_{gs}) and gate-to-drain capacitance (C_{gd}) extracted from high-frequency S-parameter measurement at corresponding bias conditions. V and I_{on} are the applied drain voltage and on-state current, respectively. Observed from such definition, the gate delay is strongly dependent on the choice of the gate bias.

Recent researches confirm that the nanogate length InGaAs-based HEMT exhibits impressive results for high performance and low power consumption. del Alamo and Kim present 60 nm InAs HEMTs [Alamo 2007], demonstrating DIBL = 93 mV/V, S = 88 mV/decade, I_{ON}/I_{OFF} ratios in excess of 10^4 and comparing the InGaAs HEMTs against state-of-the-art Si MOSFETs (including ultrathin-body and nonplanar-device architectures) shown in Figures 13.35 and 13.36. It is obvious that III–V devices can achieve comparable logic performance or even better performance than traditional Si FETs.

Kuo et al. present 80 nm InAs HEMTs [Kuo 2008b] that yield a very low intrinsic gate delay time of 0.54 ps at 0.5 V drain bias (as shown in Figure 13.37). The f_T versus DC power consumption for this 80 nm device under different drain biases is plotted in Figure 13.38, with the published data of 80 nm Si MOSFETs [Kuhn 2004] biased at 0.7 V for comparison.

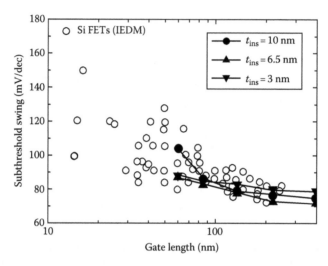

FIGURE 13.35

Subthreshold swing (*S*) versus L_g for InGaAs HEMTs and Si MOSFETs. The Si data come from reports at recent IEDM (from 1990 to 2008). (From del Alamo, J.A. and Kim, D.H., *Proceedings of the 19th International Conference on InP and Related Materials*, Matsue, Japan, 51, 2007. With permission.)

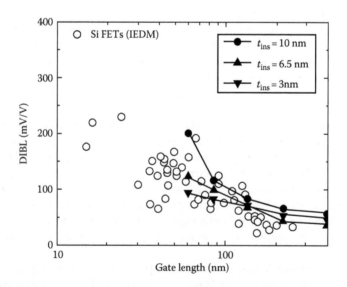

FIGURE 13.36

DIBL versus L_g for InGaAs HEMTs and Si MOSFETs. The Si data come from reports at recent IEDM (from 1990 to 2008). (From del Alamo, J.A. and Kim, D.H., *Proceedings of the 19th International Conference on InP and Related Materials*, Matsue, Japan, 51, 2007. With permission.)

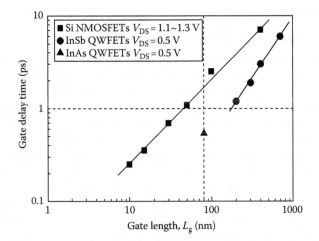

FIGURE 13.37
Gate delay of 80 nm InAs HEMTs and Si n-MOSFETs as a function of gate length. Low intrinsic gate delay time of 0.54 ps at 0.5 V drain bias was shown. (From Kuo, C.-I. et al., *Electrochem. Solid-State Lett.*, 11, H193, 2008. With permission.)

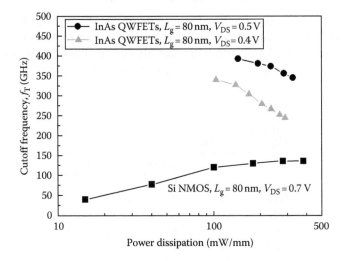

FIGURE 13.38
Cutoff frequencies of 80 nm InAs HEMTs and Si n-MOSFETs as a function of the power dissipation. InAs HEMTs can exhibit higher speed than Si n-MOSFETs under the same power consumption. (From Kuo, C.-I. et al., *Electrochem. Solid-State Lett.*, 11, H193, 2008. With permission.)

From the result of the recent III–V HEMTs researches for logic applications, we can conclude that III–V channel-based devices can achieve higher f_T than Si MOS devices under the same DC power consumption level due to their better electron transport properties. These remarkable results indicate that III–V is the material with great potential for future low-power logic applications.

13.6 Development of III–V MOS-HEMT

With high electron mobility, III–V material HEMTs exhibit superior device performance such as higher transconductance, higher current density, and higher operating frequency in comparison to Si MOSFETs. However, unlike the excellent insulating property of metal-oxide-semiconductor (MOS) structure in Si MOSFET, the Schottky gate in HEMT device suffers from high leakage current, which results in high I_{ON}/I_{OFF} ratio for the III–V devices for logic applications. In order to reduce the gate leakage current, high k insulating layers are introduced under the gate metal to form so-called MOS-HEMT structures. There are many different high-k materials that can be chosen, such as Al_2O_3, HfO_2, ZrO_2, La_2O_3, and CeO_2, etc. A comparison of insulating material properties is provided in Table 13.3. However, it is worth noting that proper surface treatment is needed before applying these insulating layers to reduce the interface trap density, which would cause the Fermi-level pinning of the carriers. These characteristics will be discussed in the following sections.

13.6.1 Performance of III–V MOS-HEMT Devices

Figure 13.39 is a cross-sectional diagram of the MOS-HEMT (which is also called buried-channel MOSFET) device from the report published by IBM in IEDM, 2008. The device contains all the processes such as effective wet-chemical cleaning, atomic layer deposition (ALD) surface self-cleaning,

TABLE 13.3

Comparison of Relevant Properties for High-k Candidates

	k	Gap (eV)	CB Offset (eV)
Si		1.1	
SiO_2	3.9	9	32
Si_3N_4	7	5.3	2.4
Al_2O_3	9	8.8	2.8 (not ALD)
Ta_2O_5	22	4.4	0.35
TiO_2	80	3.5	0
$SrTiO_3$	2000	3.2	0
ZrO_2	25	5.8	1.5
HfO_2	25	5.8	1.4
$HfSiO_4$	11	6.5	1.8
La_2O_3	30	6	2.3
Y_2O_3	15	6	2.3
a-$LaAlO_3$	30	5.6	1.8

Source: Robertson, J., *Rep. Prog. Phys.* 69, 327, 2006. With permission.

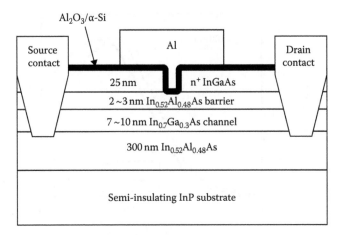

FIGURE 13.39
Schematic cross-sectional diagram of a short-channel $In_{0.7}Ga_{0.3}As$ MOS-HEMT. (From Sun, Y. et al., *IEDM Tech. Dig.*, 4796696, 2008. With permission.)

a-Si interface passivation, and post-deposition annealing (PDA). The performance of depletion-mode (D-mode) and the enhancement-mode (E-mode) buried-channel MOSFET devices with different gate lengths are both fabricated by the same process, the structure is as shown below.

The DC characteristics of a D-mode MOS-HEMT with $L_g = 100\,nm$ are shown in Figure 13.40. The device exhibits good saturation and pinch-off characteristics with a drain current of 960 mA/mm at $V_{GS} = 1\,V$, and $V_{DS} = 1\,V$. Figure 13.41 also demonstrates the extrinsic transconductance (g_m) as a

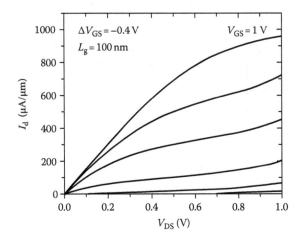

FIGURE 13.40
Output characteristics of a 100 nm D-mode $In_{0.7}Ga_{0.3}As$ MOS-HEMT. (From Sun, Y. et al., *IEDM Tech. Dig.*, 4796696, 2008. With permission.)

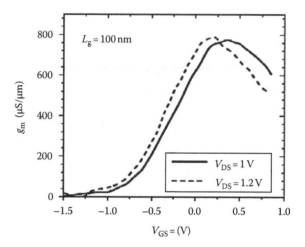

FIGURE 13.41
Transconductance of a 100 nm D-mode In$_{0.7}$Ga$_{0.3}$As MOS-HEMT. (From Sun, Y. et al., *IEDM Tech. Dig.*, 4796696, 2008. With permission.)

function of the gate bias (V_{GS}). The g_m has a peak value of 793 mS/mm at $V_{DS} = 1.2$ V, which is the highest reported value for III–V MOS-HEMT so far. Figure 13.42 shows the corresponding subthreshold characteristics. The device has a threshold voltage of −0.5 V, determined by linear extrapolation from the peak transconductance at $V_{DS} = 50$ mV. The subthreshold slope at $V_{DS} = 50$ mV is 180 mV/decade.

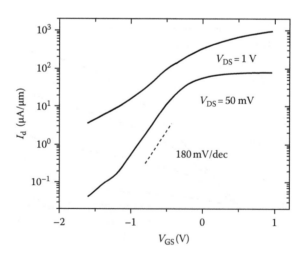

FIGURE 13.42
Subthreshold characteristics of a 100 nm D-mode In$_{0.7}$Ga$_{0.3}$As MOS-HEMT. (From Sun, Y. et al., *IEDM Tech. Dig.*, 4796696, 2008. With permission.)

The following are the output characteristics of a 90 nm E-mode HEMT. The device has a drain current of about 400 mA/mm, a threshold voltage of 0.26 V, and a subthreshold slope of 130 mV/decade as shown in Figures 13.43 and 13.44. The positive threshold voltage shift and the subthreshold slope improvements of the E-mode device are due to the thinner $In_{0.52}Al_{0.48}As$ barrier layer (2 nm) as the use of Si/Al_2O_3 process. The external peak transconductance for the E-mode device is 610 mS/mm at $V_{DS} = 1$ V (Figure 13.45). The lower g_m of the E-mode device compared with D-mode devices is

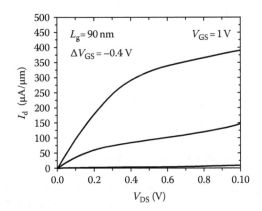

FIGURE 13.43
Output characteristics of a 90 nm E-mode $In_{0.7}Ga_{0.3}As$ MOS-HEMT. (From Sun, Y. et al., *IEDM Tech. Dig.*, 4796696, 2008. With permission.)

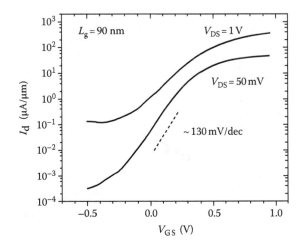

FIGURE 13.44
Subthreshold characteristics of a 90 nm E-mode $In_{0.7}Ga_{0.3}As$ MOS-HEMT. (From Sun, Y. et al., *IEDM Tech. Dig.*, 4796696, 2008. With permission.)

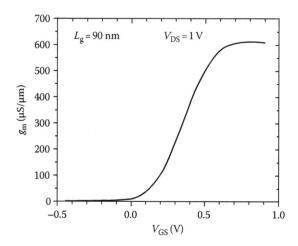

FIGURE 13.45
Transconductance of a 90 nm E-mode $In_{0.7}Ga_{0.3}As$ MOS-HEMT. (From Sun, Y. et al., *IEDM Tech. Dig.*, 4796696, 2008. With permission.)

partially due to the larger external resistance from the larger source/drain-to-gate distance.

Figure 13.46 provides the on-state resistance (R_{on}) at $V_{DS} = 50\,mV$ versus L_g. The extrapolated external resistance (R_{ext}) is $380\,\Omega\,\mu m$ for D-mode devices and $825\,\Omega\,\mu m$ for E-mode devices. The D-mode devices have a lower external resistance due to their lower gate-to-source/drain distance, which results in higher current and transconductance as previously discussed. These III–V buried-channel MOSFET devices also reveal a lower gate leakage current of

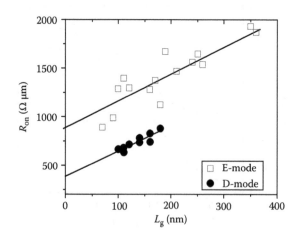

FIGURE 13.46
R_{on} versus L_g of E-mode and D-mode $In_{0.7}Ga_{0.3}As$ buried-channel MOSFET. (From Sun, Y. et al., *IEDM Tech. Dig.*, 4796696, 2008. With permission.)

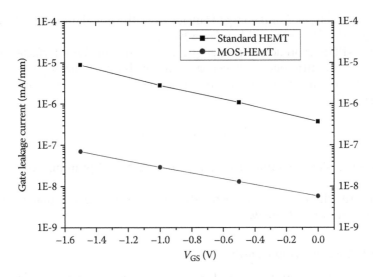

FIGURE 13.47
The gate leakage current of standard HEMT and MOS-HEMT.

up to two orders in contrast to the standard HEMT devices in Figure 13.47. It increases the breakdown voltage of the devices and enhances the device performance.

III–V buried-channel MOSFET structures are more suitable for digital applications due to their higher drive current as well as lower gate leakage. After those efforts, however, no Fermi-level pinning and low equivalent oxide thickness III–V buried-channel MOSFET devices are fabricated. It probably will become the most important technology to extend the Moore's law to a new century.

13.6.2 Challenge of Metal Gate/High-*k* Materials/III–V MOS Structure

Metal gate/high-*k* materials/III–V structure can provide the advantages of high electron mobility and low gate leakage current. However, unlike the Si MOSFET, III–V MOS structures have high trap density in the interface between high-*k* dielectrics and III–V materials. When the density of the interface states is very high ($>10^{12}$ cm^{-2}), the position of the Fermi level (work function as well) is determined by the neutral level of the surface states and becomes independent of doping concentration. This phenomenon is known as Fermi-level pinning, which is expected to cripple the device performance. In many relevant metal/high-*k* interfaces, the ability to tune the threshold voltage (V_T) and flatband voltage (V_{FB}) is limited by Fermi-level pinning, which becomes more severe with increasing *k*-values. This effect is resulted from the formation of a dipole layer at the interface, which is contributed both from interface-specific chemistry [Tung 2001] (interfacial binding, defects, etc.) and charge transfer from the metal into metal-induced gap states (MIGS) in the oxide [Heine 1965].

In the MIGS approach, the strength and nature of the interfacial dipole depend on the difference between φ_m and the charge neutrality level (CNL, φ_{CNL}) in the oxide. The effective work function φ_m^{eff} for the metal gate is given by [Yeo 2002]

$$\varphi_m^{eff} = \varphi_{CNL} + S(\varphi_m - \varphi_{CNL})$$

where S is a slope parameter ranging from zero (full pinning) to unity (no pinning).

There are three major characteristic analyses for high-k material/III–V structures described below.

13.6.2.1 Interface Trap Density

The nature of the interface trap density (or interface trap charge density) is not completely understood yet, but it can be controlled to very low values by certain techniques that will be mentioned in the following sections. For interface trap density characterization, either the high-frequency method, typically using 1 MHz C–V curve as is demonstrated in Si MOS, or the low-frequency quasistatic capacitance method, are commonly used, ignoring the frequency dispersion taking place in between.

13.6.2.2 Hysteresis

There is a flatband voltage shift that is believed to be related to charge trapping in the gate dielectric layer. It is caused by "border traps" of slow response, described by various names such as "slow traps," "near-interface oxide traps," and "slow interface states." These traps tend to decrease when annealed at high temperature.

13.6.2.3 Frequency Dispersion

In physics, dielectric dispersion is the dependence of the permittivity of a dielectric material on the frequency of an applied electric field. Because there is always a delay between changes in polarization and changes in an electric field, the permittivity of the dielectric is a complicated, complex-valued function of frequency of the electric field. This response is very important for the application of dielectric materials and the analysis of MOS systems.

A normal frequency dispersion of Si MOS capacitors is shown in Figure 13.48. Here, the capacitance is reduced with increasing frequency, which is due to the interface state capacitance relaxation; caused by the slow capture and emission processes of electrons on the interface states. It is characterized by the time constant given in the following equation:

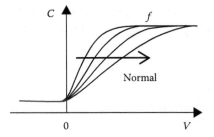

FIGURE 13.48
Schematic for classification of typical *C–V* curves of Si MOS. (From Akazawa, M. and Hasegawa, H., *J. Vac. Sci. Technol. B*, 1569, 2008. With permission.)

$$\tau_0(E) = \frac{1}{\tau_{0n}N_c v_{th}} e^{\frac{E_c - E}{kT}}$$

where
τ_{0n} is the cross section for electrons
N_c is density of states at the conduction band minimum
v_{th} is thermal velocity of electrons

On the other hand, III–V MOS capacitors tend to reveal anomalous frequency dispersion characteristics; their overall behavior could be classified into the "vertical type," shown in Figure 13.49a and the "horizontal type" shown in Figure 13.49b. The frequency dispersion is defined in the following equation:

$$\text{Frequency dispersion} = \frac{C_{10kHz} - C_{1MHz}}{C_{1MHz}} \times 100\%$$

where C_{10kHz} and C_{1MHz} are the capacitances at the frequency of 10 kHz and 1 MHz, respectively. Unfortunately, no sufficient explanation has been given about the mechanism causing these two types of frequency dispersions.

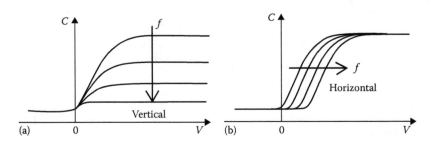

FIGURE 13.49
Schematic for classification of typical *C–V* curves of III–V MIS structures. There are (a) vertically shifting type and (b) horizontally shifting type curves. (From Akazawa, M. and Hasegawa, H., *J. Vac. Sci. Technol. B*, 1569, 2008. With permission.)

In the following sections, several methods will be described in detail to achieve an unpinning Fermi level and to improve the III–V MOS structure performance.

13.6.3 Methods of Providing an Unpinned Fermi Level at the III–V/Gate Dielectric Interface and a Low Equivalent Oxide Thickness

For the challenges for III–V MOS structures described, the interface trap density is the most important issue, which affects the device characteristics. To reduce the interface trap density and avoid Fermi-level pinning, surface treatment becomes a crucial step. The surface treatment can be divided into several parts: native oxide wet-chemical clean treatment, interface self-cleaning by atomic layer deposition, interface passivation layer (IPL), and PDA. These methods will be introduced in the following sections.

13.6.3.1 Wet-Chemical Cleaning and Sulfidization Procedures

Wet-chemical cleaning is usually adopted to etch the native oxide layer. There are researches regarding to the usage of different chemicals, such as HCl, HF, NH_4OH, $(NH_4)_2S$, etc. Table 13.4 shows that the best native oxide layer cleaning treatment combines the HCl, NH_4OH, and heated $(NH_4)_2S$ treatments. As shown in Figure 13.50a and b, from the As $2p_{3/2}$ and Ga $2p_{3/2}$ XPS spectra, the As_2O_x and Ga_2O_x can be reduced by 50% and 17%, respectively using this treatment, which is much higher in contrast to the treatments with HCl only. The 80°C $(NH_4)_2S$ will also form a thin sulfur passivation layer to protect the GaAs substrate from the oxidation reaction on the surface.

TABLE 13.4

Chemical Ratios of As $2p_{3/2}$ and Ga $2p_{3/2}$ Spectra for the Cleaned GaAs Surface after Different WCPs

	As $2p_{3/2}$ and Ga $2p_{3/2}$				
	As–As/As$_{tot}$ (%)	As$_2$O$_2$/As$_{tot}$ (%)	Ga$_2$O$_x$/Ga$_{tot}$ (%)	As–S/As$_{tot}$ (%)	Ga–S/Ga$_{tot}$ (%)
HCl only	15.4	45.6	27.8	—	—
HCl + sulfur (room temperature)	14.9	34.8	27.6	5.4	2.6
HCl + NH$_4$OH + sulfur (room temperature)	12.8	28.1	23.5	3.5	11.7
HCl + NH$_4$OH + sulfur (80°C)	12.3	22.9	23.2	2.6	12.5

Source: Cheng, C.-C. et al., *J. Electrochem. Soc.*, 155, G56, 2008. With permission.

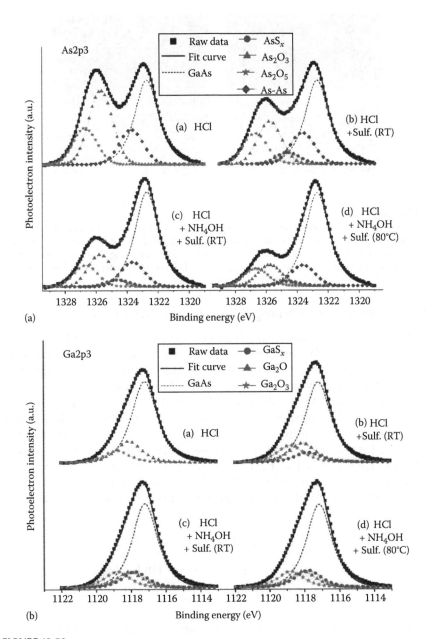

FIGURE 13.50
(a) As $2p_{3/2}$ XPS spectra of clean GaAs substrate subjected to four different WCPs. Five components were extracted: GaAs, AsS_x, As_2O_3, As_2O_5, and elemental As. (b) Ga $2p_{3/2}$ XPS spectra of clean GaAs substrate subjected to four different WCPs. Four components were extracted: GaAs, GaS_x, Ga_2O, and Ga_2O_3. (From Cheng, C.-C. et al., *J. Electrochem. Soc.*, 155, G56, 2008. With permission.)

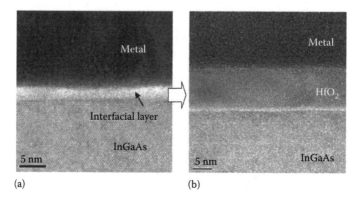

FIGURE 13.51
Cross-sectional TEM images of as-received InGaAs/GaAs capped with metal and ALD HfO_2 deposited on InGaAs/GaAs substrate capped with metal, respectively. It should be noted that HfO_2 was directly deposited on InGaAs without wet-chemical cleaning. The native oxide thickness was reduced from 2 nm to less than 1 nm after ALD HfO_2 deposition. (From Kobayashi, M. et al., *Appl. Phys. Lett.*, 93, 182103, 2008. With permission.)

13.6.3.2 Interface Self-Cleaning by Atomic Layer Deposition

ALD has many advantages, which includes the exact control of deposition thickness, low growth temperature, and superb step coverage. To date, more and more researches demonstrate that the ALD has the ability of interface self-cleaning. The TEM images in Figure 13.51a is a single InGaAs layer on GaAs substrate deposited with W and Figure 13.51b is the W/HfO_2/InGaAs structure on GaAs substrate. Without any surface treatment, a clean native oxide layer can be seen in the interface between the InGaAs substrate and metal, see Figure 13.51a. And for the case of HfO_2 layer directly deposited on the InGaAs/GaAs substrate by ALD followed by W metal deposition, the native oxide layer is reduced from 2 to 0.5 nm or less, Figure 13.51b.

The interface chemical reactions during ALD deposition are described as follows:

$$HfCl_4 + Ga_2O_x \rightarrow GaCl_3 + HfO_2$$

$$HfCl_4 + As_2O_x \rightarrow AsCl_3 + HfO_2$$

$$HfCl_4 + In_2O_x \rightarrow GaCl_3 + HfO_2$$

The oxygen atoms in the native oxide are reduced by the positive ions in the precursor. Thus, the unwanted native oxide thickness is reduced by chemical precursor.

The electron energy loss spectra (EELS) exhibit composition profiles (Figure 13.52) of the $In_{0.53}Ga_{0.47}As$/InP substrate with native oxide and HfO_2/InGaAs gate stack, respectively. The native oxide layer is significantly reduced after the ALD HfO_2 deposition.

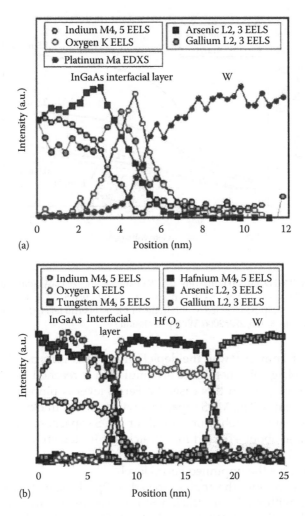

FIGURE 13.52
Cross-sectional composition information measured by EELS of as-received InGaAs substrate with native oxide and W/HfO$_2$/InGaAs, respectively. (From Kobayashi, M. et al., *Appl. Phys. Lett.*, 93, 182103, 2008. With permission.)

13.6.3.3 Interface Passivation Layer

There is another surface treatment that can be used to reduce the surface trap density: the deposition of a thin amorphous silicon or germanium layer before the high-k materials deposition. The thin layer is called IPL and it acts as a protection for the InGaAs/GaAs substrate from the atmosphere. It can effectively reduce the generation of native oxide layer.

Figure 13.53 compares the high-frequency C–V characteristics of GaAs and InGaAs MOS capacitors with various thicknesses of a-Si IPL. For IPL

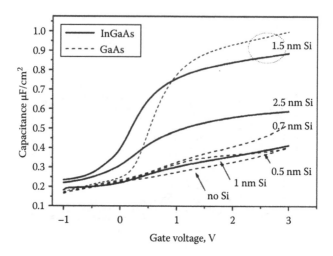

FIGURE 13.53
High-frequency (1 MHz, sweep rate 50 mV s) C–V characteristics of GaAs (dashed) and InGaAs (solid lines) MOSCaps with various thicknesses of a-Si IPL: 0, 0.5, 0.7, 1, 1.5, and 2.5 nm. IPL thickness of at least 1.5 nm is needed to obtain the unpinned Fermi level. (From Oktyabrsky, S. et al., *Mater. Sci. Eng. B*, 135, 272, 2006. With permission.)

thicknesses of 1 nm or lower, the depletion width is very poorly controlled by the gate voltage and negligible accumulation is reached even when +3 V is applied to the gate. In this case, the Fermi level is pinned by the interface bandgap states. The MOS capacitors with a-Si IPL thicknesses of 1.5 nm and above reveal C–V behavior typical for MOS capacitors with distinctive accumulation at about +1 V and low stretch-out in depletion. Thus, for both GaAs and InGaAs channel materials, at least 1.5 nm thickness of Si IPL is needed to observe the unpinned Fermi level.

Figure 13.54 provides the angle-resolved As 2p XPS spectra of GaAs channel with and without 1.5 nm thick a-Si IPL. The oxide layer can be found in the sample without Si IPL through the AsO_x peak. But the sample with a-Si IPL can reduce the oxidation reaction of As atoms. According to the XPS spectra, a-Si IPL is a sacrificial layer that forms SiO_2 and prevents As atoms from oxidation. It protects the GaAs channel material from forming the native oxide and thus improve the C–V characteristics of the MOS structure.

13.6.3.4 Post-Deposition Annealing

To decrease the interface trap density, either a wet-chemical native oxide cleaning or IPL must be present. In addition, it is predicted that annealing after the high-k materials deposition is also a key issue to achieve low defect density oxide layer. Figure 13.55 shows the C–V characteristics of Al_2O_3/ $In_{0.53}Ga_{0.47}As$ MOS capacitors at 1 MHz for $HCl + NH_4OH + (NH_4)_2S$ treatment

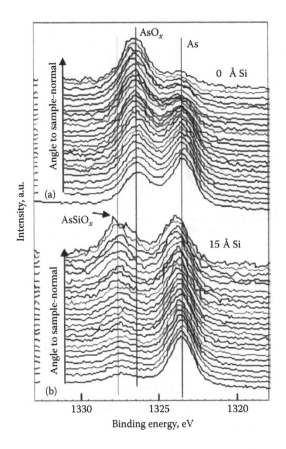

FIGURE 13.54

Angle-resolved As 2p XPS spectra of GaAs channel without (a) and with (b) 1.5 nm thick a-Si IPL after half-hour exposure to air. Angle relative to sample-normal was varied from 24° (bottom) to 82° (top). (From Oktyabrsky, S. et al., *Mater. Sci. Eng. B*, 135, 272, 2006. With permission.)

with different PDA temperatures. As the temperature rises over 500°C, a substantial enhancement of the accumulation capacitance is observed. The leakage current density is also improved for $Al_2O_3/In_{0.53}Ga_{0.47}As$ MOS while the optimized temperature is around 500°C, which is shown in Figure 13.56.

13.6.4 Characteristics of III–V Compound Semiconductor MOSFETs

Recently, considerable interest has been directly toward channel engineering by means of high-intrinsic-electron-mobility III–V materials such as GaAs, or InGaAs for the n-channel MOSFET applications. However, the lack of stable oxide on III–V compound semiconductor has been a major obstacle

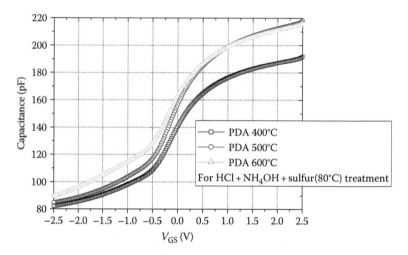

FIGURE 13.55
C–V characteristics of $Al_2O_3/In_{0.53}Ga_{0.47}As$ MOS capacitors at 1 MHz for $HCl + NH_4OH + (NH_4)_2S$ (80°C) treatment with different PDA temperatures.

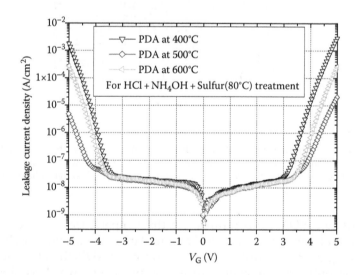

FIGURE 13.56
Plot of current density (J_g) against gate voltage of $Al_2O_3/In_{0.53}Ga_{0.47}As$ MOS capacitors with different PDA conditions.

for the fabrication of III–V MOSFETs. With researchers' recent efforts, III–V n-MOSFETs have been successfully demonstrated.

Lin reported a high-performance $In_{0.53}Ga_{0.47}As$ n-channel MOSFET integrated with a HfAlO gate dielectric and a TaN gate electrode using a self-aligned process [Lin 2008a]. Figure 13.57a shows the cross-sectional diagram of the device structure of the self-aligned n-channel In$_{0.53}$Ga$_{0.47}$As MOSFET and the TEM image of the cross section of the TaN/HfAlO/In$_{0.53}$Ga$_{0.47}$As gate stack, where the HfAlO thickness was measured to be 11.7 nm, as shown in Figure 13.57b.

Figure 13.58 shows the linear-scale drain current density as a function of gate voltage of the 4 μm-gate-length In$_{0.53}$Ga$_{0.47}$As nMOSFET at a drain

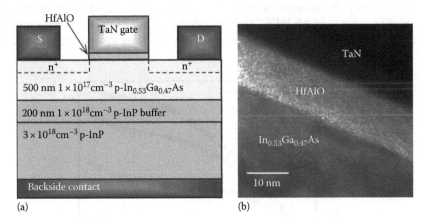

(a)   (b)

FIGURE 13.57
(a) Shows the cross-sectional diagram of the device structure of the self-aligned n-channel In$_{0.53}$Ga$_{0.47}$As MOSFET. (b) The TEM image of the cross section of the TaN/HfAlO/In$_{0.53}$Ga$_{0.47}$As gate stack. (From Lin, J.Q. et al., *IEEE Electron. Dev. Lett.*, 29, 977, 2008. With permission.)

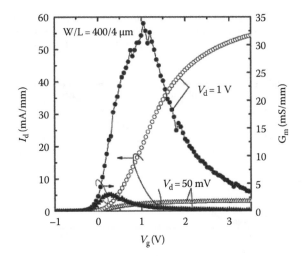

FIGURE 13.58
Linear-scale I_d–V_g of an In$_{0.53}$Ga$_{0.47}$As nMOSFET and gate transconductance versus gate bias with different drain biases. (From Lin, J.Q. et al., *IEEE Electron. Dev. Lett.*, 29, 977, 2008. With permission.)

bias of 50 mV and 1 V as well as the transconductances. The maximum gate transconductances were 3.3 mS/mm for $V_d = 50$ mV and 34 mS/mm for $V_d = 1$ V.

Meanwhile, self-aligned inversion-channel $In_{0.53}Ga_{0.47}As$ MOSFETs using ultrahigh-vacuum (UHV) deposited $Al_2O_3/Ga_2O_3(Gd_2O_3)$ (GGO) dual-layer dielectrics were demonstrated by Prof. M. Hong [Lin 2008b]. An Al_2O_3/GGO dual-layer gate dielectric was e-beam evaporated onto $In_{0.53}Ga_{0.47}As$ surface and a peak mobility of 1300 cm²/Vs has been achieved.

The DC output characteristics, I_d–V_d curves, of the self-aligned inversion channel $In_{0.53}Ga_{0.47}As$ MOSFET with a gate length of 1 μm and a gate width of 10 μm are shown in Figure 13.59, the gate bias was varied from 0 to 2 V

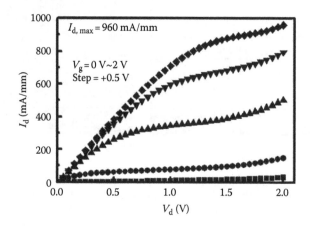

FIGURE 13.59

Output characteristics I_d versus V_d of an inversion-channel $In_{0.53}Ga_{0.47}As$ MOSFET of 10 μm gate width, and 1 μm gate length. A maximum drain current I_d of 960 mA/mm is measured at $V_g = 2$ V and $V_d = 2$ V. (From Lin, T.D. et al., *Appl. Phys. Lett.*, 93, 033516, 2008. With permission.)

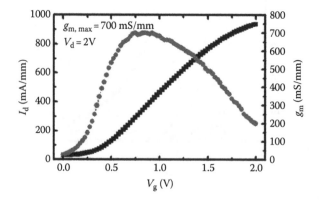

FIGURE 13.60

The transfer characteristics and transconductance g_m curve of an inversion-channel $In_{0.53}Ga_{0.47}As$ MOSFET. (From Lin, T.D. et al., *Appl. Phys. Lett.*, 93, 033516, 2008. With permission.)

in steps of +0.5 V. A maximum drain current density of 960 mA/mm was obtained under a gate bias of 2 V and a drain bias of 2 V. The transfer characteristics and transconductance curve of the device are shown in Figure 13.60. A maximum g_m of 700 mS/mm was demonstrated at $V_d = 2$ V and $V_g = 0.85$ V. The threshold voltage extracted from a linear extrapolation is about 0.3 V. The improved device performance was attributed to a high-quality gate oxide, unpinning Fermi level, and higher In content in the channel.

References

[Akazawa 2008] M. Akazawa and H. Hasegawa, *J. Vac. Sci. Technol. B*, 26 (4), 1569–1578 (2008).

[Alamo 2007] J. A. del Alamo and D. H. Kim, *Proceedings of the 19th International Conference on InP and Related Materials*, Matsue, Japan, pp. 51–54, 2007.

[Chang 2005] E. Y. Chang, T.-H. Yang, G. Luo, and C.-Y. Chang, *J. Electron. Mater.*, 34 (1), 23–26 (2005).

[Chau 2005] R. Chau, S. Datta, M. Doczy, B. Doyle, B. Jin, J. Kavalierous, A. Majumdar, M. Metz, and M. Radosavljevic, *IEEE Trans. Nanotechnol.*, 4 (2), 153–158 (2005).

[Chen 1996] K. J. Chen, T. Enoki, K. Arai, and M. Yamamoto, *IEEE Trans. Electron. Dev.*, 43, (2), 252–257 (1996).

[Cheng 2008] C.-C. Cheng, C.-H. Chien, G.-L. Luo, C.-K. Tseng, H.-C. Chiang, C.-H. Yang, and C.-Y. Chang, *J. Electrochem. Soc.*, 155 (3), G56–G60 (2008).

[Chu 2007] L. H. Chu, E. Y. Chang, L. Chang, Y. H. Wu, S. H. Chen, H. T. Hsu, T. L. Lee, Y. C. Lien, and C. Y. Chang, *IEEE Electron. Dev. Lett.*, 28, 82–85 (2007).

[Deal 2005] W. R. Deal, R. Tsai, M. D. Lange, J. B. Boos, B. R. Bennett, and A. Gutierrez, *IEEE Microw. Compon. Lett.*, 15 (4), 208–210 (2005).

[Fukui 1979] H. Fukui, *IEEE Trans. Electron. Dev.*, 26 (7), 1032–1037 (1979).

[Hacker 2004] J. B. Hacker, J. Bergman, G. Nagy, G. Sullivan, C. Kadow, H. K. Lin, A. C. Gossard, M. Rodwell, and B. Brar, *IEEE Microw. Compon. Lett.*, 14 (4), 156–158 (2004).

[Heine 1965] V. Heine, *Phys. Rev.*, 138 (6A), A1689 (1965).

[Hsieh 2007] Y. C. Hsieh, E. Y. Chang, G. L. Luo, M. H. Pilkuhn, S. S. Tang, and C. Y. Chang, *Appl. Phys. Lett.*, 90, 083507 (2007).

[Hsu 2007] H. T. Hsu, C. Y. Chang, H. S. Hsu, and E. Y. Chang, *Asia-Pacific Microwave Conference*, Bangkok, Thailand, pp. 1–4, TH A2-F4, 2007.

[Hsu 2008] H.-T. Hsu, C.-I. Kuo, and E. Y. Chang, *Asia-Pacific Microwave Conference*, Hong Kong, pp. 16–19, 2008.

[Hudait 2007] M. K. Hudait, G. Dewey, S. Datta, J. M. Fastenau, J. Kavalieros, W. K. Liu, D. Lubyshev, R. Pillarisetty, W. Rachmady, M. Radosavljevic, T. Rakshit, and R. Chau, *IEDM Tech. Dig.*, 625–628 (2007).

[Jang 2002] B. J. Jang, I. B. Yom, and S. P. Lee, *ETRI J.*, 24 (3), 190–196 (2002).

[Kim 2006] D. H. Kim, J. A. del Alamo, J. H. Lee, and K. S. Seo, *Proceedings of the 18th International Conference on InP and Related Materials*, Princeton, NJ, pp. 177–180, 2006.

[Kobayashi 2008] M. Kobayashi I, P. T. Chen, Y. Sun, N. Goel, P. Majhi, M. Garner, W. Tsai, P. Pianetta, and Y. Nish, *Appl. Phys. Lett.*, 93, 182103 (2008).

[Kuhn 2004] K. Kuhn, R. Basco, D. Becher, M. Hattendorf, P. Packan, I. Post, P. Vandervoorn, and I. Young, *VLSI Symp. Tech. Dig.*, 224–225 (2004).

[Kuo 2008a] C.-I. Kuo, H.-T. Hsu, E. Y. Chang, C.-Y. Chang, Y. Miyamoto, S. Datta, M. Radosavljevic, G.-W. Huang, and C. Ting, *IEEE Electron. Dev. Lett.*, 29 (4), 290–293 (2008).

[Kuo 2008b] C.-I. Kuo, H.-T. Hsu, and E. Y. Chang, *Electrochem. Solid-State Lett.*, 11 (7), H193–H196 (2008).

[Lin 2008a] J. Q. Lin, S. J. Lee, H. J. Oh, G. Q. Lo, D. L. Kwong, and D. Z. Chi, *IEEE Electron. Dev. Lett.*, 29 (9), 977–980 (2008).

[Lin 2008b] T. D. Lin, H. C. Chiu, P. Chang, L. T. Tung, C. P. Chen, M. Hong, J. Kwo, W. Tsai, and Y. C. Wang, *Appl. Phys. Lett.*, 93, 033516 (2008).

[Liu 1999] W. Liu, *Fundamentals of III -V Devices: HBTs, MESFETs, and HFETs/HEMTs*, John Wiley & Sons, Inc., p. 422 (1999).

[Luan 1999] H.-C. Luan, D. R. Lim, K. K. Lee, K. M. Chen, J. G. Sandland, K. Wada, and L. C. Kimerling, *Appl. Phys. Lett.*, 75 (19), 2909 (1999).

[Luo 2003] G. Luo, T.-H. Yang, E. Y. Chang, C.-Y. Chang, and K.-A. Chao, *Jpn. J. Appl. Phys.*, 42, L517–L519 (2003).

[Matsuzaki 2005] H. Matsuzaki, T. Maruyama, T. Kosugi, H. Takahashi, M. Tokumitsu, and T. Enoki, *IEDM Tech. Dig.*, 775–778 (2005).

[Mimura 1980] T. Mimura, S. Hiyamizu, T. Fujii, and K. Nanbu, *Jpn. J. Appl. Phys.*, 19, L225–L227 (1980).

[Oktyabrsky 2006] S. Oktyabrsky, V. Tokranov, M. Yakimov, R. Moore, S. Koveshnikov, W. Tsai, F. Zhu, and J. C. Lee, *Mater. Sci. Eng. B*, 135, 272–276 (2006).

[Robertson 2006] J. Robertson, *Rep. Prog. Phys.*, 69, 327–396 (2006).

[Samavedam 1997] S. B. Samavedam and E. A. Fitzgerald, *J. Appl. Phys.*, 81, 3108 (1997).

[Suemitsu 2002] T. Suemitsu, H. Yokoyama, T. Ishii, T. Enoki, G. Meneghesso, and E. Zanoni, *IEEE Trans. Electron. Dev.*, 49 (10), 1694–1700 (2002).

[Sun 2008] Y. Sun, E. W. Kiewra, J. P. de Souza, J. J. Bucchignano, K. E. Fogel, D. K. Sadana, and G. G. Shahidi, *IEDM Tech. Dig.*, 4796696 (2008).

[Sze 1990] S. M. Sze, *High Speed Semiconductor Device*, Murray Hill, New Jersey (1990).

[Takagi 2008] S. Takagi, T. Irisawa, T. Tezuka, T. Numata, S. Nakaharai, N. Hirashita, Y. Moriyama, K. Usuda, E. Toyoda, S. Dissanayake, M. Shichijo, R. Nakane, S. Sugahara, M. Takenaka, and N. Sugiyama, *IEEE Trans. Electron. Dev.*, 55 (1), 21–39 (2008).

[Tung 2001] R. T. Tung, *Phys. Rev. B*, 64, 205310 (2001).

[Williams 1991] R. Williams, *Modern GaAs Processing Methods*, Artech House, Inc., p. 221 (1991).

[Yang 2004] T. H. Yang, G. L. Luo, E. Y. Chang, Y. C. Hsieh, and C. Y. Chang, *J. Vac. Sci. Technol. B*, B22, L17 (2004).

[Yeo 2002] Y. C. Yeo, T. J. King, and C. M. Hu, *J. Appl. Phys.*, 92 (12), 7266 (2002).

[Yeon 2007] S. J. Yeon, M. Park, J. Choi, and K. Seo, *IEDM Tech. Dig.*, 613–616 (2007).

14

III–V Solar Cells on Silicon

Steven A. Ringel and Tyler J. Grassman

CONTENTS

14.1 Introduction

The monolithic integration of device-quality III–V compound semiconductor materials and their advantageous optical and electronic properties with Si substrates has been a driving force in fundamental semiconductor materials research for decades. While there are, especially in recent years, two major motivations for this effort—the addition of III–V functionality to the Si VLSI platform (e.g., optoelectronic interconnects and high-mobility CMOS channel materials) and the integration of high-efficiency III–V photovoltaics (PV) on low-cost Si substrates (versus high-cost conventional GaAs or Ge)—it is the latter objective that we discuss here. However, it is worth noting that, in addition to the economic benefits associated with the integration of III–V solar cells on Si substrates, there are important performance advantages in the use of Si based on the physical properties of Si itself and the unique epitaxial III–V bandgap profiles achievable on the Si platform.

14.1.1 Background

14.1.1.1 Motivation for III–V/Si Photovoltaics Integration

The solar cell efficiency advantages for III–V PV are well known and there are two primary enablers of this fact. First, the direct bandgaps and bandgap profiles (in the case of multijunction solar cells) of III–V compounds, alloys, and heterostructures can be tuned through well-established epitaxy techniques to advantageously match the complex solar spectrum of photons incident upon the earth, thereby converting the incoming optical energy to useful electric energy with extremely high efficiency. Second, III–V compounds possess superior electronic quality, which leads to very efficient collection of photogenerated carriers, a fact that is a product of the leverage afforded by other existing III–V epitaxial device technologies. This combination of tunable optical properties and superior electronic quality has made III–V solar cells the trend-setting ideal for highest efficiency PV technologies for decades, breaking the long-sought 40% efficiency barrier in 2007 [King 2007].

However, from the economic perspective, the targeted energy market for PV is not a high-margin market in general, i.e., the energy market is highly commoditized, noting that achievement of $1/W solar power has been a bellwether target for any successful, terrestrial solar cell technology. As a result, cost is crucial and III–V compounds are almost as notorious for high cost as they are famous for high performance. This fact has driven the III–V terrestrial PV interest toward so-called concentrator PV technologies, where large areas of expensive III–V solar materials are exchanged for low cost, optically concentrating system elements that focus sunlight from large areas down to a small area III–V solar cell, typically providing on the order of 100–500× area reduction (optical concentration). While the concentrator III–V solar cell field is a niche area of growth at present, they have become promising for specific applications where sunlight is mostly direct and for energy conversion at the utility scale now that efficiencies of 40% or more have been reached. In contrast, an area that is already dominated by III–V solar cell technology is space PV. This is due to several important factors. First, the excellent resistance to radiation-induced degradation leads to an improved power profile, greater reliability, and higher "end-of-life" performance, which is essential especially for long-term flight and space missions. Second is that the high energy conversion efficiency afforded by III–V solar cells translates directly to a reduction in array size for missions that are scaled to a constant power requirement, and therefore, lighter weight payloads with smaller stowage volume compared to other technologies are achieved. Of course, the higher efficiency also translates simply to providing more on-board power per solar array area in general.

It is clear then that if the cost of III–V PV could be reduced, the field of PV could capture the tremendous performance advantages afforded by III–V semiconductors; this has been the primary motivation behind the interest in creating a high-performance III–V/Si solar cell technology for many years. The use of Si as an alternative substrate material for III–V epitaxial solar cells in place of the common III–V PV substrate materials—GaAs and Ge—would generate significant cost savings from a variety of both obvious and less obvious perspectives. At the materials level, Si is at least 10–100× less expensive per area than the conventional substrate materials. Additionally, Si is also readily available in very large (300 mm diameter in 2009) wafer sizes compared to the typical 100–150 mm diameter wafers of Ge or GaAs. Since epitaxy and wafer-processing costs do not significantly scale with wafer size, moving to a large area substrate (Si) would translate to an additional, sizable decrease in cost per fabricated solar cell. Perhaps even more significant is that by porting the III–V solar cell technology to a Si platform, the decades mature global Si manufacturing base can be accessed, which is built upon the ideals of extreme scalability, enabling fast production ramp-up, high volume, and high throughput. With solar conversion efficiencies of 40%–50%, a lofty but achievable goal with III–V solar cell technology, compatibility with the existing Si infrastructure holds great promise to enable the rapid realization of several GW/year global solar cell production levels, with significantly faster ramp-up and reduced capital expenditures than would be expected for a technology requiring niche manufacturing.

Given the potential for cost reduction through both materials and manufacturing and the promise of III–V solar cell efficiencies nearing 50%, the case of III–V/Si is quite different from every other PV technology, mature or developing; the Si platform allows for the reduction of materials costs and the capture of large-scale incumbent manufacturing, while still providing high PV efficiency via the integrated III–V materials. Even for low-cost, lower-efficiency thin films, or for Si and III–V cells produced in a niche environment, such rapid scaling is very difficult; significant market penetration requires both high efficiency and high throughput.

14.1.1.2 Additional Photovoltaic Performance Benefits

While the purely economic benefit of integrating high-efficiency III–V materials with Si are indeed substantial, there are additional performance advantages that are often overlooked. One major such advantage is the ability to access unique, optimal bandgap profiles that are typically inaccessible from the conventional GaAs and/or Ge substrate lattice constant. In fact, at lattice constants between Si and GaAs, there exist non-Al-containing GaAsP, InGaP, and GaInAsP alloys with direct bandgaps ranging from 1.5 to 2.2 eV, which can be reached utilizing metamorphic graded buffers on Si substrates (see Figure 14.1). When integrated with active Si (or even SiGe) sub-cells, these materials enable the realization of multijunction solar cell stacks with the potential to yield extremely high conversion efficiencies, and with fewer metamorphic steps than lattice-mismatched PV stacks grown on GaAs or Ge substrates [Zdanowicz 2005].

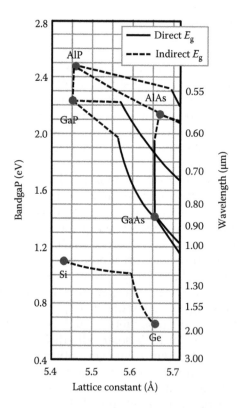

FIGURE 14.1
Bandgap versus lattice constant map for semiconductors having lattice constants between that of Si to slightly larger values than GaAs, to include additional bandgap energies of interest for III–V photovoltaics.

TABLE 14.1

Selection of Important Physical Properties of Substrate
Materials–GaAs, Ge, and Si– for III–V Photovoltaics

Substrate Qualities	GaAs	Ge	Si
Mass density (g·cm⁻³)	5.316	5.323	2.329
Fracture toughness (MPa·m$^{1/2}$)	0.4–0.5	~0.6	~0.9
Thermal conductivity (W·cm⁻¹·K⁻¹)	0.56	0.64	1.24
Thermal expansion mismatch (versus GaAs)	0%	0%–3%	70%–120%
Lattice mismatch (versus GaAs)	0%	0.08%	4.1%

A number of silicon's physical properties also present distinctive advantages over the conventional substrates, for both terrestrial and space-based PV systems, a few of which are given in Table 14.1. For space-based PV systems, cell mass is a very important parameter due to launch payload weight and stowage limitations. Silicon has a mass density less than half that of the GaAs and Ge, meaning comparably sized solar cells on Si would weigh less than half of those based on the more conventional substrate materials. Furthermore, not only is Si significantly lighter than competing substrate materials, it is also mechanically stronger, with a fracture toughness twice that of GaAs and 50% greater than Ge, allowing for the use of thinner Si substrates and thus even greater reduction in system mass. Therefore, high-performance III–V/Si solar cells with conversion efficiencies equal to or greater than that achievable by state-of-the-art III–V/Ge cells would yield an increase in specific power (W/kg) of at least 2×, but potentially upward of 4×.

For terrestrial applications, these particular properties are not eminently important, although reduced cell mass and increased mechanical strength are certainly beneficial, especially for use in such applications as large-area roof-top systems. However, in the case of terrestrial concentrator systems, where the mitigation of excess heat is paramount to high-performance operation, Si substrates possess a thermal conductivity that is approximately 2× higher than GaAs and Ge, providing for a more efficient transfer of heat out of the PV cell and into the dedicated heat sink.

14.1.2 Materials Incompatibilities

As might be expected, given the numerous substantial advantages of III–V/Si heterointegration for a large variety of semiconductor technologies, a great deal of research has been undertaken in pursuit of this goal. After decades of work, success has been growing, particularly in recent years during which several breakthroughs have occurred. The majority of the problems encountered over the years in this undertaking stem from the basic materials-related issues of the III–V/IV heteroepitaxial interface, which can

FIGURE 14.2
Dark-field (200) TEM micrograph of a GaP epilayer grown directly on Si(100) without appropriate interface control. Notice the large number of features, including stacking faults, microtwins, and anti-phase domain borders (two examples of which are indicated by arrows). (After Narayanan, V. et al., *Acta Mater.*, 50, 1275, 2002. With permission.)

be classified as a number of materials properties mismatches, including lattice constant mismatch, thermal expansion mismatch, chemical mismatch, and, perhaps most insidious, crystal symmetry mismatch (polar/nonpolar, or heterovalent, interface).

A number of crystalline defects may result from the nucleation of polar materials (e.g., zinc blende III–V semiconductors) on nonpolar substrates (e.g., diamond structure group-IV elemental semiconductors), due to the mismatch of crystal symmetries between the two types of materials. These defects, and their sources, can be divided into two basic categories: extended planar stacking sequence defects—stacking faults and microtwins (MTs)—resulting from nonoptimal III–V nucleation morphology, and subdomain defects—anti-phase domains (APDs) and anti-phase borders (APBs)—resulting from non-self-consistent deviations in III–V crystal structure registry. Both types of heterovalent interface defects are electrically active, potentially detrimental to any integrated devices, and are especially problematic for minority carrier devices, such as solar cells, as they can serve as highly efficient recombination centers, as well as current shunt paths. Figure 14.2 shows a transmission electron micrograph of a GaP/Si(100) heteroepitaxial interface, which provides a clear illustration of the problem.

Nonoptimal III–V nucleation morphology on group-IV substrates is, at least partially, a consequence of the high interfacial energy due to the truncation of the polar material, which is predicted to produce a significant electric field and resultant charge buildup [Kroemer 1980; Kroemer 2002] promoting an initial 3D growth mode, similar to strain-induced Stranski–Krastanov (SK) or Volmer–Weber (VW) growth of highly lattice-mismatched materials.

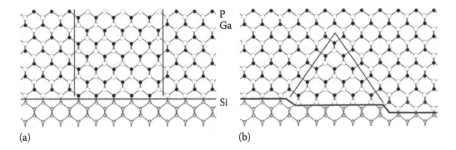

FIGURE 14.3
Diagrams of anti-phase domains formed by (a) spontaneous reordering at III–V/IV(100) nucleation interface and (b) nucleation over single atomic height steps on the group-IV surface. (After Kunert, B. et al., *Thin Solid Films*, 517, 140, 2008. With permission.)

The result is the formation of III–V islands at the high-energy heterovalent interface, stabilized by the lower surface energy {111} facets [Narayanan 2000, 2002b]. Because these facets tend to exhibit a non-ideal chemical reactivity, interruptions or errors in the crystal plane stacking sequence are often introduced, yielding stacking faults and microtwins, upon coalescence of the islands. The result is 2D crystal defects that usually extend throughout the entire remainder of the epitaxial growth.

APDs, which are widespread in all forms of III–V/IV heteroepitaxy, result from a mismatch in crystal symmetries, enabling the formation of regions of incorrect polar crystal registry. These registry errors stem from two main causes, as displayed in Figure 14.3: (1) Spontaneous reordering at the polar/nonpolar III–V/IV interface to help reduce the high interfacial energy, and (2) vertical half unit cell crystal registry shifts due to III–V nucleation over single atomic height steps on the group-IV substrate. Both mechanisms effectively result in the formation of 90° rotated subdomains, whose borders with the surrounding proper domains are electrically active because of the incorrect bonding structures (e.g., Ga–Ga and As–As bonds instead of Ga–As for the case of GaAs) and shifts in crystal polarity. APDs, depending on nucleation and subsequent growth conditions, can also extend throughout the entire epitaxial materials layer, or may self-annihilate within the epilayer.

As previously mentioned, the issue of crystal symmetry mismatch at III–V/IV interfaces is probably the most difficult problem to overcome, but it is not the only one. There is, of course, the common heteroepitaxial issue of lattice mismatch, which can lead to the creation of high densities of crystal defects through the relaxation of misfit strain in the epilayer by the formation of dislocations, both interfacial (generally benign) misfit segments and the upward extending (potentially detrimental) threading segments. In fact, the 4.1% lattice mismatch between GaAs and Si can result in threading dislocation densities (TDDs) in heteroepitaxial GaAs/Si films as high as 10^9–10^{10} cm^{-2}. Posing additional difficulty in this issue is the mismatch of thermal expansion properties, especially in the case of integration with Si substrates because of the nearly 1.5–2.0× smaller coefficient of thermal expansion versus the III–Vs. The net effect here is that the lattice mismatch between Si and the epitaxial III–V layer changes with respect to temperature, making necessary the consideration of variable strain states in growth and device designs to prevent reliability issues and epilayer cracking. And finally, there is the issue of chemical mismatch, which can result in two different negative effects. One such effect is that of chemical reactivity at the group-IV surface during III–V nucleation, especially with respect to impinging group-V elements, which can displace surface group-IV (Si and Ge) atoms, roughening the substrate surface and leading to poor nucleation morphology. The other effect is that of interdiffusion across the III–V/IV interface, leading to the unintentional autodoping of the near-interface regions (e.g., Si-doped GaP and Ga- and/or P-doped Si at the GaP/Si interface), thereby disrupting the intentional doping profiles of the grown structure.

Given the many difficulties and potential sources of detrimental defects in the monolithic integration of high-performance III–V materials with Si substrates, it is not surprising that it is still a highly active area of research, even after over 30 years of constant attention, but not yet a mainstream technology. However, much progress has been made over the years, bringing the goal ever closer to full realization. This work has led to the understanding of the nature of the numerous mismatches found at III–V/IV interfaces, and the steps necessary to circumvent them, as well as a vastly improved knowledge base regarding the basic materials science of both III–V and group-IV semiconductor materials. There have been tremendous advancements in the field of metamorphic (lattice-mismatched) growth techniques, providing scientists and engineers with the tools necessary to mitigate differences in lattice constants with a minimal density of defects. In just over the last decade, the development of a methodology for epitaxial integration of GaAs on Ge(100) substrates has paved the way for a commercializable III–V on Si technology through the use of SiGe graded buffers. And recent work has even helped break through the barriers that had been preventing a realization of direct III–V/Si integration, demonstrating high-quality, defect-mitigated epitaxy of GaP on Si(100).

We shall discuss in the remainder of this chapter some of the approaches undertaken in pursuit of the goal of integrating III–V semiconductors with Si substrates, with the ultimate goal of producing high-performance PV materials, including both epitaxial, metamorphic methods, and non-epitaxial methods. We shall also discuss some of the important parameters to the achievement of solar cell quality materials, including the effects of resultant defects on the electrical properties of the heterointegrated III–V materials. Finally, we will discuss some of the most notable work undertaken to integrate III–V solar cells on Si substrates, clearly demonstrating the vast potential for success, and the directions in which continuing research and development will embark.

14.2 III–V/Si Integration Approaches for Photovoltaics

14.2.1 Epitaxial and Metamorphic Methods

A wide variety of strategies for the epitaxial integration of III–V PV materials on Si substrates have been studied over the years, with varying degrees of success and resultant materials quality. The unifying goal, however, is the mitigation of defects related to the multiple materials properties mismatches, as discussed in Section 14.1.2, which can be detrimental to solar cell performance. Aside from the more fundamental nucleation-related studies of GaP-on-Si and GaAs-on-Ge, the vast majority of research has focused on the heteroepitaxial integration of GaAs on Si(100) substrates. In the GaAs/Si work, the mismatch-related defects receiving the most attention have been threading dislocations (TDs), a result of the large lattice constant mismatch (4.1% at room temperature, 4.3% at 600°C), and large-scale APDs, related to the crystal symmetry mismatch at the heterovalent GaAs/Si interface. We shall introduce in the following two sections some of the more common dislocation reduction methodologies attempted for the growth of GaAs on Si, with the ultimate goal of producing PV quality III–V materials; the mitigation of other heterovalent interface-related defects, including APDs and APBs, will be discussed in Section 14.2.1.3.

14.2.1.1 III–V Buffers on Si

It was realized early on that direct, high-temperature (i.e., equal to that for homoepitaxy) growth of GaAs on Si substrates led to exceedingly high defect densities in the final GaAs films, due in large part to a strain- and heterovalency-induced VW initial growth morphology. At high nucleation and growth temperatures, the GaAs forms large islands that are able to ripen, aided by the high rate of surface diffusion of the Ga and As constituents, to the point that they induce enough misfit strain to nucleate large numbers of dislocations, even before coalescence. When those islands do later coalesce, a high density of immobile and/or pinned dislocations are produced, yielding a final GaAs film with TDDs exceeding 10^9–10^{10} cm^{-2}. A representative example of a single growth step GaAs/Si structure having a very high defect density is shown in Figure 14.4.

In order to prevent the large-scale 3D growth mode that occurs at the direct GaAs/Si interface, it was deemed necessary to utilize some kind of buffer layer between the Si and the final high-temperature GaAs layer that could facilitate high-quality growth and reduce the number of defects stemming from the various interfacial and materials mismatches. The simplest such buffer is merely a thin layer of GaAs nucleated and grown directly on the Si surface at temperatures significantly below that for standard GaAs epitaxy (typically 300°C–400°C) [Akiyama 1984; Wang 1984]. The low nucleation

FIGURE 14.4
Cross-sectional TEM image of a representative GaAs/Si epitaxial heterostructure grown using a single-step process, for which extremely high defect densities are incorporated. (After Lee, H. et al., *J. Electron. Mater.*, 20, 179, 1991. With permission.)

temperature reduces the surface diffusivity of the Ga and As constituents, precluding the formation of large, defective islands, instead producing a high density of pseudomorphic (fully strained) islands. These islands can then coalesce to form a complete film before a large density of dislocations are formed, providing a more controlled dislocation nucleation process and a lower overall TDD. The temperature can then be increased for normal high-temperature, high-rate homoepitaxy. Additionally, the low-temperature nucleation process is found to help reduce large-scale interdiffusion across the III–V/IV interface, helping to relieve, at least partially, the autodoping problem. This "two-step" growth strategy remains a common base process in many forms of III–V/Si heteroepitaxy.

It is possible, of course, to utilize different materials than the final III–V layer as the heteroepitaxial buffer. A large variety of different materials have been tried for this purpose, with mixed success, including "soft" materials, such as GaSb [Uchida 1995] and ZnSe [Bringans 1992], to help trap dislocations within the buffer and prevent them from propagating into the final GaAs layer, and "hard" materials, such as Al-based III–Vs [Ueda 1991] which tend to yield smoother 2D growth morphology than with GaAs alone. GaP, which has a small lattice mismatch with Si (0.37% at room temperature), and its Al-containing alloys, has found reasonable success as a buffer layer for

GaAs [Soga 1989] on Si by separating the highly lattice-mismatched interface from the problematic heterovalent III–V/IV interface. However, regardless of the buffer material used, a simple two-step growth process alone cannot mitigate the vast majority of the resultant dislocation density, and films grown in this manner typically yield TDDs no better than 10^8–10^9 cm^{-2}.

Given the nature of dislocation motion and annihilation reactions (at least in the case of mobile dislocations), it is possible to yield lower final TDD values merely by growing the final III–V epilayer thicker, but because of the large difference in thermal expansion, film thickness (including integrated devices) must be kept to a relative minimum in order to prevent cracking. In the case of GaAs-on-Si, this "critical cracking thickness" is about 4 μm [Yang 2003] much too thin to produce significant TDD reduction by epilayer thickness alone. Therefore, it is necessary to utilize a growth process that can mitigate the large resultant defect densities in a more efficient manner, allowing for the use of thinner buffers.

An obvious extension to the two-step growth process is the application of additional thermal energy, as well as extra strain due to the thermal expansion mismatch between the III–Vs and Si, in an attempt to promote dislocation glide and beneficial interaction/annihilation. Post-growth annealing (PGA), both in- and ex-situ, has been shown to provide a reduction in final TDD [Chand 1986; Choi 1987] but the net effect tends to be rather limited and saturates quickly with respect to annealing time. Taking this concept further, thermal cyclic annealing (TCA), a process in which the sample is cyclically ramped from temperatures below to temperatures above the epitaxial growth temperature, has been demonstrated to yield substantial reduction in final film TDD [Tsaur 1982; Yamaguchi 1991]. The cyclic application of thermal energy and misfit strain induces dislocation motion and annihilation, the total magnitude of the latter depending upon the number of cycles and the difference between the high and low temperatures (see Figure 14.5), and has been demonstrated, in conjunction with the previously discussed two-step growth process, to yield final GaAs/Si TDD values on the order of mid-10^7 cm^{-2}.

Another method commonly used to further reduce dislocation densities is the insertion of strained-layer superlattices (SLSs) between the initial buffer layer and the final device-level layer. The large strain fields related to the SLSs, using such mismatched materials as GaAs/InGaAs, GaAsP/GaAs, and GaAsP/InGaAs for the GaAs/Si heteroepitaxial system, have been shown to strongly interact with propagating TDs, forcing them to bend at the SLS interfaces, resulting in a variety of beneficial interactions that provide a reduction in TDD in the final III–V epilayers [El-Masry 1988]. However, at best, when combined with an optimal TCA process, the use of SLSs seems to only be capable of yielding final GaAs/Si TDD values on the order of ~10^7 cm^{-2}. While this value represents a substantial reduction in TDD versus direct heteroepitaxy, it is still insufficient for high-quality PV applications.

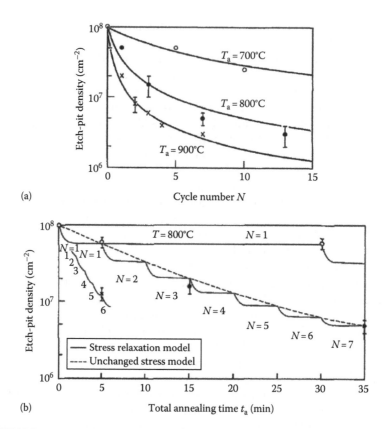

FIGURE 14.5
(a) The number of thermal cycles for different annealing times and (b) the amount of annealing time per cycle are both effective in reducing etch pit density associated with threading
dislocations for GaAs/Si heteroepitaxy. (After Yamaguchi, M., *J. Mater. Res.*, 6, 376, 1991. With
permission.)

Due to the large instantaneous interfacial misfit strain that must be accommodated at the GaAs/Si interface in any of the approaches outlined thus
far, most of the dislocations are formed within the first few nanometers
of mismatched heteroepitaxy. At most, a low nucleation temperature may
kinetically delay the nucleation of dislocations, but ultimately, the large lattice mismatch must be relieved, and even at low temperatures, this occurs
rapidly. Such a massive near-simultaneous formation of dislocations leads to
a large degree of dislocation–dislocation interactions that work to block misfit glide at the interface, producing a large number of permanent dislocation
pile-ups. Additionally, dislocation nucleation at highly mismatched (>2%
misfit) face-centered cubic (FCC) crystal interfaces leads to the formation of
large numbers of pure edge (90°) dislocations rather than the edge-screw
mixed (60°) dislocations formed at smaller misfits. While 90° dislocations are
more efficient at relieving strain than 60° dislocations (with respect to equal

misfit segment length) due to the fact that their Burger's vectors lie entirely within the interface plane, their threading segments cannot glide and are permanently stuck in place, which ends up adding to the final TDD. On the other hand, the threading segments of 60° dislocations are glissile (able to glide) and can therefore, if not hindered, produce very long misfit segments, relieving a greater amount of strain without increasing TDD [Fitzgerald 1991a]. Therefore, the degree to which TDD can truly be reduced at high misfit interfaces, even with the insertion of additional strained interfaces (SLSs) and annealing to promote dislocation glide, is fundamentally limited; reproducible III–V films with TDD values of ~10^6 cm^{-2} or below, a necessity for the production of high-performance solar cells regardless of heterointegration method, is exceedingly difficult.

Because the number of dislocations nucleated in an epitaxial film is directly related to the magnitude of the misfit strain with the underlying substrate, it follows that the reduction of the lattice mismatch would yield low final TDD. Of course, the large absolute lattice mismatch that exists between most III–V PV materials, such as GaAs, and Si cannot actually be reduced. However, it is possible to spread out the strain relief over a longer distance in a more controllable manner with the use of compositionally graded (lattice constant graded) alloy buffers between the Si and the final PV materials. Naturally, through the relief of strain by epilayer relaxation, the compositional grading has the net effect of producing a change in lattice constant; such structures are thus often denoted as "metamorphic" buffers. The main idea behind the use of metamorphic buffers is that while we cannot prevent the ultimate formation of dislocations, we can control the structure of the resultant dislocation network. This task is accomplished by promoting 60° misfit dislocation nucleation in an incremental manner, defined by the compositional grading rate (i.e., % misfit per µm), and in densities that will not result in detrimental interactions, allowing the individual dislocations to glide for long distances, providing the most efficient degree of epilayer relaxation with the lowest resultant TDD possible.

There are two basic methods of compositional grading used to achieve this effect: linear grading and step grading. In linear grading, the alloy composition is continuously adjusted throughout the buffer growth, promoting a gradual generation of dislocations, and attendant epilayer relaxation, through the entire buffer. In step grading, the composition is changed at discrete intervals, providing predetermined mismatched interfaces with just enough misfit strain to promote highly efficient dislocation nucleation and glide, thereby relaxing each layer in succession. Ideally, if designed correctly, the TD segments generated at the first mismatched interface will glide at each successive mismatched interface due to the localized strain field, effectively recycling the same set of dislocation loops throughout the entire growth, with annihilation reactions providing the potential for dislocation reduction, thereby keeping the overall dislocation density at a bare minimum.

To bridge the lattice constant gap between Si and GaAs, there exist two logical pathways for III–V-based compositional grading, $GaAs_yP_{1-y}$ and $In_xGa_{1-x}P$. Note that $AlAs_yP_{1-y}$ and $In_xAl_{1-x}P$, as well as the various related quaternaries, are available in this region, but Al-containing alloys are typically avoided due to issues with oxygen incorporation and alloy hardness, and quaternaries often exhibit unpredictable phase instabilities in strained layer growth. Both $GaAs_yP_{1-y}$ and $In_xGa_{1-x}P$ grading pathways utilize an initial GaP buffer layer on the Si surface, details of which will be discussed in Section 14.2.1.3. This particular III–V/Si integration approach, which has received a comparatively small amount of attention over the years [Hayashi 1994; Komatsu 1997] has seen a recent resurgence of activity, with demonstrations of relatively high-quality PV/optoelectronic materials, and shows promise as a pathway for high-quality III–V/Si heterointegrated materials. We will briefly revisit this topic in Section 14.4.4.

14.2.1.2 *SiGe Buffers on Si*

Given the numerous difficulties (i.e., materials mismatches) involved with III–V PV materials grown directly on Si substrates, a considerable amount of research has concentrated on breaking the problem down into less complicated systems that can be more easily solved. To this end, another group-IV material, Ge, has received much attention. While the issues of the heterovalent interface still exist for GaAs grown on Ge (discussed in the following section), the problematic lattice constant mismatch and thermal expansion mismatch are virtually eliminated, making the process of III–V/IV heterointegration a fair bit simpler. Unfortunately, the use of Ge substrates does not contribute to the goal of true III–V PV integration with Si.

Therefore, much work has also gone into the integration of Ge with Si substrates, where issues related to the heterovalent interface are completely removed, and only lattice constant and thermal expansion mismatches remain (Ge possesses a coefficient of thermal expansion nearly identical to that of GaAs; see Table 14.1). In fact, Ge/Si (and the SiGe alloys) is the prototypical system in which much of the basic science regarding lattice-mismatched heteroepitaxy of FCC materials has been studied. As with GaAs-on-Si, two main approaches to Ge/Si integration exist: direct epitaxy of Ge on Si substrates, with various methods for dislocation density reduction, and the use of compositionally graded SiGe buffers to bridge the lattice mismatch between the epitaxial Ge and the base Si substrate.

Almost identical to the case of GaAs on Si heteroepitaxy, there is a 4.2% lattice constant misfit (at room temperature) between Ge and Si, as well as >2× difference in thermal expansion coefficient (TEC), which results in massive final TDDs when grown without special defect mitigation conditions. One difference, however, is that thanks to the lack of the additional interface energy due to the GaAs/Si heterovalent interface, the Ge/Si growth mode is actually SK, rather than VW; unfortunately, the result is still an initial 3D (island) growth morphology, which coalesces to produce a highly defective film. The use of a two-step growth process and TCA, similar to that discussed previously for GaAs/Si, has been shown to yield direct-grown Ge-on-Si films with TDD in the mid-10^7–10^8 cm^{-2} range [Luan 1999]. Other techniques, such as post-epitaxial hydrogen annealing [Nayfeh 2004] and the use of surfactants (e.g., As, Sb) to prevent the SK growth mode [Wietler 2005] have been used to produce fully relaxed Ge/Si heteroepitaxial films that boast very smooth surfaces. However, regardless the growth process used, the resulting TDDs for direct Ge/Si heteroepitaxy, just like in the GaAs/Si system, are limited to no better than about mid-10^7 cm^{-2}.

Of the various Ge/Si integration strategies, that with the most currently widespread usage, and indeed the only with commercial availability, is the use of compositionally graded SiGe alloy buffers. As discussed previously for the case of III–V graded buffers, the goal here is the incremental relief of misfit in a controlled, efficient manner. While linear grades have been explored, step grades have thus far found more success. After the seminal work of the early 1990s on the realization of relaxed $Si_{1-x}Ge_x$ layers on Si substrates [Fitzgerald 1991b] rapid progress was made in the development of a step-graded SiGe buffer technology to yield low defect density Ge virtual substrates [Currie 1998]. This original buffer design, which spreads the lattice mismatch strain over a thickness of about 10 μm and utilizes a mid-growth ($Si_{0.5}Ge_{0.5}$) chemical–mechanical polish to reduce the deep crosshatch-induced surface roughness (a result of dislocation glide at misfit interfaces), as well as design elements to account for the difference in thermal expansion between the various alloy compositions and final Ge layer, yielded final TDD values of 1×10^6 cm^{-2}, a value finally reaching the limits of PV quality. An example of this is shown in Figure 14.6, for which a high-quality GaAs overlayer is also included. Numerous researchers have since further refined the SiGe buffer process, reporting final Ge layer TDDs in the 10^5 cm^{-2} range [Thomas 2003] helping to cement SiGe as the current state-of-the-art technology for III–V integration with Si substrates.

14.2.1.3 III–V Nucleation on IV Substrates

As was discussed in Section 14.1.2, there are a number of important materials incompatibilities between III–V and IV semiconductors, in addition to lattice constant mismatch, that make III–V/IV heteroepitaxy considerably more difficult and defect-prone than III–V/III–V, or even IV/IV, heteroepitaxy. Since the 1960s, a considerable amount of research has gone into the study of a variety of heterovalent heteroepitaxial systems, including GaAs/Si, GaAs/Ge, and GaP/Si, in an effort to both better understand the basic materials science involved, as well as to develop an effective growth methodology. From this work, a small set of basic, but paramount, strategies necessary for the successful integration of III–V materials on group-IV substrates were extracted, including necessary substrate orientation and preparation,

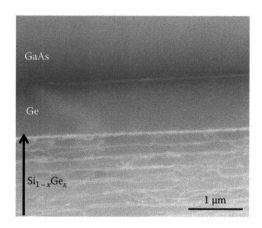

FIGURE 14.6
TEM cross-sectional image of a GaAs layer grown on an optimized, relaxed layer of Ge grown on a metamorphic, step-graded SiGe buffer on Si. TDD in the Ge and GaAs layers are equivalent and equal to 1×10^6 cm^{-2}. (After Andre, C.L. et al., *Proceedings of the 29th IEEE Photovoltaic Specialists Conference*, Piscataway, NJ, 2002. With permission.)

and the proper control of III–V nucleation on the group-IV surface. In recent years, these methods have been employed to yield PV-quality III–V materials grown on group-IV substrates.

Early research into the mitigation of APD formation in III–V/IV heteroepitaxial integration focused on group-IV substrate orientation, using higher-indexed wafer surfaces, such as (110) and (211), which possess atomic surface structures predicted to provide an electrically neutral interface with well-defined bonding order, helping to prevent the generation of APDs by eliminating the interface charge due to heterovalent bonding [Kroemer 1980; Wright 1982; Kroemer 1983; Wright 1984; Uppal 1985]. It was later suggested that suppression of APDs in III–V/IV heteroepitaxy should be obtainable on (100)-oriented substrates with fully double atomic height stepped surface reconstructions, as realized using substrates intentionally offcut in the ⟨011⟩ directions [Fischer 1985; Nishi 1985; Kroemer 1987]. Key to this approach is the lack of single-height surface steps (unavoidable with even nominally on-axis substrates), which lead to the formation of APDs by shifting the crystal structure registry by one-half unit cell, instead providing only double-height steps, which do not disrupt the III–V crystal structure registry. Both approaches have been reported to yield GaP/Si, GaAs/Si, and GaAs/Ge heteroepitaxial films generally free of large-scale anti-phase disorder, proving the efficacy and importance of proper substrate orientation.

In addition to substrate orientation, it has also been shown that the surface chemistry of the group-IV substrate is of great importance in the achievement of high-quality III–V/IV heteroepitaxy. The vast majority of possible contaminants (metals, oxygen, carbon) on Si and Ge surfaces can be effectively removed using standard chemical cleaning regimens, and the native oxides of both can be completely removed with appropriate thermal treatment (annealing in vacuum). However, carbon is an especially tenacious impurity in that while it may be removed during chemical cleaning, it is very difficult to keep it off of the substrate surface, especially during high-temperature annealing for oxide removal or double-atomic-height step reconstruction; once adsorbed onto the surface, it cannot easily be removed in vacuum. While the importance of substrate cleanliness has often been overlooked in the pursuit of III–V/Si integration, problems stemming from C contamination in Si and Ge homoepitaxy have been known for some time [Henderson 1971; Ota 1983]. More recent results have highlighted the detrimental effect of carbon contamination and the importance of Si and Ge surface preparation for III–V/IV heterointegration [Kawanami 1997; Sieg 1998a; Xu 1998; Grassman 2009a]. In order to circumvent the problems caused by C impurities at the surface—i.e., the supply of heterogeneous nucleation sites (due to its stronger bonding potential to impinging group-III and V constituents versus Si or Ge) and the disturbance of the double-atomic-height step reconstruction—homoepitaxial films of Si or Ge are grown prior to III–V nucleation, effectively burying surface impurities and providing a pristine surface reconstruction. This was shown to be necessary to yield highest-quality GaAs/Ge and GaP/Si heteroepitaxial films [Sieg 1998a,b; Grassman 2009a].

III–V/IV integration research has also focused on the importance of the III–V nucleation morphology in nearly lattice-matched systems, especially in GaP/Si, wherein initial island-like growth and subsequent coalescence (even in the absence of significant misfit strain) was indicated to be the source of planar defects commonly found in such films (stacking faults—SFs, microtwins—MTs) [Narayanan 2000, 2002b]. It was reported that a significant reduction of these defects was possible through the use of low-temperature migration enhanced epitaxy (MEE) for the GaP nucleation process [Takagi 1998] which promotes a 2D initial growth morphology, preventing the island/coalescence mechanism. The low-temperature MEE nucleation process has also been additionally useful in both the molecular beam epitaxy (MBE) growth of GaAs/Ge and GaP/Si in that the 2D nucleation morphology helps prevent the spontaneous formation of APDs at the interface [Fitzgerald 1994; Grassman 2009a; Sieg 1998a] and the low-temperature nucleation inhibits interdiffusion across the interface, preventing autodoping problems. Figure 14.7 shows examples of how growth initiation conditions affect interfacial defect structure for III–V/IV heterovalent epitaxy.

Finally, while the analogous GaP/Si and GaAs/Ge heteroepitaxial systems have much in common, they do possess a couple of important differences. One such difference, which is important in the III–V nucleation process, is the difference in surface reactivity with group-V constituents. Namely, at temperatures at which GaAs and GaP are typically nucleated (300°C–400°C) on Ge and Si substrates, respectively, phosphorous has been shown to roughen the surface through the displacement of Si atoms [Curson 2004]

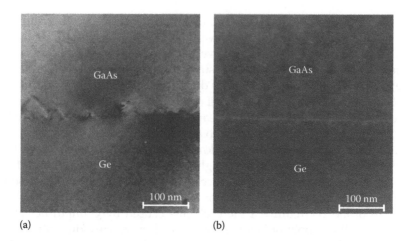

(a)   (b)

FIGURE 14.7
TEM cross-sectional images of MBE-grown GaAs/Ge interfaces (a) grown without MEE nucleation and (b) with MEE nucleation. For both cases, a two-step 500°C/620°C MBE layer was subsequently grown on each interface.

while arsenic appears to cleanly adsorb to the Ge surface. As a solution to this issue, initiating the low-temperature MEE III–V nucleation with a Ga pre-layer on the group-IV, rather than the customary group-V covered surface, has been shown to yield high-quality, APD-free epilayers [Fitzgerald 1994; Grassman 2009a; Sieg 1998a]. It should be noted, however, that there are a number of apparent contradictions in the literature with regard to the efficacy of As and P pre-layers, which may be the result of varying nucleation conditions, including substrate preparation, nucleation temperature, growth method (MBE versus CVD), and group-V species (As_4 versus As_2, PH_3 versus P_2, etc.). Also, as previously mentioned, there exists a difference in coefficient of thermal expansion ratios between GaAs/Ge and GaP/Si, which comes into play for III–V epilayers approaching critical thickness. GaAs and Ge have nearly identical coefficients of thermal expansion resulting in a constant misfit regardless of temperature, while GaP (and any of III–V grown on top of it) and Si differ by more than 50%, which means that the misfit strain state at the heterointerface is not constant with respect to temperature, and must be taken into account in growth designs to prevent detrimental issues regarding relaxation and epilayer cracking.

14.2.2 Non-Epitaxial Methods

Not all approaches for III–V/Si integration involve lattice-mismatched heteroepitaxy. Indeed, some of the earliest methods for integration involved non-epitaxial steps to transfer III–V materials to Si substrates. Of primary importance are the epitaxial liftoff (ELO) methods. ELO generally consists of III–V epitaxy on a lattice-matched substrate, followed by bonding of the epitaxial layer surface to a desired, alternative substrate, e.g., Si, which is then followed by removal or separation of the original growth substrate. Several methods have been employed to accomplish this type of integration process but all require two separate substrates, one for epitaxy and one for the final layer transfer. Substrate removal or separation can be accomplished by lateral, selective chemical etching of a release layer that is grown between the solar cell structure and the substrate, such as AlAs [Taguchi 2005] which rapidly dissolves in HF, or by physical separation such as the cleavage of lateral epitaxial films for transfer (CLEFT) method [Fan 1980]. To decrease the cost of ELO, it may be possible to re-use the original epitaxial substrate multiple times.

These ELO bonding-based methods avoid several problems related to both heterovalent epitaxy and lattice mismatch, and indeed many other technologies already utilize some aspects of wafer bonding, including various optoelectronic devices (LEDs) and silicon- and germanium-on-insulator (SOI, GeOI) substrates. However for solar cells, where relatively large area is required, wafer bonding approaches suffer from mismatch issues due to difference in the coefficient of thermal expansion (CTE) between III–V compounds and Si, giving rise to cracking. Additionally, the bonding interface itself must be uniformly free of voids, and this requires identical bonding surfaces from the perspective of smoothness, flatness, and cleanliness. This is particularly necessary for intrinsic bonding, which is preferred due to the possibility for successful bonding at low temperatures and low mechanical pressure via such interactions as large-scale van der Waals forces. Subsequent annealing and/or mechanical pressure can then further strengthen the wafer bond by driving the formation of covalent bonds between the two materials. Barring this, extrinsic bonding is possible, and this follows the same basic process, but includes the use of "foreign" materials to facilitate the attachment (such as SiO_2), potentially necessary when the bonding surfaces are not sufficiently smooth or flat and cannot be polished to become so, or the functionalization (such as oxidation or other chemical treatment) of one or both surfaces if the native bonding potential is insufficient [Gösele 1999].

More recently, with the advent of so-called Smart Cut™ technology by which layers are removed from their epitaxial host substrate via a hydrogen ion implantation process that physically weakens a thin region of material beneath the epitaxial layer to accomplish layer separation by exfoliation, III–V/Si integration has been achieved [Schone 2006]. Here, SiO_2 is deposited first on the III–V/III–V epilayer/substrate structure and also on a Si wafer, to accommodate extrinsic bonding mediated by SiO_2. Subsequent implantation, bonding, and separation effectively enables a transfer of a thin layer of GaAs to Si, and this has been recently used as a substrate for subsequent III–V solar cell epitaxy. However, final devices suffered from cracking issues that limited performance, as described above [Schone 2006]. Improvements in the H+ implant/bonding/transfer have been reported recently, yielding crack-free GaInP/GaAs dual-junction solar cell structures bonded to Si, with promising preliminary results [Archer 2008]. Ultimately, ELO processes are attractive for a variety of reasons but also generate a different set of potential limitations compared with graded heteroepitaxial integration, and more work is necessary in this regard. It is therefore interesting to note the tremendous advances currently being made by combining both ELO and metamorphic epitaxy through the so-called inverted metamorphic multijunction (IMM) approach. The IMM concept is described later as part of the solar cell discussion, but in effect optimizes the best attributes of both processes, leading to record performance solar cell devices.

14.3 Electronic Properties

The electronic properties of III–V materials integrated with Si, and in particular the electronic quality, where quality in the context of PV can be defined by possessing long diffusion lengths, long carrier recombination lifetimes, well-passivated interfaces and surfaces, are of obvious importance to solar cell applications and define the most promising methods for integration. While much attention in PV is applied to engineering of the absorption profile afforded by engineering the bandgaps of III–V compounds to optimally absorb the solar spectrum, the question of efficiently collecting electrons and holes that are subsequently photogenerated, without the availability of an applied bias voltage, is equally important for a high-performance solar cell. Hence, this section focuses specifically on the electronic properties and electronic quality of III–V/Si materials, the importance of characterizing electronic properties in addition to structural properties, and how the presence of mismatch-related dislocations results in new criteria for optimum III–V/Si solar cell design that is not shared with lattice-matched devices.

14.3.1 Minority Carrier Lifetimes and Measurements in GaAs/Si

The traditional methods used to characterize the quality of lattice-mismatched III–V/Si structures are structural in nature. Transmission electron microscopy (TEM) in cross-section (XTEM) and plan-view (PV-TEM), etch-pit density (EPD), electron beam-induced current (EBIC), and other methods are able to elucidate a variety of extended defects. While dislocations are the defect "du jour" when it comes to III–V/Si integration, it is the summation of impact of all types of defects, including point defects, on the transport of minority carriers that is of greatest consequence for PV material quality. Crystal defects can create direct electrical shorts across p–n junctions, they can trap carriers, and they generally provide pathways for carrier recombination currents that counteract the desired photocurrent of a solar cell. When considering which electronic material parameter is of most consequence regarding the collection efficiency of photogenerated minority carriers, we note that bulk carrier mobility tends to be only weakly dependent on dislocation density below values of ~10^8 cm^{-2} (this is mostly assumed for majority carriers; little information exists regarding minority carrier transport with respect to dislocations). It is the minority carrier recombination lifetime, τ, which is highly sensitive to defects and thus is of great importance when characterizing mismatched III–V/Si heterostructures for PV applications.

The first quantitative correlation between dislocation density (i.e., dislocation spacing) and carrier lifetimes in III–V materials emanated from a model of dislocation-limited diffusion length by Yamaguchi et al. [Yamaguchi 1985]. By solving the minority carrier diffusion equation for a physical system surrounding a TD, a direct dependence of diffusion length on TDD was derived, from which an expression that directly links TDD to the minority carrier lifetime, τ, could be created. The result of this treatment is

$$\tau_{TDD} = \frac{4}{\pi^3 D[TDD]} \tag{14.1}$$

where D is the minority carrier diffusion coefficient given by the Einstein relation,

$$D = \frac{k_B T}{q} \mu \tag{14.2}$$

that, in turn, connects to the minority carrier diffusion length,

$$L_D = \sqrt{D\tau} \tag{14.3}$$

Verification of Equation 14.1 requires experimental knowledge of the true TDD in addition to the minority carrier lifetime. For counting TDs, this most rigorously requires careful PV-TEM, especially for TDDs less than a few ×10^7 cm^{-2}, since X-TEM measurements become statistically limited by the field of view. Historically, PV-TEM has not been a mainstay method in the PV community and methods such as EPD, a chemical method, or EBIC, an electron beam-based imaging approach, are commonly employed. Both methods are relatively simple to use and can be quite helpful; however both rely upon how dislocations interact with the local electronic and/or chemical environment and TDs may not always be easily revealed depending on the specific material in question [Hudait 2003]. As a result, consideration should be given to the method by which TDDs are measured in the literature when correlating TDD with minority carrier lifetime. Additionally, though dislocations are the most well known of extended defects resulting from III–V/Si integration, carrier recombination lifetimes are similarly impacted by all forms of defects, and accounting for how lifetimes are impacted only by dislocations alone in a real material system can be misleading.

Fortunately, for the purpose of characterizing PV material quality, the minority carrier lifetime is a very sensitive parameter to defects of all physical structures, and this has become the standard bearer of material quality for III–V/Si PV materials. The importance of the minority carrier lifetime has long been known even for traditional crystalline Si PV, where extended defects are a very unusual event and thus carrier lifetimes are used to gauge the impact of process conditions, doping, passivation, and so forth for crystalline Si solar cell optimization. For III–V materials on Si and for III–V PV materials in general, time-resolved photoluminescence (TRPL) is a common method by which carrier recombination lifetimes are measured [Ahrenkiel 1990]. By monitoring the decay of the band-band PL peak intensity (e.g., for GaAs at ~870 nm) after appropriate pulsing of the PL light source, a TRPL decay time constant, τ_{PL}, can be obtained, which is directly related to the net sum of recombination processes contributing to the removal of excess photogenerated carriers as the carrier population drives toward thermal equilibrium. These processes include band-band or radiative recombination, Auger recombination and the defect-mediated Shockley–Read–Hall (SRH) recombination mechanism, in addition to surface or interface recombination effects. Since III–V epitaxy is amenable to growth of heterostructures, the most accurate method to perform lifetime characterization via TRPL is by using double heterostructures (DHs), an example of which is AlGaAs/GaAs/AlGaAs to characterize GaAs recombination properties. By choosing a monochromatic PL light source whose energy is greater than the bandgap energy of the lower bandgap "sandwiched" or "well" material, but less than that of the surrounding barrier materials, DH structures enable a simple separation of bulk and interface recombination processes. This is possible since the weighted contribution of the bulk and interface recombination processes on the measured TRPL lifetime can be controlled by varying the thickness of the smaller bandgap DH "well." For instance, very thin DH structures will be dominated by interface recombination if the diffusion length of the DH well material is greater than the thickness of the DH well layer, whereas for very thick well layers, bulk recombination will dominate since the majority of photogenerated excess carriers will have recombined in the bulk prior to diffusing to the barrier/well interface. A simple relationship can be generated that relates DH well thickness to the measured TRPL decay time, according to

$$\frac{1}{\tau_{PL}} = \frac{1}{\tau} + \frac{(S_1 + S_2)}{d} \tag{14.4}$$

where the bulk minority carrier lifetime, τ, is given by

$$\frac{1}{\tau} = \frac{1}{\tau_{SRH}} + \frac{1}{\tau_{rad}} + \frac{1}{\tau_{Aug}} \tag{14.5}$$

where

    τ_{SRH} is the SRH lifetime
    τ_{rad} is the radiative recombination lifetime
    τ_{Aug} is the Auger recombination lifetime

In the above expressions, S_1 and S_2 are the interface recombination velocities of the top and bottom interfaces of the DH structure, respectively. If the bottom and top interfaces of the DH structure are sufficiently identical such that $S_1 = S_2$, an assumption that is often made for ease of evaluation, but that tends to only be experimentally valid for the most ideal interfaces (e.g., AlGaAs/GaAs), is $(S_1 + S_2)/2 = S$, from which we obtain the more usual form of

$$\frac{1}{\tau_{PL}} = \frac{1}{\tau} + \frac{2S}{d} \tag{14.6}$$

From this expression, values for S and τ can be obtained simply by performing TRPL measurements on DH structures identical in every way except for varying the DH well thickness, d. Strictly speaking, exact application requires knowledge of photon recycling effects, especially for high-quality material with long minority carrier diffusion lengths, but these effects can be minimized by appropriate choice of the DH thickness [Ahrenkiel 1992]. With this treatment, TRPL measurements made on DH structures can provide a direct connection between TDD and minority carrier lifetimes according to

$$\frac{1}{\tau} = \frac{1}{\tau_{max}} + \frac{1}{\tau_{TDD}} = \frac{1}{\tau_{max}} + \frac{\pi^3 D[TDD]}{4} \tag{14.7}$$

Here τ_{max} represents the maximum minority carrier of a particular semiconductor material at a specific doping concentration from all contributions (radiative, Auger, SRH) in the absence of TDs; i.e, the homoepitaxial lifetime. For lightly doped DHs, on the order of several $\times 10^{17}$ cm^{-3} that is typical of solar cell base layer doping concentrations, radiative and Auger recombination processes are typically not rate limiting and the bulk recombination lifetime is dominated by SRH processes. The lifetime dependence on TDD can be easily transformed to reflect how the minority carrier diffusion length varies with TDD, via

$$\frac{1}{L_D^2} = \frac{1}{L_{D,max}^2} + \frac{1}{L_{D,TDD}^2} \tag{14.8}$$

where $L_{D,max}$ is defined to be the maximum diffusion length for a given material at a given doping without the influence of dislocations and

$$L_{D,TDD} = \frac{2}{\sqrt{\pi^3[TDD]}} \tag{14.9}$$

Figure 14.8 shows raw TRPL data taken at 300 K from a series of n-type (Si doped at 2×10^{17} cm^{-3}) AlGaAs/GaAs/AlGaAs DH structures grown under several different growth conditions to demonstrate the strong dependence

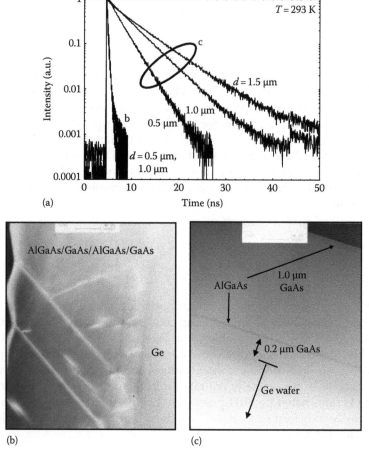

(a)

(b)   (c)

FIGURE 14.8
(a) TRPL data obtained from AlGaAs/GaAs/AlGaAs DH structures grown (b) intentionally with extended defects and (c) with suppression of extended defects, to demonstrate the sensitivity of TRPL decay lifetimes with structural defects. The groupings of PL decay data (denoted by b and c) correspond with the appropriate TEM image.

of the carrier lifetime on the presence of extended defects. For one sample, growth conditions were chosen so that high concentrations of large-scale extended defects are present, and for the other sample, growth conditions leading to ideal, low-mismatched interfaces were chosen. The defect micro-structures of these DHs are shown in the associated cross-sectional TEM images. The fast PL decay obtained for the DH in the defective structure is obvious, indicating very fast recombination via defects and very short recombination lifetimes, as compared with the long decay time for the low defect structure. These data highlight the sensitivity and direct connection with structural defects.

If growth can be controlled such that the only extended defects present in the III–V DH on Si are TDs, as was described earlier in this chapter, then Equation 14.7 should reasonably fit the measured data, assuming that bulk recombination lifetimes can be accurately extracted following Equation 14.6, above. Figure 14.9 shows the measured variation in TRPL lifetime (τ_{PL}) as a function of DH thickness, plotted according to Equation 14.6. For this particular set of AlGaAs/GaAs/AlGaAs DHs grown on Ge/SiGe/Si, a bulk minority carrier hole lifetime of 7.7 ns was obtained, along with an average interface recombination velocity of 3900 cm/s based on a series of growths with three different DH thicknesses [Sieg 1998b]. Although not discussed here, it should be noted that there is a significant dependence of both τ and S on the growth conditions, particularly of the III–V/IV nucleation conditions and cleaning processes, well beyond just the TDD value of the bulk material.

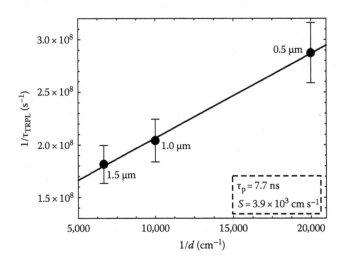

FIGURE 14.9

Measured TRPL decay lifetimes plotted as a function of GaAs well layer thickness for AlGaAs/GaAs/AlGaAs DH structures. The linearity of the data follows Equation 14.6 from which bulk minority carrier lifetimes and interface recombination velocities can be extracted. (After Sieg, R.M. et al., *Appl. Phys. Lett.*, 73, 3111, 1998. With permission.)

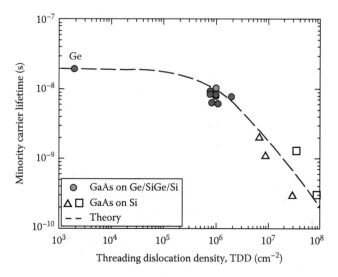

FIGURE 14.10

Compilation of measured minority carrier hole lifetimes in n-type GaAs as a function of TDD, plotted along with the theoretically expected relationship. The highest measured lifetime of 10.5 ns was obtained using SiGe interlayers to accommodate mismatch between GaAs and Si. All lifetime data for SiGe/Si substrates are grouped near a TDD = 1×10^6 cm^{-2}. Referenced data from $\triangle$ [Yamaguchi 1991], $\square$ [Freundlich 1988], $\circ$ [Carlin 2000a,b]; theory from: - - - [Yamaguchi 1985].

Figure 14.10 shows a compilation of minority carrier hole lifetime data for n-type GaAs as a function of measured TDD values that is plotted along with the theoretically expected dependence based on Equation 14.7 at an n-type doping concentration of 2×10^{17} cm^{-3}. This plot includes minority carrier hole lifetime measurements from several groups and compares GaAs integrated on Si by SiGe graded buffers, III–V buffers, TCA, and other processes [Freundlich 1988; Yamaguchi 1991; Carlin 2000a,b]. Direct epitaxy of GaAs on Si via III–V SLSs have been reported to yield lifetimes as high as 2 ns and are likely limited to the high TDD in such structures of ~7×10^6 cm^{-2} [Freundlich 1988; Yamaguchi 1988, 1989a; Ahrenkiel 1990; Wang 1998]. Other attempts to increase the minority carrier hole lifetime using thick Ge or Si$_{0.04}$Ge$_{0.96}$ interlayers on Si resulted in modest improvements up to 3 ns [Leycuras 1990; Venkatasubramanian 1991]. As seen, several data points do lie on the theoretical line, suggesting that TDDs are the primary limitation on carrier lifetime for those studies. Several data points lie somewhat off the predicted dependence, and this may be due to either inaccurate dislocation counting, the presence of other defects that are limiting the carrier lifetime, and/or the simplicity of this model. Currently, the highest measured minority carrier hole lifetime reported for n-type GaAs is 10.5 ns [Carlin 2000a,b] and, as will be discussed later in this chapter, lifetimes of this value are quite adequate for the fabrication of high-performance III–V/Si solar cells. This result was achieved using SiGe buffers, which as described earlier provide an ideal template for III–V epitaxy when combined with careful III–V nucleation to preclude the formation of APBs. Finally, very limited information on the minority carrier lifetime in other III–V materials on Si is currently available as GaAs/Si continues to be the standard for gauging the quality of III–V/Si PV integration for PV applications.

14.3.2 Interdiffusion and Polarity Control

The prior sections focused primarily on structural defects and their impact on carrier transport and recombination. However, III–V solar cells are p–n junction devices and therefore concern over autodoping through interdiffusion of group III, V, and IV elements as a result of III–V/Si integration is a serious issue, especially if inadvertent p–n junctions are formed within the solar cell structure. Inadvertent junctions can cause internal field regions that counteract the desired photovoltage, they can impede the flow of photogenerated carriers, and rampant autodoping can destroy optimum dopant profiles. That said, it is noteworthy that the primary III–V multijunction solar cell device is a lattice-matched $In_{0.48}Ga_{0.52}P/In_{0.01}Ga_{0.99}As/Ge$ triple-junction solar cell in which the bottom (low bandgap) junction in this three-junction stack is indeed intentionally formed by diffusion of As into p-type Ge substrates to form an n^+p Ge bottom solar cell that occurs during the III–V epitaxial growth. Similar notions have been considered for III–V on Si directly but the large lattice mismatch between Si and the III–V compounds having bandgap energies that are appropriate for solar energy conversion have stymied attempts thus far. Moreover, the higher temperatures necessary to remove the native oxide on the Si surface in a III–V growth chamber as opposed to the oxide removal from a Ge surface would suggest that interdiffusion and autodoping from the group V-rich background of a III–V epitaxial growth chamber could be more severe.

The initial excitement generated by the high minority carrier lifetimes of GaAs integrated on Si via SiGe led to intense interest in the integration of III–V single and multijunction solar cells on Si via SiGe graded buffers. Hence, the issue of possible interdiffusion and autodoping in the lattice-mismatched GaAs/SiGe/Si system became an area of focus. Several questions needed to be answered with respect to interdiffusion, most importantly being the potential impact of residual TDs existing throughout the III–V and upper Ge layers, and the complex network of dislocations within the graded buffer, on autodoping, since dislocations can enhance atomic interdiffusion substantially.

Figure 14.11 shows a series of secondary ion mass spectroscopy (SIMS) data obtained for GaAs grown by MBE at conventional temperatures on Ge/SiGe/Si substrates. Data taken for identical MBE growth conditions for growth on Ge substrates are shown for comparative purposes. In all cases, an initial, low-temperature nucleation step was used as this was critical to avoid

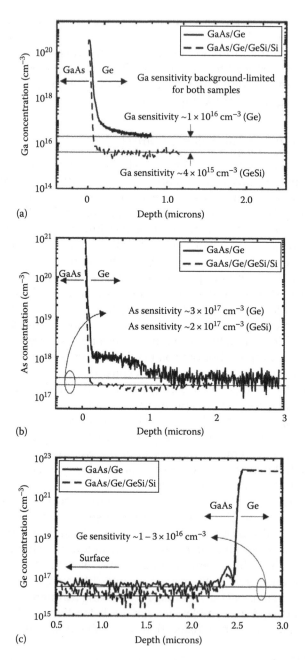

FIGURE 14.11
SIMS depth profiles showing control of interdiffusion for Ga, As and Ge across GaAs/Ge interfaces grown by MBE on Ge wafers and on Ge/SiGe/Si wafers. Nucleation conditions include an MEE initiation sequence for all cases. (After Carlin, J.A. et al., *Appl. Phys. Lett.*, 76, 1884, 2000. With permission.)

the formation of interfacial APDs, described earlier in the chapter. The data reveals that interdiffusion of Ge into the GaAs layer is below SIMS detection, as is Ga diffusion into the Ge layer. Of particular note is that the As diffusion into the Ge layer on SiGe/Si is actually less than that seen for epitaxy on Ge, and itself is below SIMS detection [Carlin 2000a,b]. These data imply that conventional MBE growth for GaAs, which occurs at ~620°C, does not enhance interfacial interdiffusion through dislocations. The important ramification of this result is that MBE growth enables total control over device polarity on SiGe/Si substrates and this means that both p+n- and n+p-configured devices can be grown and fabricated. This provides expanded design space for optimizing solar cell designs.

14.3.3 Doping Effects

As just described, the polarity of III–V solar cells may be an important design criterion, depending on application or the process technology used to fabricate the solar cell. Therefore, for III–V/Si solar cells, knowledge of how dislocations might impact minority carrier transport for n-type and p-type III–V materials is essential, since this could create an optimal choice for device polarity as well. Similar to the experiments and modeling described in Section 14.3.1 for minority carrier hole lifetimes in n-type GaAs on Si, electron minority carrier lifetimes in p-type GaAs on Si have been explored with this in mind. Figure 14.12 shows TRPL decay data obtained for p-type $(2 \times 10^{17} \text{ cm}^{-3})$ GaAs DH structures on two different SiGe/Si substrates having two different values for TDD, along with data obtained for a DH structure grown on GaAs as a homoepitaxial control sample. The impact of increasing TDD is immediately obvious with the strong, monotonic decrease in characteristic decay times [Andre 2004]. By fitting the data to the series of Equations 14.5 through 14.7 based on multiple thicknesses of DH structures, electron minority carrier lifetimes were extracted. Note that the τ_{max} was measured to be 21 ns from the homoepitaxial control and Figure 14.13 shows the electron lifetime data for the p-GaAs on SiGe/Si, plotted, along with the previously discussed minority carrier hole lifetime results for n-GaAs as a function of TDD for comparison. Looking at the measured electron and hole lifetimes at TDD of $1 \times 10^6 \text{ cm}^{-2}$, the hole lifetime is almost an order of magnitude longer than the electron lifetime for the same TDD value and, in general, a very large disparity in minority carrier lifetimes persists for all TDD values, with convergence occurring only toward very low TDD. This is a fundamental result of the large disparity in the electron mobility and the hole mobility for GaAs and can be explained by the fact that higher mobility electrons will, on average, reduce the average time it takes for carrier–dislocation interactions to occur. That is, the higher rate of "sampling" of dislocations by electrons results in a lower measured electron lifetime, compared with the slower holes. As will be seen later, this disparity in lifetime will have important consequences on device characteristics.

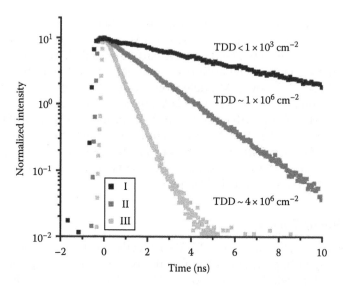

FIGURE 14.12
TRPL data measured at room temperature for p-type $Ga_{0.52}In_{0.48}P/GaAs/Ga_{0.52}In_{0.48}P$ DH structures grown on GaAs, and on two different SiGe/Si substrates having TDD values of 1×10^6 and 4×10^6 cm^{-2}. All structures possessed nominally identical doping of 2×10^{17} cm^{-3}. (After Andre, C.L. et al., *Appl. Phys. Lett.*, 84, 3447, 2004. With permission.)

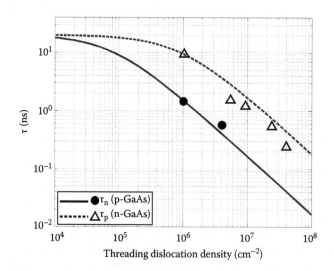

FIGURE 14.13
Plot of minority carrier electron lifetime as a function of TDD for p-type GaAs, shown along with the data for minority carrier hole lifetimes in n-type GaAs. The comparison highlights substantial differences in the impact of TDD for the recombination lifetimes of both carrier types. (After Andre, C.L. et al., *J. App. Phys.*, 98, 014502, 2005. With permission.) Referenced data (Δ) after [Yamaguchi 1989b; Carlin 2000a,b].

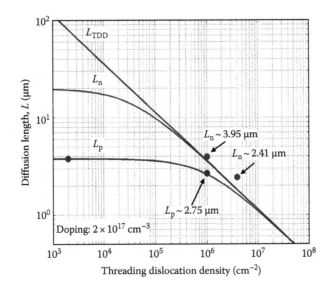

FIGURE 14.14
Comparison of minority carrier electron and hole diffusion length dependencies on TDD, with several measured data points. L_{TDD} is the average threading dislocation spacing.

If we consider the impact of conductivity type on the minority carrier diffusion length, the shorter electron lifetimes combined with the higher electron mobilities reduce this disparity somewhat, and this is seen in Figure 14.14 where for TDD values of ~10^6 cm^{-2} or greater, the difference between electron and hole transport is reduced. As TDD increases, the electron and hole diffusion lengths converge as they become limited by the average TD spacing. In contrast, for low TDD values, L_D approaches $L_{D,max}$, as described by Equations 14.8 and 14.9, and the electron diffusion length is substantially longer, as expected. From a solar cell perspective, this long diffusion length enables n$^+$p cells to be relatively thick compared to p$^+$n cells so that the maximum absorption of solar photons is possible without loss of carrier collection efficiency (i.e., current) from the base layer.

14.3.4 GaAs Diodes on Si

The ultimate measure of III–V/Si material quality for solar cell applications is based on device characteristics. Figure 14.15 shows a comparison of current density-voltage (J–V) measurements obtained from identical p$^+$n GaAs diodes grown on GaAs, Ge, and SiGe/Si substrates, the only difference being an increase in TDD from a negligible value for the GaAs substrate to 1×10^6 cm^{-2} for the SiGe/Si substrate. The detailed analysis of the I–V data reveals that the extracted reverse saturation current density, J_0, increases from 1.3×10^{-11} A cm^{-2} for the homoepitaxial diode, to 5×10^{-10} A cm^{-2} for the diode grown on the SiGe/Si substrate. While this is a very good, low value

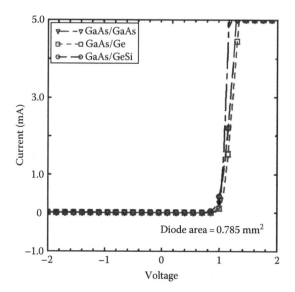

FIGURE 14.15

Current–voltage characteristics for identical p⁺n GaAs diodes grown on GaAs, Ge and SiGe/Si substrates, revealing high-quality diodes for each substrate type, in spite of increasing lattice mismatch from 0% to 4%.

for J_0, the increase is due to a decrease in the measured minority carrier hole lifetime from 22.4 ns for the diode on GaAs, to 10 ns for the diode on SiGe/Si, and is attributed to depletion region recombination resulting from the low, residual TDD in the heteroepitaxial diode. In support of this, the extracted ideality factor increased from 1.8 to 2.0 between the homoepitaxial and heteroepitaxial diodes.

The excellent match of the *I–V* characteristics predicts very good solar cell results, but before moving to that level, it is instructive to consider the diode characteristics for the opposite polarity type, i.e., n⁺p junctions. Figure 14.16a shows a comparison of the 300 K *I–V* characteristics obtained for an n⁺p diode and a p⁺n diode, each grown on SiGe/Si substrates with matching TDD values of 1×10^6 cm⁻². There is a clear reduction in the forward bias turn-on voltage for the n⁺p polarity and an increase in the reverse current density even for the same TDD value. In conjunction with this observation, detailed analysis revealed that the value for J_0 obtained for the p⁺n diode is 5×10^{-10} A cm⁻², whereas for the n⁺p diode, the value of J_0 increases by a factor of ~10 to 6×10^{-9} A cm⁻². This difference is significant and is due to the minority carrier lifetimes in the base of each type of structure being markedly different. The minority carrier hole lifetime in the n-type base of the p⁺n device is ~10 ns and the minority carrier electron lifetime in the p-type base of the n⁺p device ~1.5 ns, as described in Section 14.3.3. Note that the carrier lifetime ratios trend in a very similar fashion to the ratio of J_0 values for each device polarity. To prove that this difference is totally

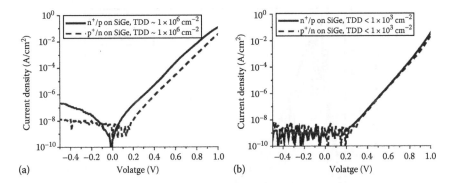

FIGURE 14.16
Log *J–V* data obtained for p+n and n+p GaAs diodes grown on (a) n- and p-type SiGe/Si substrates, respectively, each possessing nominally identical TDD values of 1×10^6 cm^{-2}, and on (b) n- and p-type GaAs substrates having TDD values of less than 1000 cm^{-2}. All measurements were taken at 300 K. (After Andre, C.L. et al., *J. Appl. Phys.*, 98, 014502, 2005. With permission.)

and only a function of the residual TDD, Figure 14.16b shows a similar plot obtained for the same p+n and n+p GaAs diode design but now grown on GaAs substrates, for which TDD < 10^3 cm^{-2}. The measured minority carrier hole and electron lifetimes are 22 and 20 ns, respectively, from TRPL data, and as seen, there is now a nearly perfect match of diode parameters and *J–V* characteristics. The conclusion from these observations is that while the presence of a low, but non-negligible concentration of TDs in GaAs on SiGe/Si is low enough to yield diodes of extremely high quality, this nevertheless creates a new design consideration that is specific to metamorphic GaAs devices [Andre 2005a].

14.4 Solar Cell Characteristics

14.4.1 Solar Cell Device Physics Overview

With the above sections describing electronic properties of GaAs/Si materials and test diodes, we can now move on to discuss solar cells explicitly. First, we briefly provide an overview of solar cell device physics, to provide context to the remainder of this chapter. There are several excellent texts that provide a rigorous treatment of solar cell device physics, and the reader is referred to these for in-depth treatments [Fonash 1981; Fahrenbruch 1983; Nelson 2003].

The incident solar spectrum on a solar cell depends significantly on the solar cell location. In space, the solar spectrum has a total incident integrated power density of 1353 W/m^2, which is referred to as the solar constant. The energy distribution of AM0 illumination is shown in Figure 14.17.

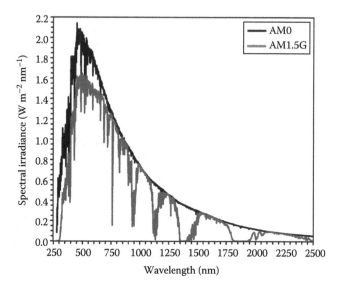

FIGURE 14.17
Irradiance for the AM0 and AM1.5G solar spectra from the ASTM G173-03 reference standard. (Courtesy of http://rredc.nrel.gov/solar/spectra/)

As sunlight passes through the earth's atmosphere, a variety of losses occur due molecular absorption and Rayleigh scattering, the magnitude of which depends on the optical path length through the atmosphere. The resultant losses substantially reduce the input power density on earth and dramatically alter the spectral content. Shown along with the AM0 radiation spectrum in Figure 14.17 is the AM1.5 terrestrial spectrum, having an integrated power density of ~1000 W/m². The "AMn" nomenclature is a measure of the optical path length through the atmosphere, with $n = 1/\cos \theta$ and θ is the angle of the incident sunlight with respect to earth normal. The shortest path length is for $\theta = 0$, or AM1. The accepted spectrum for North America is AM1.5G, where G refers to global radiation (AM1.5D, where D refers to the direct spectrum, is important for concentrator solar cell measurements; AM1.5 is often written to denote AM1.5G and this convention will be followed in the text below).

Solar cells based on III–V compounds are well suited to convert the sun's energy into electricity due to the wide range of direct bandgaps giving rise to very efficient absorption of the solar spectrum in only a few microns of material, the availability of lattice-matched heterostructures to create bandgap profiles for so-called multijunction solar cells, and the presence of an existing and vibrant III–V optoelectronics industry that provides important practical leveraging of fabrication tools and processing. The ideal bandgap for a single-junction solar cell is approximately 1.4–1.6 eV for AM0 and AM1.5 incident spectra, as this range of bandgaps represent the optimum trade-off

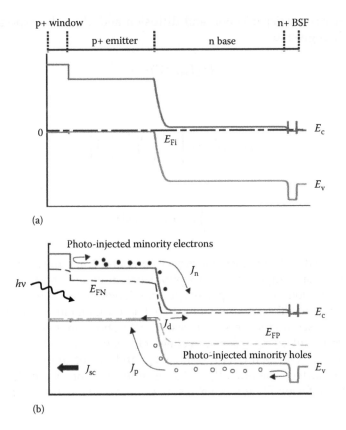

(a)

(b)

FIGURE 14.18
Simplified band diagram of a p⁺n solar cell (a) in equilibrium and (b) during illumination highlighting photocurrent flow and quasi-Fermi level separation resulting from the light biasing effect.

between current generation, which increases with decreasing bandgap, and photovoltage, which increases with increasing bandgap. Hence, GaAs, with its bandgap of 1.42 eV at 300 K, is nearly ideal. The maximum theoretical solar cell efficiency at the GaAs bandgap is calculated to be ~30% for AM1.5 illumination and ~25% for AM0 [Möller 1993]. The highest efficiencies achieved for single-junction GaAs cells are 25.1% (AM1.5) and 23.0% (AM0) [Green 2004].

To understand the performance of a solar cell under illumination, consider the simplified band diagram of a p⁺n homojunction solar cell shown in Figure 14.18. The photocurrent is provided by the collection of minority carrier holes that cross the junction from the n-type base and collection of photogenerated electrons crossing the junction from the p-type emitter, along with electron and hole components due to absorption in the depletion region. The photocurrent opposes the standard diode current that results from majority carrier injection and diffusion and, thus, the expression for this diode is given by

$$J = J_{\text{diode}}(V) - J_{\text{L}} \tag{14.10}$$

where

$$J_{\text{diode}}(V) = J_0 \left[\exp\left(\frac{qV}{nk_{\text{B}}T} \right) - 1 \right] \tag{14.11}$$

and

$$J_{\text{L}} = \int (J_{\text{n}} + J_{\text{p}} + J_{\text{d}}) d\lambda \tag{14.12}$$

The value for the ideality factor, n, is between 1 and 2, and the terms J_{n}, J_{p}, and J_{d} represent the components of the photocurrent emanating from minority carrier generation in the p-type, n-type, and depletion layers of the p–n junction, respectively. These expressions do not account for either series or shunt resistance losses, and for this ideal case, $J_{\text{L}} =$ the short circuit current density, J_{sc}. The resultant current density-voltage characteristic is shown in Figure 14.19, where J_{sc}, the open-circuit voltage, V_{oc}, and the maximum power point, P_{max} are indicated. The ratio, $P_{\text{max}}/V_{\text{oc}}J_{\text{sc}}$ is an important solar cell parameter called the fill factor, FF, and is a measure of how well the diode behavior can

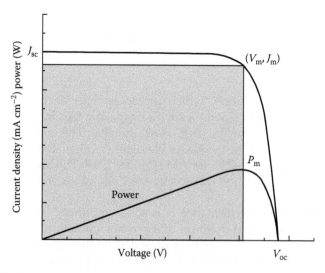

FIGURE 14.19
An example of illuminated current–voltage (lighted *I–V*) characteristics of a solar cell showing V_{oc}, J_{sc}, and P_{max}. The ratio of the shaded area and the total area under the *I–V* curve provides the value for the fill factor.

approximate the ideal maximum power generation. Typical values of FF for high-performance solar cells are between 80% and 90%. The V_{oc} can be easily measured and rearranging Equations 14.10 and 14.11, we can obtain

$$V_{oc} = \frac{k_B T}{q} \ln\left(\frac{J_L}{J_0} + 1\right) \approx \frac{k_B T}{q} \ln\left(\frac{J_{sc}}{J_0} + 1\right) \qquad (14.13)$$

From these expressions, we can finally write the solar cell conversion efficiency, η, as

$$\eta = \frac{P_{max}}{P_{in}} = \frac{J_{sc} V_{oc}}{P_{in}} \text{FF} \qquad (14.14)$$

Here, P_{in} is determined by the particular incident solar spectrum (e.g., AM0, AM1.5, etc.). More rigorous treatments than summarized here would reveal the direct connection between carrier lifetimes, interface recombination velocities, dislocation densities, and diffusion lengths on the fundamental components of V_{oc}, FF, and J_{sc}, but such treatment is beyond the scope of this brief overview.

While the nearly ideal bandgap of GaAs was the primary cause for enormous research and development of III–V PV initially, more recently, the realization that by conventional III–V lattice-matched heteroepitaxy, vertically integrated solar cells containing multiple bandgaps could be grown greatly extends the efficient utilization of the solar spectrum across a much larger range of energies. This multijunction concept is premised on the vertical "stacking" of solar cells with higher bandgap cells on top of lower bandgap cells. In this way, higher energy photons are absorbed by upper cells, and these cells transmit photons having energies less than their bandgaps to cells beneath, which have lower bandgaps. Hence, the solar spectrum is effectively divided amongst sub-cells, with the efficiency improvement due to the minimization of transmission losses for low-energy photons, and the minimization of thermal losses for photons that have energies much greater than the bandgap. The most convenient multijunction III–V cell has been based on lattice-matched $Ga_{0.52}In_{0.48}P/In_{0.01}Ga_{0.99}As/Ge$, where the two III–V cells are lattice matched to each other and in turn to an underlying Ge substrate. Due to diffusion of As into a p-type Ge substrate to create an integrated Ge "bottom" cell, a triple-junction cell having a bandgap profile of ~1.85 eV/1.42 eV/0.67 eV can be created. While these bandgap values are not aligned with the optimum arrangement for a three bandgap multijunction cell, very high efficiencies have been demonstrated due to the much improved spectral utilization compared with single-junction cells. At this time, the highest efficiency obtained for this triple-junction cell is ~32% at AM1.5, which under concentrated sunlight reaches in excess of 40% [King 2007]. This technology is the first PV technology to break the 40% efficiency barrier. Recently, these extraordinary efficiencies have been exceeded by using metamorphic grading to obtain a more ideal split of the solar spectrum amongst the three bandgaps of a triple-junction cell, and using a bandgap profile for GaInP/InGaAs/Ge of ~1.67 eV/1.18 eV/0.67 eV, a 41.1% solar cell was demonstrated at 454× solar concentration [Guter 2009]. The recently reported inverted metamorphic approach, where the solar cell stack is grown in reverse, with the high bandgap top cell initially grown to a lattice-matched substrate, followed by the growth of subsequently lower bandgap sub-cells using metamorphic buffers to tune the bandgaps of these cells, has resulted in III–V triple-junction cells with 1-sun efficiencies as high as 33.8% at AM1.5G [Geisz 2008]. Key to the successes of these various approaches is the minimization of TDD, much as in the case for III–V/Si solar cells. As will be explained below, a future for such technologies is their integration with Si, as this directly addresses cell costs and also enables creative bandgap profiles.

14.4.2 III–V Solar Cells on Si Using SiGe Buffers

14.4.2.1 *Single-Junction GaAs Solar Cells on SiGe/Si*

To date, the integration of GaAs on Si for PV applications using SiGe interlayers to minimize the residual density of TDs has met with greatest success from the perspective of electronic material quality, as described in Section 14.3, and by device performance. This section describes the properties of single-junction GaAs cells grown on Ge/SiGe/Si substrates, extending to GaInP/GaAs dual-junction cells on Ge/SiGe/Si substrates.

As described in Section 14.3.4, GaAs diodes grown on Si via SiGe metamorphic buffers display extremely well-behaved dark J–V characteristics, with very low reverse bias current densities and turn-on voltages that match what is obtained for homoepitaxial diodes. One would thus expect that solar cell characteristics would display similar results. Figure 14.20 shows the schematic of a typical GaAs single-junction solar cell using a p⁺n configuration (p-type emitter and n-type base) grown on Ge/SiGe/Si and its corresponding cross-sectional TEM image. The lack of TDs and other extended defects in the device layers in the cross-sectional TEM image, with its somewhat limited field of view, is evident, and a combination of EBIC[1] and plan-view TEM reveal TDD values of ~1×10^6 cm⁻² for these cells. The matching of the TDD in the solar cell layers with that of the final Ge layer in the step-graded buffer demonstrates the robustness of the defect-reduction approach against standard solar cell epitaxy conditions and device fabrication. The lighted I–V results obtained from this cell structure are shown in Figure 14.21, measured under an NREL standard AM1.5G spectrum. Record-high V_{oc} values for GaAs/Si solar cells were achieved, reaching 981 mV, with a total area efficiency of 18.1% [Andre 2005b]. Correcting for the relatively high percentage of the front surface covered with ohmic grid fingers for this cell (10%) by

p^{++} GaAs contact layer (1000 Å)	~1×10^{19} cm^{-3}
p^{+} In$_{0.49}$Ga$_{0.51}$P window (500 Å)	~1×10^{18} cm^{-3}
p^{+} GaAs emitter (5000 Å)	~2×10^{18} cm^{-3}
n GaAs base (2.0 µm)	~1×10^{17} cm^{-3}
n^{+} In$_{0.49}$Ga$_{0.51}$P back surface field (1000 Å)	~1×10^{18} cm^{-3}
n^{+} GaAs buffer (1000 Å)	~1×10^{18} cm^{-3}
Low-temperature n^{+} GaAs buffer (1000 Å)	~2×10^{18} cm^{-3}
SSMBE Ge buffer layer (300 Å)	uid
Ge termination layers (~1.0 µm)	~1×10^{18} cm^{-3}
n SiGe step graded buffer layers (~10 µm)	~1×10^{18} cm^{-3}
n Si substrate	~0.5–$2\ 1 \times 10^{18}$ cm^{-3}

(a)

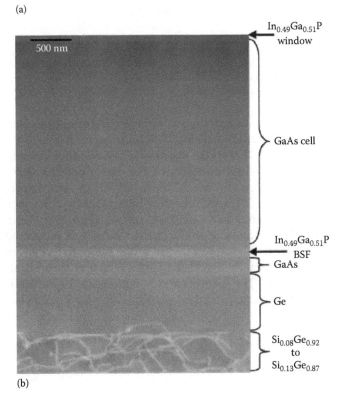

(b)

FIGURE 14.20

(a) Schematic of a p^{+}n single-junction GaAs cell on SiGe/Si and (b) the corresponding TEM cross-sectional image showing the network of dislocations within the SiGe buffer, that do not extend into the cell layers. (After Andre, C.L. et al., *IEEE Trans. Electron. Dev.*, 52, 1055, 2005. With permission.)

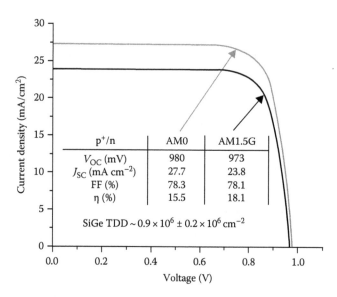

FIGURE 14.21
Illuminated *J–V* characteristics for a single-junction p⁺n GaAs solar cell grown on SiGe/Si substrates, under AM0 and AM1.5G spectra measured by NREL. The measured cell performance parameters are listed in the figure inset. (After Andre, C.L. et al., *IEEE Trans. Electron. Dev.*, 52, 1055, 2005. With permission; Andre, C.L. et al., *J. Appl. Phys.*, 98, 014502, 2005. With permission.)

calculating the projected current density obtained for more standard 4% grid coverage, the efficiency reaches 18.8%. The high performance of these cells results from the lower TDD and the suppression of all other extended defect modes leading to the >10 ns minority carrier base lifetimes, lower diode saturation current density, and much improved V_{oc}. In comparison, typical values for V_{oc} obtained for GaAs cells on Si integrated by other means have been reported in the range from ~750 to ~900 mV [Fan 1981; Vernon 1988; Yamaguchi 1989b]. Figure 14.22 shows a distribution of measured V_{oc} for GaAs/Si cells obtained at 1-sun conditions, along with other relevant experimental data for intentionally dislocated GaAs cells and theoretical calculations of V_{oc} versus TDD. The theoretical calculations are based on the model for minority carrier lifetime described earlier, which for V_{oc} translates to (for GaAs)

$$V_{oc} = \frac{k_B T}{q} \ln\left(\frac{J_{sc}}{J_{02}} + 1\right) \tag{14.15}$$

where

$$J_{02} = \frac{q n_i W_D}{2}\left(\frac{1}{\tau_{base}}\right) \tag{14.16}$$

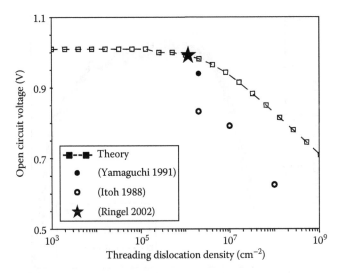

FIGURE 14.22

Measured and modeled V_{oc} for p⁺n GaAs single-junction solar cells on Si substrates versus measured TDD for 1-sun AM0 illumination. The line is the theoretical V_{oc} dependence on TDD calculated using Equations 14.14 and 14.15. (After Ringel, S.A. et al., *Progress Photovolt. Res. Appl.*, 10, 417, 2002. With permission.) Referenced data from: ○ [Itoh 1988], ● [Yamaguchi 1991].

and $\tau_{base} = \tau$ from Equation 14.7, specifically for the n-type base layer of the solar cell in Figure 14.20.

Here, it is noted that the dominant mode for the dark diode current transport of GaAs is via depletion region recombination, thus the expression (14.16) is used in the V_{oc} calculation. From Figure 14.22, one can see the proximity of the GaAs/Ge/SiGe/Si V_{oc} to the plateau of V_{oc} and the measured homoepitaxial control data. Hence, for p⁺n cells, little improvement in V_{oc} is expected for TDD values less than mid-10^5 cm⁻² range. However, calculations and experimental data are also shown for n⁺p configured GaAs cells, and the clear reduction in V_{oc}, which is related to the dependence of minority carrier lifetimes on the conductivity type for GaAs already discussed, will be the focus of the subsequent section of this chapter.

To demonstrate the similarities of identical p⁺n GaAs cell structures grown on GaAs, Ge, and Ge/SiGe/Si substrates from the perspective of carrier collection efficiency, diffusion length, and overall current collection, Figure 14.23 shows a comparison of external quantum efficiency measurements on all three types of substrates, going from homoepitaxy, to nearly lattice-matched heteroepitaxy, to lattice-mismatched metamorphic heteroepitaxy. The nearly identical quantum efficiency magnitudes as a function of wavelength show that identical current contributions can be expected from the p-type emitter, n-type base, and depletion regions of the cell, with no evidence of increased carrier losses at either the front window/emitter or base/back surface field heterostructures interfaces. Hence, the "porting" of GaAs

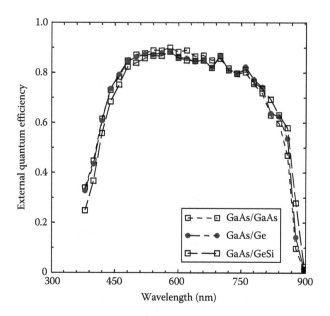

FIGURE 14.23
External quantum efficiency measurements of identical GaAs solar cells grown on GaAs, Ge, and SiGe/Si substrates indicating nearly identical collection efficiency that is independent of substrate material and lattice mismatch. (After Carlin, J.A. et al., *Proceedings of the 28th IEEE Photovoltaic Specialists Conference*, Anchorage, AL, 2000. With permission.)

cell technologies to Si via the use of SiGe interlayers maintains the necessary carrier collection efficiency, consistent with the V_{oc} and earlier materials characterization results.

A key concern for III–V/Si integration in general, and for III–V/Si solar cells in particular, is the issue of differences in coefficients of thermal expansion (CTE) between GaAs and Si. TCE mismatches can lead to microcracking and thickness thresholds have been the subject of recent studies [Yang 2003]. The effect of microcracking in large area solar cells is to create small, isolated "dead zones" within the cell, wherein the microcracks prevent any generated current from being collected by the solar cell grid fingers. From an *I–V* perspective, the effect is to reduce the magnitude of collected current, and if particularly problematic, can create unwanted shunt paths that reduce V_{oc}. The CTE mismatch can be managed by incorporating a residual compressive strain during growth to partially offset the tensile strain resulting from CTE differences during the cool-down stage from growth conditions to room temperature. Figure 14.24 shows a comparison of lighted *J–V* characteristics for three different area GaAs cells from 0.36 to 4 cm² of the same cell design, grown on identical Ge/SiGe/Si substrates. The TDD for each of these substrates was 2×10^6 cm⁻². The similarities in the results reveal that microcracks indeed can be controlled and prevented by proper growth conditions that compensate for strain introduction via CTE mismatch. This has

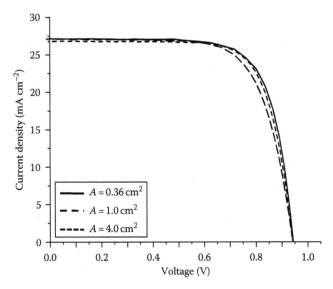

FIGURE 14.24

AM0 lighted *I–V* characteristics obtained for p⁺n GaAs cells on Si using SiGe buffers as a function of cell area from 0.36 to 4 cm², revealing a very good match in characteristics over a 10-fold increase in cell area. The TDD in each case was 2×10^6 cm⁻². (After Andre, C.L. et al., *IEEE Trans. Electron. Dev.*, 52, 1055, 2005. With permission; Andre, C.L. et al., *J. Appl. Phys.*, 98, 014502, 2005. With permission.)

been demonstrated quite dramatically for InGaP/InGaAs metamorphic dual junctions on SiGe/Si in recent work as well [Wilt 2005].

14.4.2.2 Polarity Effects for III–V Cells on SiGe/Si

In Sections 14.3.3 and 14.3.4 the impact of changing conductivity type on carrier recombination properties and diode characteristics was described. For an ideal solar cell in which the effect of incident solar photons is to simply add a voltage-independent photocurrent to the diode *I–V* characteristics, one would expect such effects to manifest in solar cells as well. Figure 14.25a shows the impact of cell polarity rather dramatically. As seen, the primary effect on the solar cell *I–V* characteristics is a reduction in the V_{oc} for n⁺p compared to p⁺n cells at the same value of TDD grown on Ge/SiGe/Si substrates. Note that the cells have the same total area and this effect is only related to an increase in J_0 resulting from increased recombination rates within the p-type side of the depletion region, where the higher mobility of minority carrier electrons creates a condition of faster recombination mediated by the residual TDD in the device. The light *I–V* characteristics in Figure 14.25b, which compares homoepitaxial cells of both p⁺n and n⁺p configurations with nearly perfectly matched cell characteristics, confirm that the effect is related to the presence of dislocations that mediate depletion recombination.

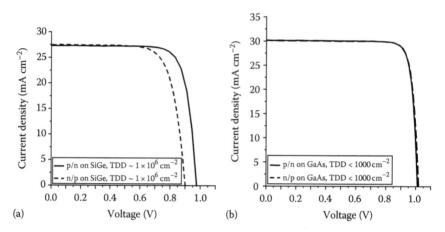

FIGURE 14.25
AM0 lighted *I–V* characteristics comparing p⁺n and n⁺p configured GaAs cells on (a) SiGe substrates with equivalent TDD of 1×10^6 cm⁻² and on (b) low dislocation density GaAs substrates.

Moreover, the variation of V_{oc} with TDD for the n⁺p cells tracks the theoretical calculations of the V_{oc} versus TDD model discussed above and shown in Figure 14.22.

The general conclusion from these finding is that while GaAs integrated on Si possesses very similar properties to homoepitaxial GaAs, subtle effects of the low concentration of residual dislocations on minority carrier transport in devices demonstrates that GaAs/Si solar cells are not equivalent to their non-metamorphic counterparts when it comes to choice of device polarity, and that this is due to the large disparity of electron and hole mobility. For materials having more similar electron and hole mobility values, such a polarity preference in the presence of dislocations would not be expected.

14.4.2.3 Dual-Junction GaInP/GaAs Solar Cells on SiGe/Si

The natural extension of single-junction III–V cells on Si is toward multijunction III–V cells on Si. Efforts in this direction are currently at an earlier stage, but already great promise has been reported. Figure 14.26 shows cross-sectional TEM images of a p⁺n configured GaInP/GaAs lattice-matched dual-junction cell having a bandgap profile of 1.9 eV/1.42 eV, with a zoomed-in image to reveal the upper cell and tunnel junction region. Note that this work utilized MBE growth, whereas the single-junction cell data described in the sections above was for MOCVD-grown material. As in the case for the single-junction cells, the TEM reveals very high-quality material throughout what is now a far complex III–V growth structure, and includes not only

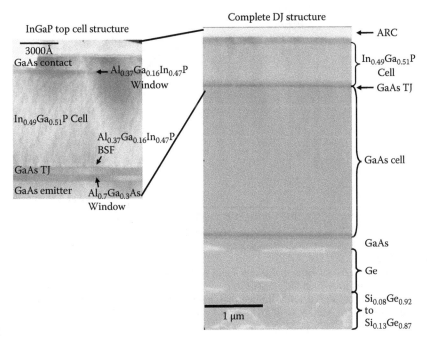

FIGURE 14.26
TEM cross section of a GaInP/GaAs dual-junction (DJ) solar cell on SiGe/Si, with a magnified image of the top cell and tunnel junction (TJ) region showing the presence of very sharp interfaces.

GaAs and GaInP cells, but a thin GaAs p^+/n^+ tunnel junction that provides a low-resistance connection between the series-connected sub-cells. The light *I–V* results are shown in Figure 14.27 with the corresponding external quantum efficiency shown in Figure 14.28. The 1-sun V_{oc} value of 2.21 V under AM0 illumination is a sum of the V_{oc} generated by the GaAs sub-cell and the GaInP sub-cell. Assuming the GaAs sub-cell provides ~0.98 V from the single-junction experiments, the GaInP is providing ~1.23 V. Note that this is an underestimate for the GaInP cell V_{oc} since the spectrum incident on GaAs is now at a much lower incident flux due to the presence of the GaInP top cell, which in reality will slightly reduce the GaAs V_{oc} value. The overall high value for the V_{oc} of the dual-junction cell indicates that very good material quality has been maintained throughout the dual-junction structure. This was confirmed by the reasonable match with the V_{oc} value obtained for identical dual junctions grown on GaAs and processed side-by-side, which produced 2.34 V under AM0 conditions. While the overall efficiency of the dual-junction cell on Si was 16.8% measured using a calibrated AM1.5G solar simulator (15.3% under AM0), the performance is limited by several external factors, including a large, 10% grid obscuration, the use of a GaAs tunnel

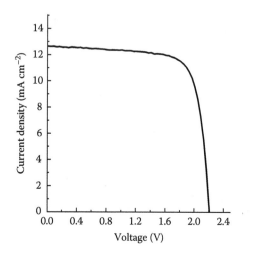

FIGURE 14.27
Light *I–V* characteristics measured under 1-sun AM1.5G illumination of a GaInP/GaAs dual-junction solar cell grown on SiGe/Si substrates by MOCVD. An efficiency of 16.8% was measured. This cell was significantly limited by large grid shadowing and a poor antireflection coating. (After Lueck, M.R. et al., *IEEE Electron Device Lett.*, 27, 142, 2006. With permission.)

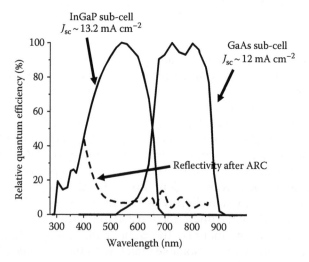

FIGURE 14.28
Quantum efficiency characteristics of GaInP/GaAs dual-junction cell on SiGe/Si. The measured surface reflectivity reveals a nonoptimum antireflection coating. The integrated current density from each sub-cell is nearly matched at ~13 mA/cm² for the AM0 spectrum. (After Lueck, M.R. et al., *IEEE Electron Device Lett.*, 27, 142, 2006. With permission.)

junction as opposed to a wider bandgap tunnel junction, and a relatively poor antireflection coating as can be seen by the high measured reflectivity in Figure 14.28, all of which combine to reduce the current output by several mA cm^{-2}. Correcting such issues as this very promising technology moves away from the prototype stage should realize 1-sun efficiencies well in excess of 20% on Si for only two active junctions. Future work can add a third, Ge sub-cell using the top layer of the SiGe graded buffer, consistent with GaInP/GaAs/Ge triple junctions that are state-of-the-art but are grown on high-cost Ge substrates.

14.4.3 III–V/Si Solar Cells for Space Applications

There is great interest in III–V/Si solar cells for space power applications. The primary issues for space power are related to the finite area available for solar panels, the need for tolerance to radiation-induced defects and large temperature swings in orbit, the need for robustness, and the requirement for minimizing total weight (mass density). As launch costs for payloads are quite expensive, as are the costs of satellite and planetary missions in general, the solar power systems should never be the limitation on mission success and life expectancy, ideally. Comparing Si to Ge for III–V solar applications as a substrate material, Si possesses vastly superior physical properties. Its mass density of 2.33 g/cm^3 is less than half that of Ge or GaAs, and since the semiconductor substrate in effect comprises the weight of a solar cell, substitution of Si for Ge automatically reduces weight by more than a factor of two, and this translates to more than a factor of two increase in the specific power, measured as W/kg for the same efficiency solar cell. Moreover, the fracture toughness of Si compared to Ge is increased by almost 50% and this can further increase the specific power since Si can be thinned to significantly lower thicknesses and at the same time be more robust against mechanical breakage, improving reliability. The thermal conductivity of Si at 1.3 W/cm K is almost twice that of Ge and so for space applications where illuminated panels can routinely reach temperatures in excess of ~80°C–100°C, heat dissipation is important. Finally, integrating III–V solar cells onto Si maintains the excellent radiation resistance of III–V compounds, which is already an important advantage over competing solar cell technologies for space applications. These technical advantages for using Si substrates do not even include the tremendous cost savings potential in switching from a high cost, niche substrate to a low cost, already scaled manufacturing base using Si. Summarizing, the high efficiency of III–V solar cells translates to optimum use of available area for providing maximum power, the low mass density provided by Si substrates directly increases the specific power, decreasing payload weight, and the mechanical robustness of Si addresses reliability in flight.

FIGURE 14.29
Photograph of the MISSE-5 experiment (circled) connected to the International Space Station, which contained prototype GaAs/SiGe/Si solar cells. (After Wilt, D.M. et al., *Proceedings of the 4th IEEE World Conference on Photovoltaic Energy Conversion*, Waikoloa, HI, 1915, 2006. With permission.)

The result of solar cell successes for GaAs on Si cells using SiGe interlayers advanced to their inclusion on the Forward Technology Solar Cell Experiment (FTSCE) that was part of the NASA-led Materials on the International Space Station Experiment number 5 (MISSE-5) mission in 2006–2007. Figure 14.29 shows a photograph of the International Space Station with the MISSE-5 experiment plate circled. A goal was to study how temperature cycling and other flight effects (the ISS orbit is not in a high radiation environment) might impact the prototype GaAs/Si cells due to their great promise. The cells initially received extensive ground-based thermal cycling testing to determine their suitability for flight, for which there were no differences in degradation between the heteroepitaxial GaAs/SiGe/Si cells and homoepitaxial control cells. Testing pre- and post-flight data after one year in space, however, revealed slight degradation in several parameters, but as there was no degradation observed in pre-flight testing that was done at even more extreme temperature cycles, the post-flight degradation has been suggested to be related to issues associated with adhesive mounting materials used for the flight experiments, since the source of degradation appears to be minor microcracking in the largest area cells tested (4 cm^2) and not evident in the 1 cm^2 cells [Wilt 2008]. Regardless, there is no indication of degradation in material properties and III–V/Si PV are of great promise for future aerospace missions.

14.4.4 Future Developments

Since the early 2000s, the dominant, and most successful, III–V/Si integration approach has been the SiGe step-graded buffer with GaAs/Ge integration interface, as evidenced by the preceding text. This is due not only to the efficacy of this materials system, but also due to the more modest level of success via other avenues. It is clear by even a cursory glance at the III–V lattice constant map (see Figure 14.1) that there are other potential approaches for the integration of GaAs (or other III–V PV of interest) with Si substrates, namely that of III–V metamorphic buffers, most likely via the $GaAs_yP_{1-y}$ or $In_xGa_{1-x}P$ alloys. Unfortunately, as discussed in Sections 14.1.2 and 14.2.1, attempts at the direct integration of III–V materials with Si substrates has been hindered by significant materials incompatibilities and mismatches, including lattice constant mismatch, interfacial heterovalency, and thermal expansion mismatch.

Following decades of research with mixed success, though fundamentally important results, recent work has led to the development of a process for the direct growth of GaP on Si(100) substrates, with the resulting GaP materials free of detrimental heterovalent nucleation related defects, including APDs, stacking faults, and MTs [Grassman 2009a]. Such a III–V/Si integration system not only opens the door to new pathways for high-quality GaAs-on-Si, but also to a whole new region of III–V accessibility, enabling the growth of direct bandgaps that were previously only achievable through the use of Al- or N-containing alloys. Additionally, the use of $GaAs_yP_{1-y}$ or $In_xGa_{1-x}P$ metamorphic grading, rather than $Si_{1-x}Ge_x$, enables the production of true multijunction III–V/Si PV thanks to the optical transparency of the III–V metamorphic buffer (i.e., increasing bandgap with decreasing lattice constant) to photons transmitted through the top III–V subcell(s), making them available for absorption in a Si bottom cell. III–V/Si multijunction solar cell bandgap profiles boast some of the highest theoretical PV efficiencies possible, providing a strong impetus for investigation.

Such an approach has been attempted sparingly in the past, but is now receiving significantly increased interest. Recent demonstrations of $GaAs_yP_{1-y}$ compositional grading on Si substrates indicate a high level of metamorphic efficiency, allowing for the accommodation of large amounts of lattice constant misfit, such as that between GaAs and Si, with very thin buffers and low TDD; see Figure 14.30 for an example. Efficiently relaxing buffers are highly desirable as they help prevent the cracking issues that can occur with thicker metamorphic growths due to the thermal expansion mismatch between Si and Ge/SiGe and/or III–Vs. Recent results involving early stage GaAsP solar cells on Si substrates offer very promising performance, with plenty of room for improvement and clear pathways for achievement [Geisz 2006; Grassman 2009b]. Much more work in this area is to come, offering great promise for future high-efficiency III–V/Si multijunction solar cells.

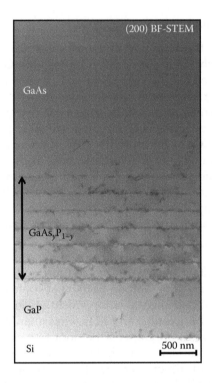

FIGURE 14.30
Bright field (200) STEM cross-section image of anion-graded $GaAs_yP_{1-y}$ step-graded metamorphic buffer grown on Si. The image extends from the Si to GaAs lattice constant, a 4% mismatch, using a comparatively aggressive grading rate of ~3% μm^{-1} for GaP to GaAs. All layers are nearly fully relaxed and the final TDD is on the order of 1×10^7 cm^{-2}. There is no evidence for phase decomposition in this graded anion-based metamorphic buffer.

References

[Ahrenkiel 1990] R. K. Ahrenkiel, M. M. Al-Jassim, B. Keyes, D. Dunlavy, K. M. Jones, S. M. Vernon, and T. M. Dixon, *J. Electrochem. Soc.* **137**(3), 996 (1990).

[Ahrenkiel 1992] R. K. Ahrenkiel, B. M. Keyes, G. B. Lush, M. R. Melloch, M. S. Lundstrom, and H. F. Macmillan, *J. Vac. Sci. Technol. A* **10**(4), 990 (1992).

[Akiyama 1984] M. Akiyama, Y. Kawarada, and K. Kaminishi, *Jpn. J. Appl. Phys.* **23**(11), L843 (1984).

[Andre 2004] C. L. Andre, J. J. Boeckl, D. M. Wilt, A. J. Pitera, M. L. Lee, E. A. Fitzgerald, B. M. Keyes, and S. A. Ringel, *Appl. Phys. Lett.* **84**(18), 3447 (2004).

[Andre 2002] C. L. Andre, A. Khan, M. Gonzalez, M. K. Hudait, E. A. Fitzgerald, J. A. Carlin, M. T. Currie, C. W. Leitz, T. A. Langdo, E. B. Clark, D. M. Wilt, and S. A. Ringel, *Proceedings of the 29th IEEE Photovoltaic Specialists Conference*, Piscataway, NJ (2002).

[Andre 2005a] C. L. Andre, D. M. Wilt, A. J. Pitera, M. L. Lee, E. A. Fitzgerald, and S. A. Ringel, *J. Appl. Phys.* **98**(1), 014502 (2005).

[Andre 2005b] C. L. Andre, J. A. Carlin, J. J. Boeckl, D. M. Wilt, M. A. Smith, A. J. Pitera, M. L. Lee, E. A. Fitzgerald, and S. A. Ringel, *IEEE Trans. Electron. Dev.* **52**(6), 1055 (2005).

[Archer 2008] M. J. Archer, D. C. Law, S. Mesropian, M. Haddad, C. M. Fetzer, A. C. Ackerman, C. Ladous, R. R. King, and H. A. Atwater, *Appl. Phys. Lett.* **92**(10), 103503 (2008).

[Bringans 1992] R. D. Bringans, D. K. Biegelsen, L. E. Swartz, F. A. Ponce, and J. C. Tramontana, *Appl. Phys. Lett.* **61**(2), 195 (1992).

[Carlin 2000a] J. A. Carlin, M. K. Hudait, S. A. Ringel, D. M. Wilt, E. B. Clark, C. W. Leitz, M. Currie, T. Langdo, and E. A. Fitzgerald, *Proceedings of the 28th IEEE Photovoltaic Specialists Conference*, Anchorage, AL (2000).

[Carlin 2000b] J. A. Carlin, S. A. Ringel, E. A. Fitzgerald, M. Bulsara, and B. M. Keyes, *Appl. Phys. Lett.* **76**(14), 1884 (2000).

[Chand 1986] N. Chand, R. People, F. A. Baiocchi, K. W. Wecht, and A. Y. Cho, *Appl. Phys. Lett.* **49**(13), 815 (1986).

[Choi 1987] C. Choi, N. Otsuka, G. Munns, R. Houdre, H. Morko, S. L. Zhang, D. Levi, and M. V. Klein, *Appl. Phys. Lett.* **50**(15), 992 (1987).

[Currie 1998] M. T. Currie, S. B. Samavedam, T. A. Langdo, C. W. Leitz, and E. A. Fitzgerald, *Appl. Phys. Lett.* **72**(14), 1718 (1998).

[Curson 2004] N. J. Curson, S. R. Schofield, M. Y. Simmons, L. Oberbeck, J. L. O'Brien, and R. G. Clark, *Phys. Rev. B* **69**(19), 195303 (2004).

[El-Masry 1988] N. A. El-Masry, J. C. Tarn, and N. H. Karam, *J. Appl. Phys.* **64**(7), 3672 (1988).

[Fahrenbruch 1983] A. Fahrenbruch and R. Bube, *Fundamentals of Solar Cells: Photovoltaic Solar Energy Conversion* (Academic Press, New York, 1983).

[Fan 1980] J. C. C. Fan, C. O. Bozler, and R. P. Gale, *Proceedings of the 14th IEEE Photovoltaic Specialists Conference*, San Diego, CA, p. 534 (1980).

[Fan 1981] J. C. C. Fan, C. O. Bozler, and R. W. McClelland, *Proceedings of the 15th IEEE Photovoltaic Specialists Conference*, Orlando, FL, p. 666 (1981).

[Fischer 1985] R. Fischer, N. Chand, W. Kopp, H. Morko, L. P. Erickson, and R. Youngman, *Appl. Phys. Lett.* **47**(4), 397 (1985).

[Fitzgerald 1991a] E. A. Fitzgerald, *Mat. Sci. Rep.* **7**(3), 87 (1991).

[Fitzgerald 1991b] E. A. Fitzgerald, Y. H. Xie, M. L. Green, D. Brasen, A. R. Kortan, J. Michel, Y. J. Mii, and B. E. Weir, *Appl. Phys. Lett.* **59**(7), 811 (1991).

[Fitzgerald 1994] E. A. Fitzgerald, J. M. Kuo, Y. H. Xie, and P. J. Silverman, *Appl. Phys. Lett.* **64**(6), 733 (1994).

[Fonash 1981] S. J. Fonash, *Solar Cell Device Physics* (Academic Press, New York, 1981).

[Freundlich 1988] A. Freundlich, M. Leroux, J. C. Grenet, A. Leycuras, G. Neu, and C. Verie, *Proceedings of the 8th European Community Photovoltaic Solar Energy Conference*, Florence, Italy, p. 1522 (1988).

[Geisz 2008] J. F. Geisz, S. R. Kurtz, M. W. Wanlass, J. S. Ward, A. Duda, D. J. Friedman, J. M. Olson, W. E. McMahon, T. E. Moriarty, J. T. Kiehl, M. J. Romero, A. G. Norman, and K. M. Jones, *Proceedings of the 33rd IEEE Photovoltaic Specialists Conference*, San Diego, CA (2008).

[Geisz 2006] J. F. Geisz, J. M. Olson, M. J. Romero, C. S. Jiang, and A. G. Norman, *Proceedings of the 4th IEEE World Conference on Photovoltaic Energy Conversion*, Waikoloa, HI, pp. 772–775 (2006).

[Gösele 1999] U. Gösele, Y. Bluhm, G. Kästner, P. Kopperschmidt, G. Kräuter, R. Scholz, A. Schumacher, S. Senz, Q. Y. Tong, L. J. Huang, Y. L. Chao, and T. H. Lee, *J. Vac. Sci. Technol. A* **17**(4), 1145 (1999).

[Grassman 2009a] T. J. Grassman, M. R. Brenner, S. Rajagopalan, R. Unocic, R. Dehoff, M. Mills, H. Fraser, and S. A. Ringel, *Appl. Phys. Lett.* **94**(23), 232106 (2009).

[Grassman 2009b] T. J. Grassman, M. R. Brenner, A. M. Carlin, S. Rajagopalan, R. Unocic, R. Dehoff, M. Mills, H. Fraser, and S. A. Ringel, *Proceedings of the 34th IEEE Photovoltaic Specialists Conference*, Philadelphia, PA (2009).

[Green 2004] M. A. Green, K. Emery, D. L. King, S. Igari, and W. Warta, *Progress Photovolt. Res. Appl.* **12**(1), 55 (2004).

[Guter 2009] W. Guter, J. Schöne, S. P. Philipps, M. Steiner, G. Siefer, A. Wekkeli, E. Welser, E. Oliva, A. W. Bett, and F. Dimroth, *Appl. Phys. Lett.* **94**(22), 223504 (2009).

[Hayashi 1994] K. Hayashi, T. Soga, H. Nishikawa, T. Jimbo, and M. Umeno, *Proceedings of the 1st World Conference on Photovoltaic Energy Conversion*, Waikoloa, HI, pp. 1890–1893 (1994).

[Henderson 1971] R. C. Henderson, R. B. Marcus, and W. J. Polito, *J. Appl. Phys.* **42**(3), 1208 (1971).

[Hudait 2003] M. K. Hudait, Y. Lin, D. M. Wilt, J. S. Speck, C. A. Tivarus, E. R. Heller, J. P. Pelz, and S. A. Ringel, *Appl. Phys. Lett.* **82**(19), 3212 (2003).

[Itoh 1988] Y. Itoh, T. Nishioka, A. Yamamato, and M. Yamaguchi, *Appl. Phys. Lett.* **52**(19), 1617 (1988).

[Kawanami 1997] H. Kawanami, S. Ghosh, I. Sakata, and T. Sekigawa, *Mater. Res. Soc. Symp. Proc.* **441**, 33 (1997).

[King 2007] R. R. King, D. C. Law, K. M. Edmondson, C. M. Fetzer, G. S. Kinsey, H. Yoon, R. A. Sherif, and N. H. Karam, *Appl. Phys. Lett.* **90**(18), 183516 (2007).

[Komatsu 1997] Y. Komatsu, K. Hosotani, T. Fuyuki, and H. Matsunami, *Jpn. J. Appl. Phys.* **36**(9A), 5425 (1997).

[Kroemer 1983] H. Kroemer, *Surf. Sci.* **132**(1–3), 543 (1983).

[Kroemer 1987] H. Kroemer, *J. Cryst. Growth* **81**(1–4), 193 (1987).

[Kroemer 2002] H. Kroemer, *Int. J. Mod. Phys. B* **16**(5), 677 (2002).

[Kroemer 1980] H. Kroemer, K. J. Polasko, and S. C. Wright, *Appl. Phys. Lett.* **36**(9), 763 (1980).

[Kunert 2008] B. Kunert, I. Németh, S. Reinhard, K. Volz, and W. Stolz, *Thin Solid Films* **517**(1), 140 (2008).

[Lee 1991] H. Lee, X. Liu, K. Malloy, S. Wang, T. George, E. Weber, and Z. Liliental-Weber, *J. Electron. Mater.* **20**(2), 179 (1991).

[Leycuras 1990] A. Leycuras, M. F. Vilela, J. C. Grenet, G. Strobl, M. Leroux, G. Neu, and C. Verie, *Proceedings of the 21st IEEE Photovoltaic Specialists Conference*, Orlando, FL, p. 95 (1990).

[Luan 1999] H.-C. Luan, D. R. Lim, K. K. Lee, K. M. Chen, J. G. Sandland, K. Wada, and L. C. Kimerling, *Appl. Phys. Lett.* **75**(19), 2909 (1999).

[Lueck 2006] M. R. Lueck, C. L. Andre, A. J. Pitera, M. L. Lee, E. A. Fitzgerald, and S. A. Ringel, *IEEE Electron. Dev. Lett.* **27**(3), 142 (2006).

[Möller 1993] H. J. Möller, *Semiconductors for Solar Cells* (Artech House, Boston, MA, 1993).

[Narayanan 2002a] V. Narayanan, S. Mahajan, K. J. Bachmann, V. Woods, and N. Dietz, *Acta Mater.* **50**(6), 1275 (2002).

[Narayanan 2002b] V. Narayanan, S. Mahajan, K. J. Bachmann, V. Woods, and N. Dietz, *Philos. Mag. A* **82**(4), 685 (2002).

[Narayanan 2000] V. Narayanan, S. Mahajan, N. Sukidi, K. J. Bachmann, V. Woods, and N. Dietz, *Philos. Mag. A Phys. Condens. Matter Struct. Defect Mech. Prop.* **80**(3), 555 (2000).

[Nayfeh 2004] A. Nayfeh, C. O. Chui, K. C. Saraswat, and T. Yonehara, *Appl. Phys. Lett.* **85**(14), 2815 (2004).

[Nelson 2003] J. Nelson, *The Physics of Solar Cells* (Imperial College Press, London, U.K., 2003).

[Nishi 1985] S. Nishi, H. Inomata, M. Akiyama, and K. Kaminishi, *Jpn. J. Appl. Phys.* **24**(6), L391 (1985).

[Ota 1983] Y. Ota, *Thin Solid Films* **106**(1–2), 1 (1983).

[Ringel 2002] S. A. Ringel, J. A. Carlin, C. L. Andre, M. K. Hudait, M. Gonzalez, D. M. Wilt, E. B. Clark, P. Jenkins, D. Scheiman, A. Allerman, E. A. Fitzgerald, and C. W. Leitz, *Progress Photovolt. Res. Appl.* **10**(6), 417 (2002).

[Schone 2006] J. Schone, F. Dimroth, A. W. Bett, A. Tauzin, C. Jaussaud, and J. C. Roussin, *Proceedings of the 4th IEEE World Conference on Photovoltaic Energy Conversion*, Waikoloa, HI, p. 776 (2006).

[Sieg 1998a] R. Sieg, S. Ringel, S. Ting, E. Fitzgerald, and R. Sacks, *J. Electron. Mater.* **27**(7), 900 (1998).

[Sieg 1998b] R. M. Sieg, J. A. Carlin, J. J. Boeckl, S. A. Ringel, M. T. Currie, S. M. Ting, T. A. Langdo, G. Taraschi, E. A. Fitzgerald, and B. M. Keyes, *Appl. Phys. Lett.* **73**(21), 3111 (1998).

[Soga 1989] T. Soga, S. Nozaki, N. Noto, H. Nishikawa, T. Jimbo, and M. Umeno, *Jpn. J. Appl. Phys.* **28**(12), 2441 (1989).

[Taguchi 2005] H. Taguchi, T. Soga, and T. Jimbo, *Sol. Energy Mater. Sol. Cells* **85**(1), 85 (2005).

[Takagi 1998] Y. Takagi, H. Yonezu, K. Samonji, T. Tsuji, and N. Ohshima, *J. Cryst. Growth* **187**(1), 42 (1998).

[Thomas 2003] S. Thomas, S. Bharatan, R. Jones, R. Thoma, T. Zirkle, N. Edwards, R. Liu, X. Wang, Q. Xie, C. Rosenblad, J. Ramm, G. Isella, and H. Von Känel, *J. Electron. Mater.* **32**(9), 976 (2003).

[Tsaur 1982] B.-Y. Tsaur, J. C. C. Fan, G. W. Turner, F. M. Davis, and R. P. Gale, *Proceedings of the 16th IEEE Photvoltaics Specialist Conference*, San Diego, CA, p. 1143 (1982).

[Uchida 1995] H. Uchida, M. Umeno, T. Soga, H. Nishikawa, and T. Jimbo, *J. Cryst. Growth* **150**(1–4), 681 (1995).

[Ueda 1991] O. Ueda, K. Kitahara, N. Ohtsuka, A. Hobbs, and M. Ozeki, *Mater. Res. Soc. Symp. Proc.* **221**, 393 (1991).

[Uppal 1985] P. N. Uppal and H. Kroemer, *J. Appl. Phys.* **58**(6), 2195 (1985).

[Venkatasubramanian 1991] R. Venkatasubramanian, M. L. Timmons, J. B. Posthill, B. M. Keyes, and R. K. Ahrenkiel, *J. Cryst. Growth* **107**, 489 (1991).

[Vernon 1988] S. M. Vernon, S. P. Tobin, V. E. Haven, C. Bajgar, T. M. Dixon, M. M. Al-Jassim, R. K. Ahrenkiel, and K. A. Emery, *Proceedings of the 20th IEEE Photovoltaic Specialists Conference*, Las Vegas, NV, p. 281 (1988).

[Wang 1998] G. Wang, G. Y. Zhao, T. Soga, T. Jimbo, and M. Umeno, *Jpn. J. Appl. Phys.* **37**, L1280 (1998).

[Wang 1984] W. I. Wang, *Appl. Phys. Lett.* **44**(12), 1149 (1984).

[Wietler 2005] T. F. Wietler, E. Bugiel, and K. R. Hofmann, *Appl. Phys. Lett.* **87**(18), 182102 (2005).

[Wilt 2005] D. M. Wilt, A. M. T. Pal, N. F. Prokop, S. A. Ringel, C. Andre, M. A. Smith, D. A. Scheiman, P. P. Jenkins, W. F. Maurer, B. McElroy, and E. A. Fitzgerald, *Proceedings of the 31st IEEE Photovoltaic Specialists Conference*, Orlando, FL, p. 571 (2005).

[Wilt 2008] D. M. Wilt, A. T. Pal, S. A. Ringel, E. A. Fitzgerald, P. P. Jenkins, and R. Walters, *Proceedings of the 33rd IEEE Photovoltaic Specialists Conference*, San Diego, CA (2008).

[Wilt 2006] D. M. Wilt, S. A. Ringel, E. A. Fitzgerald, P. P. Jenkins, and R. Walters, *Proceedings of the 4th IEEE World Conference on Photovoltaic Energy Conversion*, Waikoloa, HI, p. 1915 (2006).

[Wright 1982] S. L. Wright, M. Inada, and H. Kroemer, *J. Vac. Sci. Technol.* **21**(2), 534 (1982).

[Wright 1984] S. L. Wright, H. Kroemer, and M. Inada, *J. Appl. Phys.* **55**(8), 2916 (1984).

[Xu 1998] Q. Xu, J. Hsu, S. Ting, E. Fitzgerald, R. Sieg, and S. Ringel, *J. Electron. Mater.* **27**(9), 1010 (1998).

[Yamaguchi 1991] M. Yamaguchi, *J. Mater. Res.* **6**(2), 376 (1991).

[Yamaguchi 1985] M. Yamaguchi and C. Amano, *J. Appl. Phys.* **58**(9), 3601 (1985).

[Yamaguchi 1988a] M. Yamaguchi, C. Amano, Y. Itoh, K. Hane, R. K. Ahrenkiel, and M. M. Al-Jassim, *Proceedings of the 20th IEEE Photovoltaic Specialists Conference*, Las Vegas, NV, p. 749 (1988).

[Yamaguchi 1989a] M. Yamaguchi, T. Nishioka, and M. Sugo, *Appl. Phys. Lett.* **54**(1), 24 (1989).

[Yamaguchi 1989b] M. Yamaguchi, C. Amano, and Y. Itoh, *J. Appl. Phys.* **66**(2), 915 (1989).

[Yang 2003] V. K. Yang, M. Groenert, C. W. Leitz, A. J. Pitera, M. T. Currie, and E. A. Fitzgerald, *J. Appl. Phys.* **93**(7), 3859 (2003).

[Zdanowicz 2005] T. Zdanowicz, T. Rodziewicz, and M. Zabkowska-Waclawek, *Sol. Energy Mater. Sol. Cells* **87**(1–4), 757 (2005).

Index

9 780367 383268